Interdisciplinary Perspectives on Sustainable Development: Achieving the SDGs through Education, Wellbeing, and Innovation

CRC Press
Taylor & Francis Group
Boca Raton London New York

CRC Press is an imprint of the
Taylor & Francis Group, an **informa** business

First edition published 2024
by CRC Press
4 Park Square, Milton Park, Abingdon, Oxon, OX14 4RN

and by CRC Press
2385 NW Executive Center Drive, Suite 320, Boca Raton FL 33431

CRC Press is an imprint of Informa UK Limited

ISBN: 978-1-032-60104-5 (pbk)
ISBN: 978-1-003-45761-9 (ebk)

DOI: 10.4324/9781003457619

Typeset in Sabon LT Std
by Ozone Publishing Services

About

ISC 2022 is dedicated to the Niti Aayog policies to promote sustainability through exchange of ideas emerging out of the academia. The ISC is an annual conference that would be held in virtual mode until COVID restrictions on travel exist.

Vision

The vision of the conference is to capacitate Academia with the necessary ideas that will provide insights of the grassroot level development to various stakeholders of the Niti-Aayog policies. Towards this goal, the conference will create a conjunction of various stakeholders of Niti-Aayog policies that include- academic institutions, government bodies, policy makers and industry.

Mission

The ISC organizers will make concerted efforts to promote academic research that would provide technological, scientific, management & business practices, and insights into policy merits & disruptions. The framework of exchange of ideas will be geared towards adoption of deep technologies, fundamental sciences & engineering, energy research, energy policies, advances in medicine & related case studies. This framework will enable the round table discussions between the academia, industry and policy makers through its range of plenary and keynote speakers.

Contents

Editor's Biography......ix

Foreword......xii

Preface......xiii

Introduction......xiv

Details of Programme Committee......xv

1. Power and Discourse in La Fontaine's Beast Fables......1
2. Lockdown: Real-life Paradoxical Experience to Sustain Human Relationships, Healthy Lives, and Well Being......5
3. Effect of Parenting Stress on Parenting Sense of Competence Among Mothers of Children with ADHD......10
4. Sustainable Development Goals and Juvenile Justice System: A Comparative Analysis......15
5. Influence of Karma at Workplace: With Special Reference to Higher Education Institutions in South Gujarat Region......21
6. Cross-Walk of Professional Competencies for Social and Emotional Wellbeing to Cater Mental Health Problems in Schools......26
7. Emotional Intelligence Manages Sustainable Development for an Organization - the Contribution of Psychological Well-Being......31
8. The Enactment of Social Sustainable Goals in IT Organizations......36
9. Role of Higher Education in Achieving the Sustainble Development Goals (SDGs)......41
10. "Happiness Engineering": Acceptance and Commitment Therapy for University Students' Classroom Engagement, Mental Health, and Psychological Flexibility......45
11. Work-Life Balance and Its Challenges for Medical Professionals in the Health Care Sector......50
12. Political Representation of Aesop's Beast Fables in Augustan Age......57
13. Miro Application of Web Whiteboard for Sustainable Development in Teaching and Learning Research......61
14. A Study of Female Identity and Marital Discord in the Selected Works of Anita Desai......67
15. Role of Digital Competency in Sustainable Quality Education......73
16. Digital Infrastructure for SHGs of Tribal Women in Odisha: Means for MSMEs to Achieve SDGs......77
17. Fetishism: Paradoxing the Narratives of Sustainable Development Goals......84

18. A Systematic Review Study on the Quality of Life Associated with Depression Among the Elderly in India 88

19. Influence of Personality and Sector of Employment on Perceived Social Support and Work Family Conflict........ 93

20. Prophesying the Future Retailing Model of Emerging Markets with Special Focus on India 98

21. Family Conflict and Rivalry in The Shipwrecked Prince and King Lear: A Comparative Study 102

22. The Subtle Warnings Signs of Suicidal Thought and Behaviour Exhibited by Hannah Baker in "Thirteen Reasons Why" by Jay Asher 107

23. Demographic Variables and Job Satisfaction Among College Lecturers 112

24. Sustainable Crisis: Psychoanalytical Reading of Populism and Trauma in Select War Narrative........ 115

25. Impact of Problematic Internet Use on Psychological Well-Being, Hyperventilation and Chronic Fatigue Syndrome Among Youth 120

26. Psychological Distress Among IT Sector Employees During COVID-19 Pandemic in India 125

27. Triangulation Study on LGBTQ Inclusion with Sustainable Development Goal 10 using Twitter Data and Topic Modelling........ 131

28. Community Participation in Public Space Planning and Management: Cases of Indian Cities 136

29. Surveying Interest and Engagement in Political Discourse 142

30. Opinion Mining of National Education Policy 2020 to Improve Its Implementation for Women Empowerment 146

31. Factors Affecting Entrepreneurship Intention: An Empirical Study with Reference to Indian University Students 151

32. Criminogenic Cognition of Juveniles in Conflict with the Law and Use of the Internet with the Victim-Offender Overlap........ 156

33. Testifying Legal Admissibility: Germline and Embryo Editing Focusing on SDG 15-Life on Earth........ 160

34. Exploring Psychological Wellbeing of College Students in Relation to Their Demographic Identity: Predictors and Prevalence........ 165

35. India and Nepal Bridging the Gap with Hydropower Project Enhancing Science and Technological Partnership........ 169

36. Systematic Literature Review of Interlinkages between Sustainable Development & Human Development 172

37. Impact of COVID-19 on Domestic Workers with Special Reference to Pune Region 177

38. Entrepreneurial Education and Entrepreneurial Intentions: Mediation of Entrepreneurial Mindset and Moderated Mediation of Creativity 182

39. Role of Corporate Social Responsibility in Achieving Sustainable Development Goals..........188

40. Feasibility of DREAMS Afterschool Intervention to Implement SDG – 4, 5, and 11 in Rural India..........192

41. Crowdsourcing: A Technique to Sustain the Educational Industry..........197

42. Identity of Scheduled Tribes in India - A Systematic Review..........204

43. Strategies Employed to Acquire and Reflect Political Knowledge..........209

44. Impact of Select Vocabulary Learning Strategies (VLS) on Vocabulary Acquisition of Tertiary Level Learners..........213

45. Socialization of Culture: Sociopolitical and Sociocultural Contexts Ensuing Cultural Transition and Hybridity..........219

46. Parables of the Lost and Found: A Semiotic Dissection of Religious Discourse..........224

47. Incorporating Research-based Pedagogical Implications in Grammar Through the Android Application: An Experimental Study..........229

48. Legacy and Evolution of Panchayati Raj Institutions and Tribal Self-Governance in India..........235

49. Prospects of PESA Act and Inhibitions in its Implementation in Scheduled Areas of Jharkhand, India..........240

50. Exploring Health Information Seeking Behaviour Among Young Oraon Women in Jharkhand..........247

51. Evolution and Implementation of Land Acquisition Legislations in India..........252

52. Change in Gender Relations: Re-Visiting Gender-Based Violence in Tribal Communities of India..........257

53. "Sarna Adivasi" Religion Code: Contextualizing Religious Identity of Tribals in India..........262

54. Sustainable Fashion: "Form Leisure"- Deconstructing Men's Formal Shirts Into a Women's Wear Collection..........267

55. Kondapalli Toys: White Woodcraft of Andhra Pradesh..........273

56. Role of Consumer Perception on Genderless Fashion in Deconstructing Gender Stereotypes in Indian Society..........278

57. Denial of Human Right to Water During Pandemic: Experience of Indian Slum..........283

58. Dynamics of the Demographic Transition on Economic Development: Evidence From SRS Data in India..........290

59. From Function to Fashion, Face Masks as a Flourishing New Product..........296

60. Cartoons and their Visual Aspects Affecting Children..........302

61. Ergonomic issues faced by transporters of LPG gas cylinders..........308

62. Proposed Concept for Mysore Pak Packaging..........313

63. Livelihood Experiences of Working Women with Disability during COVID-19: Predicament and Prospect..........320

64. Developing Storytelling as a Method of the Design process in Bachelors of Interior Design Education....327

65. Review on Incorporating Visual Storytelling as a Method of the Design Process in Design Education....332

66. Does Musically Responsive School Curriculum enhance Reasoning Abilities and Helps in Cognitive Development of School Students?....337

67. Music and Its Effect on Mathematical and Reading Abilities of Students: Pedagogy for Twenty-First Century Schools....342

68. An Exploration of the Complexities Involved in the Regulation of Green Buildings....347

69. The Makers and Users of Fashion, a Study of Contrast....352

70. Beyond Classroom: Impact of Covid-19 on Education System....357

71. Implementation of Rawls Theory of Justice in the Present Indian Reservation System....365

72. Financial Inclusion: Conceptual Understanding to Indian Report Card....370

73. Contemplating the Problems and Issues Related to Corporate Social Responsibility in India....376

74. Untapped Power of Music-Integrated Pedagogy: Its Role in Enhancement of "Behaviour and Self-Confidence" among School Students....383

75. Casualty of Dignity and Other Rights of Children Born Out of Casual Relationship: A Legal Conundrum....388

76. Women in Civil Engineering Profession: Career Profile of Indian Women....395

77. Factors That Make Public-Private Partnerships Appealing for Highway Projects in Gujarat....401

78. Risk Management in Public–Private Partnership-Based Infrastructure Projects: A Critical Analysis....406

79. Construction Safety Practices: An Analysis....410

80. The Relevance of Kitchen Vastu Guidelines in Relation to Architecture....415

81. Drone Rules 2021: Analysis and Implications for India's UAV Programme....420

82. Toy Companies Using Unconventional Methods to Stay Relevant and Reach Evolving Minds of the Parents and Children....428

83. Behavior of Speed Breaker in Urban Context....433

84. Aesthetics of Distortion and the Absurd: Fusing Redemptive Existentialism and Berkeley's Metaphysics in Beckett's Plays....438

85. Explicit and Implicit Self-Esteem of Narcissists and Non-Narcissists....443

Editor's Biography

Proceedings of the 1st International Sustainability Conference (ISC 2022)

Edited by

Dimitrios A Karras

Bio: Dimitrios A. Karras received his Diploma and M.Sc. Degree in Electrical and Electronic Engineering from the National Technical University of Athens (NTUA), Greece in 1985 and the Ph. Degree in Electrical Engineering, from the NTUA, Greece in 1995, with honors. From 1990 and up to 2004 he collaborated as visiting professor and researcher with several universities and research institutes in Greece. Since 2004, after his election, he has been with the Sterea Hellas Institute of Technology, Automation Dept., Greece as associate professor in Intelligent Systems-Decision Making Systems, Digital Systems, Signal Processing, till 12/2018, as well as with the Hellenic Open University, Dept. Informatics as a visiting professor in Communication Systems (the latter since 2002 and up to 2010). Since 1/2019 is Associate Prof. in Intelligent Systems-Decision Making Systems, Digital Systems and Signal Processing, in National & Kapodistrian University of Athens, Greece, School of Science, Dept. General as well as Adjunct Assoc. Prof. Dr. with the EPOKA university, Computer Engineering Dept., Tirana (1/10/2018-25/9/2020). He has published more than 70 research refereed journal papers in various areas of intelligent and distributed/multiagent systems, Decision Making, pattern recognition, image/signal processing and neural networks as well as in bioinformatics and more than 185 research papers in International refereed scientific Conferences. His research interests span the fields of intelligent and distributed systems, Decision Making Systems, multiagent systems, pattern recognition and computational intelligence, image and signal processing and systems, biomedical systems, communications and networking as well as security. He has served as program committee member as well as program chair and general chair in several international workshops and conferences in the fields of intelligent Systems-Decision Making Systems, signal, image, communication and automation systems. He is, also, former editor in chief (2008-2016) of the International Journal in Signal and Imaging Systems Engineering (IJSISE), academic editor in the TWSJ, ISRN Communications and the Applied Mathematics Hindawi journals as well as associate editor in various scientific journals, including CAAI, IET. He has been cited in more than 2220 research papers (https://scholar.google.com/citations?user=IxQurTMAAAAJ&hl=en), (Google Scholar) and 1626 (ResearchGate Index, after recent recounting based only on ResearchGate database, with more than 9105 ResearchGate Index Reads metric) citations, as well as in more than 1101 citations in Scopus peer reviewed research database, with H/G indexes 20/48 (Google Scholar), Scopus-H index 15, RG-index 30.96 and RG-Research Interest index 886.1 (https://www.researchgate.net/profile/Dimitrios_Karras2/).

Sai Kiran Oruganti

Indian Institute of Technology Patna India.

ORCID: https://orcid.org/0000-0003-4601-2907

Areas: Wireless Power Transfer, Wireless technologies, IoT, Radio Science, Electromagnetics & applications.

Profile Summary

Prof. Dr. Sai Kiran Oruganti is with the School of Electrical and Automation Engineering, Jiangxi University of Science and Technology, Ganzhou, People's Republic of China as a full Professor since October 2019. He is responsible for establishing an advanced wireless power transfer technology laboratory as a part of the international specialists team for the Center for Advanced Wirless Technologies. Between 2018-2019, he served as a senior researcher/Research Professor at Ulsan National Institute of Science and Technology. Previously, his PhD thesis at Ulsan National Institute of Science and Technology, South korea, led to the launch of an University incubated enterprise, for which he served as a Principal Engineer and Chief Designer in 2017-2018. After his PhD in 2016, he served Indian Institute of Technology, Tirupati in the capacity of Assistant Professor (Electrical Engineering) between 2016-2017.

Research

Prof. Dr. Oruganti, prime research focus is in the development of Wireless Power Transfer(WPT) for applications- Internet of Things (IoT) device charging, Agriculture, Electric Vehicle Charging, Biomedical device charging, Electromagnetically induced transparancy techniques for military and defence applications, Secured shipping containers, Nano Energy Generators.

Achievements

Prof. Dr. Oruganti has more than 21 patents pending on his credit and with several of those patent applications passing the NoC stage. As of 2021, 16 of 21 patents have been granted. He is credited with the pioneering work in the field of Zenneck Waves based Wireless Power Transfer system. Most notably, he has been regarded as one of the only few researchers in the field of WPT to be able to conduct power and signal transmission across partial Faraday shields. His recent paper accepted by Nature Scientific Reports has generated a lot of interest and excitement in the field. International Union of Radio Science(URSI) recognized his research efforts and awarded him Young Scientist Award in 2016. He is also recipient of IEEE sensors council letters of appreciation.

Sudeshna Ray

Dr. Ray's research at the interface of Chemistry and Material Science is focused on the development of Novel Inorganic based Luminescent materials for the application in white Light Emitting Diodes (LEDs) and as Spectral Converters in Solar Cell. The novel materials can be a single composition multi-centred phosphor or near UV/blue excitable blue, green, yellow and red emitting phosphor. The utilization of 'Green' solution based Synthesis Methodology for the precise control of the composition of the phosphors and achievement of a homogenous distribution of small amounts of activators in the host compounds is the main paradigm of my Research. In addition, to my previous focus on Synthesis, Characterization and Optical studies of size and shape tuned nanocrystalline Y2O3, YVO4 and YPO4 based phosphors, I am extensively involved into the research for the development of Advanced Luminescence Materials so called Quantum Cutters for the application in Solar Cell. A unifying theme of my research is the compositional tuning of the properties of extended solids through solid solution; sometimes referred to as the game of x and y, as, for example, in Sr2(1-x-y/2)Eu2xLaySi1-yAlyO4. Design of New phosphor for LEDs and fabrication of LEDs using the phosphors is an integral part of my Research. Currently, I am involved in the synthesis of Persistent Phosphors for the fabrication of Glow Bullet for Defense Application. RESEARCH INTERESTS • Exploration of New Phosphors by Mineral Inspired Methodology • Development of water soluble silicon compound by alkoxy group exchange reaction • Synthesis of Eu2+ and Ce3+ doped silicate phosphors using water soluble silicon compound • Solution Synthesis of 'Size' and 'shape' tuned Nanomaterials • Characterization of Nanocrystalline phosphors by XRD, TEM, FE-SEM and Raman spectroscopic measurement. • Study of 'Up-conversion', 'Down-conversion', 'Down-shifting' phenomena by steady state photoluminescence and lifetime measurement and analysis. • Measurement of 'Quantum Efficiency' and Thermal stability of phosphors. • Development and optical study of 'Quantum Cutting' Materials as Spectral Converter for Solar cell

RESEARCH EXPERIENCE

- Postdoctoral Fellow Phosphor Research Laboratory, Department of Applied Chem. September (2012) –July (2013) National Chiao Tung University, Taiwan Advisor: Prof. Teng Ming Chen
- Study of Energy Transfer from sensitizer to activator by lifetime analysis.
- Measurement of Quantum Efficiency

Foreword

I had the privilege to serve as the convenor for the first ISC2022 (formerly ICTSGA-1) which is dedicated to the realization of Niti-Aayog policies of the government of India. As a chairman for the Technology Innovation Hub IIT Patna whose National Mission on Interdisciplinary Cyber-Physical Systems is dedicated to Analytics, I have always felt the need to make an outreach to the Indian as well as international academia. ISC2022 is a prelude to the larger vision to capacitate Academia with the necessary ideas that will provide insights of the grassroot level development to various stakeholders of the Niti-Aayog policies. Towards this goal, the conference has aimed to create a conjunction of various stakeholders of Niti-Aayog policies that include- academic institutions, government bodies, policy makers and industry.

I hope that the contents of this series will serve as a guide for young researchers and policy makers for their future endeavours towards a holistic growth of the human society.

Trilok Nath Singh, PhD
Professor & Director, Indian Institute of Technology Patna.
Chairman and Board of Directors, Technology Innovation Hub,
Indian Institute of Technology, Patna-INDIA.

Preface

ISC 2022 (Formerly ICTSGA-1) is dedicated to the Niti Aayog policies to promote sustainability through exchange of ideas emerging out of the academia. The ISC is an annual conference that would be held in virtual mode until COVID restrictions on travel exist. The conference featured several plenary and keynote speeches from UN, African Environmental Sustainability, Mahatma Gandhi University, Universidad Politécnica de Valencia, Technical University of San Luis Potosí Mexico, Delhi Technological University, University at Johannesburg. The sessions were divided into two major sections: (a) Sciences & Engineering (b) Management & Humanities. In addition, a dedicated session on women in sciences, and academia was held. The sessions included ~50% representation from Women in academia, and research.

Statistics

The ISC 2022 received 816 abstract submissions of these only 340 were selected.

- Total Plenary Talks: 4
- Total Keynote Talks: 8
- Total Invited Talks: 7
- Total Oral Talks spread across two days: 35
- Total Women in SDG talks: 26

Conference Chairs

Ravindra Pratap Singh,
IGNTU Amarkantak, India

Dimitrios A Karras
National and Kapodistrian University of Athens, Greece.

Amarendra Pratap Singh
IGNTU Amarkantak, India

Introduction

Vision

The vision of the conference is to capacitate Academia with the necessary ideas that will provide insights of the grassroot level development to various stakeholders of the Niti-Aayog policies. Towards this goal, the conference will create a conjunction of various stakeholders of Niti-Aayog policies that include- academic institutions, government bodies, policy makers and industry.

Mission

The ISC organizers will make concerted efforts to promote academic research that would provide technological, scientific, management & business practices, and insights into policy merits & disruptions. The framework of exchange of ideas will be geared towards adoption of deep technologies, fundamental sciences & engineering, energy research, energy policies, advances in medicine & related case studies. This framework will enable the round table discussions between the academia, industry and policy makers through its range of plenary and keynote speakers.

Details of Programme Committee

Organizing Committee

T. Sunder Selwyn, PhD
Prince Dr. K. Vasudevan College of Engineering and Technology, Chennai India

K.K. JACOB, PhD
Faculty of Applied Sciences and Technology, Universiti Tun Hussein Onn Malaysia.

Mengistu Tulu Balcha, PhD
Ambo University Ambo, Ethiopia.

Kannaiya Raja, PhD
Ambo University Ambo, Ethiopia.

Karthikeyan Kaliyaperumal, PhD
Ambo University Ambo, Ethiopia.

Publishing Committee:

Dimitrios A Karras, PhD
University of Athens (NKUA), Greece

Karthikeyan Kaliyaperumal, PhD
Ambo University Ambo, Ethiopia.

Sudenshna Ray, PhD
RNTU-AIISECT University Bhopal, India

Power and Discourse in La Fontaine's Beast Fables

Sanaa Parween,[*,a] Dr. Jayatee Bhattacharya,[b] and Dr. Simi Malhotra[c]

[a]Amity University of Noida, India
[b]Amity University of Noida, India
[c]Jamia Milia Islamia of New Delhi, India
E-mail: [*]pseudoliterarian@gmail.com

Abstract

Storytelling originated from the need of shaping minds desirably and creatively. A genre that successfully established itself in the didactic field of storytelling is that of fable. An ancient narrative form that engaged with moral, political, and philosophical instruction to bring forth a unique understanding of human nature by using beasts. Seventeenth century witnessed civil and political conflicts and it was reflected in the literary works of that period, the fables of Jean de La Fontaine are counted among the literary treasures of this period. This paper aims to discuss two fables of La Fontaine for their social and political significance in terms of power. In addition, the above discussion, this research paper will attempt to highlight the discourse of animal tales and their significance with help of Foucault's theory of knowledge and power.

Keywords: Fables, La Fontaine, Politics, Power, Discourse, Beast Fables

Introduction

The tradition of fables has always been an elegant source of wisdom and didactic derivations. It can be found both in prose and verse form, existing mostly in collections. But the distinguishing characteristic of a fable comes from its fabulist, even with the adaptation of ancient fables by modern fabulists, the factor of uniqueness remains. Within the framework of studying or analyzing the genre, it becomes evident that each literary period has its take on fables. The frequent adaptations and translation cause the overlapping of number of tales, which is also because of the oral transmission of the tales. Stories across the globe can be traced back to the rich history of wisdom tales, oral tales from India, China, Arab, Europe, etc.

A technical categorization of fable that often helps in a scholarly distinction can be found in Newbigging's book. First, a *fabulae* is the ancient Scandinavian and Grecian gods around which the legends and mythological lore are woven, the tales of remote past laced with imagination and superstitions. Second category is a *fabellae,* a significant branch of literature that involves fabulists and the subject of reasoning combined with imaginative faculty (Newbigging 1895). This category exercises a narrative relationship between the reader and the fable writer, this accepted narrative relationship is based on tales of human experience, wisdom, and knowledge.

Thomas Newbigging also mentions in his book Joseph Jacob's definition of beast fable as "a short humorous allegorical tale, in which animals act in such a way as to illustrate a simple moral truth or inculcate a wise maxim" (Newbegging 1895). This short humorous tale can be found both in verse and prose form and the storyline is often focused on the reader. Often the elements of rationality and irrationality seem to coincide in the structure of the fable, but it is from the experience and knowledge of the reader that the essence of fable can be enhanced.

English and French fabulist have used this genre for its application in political propaganda as well as for general movements in social contexts to reach the desired outcome. In this attempt to use fables for passing on ideas or to promote an objective through fictional narrative, the genre gained new importance during seventeenth century. It is the rationality and factor of truth that make the metaphorical nature of the fable an acceptable form of narrative.

Annabel Patterson establishes in her work the significance of fable by exploring the application of political fabling to early Tudor England. Patterson (1991) The verse fables by Lydgate is the main focus of Patterson while she points out the narrative of English tradition in the fable as a form of resistance to unjust power relations. Political consciousness in fables can be perceived through the reading of fable with an analytical lens, this consciousness extends to the use of veiled propositions (Patterson 1991).

The fundamental nature of literary fiction that enables its reading and interpretation by placing characters and plots in a certain order that it resembles a certain truth is a form of discourse in itself. The lesson intended by the fabulist can be interpreted in numerous ways, also the same fable can be told and understood from several other angles. This factor involves various discourse strategies, and rhetorical devices within a certain narrative structure to deliver a tale that seems relevant throughout human history and present. The stories can be analyzed for their use of characters and their precise ordering of events.

The origin of fables is embedded with slave narrative, before the written and printed versions of instruction, it began with oral narrative of Phaedrus a slave from the time of Tiberius, set free by emperor Augustus, Aesop a slave from Samos. These oral narratives were later recorded and developed by Greek and Roman scholars. This fact about fables beginning as slave narratives is significant while reading a fable because it harks back to a time when speaking truthfully based on facts to someone in a more powerful position than the one who is telling the story may result in death. The slave narrative began with the objective of passing a controversial political idea veiling it with blandishment to avoid offending the higher officials.

A major theme recurring in the fables is that of warning or lesson of injustice where the weak suffer. This lesson can be found throughout the tradition of fables, it can also be seen as social observation that each generation of readers can read and relate to. In the narrative discourse, it becomes an important section to focus on where the power distribution has always been uneven, as a result of which injustice prevails. This hierarchy and power relation in the fables seem to be analogs to human world of every time as it was to the slaves of Greece at that time (Clayton, 2008).

Jean de La Fontaine began his work on fables with the translation and compilation of Aesop's fables along with that of Phaedrus and other slave fables. that also influenced many of his fables. But in due course of time, he began working on fables more seriously, his early works or Book 1 can be categorized as fables centered more politically and socially. Along with the element of irony, there is a subsequent rhetoric style that engages the reader. Even in the translated fables, the essence of La Fontaine's style of narration remains. The oppositional narrative. The fables in this work are going to be from the translation of selected fables by Christopher Betts.

The aspect of power can be distinguished in two ways. The power relation is intrinsic to the fables, as depicted within the narrative of fable, and another is the extrinsic knowledge of power structure that shapes the understanding of the reader. The second is influenced by knowledge, through individual experience of society, culture, and tradition.

Fable 1 - The Heifer, The Goat, and The Sheep, in Alliance with The Lion

This verse fable of La Fontaine is from some of the early works in the collection of selected fables, sixth fable of book 1 in the collection of selected fables. It depicts the author's hold on the narrative as the tale being a reflection of how power prevails. The fable begins with the alliance of a heifer, a goat, and a sheep with the lion. This unlikely alliance can be seen as a political remark of the period, the increasing centralization of power by complete monarchy, and the eagerness of nobles from all around. This reflection can be further understood by the discourse used by fable to replace humans with animals, this literary device enables the reader to associate the knowledge of animal characteristics and that of humans.

This fable of power and ultimate reign of unlawfulness that comes with siding with an immoral yet powerful king plunge to the conclusion of a bad ending. Although the word "alliance" might put a strict tone to the entire exchange, it is ultimately the advantageous individual in power that will claim the profit. The fable ends with the lion claiming all the shares without leaving any to the rest of the members that helped him hunt by stating, "it's my reward because I am a lion" (La Fontaine, 2014).

We have a particular understanding with respect to the animal world and the fact that knowledge has arisen from many domains of knowledge's discourses. Even if there was no awareness that the emergence was only a possibility, a certain understanding of animals would have arisen (Johnson, 2012). This knowledge helps the reader of a fable to decipher the relevance to a social scenario and to implement the wisdom behind the beast fable.

Fable 2 - The Wolf and The Lamb

This particular fable can be categorized as a didactic as well as social commentary. The fable began with reasoning through example and observation, where a lamb comes face to face with a wolf while quenching his thirst. The wolf began accusing the lamb of polluting the river, the drinking place of the wolf. The tone of the wolf here is authoritative and claiming ownership of the river as his own, is a point to focus on while analyzing the fable for power relations. The brazen claim that makes the wolf entitled in accusing the lamb of quenching his thirst sends a clear streak of injustice.

The cruelty of wolf then proceeds further to blame the lamb for speaking ill of the wolf, when the lamb politely denies doing so, the wolf boldly stretches the blame to his family and extended family (La Fontaine, 2014)

It is evident from the ending of the fable that the wolf made up his mind to eat the lamb the moment he came across it drinking water, the elaborate conversation can be seen as a blatant way of the wolf trying to justify his action through the meaningless argument, trying to make sense of the heinous act. It is also from the characteristics of the Aesopic tradition that the warning in the fable is mixed with the rhetoric of mockery. This underlying characteristic is sometimes subtle, especially during the seventeenth century when the king was authoritative and the ones in charge of the courtly affairs were given power to exercise as they please.

The social commentary can be seen as a parallel to the reigning power in France that was expanding and engulfing with pomp and show, leaving behind a trail of social injustice. The vast organization of monarchy, church, and nobility was doused with an uneven distribution of authority and control over the weaker class. These were at the receiving end of a troubling tax-paying class. This unrest with the addition of other factors will lead to the French revolution in the upcoming decades.

A question of seventeenth-century totalitarian and centralized power of the king was posed to Foucault, to which he replied that the power is not singular. The power structure is an intricate network of authority, social relations, family and kinship, education system, administration, etc. (Foucault, 1984). And it does not stop at that, it is the knowledge system that controls and exerts power. This knowledge system extends to how our understanding is shaped within a state apparatus and how we happen to interpret or reason the 'truth' and 'experiences' around us. The objective reality of human interaction in the complex social structure is what makes it unique for each reader to decipher the intended meaning of the tale, even when the author specifies the circumstances and characters, there is a window of individual perception.

Conclusion

H. J. Blackham sums up the understanding of fable structure into three elements, idea, image, and expression. According to him, the idea is reflected in the image and the image is a metaphorical expression that is developed through narrative. Methodologically, a fable as fiction is an imagined action as well as a metaphor, this action resembles a certain 'truth' in the mind of the fabulist and the reader. This 'truth' is highlighted through a direct statement and left for interpretation (Blackham (1985).

It is critical to observe the tradition of speaking animals, tracing the classical Greek fables as well as the Indian, Arabian and Chinese tales, the role of the animal in political and social commentary is unique to the genre.

It is often debated as to why fable writers use beast who is incapable of speech in reality as characters in fables. A straightforward answer by Deborah Steiner "is that the very silence of animals makes them ideal ciphers and spokesmen for humans" (Steiner 2012).

The fact that fabulists can fill this silence with whatever mode of discourse, poetics, or ideology they chose implies the coerced identification of differences and similarities between the two. Depending on the cognitive understanding of a reader, the animals in a fable can be discerned according to their familiar properties. These properties can be paralleled with human traits, thus drawing a common understanding based on the existing knowledge system. The simplest definition of fable cannot hold the endless possibility of framing a narrative, it is from the treasures of literary genre that can encompass several qualities of a fictional tale only adding to the richness of its rhetoric within a fascinatingly compact plot.

References

Blackham, H. J. (1985). Fable as Literature. London: The Athlone Press.

Clayton, Edward (2008). Aesop, Aristotle, and Animals: The Role of Fables in Human Life. Humanitas: Interdisciplinary Journal, 21(2), 182.

Foucault, Michel (1984). The Foucault Reader. Paul Rainbow (ed.). New York: Pantheon Books.

La Fontaine, Jean De. (2014) Selected Fables. (trans) Christopher Betts. Oxford University Press.

Newbigging, Thomas. (1895). Fables and Fabulists: Ancient and Modern. New York: Frederick A. Stokes Company Publishers.

Patterson, Annabel M. (1991). Fables of Power: Aesopian Writing and Political History. Durham: Duke University Press.

Steiner, Deborah. (2012). Fables and Frames: The Poetics and Politics of Animal Fables in Hesiod, Archilochus, and the Aesopica. *Arethusa.* 45(1), 36. https://www.jstor.org/stable/26322720

Lockdown: Real-life Paradoxical Experience to Sustain Human Relationships, Healthy Lives, and Well Being

Thomas A. Mattappallil,[*,a,b] *Saji Varghese,*[c]
Vineeth Sahadevan,[d] *and Ashok Jacob Mathew*[e]

[*,a,c]Christ Deemed to be University, Bengaluru, Karnataka, India
[b]Rajagiri College of Social Sciences (Autonomous), Cochin, India
[d,e]St. Claret College, Bengaluru, Karnataka, India.
E-mail: [*,a]thomas.mattappallil@res.christuniversity.in, [b]thomasamp@rajagiri.edu,
[c]saji.varghese@christuniversity.in, [d]vineeth@claretcollege.edu.in, [e]ashok@claretcollege.edu.in

Abstract

Lockdown is a paradoxical situation executed by the government to regulate the pandemic where the citizens live in their houses and go out only for essential needs. It argues for people's spiritual dimension of good health dxcdc and well-being during the lockdown to sustain human relationships in a paradoxical experience. Lockdown as a paradox is a concept or theory used to analyze the revitalizing spiritual impact on human relationships, good health, and well-being. The qualitative method of interviewing and textual analysis explores the real-life paradoxical experiences of people. Interviews and recordings of people's real-life experiences and literary works like *Letters from Lockdown literature* by Natasha Kaplinsky, *Lock tales* Edited by Bhavani S and Meenakshi Shivram are the primary sources for the research. The research examines good health and well-being through anecdotes and people's real-life experiences in the lockdown. Meanwhile, it explores the influence of lockdown as an opportunity for spiritual development.

Keywords: Lockdown, paradox, human relationships, good health, and well being

Introduction

Lockdown is an offshoot of the pandemic period that has rehabilitated human lives and relationships quickly. Human life has been facing contradictory experiences from lockdown till now. This paper intends to explore and examine people's spiritual dimension of good health and well-being through human relationships during the lockdown for the world's sustainable development. The contradictory experiences of people are shared through anecdotes, testimonials, and stories from their real-life experiences read in the literary works and heard in the interview. It aims to explore the influence of lockdown as an opportunity for the spiritual development of human relationships, good health, and well-being of people for a future prosperous and peaceful life on the planet. Human relationships constitute the development of society by eradicating poverty, hunger, and illness. The spiritual dimension of people has brought a radical shift in human relationships, health, and well-being during the pandemic and lockdown. People from their real-life experiences have listened to lockdown as a spiritual opportunity to reflect on their souls, life, and other people's lives. The plagues in human history reiterate the rise of a new spiritual mentality and human relationships with God. The first plague brought a paradigm shift in people's thinking and relationship with God (Anasuya, 18). Consequently, it alters our way of thinking about spiritual and health issues. The plague of Athens in the fifth century BC emphasizes similar examples of people rethinking the spiritual dimension of good health and well-being in society (Boas, 2).

Similarly, lockdown influences people to rethink their well-being and good health. People's reaction towards lockdown as a contradictory real-life experience offers a solid contradiction with robust mindset, reasoning, and thinking. People's attitude also determines the sustainability of human relationships, healthy lives, and well-being of people in the lockdown. Paradoxes offer apparent contradiction based on the psychological reactions of people within acceptable premises (Paul, 377). People's psychological behavior and mindset assist them in overcoming contradictory situations in life. People have adopted meditation or breathing techniques to help lighten their problems like worrying in their life or other psychological issues. It relieves the tense situation of any individual and creates a fresh mind (Pal et.al., 210).

Nature and Human Relationships

Nature heals the paradoxical experiences of people to balance their healthy life, relationships, and well-being during the lockdown. The bond with nature relieves the human mind, body, and soul from all sorts of frustrations and distress and helps them to sustain and exist in life. It is said that individuals with stronger or weaker connections with nature have a positive insight into the pandemic (Hassova et al., 2020). Lockdown is a golden opportunity that reminds the significance of health. The word "health" means the state of emotional and social well-being. People need good health to manage stress and live a long and active life (Pal et al., 210–211). Al Gore, an environmental activist, in her letter to Arlo and Kika in the book *Letters from Lockdown literature* by Natasha Kaplinsky, says that "we have been able to spend outside in nature during this global crisis has been a powerful healing force" "letters from lockdown (LFL, 20)." In the initial days of lockdown, she was baffled and disoriented about life and human relationships. She experienced healing from nature because she believed in God's presence in nature's silence. It gave her the strength from nature to survive in the lockdown period. According to Al Gore, her experience of God in nature itself is a kind of prayer where she voluntarily surrenders herself to be in communion with God. It results in the growth of her faith and charitable deeds of love within the society during the lockdown (Louw, 5). The attitude of Al Gore is relevant in surviving and overcoming the lockdown and pandemic.

Friendship and Family

Dame Mary Berry is a cookery writer and TV personality, and she said that "lockdown for her was a time to be at home, much more than I usually would, and to spend more time in the garden when the sun shone and I could appreciate nature and pottering around" (LFL, 25). During this time, she called her friend who was experiencing loneliness. Her life shows that sustaining human relationships in lockdown requires immense patience and contentment. She deliberates on the significance of preserving values like respecting other feelings, expressions, and desires in maintaining human relationships in the lockdown situation. Dame Mary has genuinely understood the importance of the soul, the value of human relationships, and its connection with the universe in this modern world '(Phaedra, 268)'. The significance of cultivating people's healthy lives and well-being starts with understanding other people's problems and is evident in the life of Dame Mary Berry. The life example of Dame Mary Berry accords with the fact that the actual truth exists indeterminately with the complementary relations (Kimmelman, 151).

In an interview with Mr. Roy, he talks of his real-life lockdown experiences of lockdown. His experiences prove that valuable human relationships born in paradoxical situations sustain many relationships. His relationships with friends have created healthy living conditions themselves. Mr. Roy A Parekatt is a retired professor in botany from K. E. College, Kerala, and he says that his close acquaintance with nature and friends has strengthened his soul. Since he has been diagnosed with cancer, he returned home after treatment during the lockdown. Mr. Roy has experienced the presence of God from his friends in his soul during this lockdown. He has had cancer and its pain by serving God in nature and his friends in his day-to-day life (Chu, 29). Initially, he has more difficulty maintaining friendships, especially in face-to-face interactions. He said he had kept their company through regular phone calls in the later stage. As a result of it, they have created a good bond among themselves by doing some farming and sharing that news among themselves. The blessing, contentment, enjoyment, and happiness have strengthened his bond with nature and friends. His continuous acquaintance with God is a contingent factor in sustaining his relationships with himself, his friends, and nature.

Dr. Deepali Mallya, Assistant Professor, Christ University, Bangalore, Karnataka, India, shared her life experience during the lockdown. She said that 'Covid 19 means Family time, happiness, rejuvenation, and Peace!' in the book *Lock Tales* (Locktales 120). She has experienced a tedious time; however, she is happy with her life and family during the lockdown period. She did creative work like painting pots and glasses with her mother (Locktales, 120). Dr. Deepali spent her time caring for her mother, which is relevant to Pope France's observation, "we need to think of ourselves more and more as a single-family dwelling in a common home" (Pope, 17). She has cared for her mother, which helped her to celebrate Covid19 as a time to rejoice with family. This optimistic behavior has benefited her to overpower the eternal existence of lockdown by being with her mother. The consciousness of caring for her mother and others is evident in Dr. Deepali's life. It has to be practiced in everyone's life to solve hunger, poverty, and sickness of the poorest people in relevance to the first five sustainable development goals in the UN plan for 2030.

Migrant Labourers

An interview with Mr. Ahmed Khan, a migrant laborer from Bihar, was conducted to know the extremely contradictory experiences and paradoxical situation of migrant laborers during this lockdown. He said he was with four other friends in a single room near Gunjur, Chikka Bellandur, and Bangalore. They were not allowed to go out to buy vegetables, and we were beaten by the police every time during the first lockdown and had food once a day. He has reiterated that being with friends greatly supports overcoming this lockdown time. In the second lockdown, while traveling back to their native by train, they do not have the essential food items to satiate their hunger, as mentioned by Ahmed khan. Listening to and watching songs and movies was the only relief they had to overcome the lockdown period. The situation of Ahmed Khan and his friends is similar to all migrant laborers in lockdown. They have to walk a long distance to see their family in their native. Few state boundaries were not ready to accept the migrant workers as they could be "carriers" of the dreaded virus (Prakash et.al., 42).

Sustaining the life and lives of people in the family during the lockdown was an excellent task for the migrant laborers. In a lockdown situation, people in various families died from Covid-19, and the deceased members didn't get the last rites of cremation or burial (Prakash et al., 2). The real-life paradoxical situation of Ahmed khan and other migrant laborers was at its extreme level of desire to sustain their healthy lives and well-being with their families. He and his friends have undergone intense painful experiences of hunger and sickness detached from families and poverty, and they become the poorest in the world during the pandemic. Ahmed Khan and his friends have experienced this precarious situation as a risky and challenging opportunity to overcome hunger, sickness, and poverty and sustain their relationships with family. They have been trapped in the migrated lands, far away from their families and loved ones, with minimal essential needs. It is understood that they have no job and money in their hands to survive (Bhagat, 705–718). The government has instructed the employers to pay even for their food bills, which were deducted from their salary (Prakash et.al., 42). It shows the government's passive attitude in understanding the people's fundamental problems. As a result, a sense of alienation or isolation has been created in their minds. Still, a sense of community, attachment, and belongingness towards their family made them exist in this precarious situation without fear (Gatti et al., 4385). The life of migrant laborers shows that solid willpower and desire are necessary for executing various measures for the world's sustainable development.

Local Communities

Mr. Sachin K B is a Media Production Associate from Bangalore, Karnataka. An interview was conducted with him to know his lockdown experiences. His father has found it challenging to manage the business during the lockdown. He says that he has joined a job in a degree college after his degree, which has helped him take care of his family during the lockdown time. He is being forced to do work for the promotion of the college. He wished to return to everyday situations because there are no restrictions on being with family and friends. He learned many new things during the lockdown, which was a memorable moment. During the lockdown, they have free time to talk among themselves during lunch. In his observation, "lockdown has made rich richer and poor poorer". His friends called him to get some relief because they were restless. Lockdown has helped him and his friends to strengthen their relationships through phone calls. The government insisted on wearing the mask, which was healthier for his life, but at the same time, their stamina went down. The life experience and anecdotes of Mr. Sachin show the life of a middle-class family during the lockdown time.

These life experiences and anecdotes tell that he has developed a sense of hope that he can exist even in such a contradictory state by protecting each other. It is further understood that the purpose of community creates a "we" feeling during the lockdown. The "we" represents the collective subjective realities of people, and it depicts the historical and political facts and memory from the real-life paradoxical and spiritual experiences of lockdown (Dwivedi et al., 2). This sense of community or purpose of "we" is essential in eradicating poverty, hunger, and other problems by providing the necessary help to the poorest people in the world. In the case of Sachin, his family, and friends, their collective subjective experiences of lockdown have explored the possibility of overcoming lockdown as a real-life paradoxical experience in a local community. The social, religious, and friend's community has extended its hands for basic essential needs. The support and help from the local community have given Sachin the confidence and power to provide a healthy life and well-being for his family and friends. This power is required to develop the poorest people within our local communities. It coincides with Kierkgard's analysis of human existence, where he says that freedom gives power to people to choose things wisely in their life (Giese, 59–60).

It is understood that the local community around Sachin and his family has provided them with essential food items daily during the lockdown. It has helped his family to survive during the lockdown. His family has helped the needy in their local community with what they have and live with happiness and contentment. His sense of community has helped him listen to his friend's grief and help them be happy. He has made a great effort to take care of his family and friends by adjusting himself to the lockdown situation. Since he has that strong will and determination to withstand all his relationships with hope and wish that everyone's life will return to normal position. Local relationships and supportiveness have been the driving force for the existence of people in stressful and disruptive conditions for spiritual development of people (Gatti et al., 4385).

Conclusion

The research proves that lockdown is an opportunity for everyone to reflect on their lives. Al Gore witnessed the power of God in the serene, calm nature, which strengthened her faith to heal herself and her friend's problems. In the case of Dame Mary, though she underwent the antithetical state, she tried to find her connection with their soul and the people in this world. Her life is an example of the actual existence indeterminately with complementary relations, and it is fundamental for peaceful coexistence.

The life of migrant laborers depicts the harsh realities of suffering from poverty, famine, and sickness during the lockdown. It has been discovered that their strong willpower and desire are necessary to consistently provide the basic amenities and facilities for their healthy lives and well-being. Lockdown has made a paradigm shift in people's mindsets, attitudes, and behavior. Lockdown as a paradoxical and spiritual experience can be studied with various theories in theory and philosophy. The study shows adverse and beneficial experiences of the people affected in sustaining the human relationships, healthy lives, and well-being of people in the lockdown period. Lockdown in its spiritual sense has made people aware of care, community, freedom, blessing, contentment, enjoyment, and happiness in their lives. Through the lives of people like Mr. Sachin, Mr. Roy, Mr. Ahmed Khan, Dame Mary, and Al Gore, we can understand that people have started adopting all these values in their life. The lockdown was a time to regain our old normal in the new normal conditions.

Acknowledgments

Thankful for the support and cooperation from my supervisor Dr. Saji Varghese, Associate Professor, Department of English, CHRIST Deemed to be University, Bengaluru, Karnataka, India, and from the management and Department of Languages, Rajagiri College of Social Sciences (Autonomous), Cochin, Kerala, India.

References

Bhagat, R. B., Reshmi, R. S., Sahoo, H., Roy, A. K., and Govil, D. (2020). The COVID-19, Migration and Livelihood in India: Challenges and Policy Issues: Challenges and Policy Issues. Migration Letters, 17(5), 705–718.

Calida, C. (2021). Theology of the pain of God in the era of COVID-19: the reflections on sufferings by three Hong Kong churches through online services. Practical Theology. 14(1-2):22 34.

Anasuya, D. (2021). Xenophobia And Covid-19 From Literary: And Media Perspective. Pandemic and Xenophobia: An Analysis of The Eyes of Darkness. The Central University of Kerala, 18.

Pope, Francis. (2020). *Fratelli Tutti: On the Fraternity and Social Friendship*, Rome: Libreria Editrice, Vaticana, 17.

Gatti, F., and Procentese, F. (2021). Local community experience as an anchor sustaining reorientation processes during the COVID-19 pandemic. Sustainability. 13(8), 4385.

Giese, I. (2011). Kierkegaard's Analysis of Human Existence in *Either/Or*: There is No Choice Between Aesthetics and Ethics, Int. J. of Phil. Stud. 19(1), 59–73.

Haasova, S., et al. (2020). Connectedness with nature and individual responses to a pandemic: an exploratory study. Front. in Psycho. 2215.

Kaplinsky, N. (2021). Letters from Lockdown, London: Wren & Rook.

K.B, Sachin (2022). Telephonic interview.

Khan, Ahmed (2022). Personal interview.

Kimmelman, B. (2003). George Oppen's Silence and the Role of Uncertainty in Post-War American Avant-Garde Poetry." Mosaic: A Journal for the Interdisciplinary Study of Literature. 36 (2):145 162.

Pablo, N. (1974). Keeping Quiet, Pablo Neruda. [online] Bu edu.

Parekatt, A. R. (2022). Personal interview.

Aseem, Prakash., et al. (2022). Pandemic Precarity Life, Livelihood, and Death in the Time of the Pandemic. Economic and Political Weekly. 57(50). 40–45.

Saka, Paul (2013). Mind and paradox: Paradoxes depend on Minds. J. of Experi. and Theor. Artifi. Intel. 25 (3). 377–387.

Sanjeeviraja, B., et al. (2021). Locktales, Bengaluru: Darpan and Imprint of Prism Books.

Surya, K. P., and Ashok, K. P., (2021) The impact of increase in COVID-19 cases with exceptional situation to SDG: Good health and well-being, J. of Stat. and Man. Sys. 24(1), 209–228.

United Nations. (2022). Transforming Our World: the 2030 Agenda for Sustainable Development.

Effect of Parenting Stress on Parenting Sense of Competence Among Mothers of Children with ADHD

Jonah Angeline,[*,1] *and Maya Rathnasabapathy*[2]

[1,2]School of Social Sciences and Languages, Vellore Institute of Technology, Chennai – 600127, Tamil Nadu, India.
E-mail: [*]maya.r@vit.ac.in

Abstract

Attention deficit hyperactive disorder is a neurodevelopmental disorder that affects children's behavior, socialization, and academic behavior. Parenting requires various skills and healthy parenting may lead to healthy parent-child interaction and supports the child's positive behavior. Parenting children with ADHD may lead to parenting stress. Parenting stress may affect the parenting sense of competence. This paper attempts to see the effect of parenting stress and parenting sense of competence among mothers of child with ADHD. The samples include 90 mothers of children who have ADHD. The instruments are composed of the parenting stress index (PSI) to measure parenting stress and Parenting sense of competence scale to measure parenting self-efficacy. Results indicated that there was a significant relationship between parenting stress and the parenting sense of competence of mothers of children with ADHD. The higher the parenting stress, the lower the parenting sense of competence. The detailed results are discussed.

Keywords: Parenting stress, Attention deficit hyperactive disorder, ADHD, Parenting sense of competence

1. Introduction

It is perceived that any neurodevelopment disorder influences the entire family members (Earnhart, 2015), and the existence of small children with a Neurodevelopmental disorder affects the versatile instruments of the family (Minnes, 1998). Caregivers of the children impact their child's well-being and security, their prosperity is fundamental to upkeep a positive family environment. So to know these reasons, the latest pattern is to shift the attention in research from each individual to all members of family of child with disorder (Gardiner and Iarocci, 2012), particularly understanding anxiety and the knowledge of the parental skill. Bringing up a child is hard, however, bringing up a child with developmental disabilities can give more difficulties for the parents. The level of anxiety in parents whose child have developmental disabilities is essentially greater than those of parents of children without developmental disabilities (Yoong and Koritsas, 2012).

1.1. Parenting

Specifically, negative parenting ways of behaving contain conflicting discipline, discipline harshly, unfortunate checking and oversight, and have been more than once connected to child externalizing) issues (Pinquart, 2017) and show proof of congruity across ages (Bailey and Hawins, 2009; Tung and Lee, 2012). Negative parenting ways of behaving give a negative model of conduct, neglect to advance favorable to social youngster conduct, and obstruct improvement of versatile social-mental abilities (Trivette, 2010). Such deficiencies place child in danger of creating externalizing messes during their growth (Cooper, 2011).

1.2. Positive Parenting

Positive parenting (e.g., warmth, assertive discipline, realistic expectation, creating a positive environment, and parental inclusion) provides a way for healthy child development (Stach and Ruttle, 2010) that is directly related to minimizing child behavior problems (Kotchick and Forehead, 2002). Positive parenting is the key essential for self-regulatory skills of child, and for developing a positive and healthy mentality (Biglan and Sandler, 2012). Children who grow in positive and healthy environment and with positive authoritarian parenting shows low when they face discrimination, financial problems, and negligence (Odgers and Moftt, 2012). Advancing positive nurturing ways of behaving is a valuable system in working on the government assistance and psychosocial improvement of youngsters (Sanders, 2008; Rincon and Bustos, 2018).

1.3. Parenting Stress

During child-rearing parenting behavior due to parenting stress and the influence of child behavior outcomes are considered. Parenting stress has adverse effects on parent-child interactions and the quality of life of the parents. Parenting stress and child behavior are interrelated, due to the child's behavior parents get stressed and because of the parents' stressful behavior, child's behavior gets affected. Parenting stress leads to negative parenting and poor parent-child interaction.

1.4. Parenting Sense of Competence

Parenting sense of competence is also another important factor that impacts the parent's behavior and parent-child interaction. Satisfaction and self-efficacy are two main dimensions of parenting sense of competence where the belief of a parent in their successful parenting is termed as parenting stress and the level at which parents feel happy and satisfied in their role of parenting. Studies show a great level of parenting sense of competence.

2. Method

2.1. Hypothesis

H1: There will be a significant relationship between Parental Stress and Parenting Sense of Competence.

2.2. Participants

90 mothers of children with ADHD in and around Chennai, India are selected to participate in the research. For the research special schools were approached and mothers of children with ADHD who volunteered for the research were approached. Mothers filled in the demographic data. Only mothers of children with age 6 to 12 are included in the study. Among the 90 children, 56 are boys and 34 are girls children. The age ranges of mothers were from 22 to 46 years.

2.3. Procedure

Mothers of children with ADHD who come for special education were identified. The researcher got permission from concern school and parents before the research. The parents were well informed about the purpose of the research and parents expressed their willingness by signing the consent. The scales were explained clearly and parents took an average of 15 to 20 minutes to fill out the form.

2.4. Measures

A demographic questionnaire was completed by parents (age, educational qualification of the mother, family income status, family type, gender of the child, birth order of the child, and milestone development of the child)

2.4.1. Parental Stress Scale (PSS)

Parenting Stress Scale (Berry and Jones, 1995) is an 18-item self-report scale designed to measure with 5-point rating scale (1-Completely disagree to 5 completely agree).

2.4.2. Parenting Sense of Competence Scale (PSOC)

Parenting Sense of Competence Scale (PSOC) (Gibaud-Wallston and Wandersman, 1978) is a 17-item scale.

2.5 Analysis

T-test analysis, Descriptive Analysis, and Pearson correlations analysis were used to analyze of data.

3. Results

Table 1. Socio-demographic data of mothers of ADHD children (n = 90)

	Category	Frequency	%	Valid %	Cumulative %
Age	20–30	29	32.2	32.2	32.2
	30–40	61	67.8	67.8	100.0
Education Level	High school Education	3	3.3	3	3.3
	Secondary class Education	9	10.0	10.0	13.3
	College Education	45	50.0	50.0	63.3
	Post Graduate Education	33	36.7	36.7	100.0
Family Type	Joint Family	40	44.4	44.4	44.4
	Nuclear Family	50	55.6	55.6	100.0
Socio-economic status	Lower	7	7.8	7.8	7.8
	Middle	74	82.2	82.2	90.0
	Upper	9	10.0	10.0	100.0
Total Number of Children	1	31	34.4	34.4	34.4
	2	54	60.0	60.0	94.4
	3 and above	5	5.6	5.6	100.0
Place of Birth	Rural	14	15.6	15.6	15.6
	Urban	76	84.4	84.4	100.0
Other Comorbidity	Nil	58	64.4	64.4	64.4
	Learning disability	32	35.6	35.6	100.0

Table 1 shows the socio-demographic data of Mothers of ADHD children. There are 32.2% (29) of 20–30 age range and 67.8% (61) of 30–40 range Mothers of ADHD children participated in this study. Mothers of ADHD children preferably stay in Nuclear families (55.6%) compared to joint families (44.4%). There is a big difference in the place of birth in Mothers of ADHD children who stay in urban having ADHD children (84.4%) compare to rural areas (15.6%). However, Second child (60%) of the parents were mostly affected with ADHD. 35.6% ADHD children have other comorbidity issues. Full-term pregnancy (65.6%) child also affected with ADHD and the pre-mature delivery percentage is 34.4%.

Table 2. Relationship between Parenting Stress and Parenting Sense of Competence of Mothers of children with ADHD (n=90)

Variable	Mean	Std. Deviation	Pearson Correlation
PS_total	49.29	9.387	$-.396^{**}$
SOCTotal	60.14	13.073	

Note: ** 0.01 level

Table 1, Pearson correlation of parenting stress (mean 49.29) significantly ($p < .01$) negatively related to the Parenting sense of competence (mean 60.14). So, if parenting stress increases mothers of ADHD children's Parenting sense of competence decreases. So the hypothesis, "There is significant relationship between Parental Stress and Parenting Sense of Competence" is accepted. Comparatively, Sense of competence (60.14) of mothers of ADHD children was more than their stress level.

4. Discussion

Previous studies reported that parenting sense of competence is crucial for the confidence level of parents in growing up their children, especially in parents of children with ADHD. This is because, for most parents, their level of satisfaction with their role as parents significantly influences their level of satisfaction with life. (Langley et al., 2020; Arakkathara and Bance, 2019). Parents' well-being is highly correlated to children's well-being and existing studies reported that parents' sense of competence has an impact on the quality of care given by parents, which significantly affects children's adjustment to their parenting role. (Coleman and Karraker, 1997; Greenspan, 2003; Hudson, Elek, and Fleck, 2001). So this study focused to assess the role of parental stress on sense of competence of mothers of ADHD children.

Correlational analysis results clearly show parenting stress is negatively related to Parenting sense of competence of mothers of ADHD children whereas parents who have a higher level of sense of competence estimated them to have less level of stress. Previous studies also supported these findings (Lazarus and Folkman, 2004) stress negatively affects self-esteem, competence, and self-confidence. Likewise, parents who feel not controlled in their children's lives are likely to feel less competent as parents and in other spheres of life. They also feel low receptive to their child's needs (Rybski and Israel, 2017; Zukosky, 2009).

5. Findings

Parents who feel efficient and satisfied consider themselves in efficient their parenting behavior and considered themselves as competent parent. (Rybski and Israel, 2017; Sevngy & Loutzenhised, 2010). As stress is a tension arouse in mind that affects mind and body, it affects various psychological factors. From the result, it is shown Mothers of ADHD children experience higher feelings of competence that have been associated with low levels of Parenting stress that eventually increase the psychological well-being of parents. This result was supported by others studies that were discussed in the discussion part. As mothers of children with ADHD undergo a challenge in the satisfaction and efficiency level of their parenting, parenting stress affects these efficacy and satisfaction level of their parenting. So it is important to address parenting stress by providing parenting training, parental counseling, and social support to reduce the parenting stress so that their efficacy and satisfaction level will be increased and that leads to positive and successful parenting which in turn enhances parent–child relationship and overall enhance the psychological well-being of both mother and the child.

6. References

Arakkathara, J. G., and Bance, L. O. (2019). Promotion of well-being, resilience and stress management (POWER): An intervention program for mothers of children with intellectual disability: A pilot study. *Indian Journal of Positive Psychology*, 10(4), 294–299. [Google Scholar] [Ref list]

Bailey, J. A., Hill, K. G., Oesterle S., and Hawkins, J. D. (2009). Parenting practices and problem behavior across three generations: Monitoring, harsh discipline, and drug use in the intergenerational transmission of externalizing behavior. Developmental Psychology, 45(5), 1214–1226.

Bayer, J. K., Sanson, A. V., and Hemphill, S. A. (2006). Parent influences on early childhood internalizing difficulties. Journal of Applied Developmental Psychology, 27, 542–559.

Biglan, A., Flay, B. R., Embry, D. D., and Sandler, I. N. (2012). The critical role of nurturing environments for promoting human well-being. American Psychologist, 67(4), 257–271.

Bunga, D., Manchala, H. G., Tondehal, N., and Shankar, U. (2020). Children with intellectual disability, impact on caregivers: A cross-sectional study. Indian Journal of Social Psychiatry, 36(2), 151. 10.4103/ijsp.ijsp_81_19

Cabaj, J. L., McDonald, S. W., and Tough, S. C. (2014). Early childhood risk and resilience factors for behavioral and emotional problems in middle childhood. BMC Petiatory, 14(166), 1.

Chavira, V., Lopez, S. R., Blacher, J., and Shapiro, J. (2000). Latina mothers' attributions, emotions, and reactions to the problem behaviors of their children with developmental disabilities. Journal of Child Psychology and Psychiatry, 41(2), 245–252. 10.1111/1469-7610.00605 [PubMed] [CrossRef] [Google Scholar] [Ref list]

Coleman, P. K., and Karraker, K. H. (1997). Self-efficacy and parenting quality: Findings and future applications. Developmental Review, 18(1), 47–85. https://doi.org/10.1006/drev.1997.0448

Cooper, P., and Jacobs, B. (2011) Evidence of best practice models and outcomes in the education of children with emotional disturbance/behavioral difficulties: an international review. NCSE, Co. Meath. https://ncse.ie/wp-content/uploads/2016/08/Resea rch_Report_7_EBD.pdf

Earnhart, C. L. (2015). Evidence-Based Recommendations for Parents of Children with Developmental Disabilities: A Best Practice Approach, Electronic thesis, University of Arizona. Retrieved 10. March, 2017 from http://arizona. openrepository.com/arizona/handle/10150/595032

Elek, S. M., Hudson, D. B., and Bouffard, C. (2003). Marital and parenting satisfaction and infant care self-efficacy during the transition to parenthood: The effect of infant sex. Issues in Comprehensive Pediatric Nursing, 26(1), 45–57. https://doi.org/10.1080/01460860390183065

Sustainable Development Goals and Juvenile Justice System: A Comparative Analysis

Dr. Bhavana Sharma,[1,a] *and Ms. Neha Garg*[2,b]

[a]Associate professor, Birla School of Law, Birla Global University, Bhubaneswar, Odisha, India,
[b]Assistant Professor Department of Law Bharati Vidyapeeth (Deemed to be University) Institute of Management and Research, New Delhi, India
E-mail: [1]sharmabhavana44@gmail.com; [2]neha.shrutika.jain1@gmail.com

Abstract

Juvenile justice is a philosophy that has evolved over years and plays a vital role in the reformation of society. Juvenile Justice is self-explanatory, as a juvenile is a person who has not reached adulthood, and justice is the concept of equitable treatment. Because it involves children, who are the most vulnerable group. The United Kingdom and India refuse to abandon these children, they remain committed to the reformative and rehabilitative approach. By 2030, the 17 Sustainable Development Goals set forth by the United Nations are intended to create a world that is more prosperous, more egalitarian, and more secure. This article attempts to provide a quick overview of the Juvenile Justice Systems in India, the United Kingdom, and the United States and will also analyze the impact of sustainable development goals on juvenile and their rights.

Keywords: Juvenile, Justice, Sustainable Development Goals, No Poverty, Peace, Justice, strong institutions

Introduction

> "I believe the purpose of Juvenile Law is to provide a structure where troubled kids can reform themselves and mature into better citizens."
>
> – Judge Sim Eunseok

We've all heard the phrase "Children are society's future assets." Though the idea may appear simple on the surface, it plays an important role in the community since a lack of sufficient nurturing can lead to juvenile delinquency and other harmful repercussions. As a result, when it comes to the juvenile justice system, the burden of proof is heavier. It serves two purposes: one is to connote the idea of juvenile delinquency and to hold individuals responsible for the behavior of children involved in it, and the other is to protect those who require care and protection. However, in a broad sense, the juvenile justice system around the world deals with child offenders and how it regulates, supervises, and, in some cases, corrects the behavior of children toward responsible adults.

Juvenile justice is a law and a system created to safeguard and protect the rights of young/ children. This branch of the law generally deals with underage or minors who have been neglected/abandoned or who have been accused of any crime. This approach tries to rehabilitate these children and assist them in changing their ways since it believes that children can be reformed.

Juvenile Justice Systems in India

In India, there has long been a tradition of treating juvenile offenders differently. Lord Cornwallis established Ragged School for such youngsters in the year 1843, during the colonial period. In the year 1850, the Apprentice Act was enacted, which dealt with juvenile legislation. Then followed the 1876 reformatory School Act. Following this, the reformatory school Acts of 1876 and 1897 indicated a change in criminal ideology from punitive to reformatory measures, with the primary objective now being to reform juveniles rather than subjecting them to punitive measures.

Indian Penal Code was enacted a decade later. Although the Code does not directly mention juvenile offenders, it does contain provisions that apply to minor criminals. Children under the age of seven are doli incapax, meaning they are incapable of committing a crime, according to section 82 of the IPC as they lack men's rea, or the intention to commit a crime. Section 83 is mostly concerned with children aged seven to twelve. These youngsters are punishable if they comprehend the nature of the crime while committing it. Young offenders are also addressed in sections 27 and 360 of the Code of Criminal Procedure, 1973.

The current juvenile justice system is governed by several international treaties. UN Standard Minimum Rules for the Administration of Justice and UN Convention on the Rights of the Child (CRC) are two examples (Beijing Rules). In addition, India's Constitution, Article 15(3), makes special provisions for children. This article expressly provides for the protection of children in the constitution of India. Furthermore, children have access to Articles 21, 23, and 24, which deal with fundamental rights. Children are also included in Articles 39(e) and (f), as well as Article 45 of the Indian Constitution.

In 1974, the National Policy for Children was released, which addressed training and rehabilitation, destitution, and neglected and exploited children. The Juvenile Justice Act, which covers all aspects of juvenile justice, was passed in 1986. A juvenile is defined by the Act as a child who has not reached the age of eighteen. It was repealed in 2015 by the Juvenile Justice (Care and Protection of Children Act), 2015. The Juvenile Justice (Care and Protection of Children) Act, 2015 was enacted to codify and amend the law relating to juveniles in conflict with the law and children in need of care and protection by providing the proper care, protection, and treatment that is necessary for their development by adopting a child-friendly approach in the adjudication and disposition of matters in the best interest of children and for their ultimate rehabilitation through various means.

A juvenile/child is a person who is under the age of 18 according to the Juvenile Justice (Care and Protection of Children) Act 2015. The term "child" is further divided into two categories by this legislation:

1. "A child in conflict with law"
2. "A child in need of care and protection"

According to section 14 of the JJ Act, 2015, a child who has been exploited, abused, or abandoned is referred to as a child in need of care. While young people charged with crimes fall under the category of "children in conflict with the law."

Child in Conflict With the Law

It refers to a child who is accused or found guilty of committing a crime and who is under the age of eighteen at the time the crime was committed.

According to the 2018 "Crime in India" report, about 85 percent of detained minors lived with their parents, indicating a systemic failure in terms of nurturing future generations. Furthermore, IPC-related crimes accounted for 92 percent of juvenile cases. Almost 40% of the offenses involved damaging the human body, including bodily harm and grave bodily harm, rape, and attack on women to insult their modesty. It's worth noting that 91 percent of these children have completed at least a primary school education. This metric reflects the state's consistent inability to meet its goal of providing high-quality education.

Child in Need of Care and Protection

A child who is found working in violation of labor laws, who is at risk of marrying before reaching the legal age, who resides with a person who has or threatens to injure, abuse, exploit, or neglect the child or violate any law, or a child whose parents or guardians are unfit to care for him or her are all included.

This Act also established a clear boundary between petty, serious, and heinous offenses. It stated that in the case of heinous crimes allegedly committed by a child who has reached or exceeded the age of 16, a preliminary evaluation would be done to examine his mental and physical capacity to commit such crimes, and the child will subsequently be tried as an adult.

The JJ Act, 2015 acknowledged that the rights of juvenile accused are just as important as those of victims, and hence special provisions were proposed to address heinous crimes committed by people aged 16 to 18. The Children's Court will ensure that any child found guilty of a heinous crime is put in a safe facility until they reach the age of 21, after which they will be sent to jail.

The Juvenile Justice (Care and Protection of Children) Act 2015 also made it easier for abandoned, orphaned, and surrendered children to be adopted. It established the Child Adoption Resources Authority as a legal entity (CARA). It also requires all institutions that provide child care to register.

Juvenile Justice Systems in the United States

When it came to the treatment of juveniles in the United States, the nineteenth century saw a significant shift. New York and Chicago founded the New York House of Refuge in 1825 and the Chicago Reform School in 1855, respectively, to keep children separate from adult dangerous criminals. In Cook County, Chicago, Illinois, the first juvenile court in the country was established in 1899. After that, most states in the United States built juvenile court systems. Instead of applying punitive and penal measures, these early juvenile court systems emphasized rehabilitation and reform of the criminal.

Juveniles require different treatment than adults in order to be corrected, hence their correctional methods should change as well. Majorly people believe that juveniles should be held less liable because crimes can be done owing to youth's impulsivity or malleability. Impulsivity is thought to contribute to incapacity by interfering with the ability to consider the implications of one's actions, whereas malleability makes juveniles more susceptible to negative influences, particularly from peers. Because the US has a federal framework that permits states to set their laws, the age factor varies from state to state. The establishment of juvenile courts and other measures to address juvenile delinquency has generally been recognized by courts as a reasonable enlargement of state police authority to protect the welfare and safety of children. *Parens patriae* is the idea that the state has the right to pass laws governing the custody, care, protection, and maintenance of any children under its legal control. The juvenile court may transfer adolescents to adult court if it waives or renounces its jurisdiction.

The most significant federal law about juvenile justice is "The Juvenile Justice and Delinquency Prevention Act" (JJDP) in the United States. JJDP supports the delivery of community-based services to youth who are at risk of becoming delinquent by states and local communities, the training of individuals for jobs in these fields, and the provision of technical support. There are juvenile courts in each of the 50 states and they are governed by local authorities. Around 20 states in the United States have adopted the 'Balanced and Restorative Justice (BARJ) ideology and have drafted Standard Juvenile Court Act. Around 11 states prepared their delinquency law in "Legislative Guide for Drafting Family and Juvenile Court Acts 1969." Juvenile Courts have original jurisdiction over child delinquency proceedings in the majority of states in the United States. A minor under the age of 18 is either arrested or referred to juvenile court. Procedural disparities can be evident in the United States since laws range from state to state. The juvenile justice system offers rehabilitation for individuals who are not of maturity age.

Juvenile Justice Systems in the United Kingdom

Modern Scenario requires much-needed modifications as per the changing scenario. Although the legal system in the United Kingdom has protections for children concerning their rights to employment and rehabilitation, it also featured punitive measures against minors in 1889. A statutory agreement that protected children from cruelty in public life and gave the state the ability to interfere in violent home situations through its instruments was the Children's Charter, which was introduced in 1889. Children Act of 1988 was also introduced which provided certain measures to prevent children from cruelty in their families and it also provided for the establishment of Juvenile Courts along with various other provisions. Later establishment included the Children and Young Persons Act of 1933 which provided for the role of juvenile courts in dealing with juvenile matters.

Pertaining to the criminal liability of a child, UK law provides that a child under the age of 10 years cannot be arrested under The Children and Young Person Act and he should be released immediately under section 34 (2) of the Police Act and Evidence Act. The child can be kept in police custody for 72 hours but if a Juvenile is charged with any offense, he must be brought to the magistrate under Section 46 (1) on the next day on which the charge has been established and not later than that. The UK Court considers that a child who is between the age of 10 to 14 years is incapable of knowing the difference between right and wrong and lacks requisite Mens Rea. But as per the recent report of the 2012 European Commissioner for Human Rights Thomas Hammer Berg has observed that the practice of United Kingdom juvenile justice is very punitive and focus should be made on rehabilitation and reformation of children rather than penal provisions. This is a serious concern because the youngsters who are kept in prison when they release, tend to re-offend the other people thereby increasing crime so United Nations must adopt policies that imprisonment of Juvenile children should be kept as the last resort and the period of imprisonment should be kept short so that the reformatory measures may be adopted to prevent Juvenile violence.

Juvenile Justice and Sustainable Development

Sustainable Development Goal 17 strives for peace, justice, and strong institutions. United Nations Development Program granted human rights and also provided for the establishment of stability and peace among various societies and the implementation of the Rule of Law simultaneously so that a stable and good life may be enjoyed by all. As per the sustainable development goals, the significance of sustainable development is diminished when there is an abundance of violence, exploitation, an increase in crimes, and juvenile abuse. Sustainable development aims to eliminate all forms of violence, exploitation, and racial discrimination as these issues are the major causes of its non-attainment. Women and Children constitute the vulnerable part of society. According to United Nations, young men face the highest risk of becoming murder victims and suspected perpetrators. Sustainable development goal 16.3 provides for the establishment and promotion of rule of law not only at the national level but also at the international level and it also ensures to provide equality of justice to all, this is the major sustainable development goal that marks the underline of the Juvenile justice system. The Juvenile justice system is directly related to the health rights which are being provided under goal 3 and also for the climate change goals that are aimed to provide sustainable development goal 13 and the states should deliver such services to ensure that the rights of the child may not be hampered. The sustainable development goals also aim to provide Human Rights in the Juvenile justice system and also to resolve the dispute amicably and prevent the control abuse of power by the government or the other controlling agencies so that a transparent and efficient and accountable system may be developed that available for all at affordable price. It also includes the elimination of discrimination against children and providing equal justice access to justice in the legal systems throughout the world.

Conclusion and Suggestion

The criminal liability of a child depends upon the history of a country and nowadays a child can be held responsible for aggressive behavior. The Juvenile Justice Act, of 2015 reformulated the approach of rehabilitation and reformation and these measures may be resorted to instead of taking punitive measures against the children. The judicial legislation needs to work on the efficacious development of children and proper drafting of the relevant provision is the need of the hour so that the correction in the legislation can be achieved and a bureaucratic setup must be established leading to the development of child efficacious system. The JS Verma Committee provided an innovative solution to deal with the children by lowering the age to 16 years and providing that if a juvenile of the age of 16 years is sent to imprisonment and is released at around the age of 30 then there will be a reformatory person who will not commit the crime for which he was imprisoned. What the counter may happen, as fake birth certificates can be made; the criminals might come with psychotic disorders which are against the principle of natural justice in a civilized society. So, the need of the hour is to adopt a balanced approach between punitive and reformatory measures so that sustainable development goals may be achieved simultaneously. The Juvenile Justice and sustainable development goal can be achieved by providing access to juvenile justice and relevant information pertaining to it. It should also aim to provide justice that may be accessed equally by all juveniles and there should be a policy of non-discrimination enforced at the municipal level itself. To attain the sustainable development goals pertaining to the juvenile systems, the establishments of domestic institutions must be formed as per the international guidelines and the countries must abide by them through a valid written agreement between them. It should also establish public institutions that provide for the development of juveniles running with government grants so that sustainable development is not hampered due to a lack of financial support. There is also a requirement of complete data scrutinizing pertaining to juveniles in order to have the number of juveniles imprisoned for which crimes. The public official must be appointed in the institutions and he should address the problem of such juveniles, and the motive behind crime and efforts must be made to resolve them. The establishment of the criminal justice system providing justice as per international standards must also be developed. There is a requirement for policy generation by the countries that should follow a particular model so that sufficient funding and infrastructure facilities for the marginalized and vulnerable juveniles. The sustainability of juvenile justice may be achieved when such courts are established at the local level so that the matters are resolved at the lower level and the Higher judiciary which is already overburdened with the cases does not have to deal with such matters. There should be social welfare agencies established at the municipal level so that facilities may be provided to left-out sectors of society. There should be trained social workers and police officers and the appointment of such staff who are well versed and trained. There should be the establishment of supervising development officer for social causes that operate at the district level in consonance with the family courts and other additional authorities may also be appointed to form efficient Juvenile sustainable development at the grass root level. A Juveniles information system must be established that collects information about the courts dealing with juvenile cases dealing with the activities for which the juveniles are imprisoned or kept under detention and policy reformulation may be done to mitigate such crimes

References

Bajpai, G. S. (2006). Making it Work: Juvenile Justice in India, http://www.forensic.to/webhome/drgsbajpai/lcwseminar.pdf

Dr. Anuradha. (2012). Juvenile Justice: critically juxtaposing the models in India and Singapore. Asian Law Institute.

Dr. S. Guruswamy. (1993). Juvenile Delinquency: Occurrence, Care and Cure, 32 Social Defence, and http://family.jrank.org/pages/1006/Juvenile-Delinquency-FamilyStructure.html. Accessed on 29th June 2022

Fauzia Khan. Juvenile Justice Law need to to uphold the twin objective of Justice and Deterrence. https://thewire.in/law/juvenilejustice-laws-need-to-uphold-the-twin-objectives-of-justice-and-deterrence. Accessed on 29th June 2022.

Gardner, Martin R. (1989). The right of Juvenile Offenders to be Punished: Some Implications of Treating kids as Persons, 68 NEB.L.REV, 182, 191.

Bradley, K. (2008). Juvenile Delinquency and the evolution of the British Juvenile Courts. c.1900–1950. History in Focus. Archives.

Panakal, J. J. (1961). Special Training of Police for Prevention of Juvenile Delinquency.

The Indian Journal of Social Work and K. A. Shukla (1983). Juvenile Delinquency in India: Research Trends and Priorities.

Sutherland, E. H. (1939). Principles of criminology. Chicago, Philadelphia: J. B. Lippincott Company.

Kumari, Ved (2004). The Juvenile Justice System in India -From Welfare to Rights, Oxford University Press.

Influence of Karma at Workplace: With Special Reference to Higher Education Institutions in South Gujarat Region

Hiteshkumar Patel,[*,a] *Vinod Patel,*[b] *and Manish Sidhpuria*[c]

[a]UCC & SPBCBA & SDHGCBCA & IT, Veer Narmad South Gujarat University, India
[b]DBIM, Veer Narmad South Gujarat University, India
[c]DBIM, Veer Narmad South Gujarat University, India
E-mail: *hitesh8383@gmail.com

Abstract

Karma aims to work selflessly for others without anybody's expectation. A person's actions create their future. A person's future cannot already be written; that would depend on karma. Therefore, a need for study is arisen to know how a person works at the workplace and how karma and its consequences affect it. The conceptual framework for this study is to understand how karma influences the workplace and what people think about karma and its implementation. The descriptive research design has been used in the literature review, and the research methodology used was primary. The research was conducted among 457 faculty members from the South Gujarat Region through structured questionnaires. This research shows that people believe in karma, and many believe that karma can influence the workplace. The concept of karma is presented in this article as a method for enhancing organizational performance.

Keywords: Karma, Workplace, Framework, Organisational performance

Introduction

Karma is like a multi-colored chameleon. It is not always visible, but it is a form of energy that moves from good to evil to neutral. This energy is often formed from our activities - what we say and do. There is no predefined time limit in which an individual's karma yields fruit. The results of one's activities may surface immediately or over time. It might happen in this lifetime or a future birth in the latter situation. This mechanism explains why our lives experience abrupt and unexplainable ups and downs. Lord Krishna states in Gita:

"Karmanyeva Adhtkaraste Ma Phaleshu Kadhachana
Ma Karma- phala-heturbuhu Ma The' Sangab Asthu Akarmani"

A believer in karma would not blame anybody or anything for their problems. He understands that his actions and intentions shape his existence. He also realizes that although he cannot alter the past, he can lessen the effects of his karma and create a new future for himself by seeking God's blessing. It gives him hope for the future and awareness of his present circumstances. In addition to broadening his viewpoint, it allows him to perceive himself and his existence in terms of not just this life but many lives spanning millions of years.

Your job activities influence your karma. Furthermore, if you perform well, you will be rewarded by helping society or your business grow. Every action has an equal and opposite response. Taking a bribe is also a kind of karma. It is punishable by law. Punishment is the consequence of this karma. Karma thus is not a kind of punishment.

As a result, all actions are karma. Moreover, every action has a result. Therefore, it implies that karma influences our actions. We may also believe karma controls our entire fate. Every choice we make in life sets us on a particular path. Therefore, karma and its consequences are life.

Managers must incorporate karma into the workplace and meet the everyday needs of workflow and procedures. It may be accomplished by just establishing certain principles.

Review of Literature

(Badrinarayanan and Madhavaram, 2008), Emphasized spirituality in the workplace is an increasing trend in spirituality studies. We provide a conceptual framework based on current theoretical foundations to show how important workplace spirituality is in sales firms. It provides test hypotheses for research, guidelines for research, and practical implications.

(Pradhan, 2016), No matter what, Karma-Yoga (Spirit at Work) must work (Nishkama Karma). Unlike Protestant labor ideals, Karma-Yoga emphasizes the intent of the doer (Karta) rather than the outcome (Mukti) (Sansara). Westerners labor for life, while Easterners work for money (Aisharvya Jeevan). This research describes karma yoga in five aspects. The author analyses the links between karma yoga's five components and work happiness, dedication, and quitting intention. Testable hypotheses allow researchers to explore the relationship between karma-yoga and various work attitudes, gaining new insights into this wonderful yet esoteric philosophy of labor.

(Singh & Singh, 2012), Salespeople are prone to spiritual ideas because of their profession, which enables them to see their work with a feeling of responsibility that enhances pleasure, dedication, and fulfillment. The Indian philosophy suggests that one's deeds may create pleasure or misery. Here we present a new idea, Karma Orientation salesman. It affects their sales abilities and drives. Self-awareness enhances sales productivity, ethical behavior, and spiritual well-being while enhancing the feeling of responsibility of salespeople. Karma prospect promotes ongoing work without awaiting a reward. These may be personal, organizational, or supervisory values.

(Chadha, 2021), The goal is to discover how significant eastern and western concepts are in reducing stress and increasing happiness. This article focuses on the importance of karma in Indian religious traditions for happiness. First, the Bhagavad Gita advises not to worry about the "past" or "future" but rather live in the "now." Second, the task is admiration, not the result. The karma theory principles and ideas are therapeutic and are similar to Positive Psychology constructs. This study is part of a growing body of work that seeks to link Indian karma theory philosophy and pleasure.

Gautam and Arora (2019), Stressed the importance of karma yoga, often known as "yoga of action," which involves working selflessly for others without expecting anything in return. The karma yoga paradigm for organizational success is discussed in detail in this article. This study aimed to examine the literature on workplace spirituality and karma yoga practice to determine how karma yoga incorporates spirituality. This paper describes what karma yoga is, how it may be practiced in the workplace, and its benefits. Individual, team and organizational characteristics enhance the organization's overall performance.

Objective of Study

- To know what people think about karma and whether it is real.

- To know karma influences the events that happen in the life of faculty members of higher education institutions.
- To assess whether workplace activities influence the karma of faculty members of higher education institutions.

Research Methodology

Research Design: The study is based on descriptive research in nature.

Sampling Design: Convenient sampling is used due to the lack of time and resources available for data collection. The sample size was restricted to 457 respondents.

Source of Data Collection: The questionnaire method has gathered primary data. The respondents were asked to fill out the provided questionnaire without any briefing about the concepts of karma. It avoided biases in responses.

Sampling Frame: UG and PG Higher Educational Institutions in South Gujarat Region Faculty members. Sampling Technique: Non-Probability Convenient Sampling.

Data Analysis

Table 1. Gender distribution of respondents

Gender	Response Number	Response %
Male	298	65%
Female	159	35%
Total	**457**	**100%**

Table 2. Distribution of karma and whether it is real or not.

Question	Option	Frequency (%)
Do you think that karma is something real?	Yes	90%
	No	7%
	May be	3%
Do you believe that karma is a part of your religion?	Yes	61%
	No	31%
	May be	8%
Do you believe that thoughts and feelings affect karma?	Yes	61%
	No	26%
	May be	13%
Do you believe that after people die, they are reborn in a new body?	Yes	68%
	No	13%
	May be	19%
Is there such a thing as rebirth or reincarnation?	Yes	68%
	No	13%
	May be	19%
Does the soul carry karma during reincarnation from one life to another?	Yes	72%
	No	14%
	May be	14%

Table 2: Shows that over 90% of individuals think that karma is something real, while 7% don't. 61% of individuals believe that karma is a part of their religion, while 31% don't. 68% of individuals believe that they are born in a new body after they die. 68% of the respondents believe that there is something like rebirth or reincarnation and 72% think karma carries the soul from one life to another.

Table 3. Distribution of influence of karma on the work activities performed by Respondents

Questions	Options	Frequency		P value
		n	%	
My spiritual values guide my decision at work.	Strongly Agree	41	9	0.047
	Agree	59	13	
	Neutral	354	77	
	Disagree	44	10	
	Strongly Disagree	28	6	
While working, I am completely focused on the task at hand.	Strongly Agree	152	33	0.003
	Agree	132	29	
	Neutral	126	28	
	Disagree	31	7	
	Strongly Disagree	16	4	
I gladly perform all duties which are allotted to me.	Strongly Agree	171	37	0.005
	Agree	113	25	
	Neutral	31	7	
	Disagree	43	9	
	Strongly Disagree	99	22	

I believe volunteering to help someone is very rewarding.	Strongly Agree	70	15	0.037
	Agree	126	28	
	Neutral	77	17	
	Disagree	92	20	
	Strongly Disagree	92	20	
I help others when they are in trouble at the workplace.	Strongly Agree	165	36	0.003
	Agree	109	24	
	Neutral	76	17	
	Disagree	71	16	
	Strongly Disagree	36	8	
I put conscious efforts into finding a viable solution to others' problems at work.	Strongly Agree	104	23	0.035
	Agree	79	17	
	Neutral	180	39	
	Disagree	26	6	
	Strongly Disagree	68	15	
I feel I must contribute to others.	Strongly Agree	41	9	0.017
	Agree	59	13	
	Neutral	354	77	
	Disagree	44	10	
	Strongly Disagree	28	6	

Major Findings

- Out of 457 respondents, 35%, that is, 159 are female. And 65%, that is 298, are male. So here inferred that males are more than compared to females.
- More than 70% of the respondents believed that karma is something real, and their beliefs and thought can also affect karma. However, the rest of the respondent's beliefs indicate the contrary.
- More than 93% of respondents experience an association of karma with events happening in life among various higher education faculty members belonging to different categories of Institutes. However, the rest of the respondent's beliefs indicate the contrary.
- More than 93% of respondents experience an association of karma with activates performed at the workplace among various higher education faculty members. However, the rest of the respondent's beliefs indicate the contrary.

Conclusion

It is found many people believe in karma. They believe that karma can have an impact on the workplace or even their personal lives. Good and bad deeds are somehow responsible for ensuring fairness and maintaining overall balance in a person's life. The concept of karma is presented in this article as a method for enhancing organizational performance. Nowadays, higher education institutions find it difficult to make their employees happy and involve them in a meaningful way to make them satisfied and loyal to their workplaces. The study further explained that one who practices karma focuses on their work rather than the outcome and thinks for the entire society in both adverse and favorable conditions. Managers in the organizations are also practicing Karma Yoga and can very easily practice Transformational Leadership.

Scope for Future Research

The study focuses solely on the influence of karma at the workplace on faculty members of higher education institutions. It can also be extended to elementary, intermediate, secondary, and higher secondary school teachers. Further study can also be conducted at the national and international level for the education and corporate and public sectors workplace.

References

Badrinarayanan, V., and Madhavaram, S. (2008). Workplace spirituality and the selling organization: A conceptual framework and research propositions. Journal of Personal Selling and Sales Management, 28(4), 421–434.

Chadha, N. (2021). Karma Theory And Positive Psychology: An Overview Psychology. Indian Journal of Psychiatry, 55 (Suppl 2), S150–S152.

Gautam, D., and Arora, N. (2019). Karma yoga: The practical way to integrate spirituality in workplace. Journal of Advanced Research in Dynamical and Control Systems, 11 (6 Special Issue), 1536–1540.

Pradhan, S. (2016). Karma-Yoga (Spirit at Work) and Job Attitudes: Applying Indian Wisdom to Today' s Business Problems. July, 1–13.

Singh, R., and Singh, R. K. (2012). Salesperson's Karma Orientation: A Conceptual Framework and Research Propositions. SSRN Electronic Journal. https://doi.org/10.2139/ssrn.1626699

Cross-Walk of Professional Competencies for Social and Emotional Wellbeing to Cater Mental Health Problems in Schools

Ashraf Alam[*,a], *Atasi Mohanty*[a]

[a]Rekhi Centre of Excellence for the Science of Happiness,
Indian Institute of Technology Kharagpur, India
E-mail: [*]ashraf_alam@kgpian.iitkgp.ac.in

Abstract

This investigation explores teachers' and counselors' perceptions about schools' role in supporting students' mental health, thereby examining their perspectives on the actions taken by schools to offer support. Between July 2021 and March 2022, 91 secondary school teachers and 83 school counsellors ($M_{age} = 39.46$) from New Delhi, Pune, Kolkata, and Bengaluru replied to open-ended questions. Providing support to kids at the right time, working in team, taking lead by being on the frontline, and education, were the important themes that emerged from this research, albeit there were notable differences in perceptions and opinions across school counselors and teaching staff. School counselors showed vigorous support for evidence-based initiatives in schools that specifically addressed students' mental health and wellbeing. Findings of the study suggest that effectively addressing the menace of mental health-related issues among school students would require clearly defined professional responsibilities and a coordinated effort by all the stakeholders.

Keywords: Secondary Schools, Support, Mental Health, Student, Teacher, School Counsellor, Curriculum, Pedagogy, Teacher Education

Mental Health of Students in Indian Secondary Schools

Studies in Indian settings have revealed that young people frequently obtain assistance from counselors and other mental health specialists through their schools. By incorporating mental health education directly into the curriculum, schools can lower access-related obstacles. Secondary schools are in a better position to address students' mental health issues because of their capacity for teaching and spotting psychological changes in students. Internationally, public health organizations have identified the prevention of mental illnesses as a top concern. Additionally, whole-school approaches to mental health and multi-tiered care delivery models are congruent with prevention science. These methods place a strong emphasis on the value of providing treatment for the entire spectrum of mental health. Economic cost evaluations show that prevention measures can lower the direct costs of mental health issues, such as health care, and indirect expenses, such as unemployment (Alam, 2022a).

Secondary school counselors frequently find themselves delivering reactive mental healthcare due to time constraints that prohibit them from focusing on tailored therapy, continuous treatments, or prevention measures. The implementation of proactive and preventative actions by school counselors is further hampered by the lack of resources in schools and the diversity of their tasks, such as getting engaged in teaching primary class kids and performing school administrative work. In India, only a small number of secondary schools have the resources to implement a global preventive strategy.

Perspectives on Secondary Schools' (and Their Staffs') Contributions to Students' Mental Health

Instructors' and School Counsellors' Functions

In the United States, Australia, Europe, and the United Kingdom, educators and school counselors concur that shielding students' mental health is a crucial aspect of their job. They must recognize students' mental health issues, offer an inclusive learning environment, and instruct pupils on mental health. Instructors typically think that counselors have a bigger part in assessing students for mental health issues, doing examinations, giving psychological therapy, and making referrals to other agencies or mental hospitals. In India, despite wishing to concentrate on counseling and advising, counselors often spend most of their time assisting students with academic and administrative work (Alam, 2022b).

Teachers frequently express a need for greater mental health education during pre-service and in-service teacher training programs in order to improve their own literacy and competency. Numerous teacher preparation programs put a strong emphasis on students' mental health (Alam, 2022c). Teachers believe that teacher-counselor collaboration is crucial for facilitating appropriate referrals.

Uncertainty and Conflicting Roles

The way that teachers and counselors help in curing students' mental health is impacted by role ambiguity and conflict. When employment expectations are not categorically explained, role ambiguity results (Alam, 2022d). It can be difficult for school counselors to know what exactly is expected of them to help the pupils, which can lead to role conflict and decreased work satisfaction.

There is an overlap between the job roles of teachers and school-based mental health professionals like school counselors. Teachers acknowledge overlapping duties and emphasises the value of cooperation and a shared objective. Although this idea may help diverse professionals in and out of education to promote students' mental health, there is still room for improvement in how well this is done in practise.

The Present Research

Richly detailed descriptions and in-depth comprehensions of individual responsibilities within the school setting were gathered. The two objectives of this study were: (1) to scrutinize the similarities and differences in counselors' and secondary school teachers' perceptions of the role of the school in supporting students' mental health, and (2) to examine teachers' and counselors' perspectives on the actual actions that schools take to support students' mental health.

This research intends to create a comprehensive knowledge body of the responsibilities of educators and school counselors in students' mental health with a special focus on the Indian setting. Compared to other nations like the US, UK, and Canada, very little study has been done on the viewpoints of school staff in Indian school settings. Examining Indian perspective will expand our current understanding in this field and offer insightful information on school-based student mental health strategies that are applicable in an Indian context.

Method

Recruitment, Participation, and Design

Online surveys of teachers and school counselors from secondary schools in Delhi, Pune, Hyderabad, Kolkata, Bengaluru, and Chennai were carried out in this qualitative investigation. Although school principals were also recruited and their replies were collected, this analysis does not include them due to their scant relation to the study's research objectives. Snowball sampling technique was employed for recruitment of key informants. Additionally, flyers were distributed online on social media websites (e.g., Facebook, LinkedIn, Instagram, and Twitter). The study's key informants were urged to spread the word about it on their networks. Participants who met the eligibility requirements were the ones who worked as teachers, school counselors, or school psychologists at secondary schools in India. The study was endorsed online and recruiting emails were sent out on critical occasions (including the start of school terms) between January 2019 and September 2020.

Participant Statistics

The open-ended questions received responses from a total of 97 instructors (24.7 percent men) and 93 school counselors (10.8 percent men). The average age of teachers was 37.9 years, and they have been teaching for an average of 7.2 years. The average age of school counselors was 41.3 years, and they had been in the position for an average of 7.1 years. The majority of instructors (73.6%) worked in public schools, whereas private institutions employed 91.2% of school counselors. Instructors and counselors were more likely to work at coeducational schools (89.6% and 79.2%, respectively), while schools in capital cities employed more counselors than teachers (93.6 percent and 19.8 percent, respectively). Of these respondents, only 68 school counselors and 72 instructors submitted information that were directly related to the current investigation.

Procedures and Measures

The surveys included qualitative questions to collect data on demographics, work history, and current educational status. Specifically worded, open-ended questions, such as (1) "What do you believe is the function of the school in preserving students' mental health?" (question for educators only); (2) "What do you believe is the function of school psychologists and/or counselors in preserving kids' mental health?" (question for school counselors only); (3) What policies or procedures does your school already have in place to oversee students' mental health? (question for both teachers and counselors); and (4) Have you ever utilized or suggested to your pupils' online therapy programs? (question only for counselors).

Analysis

Rigour in Research

By paying close attention to research procedures, rigor in theme and content analysis was addressed. The use of numerous code talks with experts from eight Indian universities, and continual interaction with the text data were the various methods that were used to address procedural rigor.

Thematic Examination

The pioneering six-stage theme analysis guidelines developed by Clarke and Braun were largely used to evaluate qualitative data. This method was used to examine the data from questions 1 and 2. Thematic analysis makes it possible to find, understand, and report on recurring patterns of meaning in data. Due to its adaptability, rigor, and ability to consider the reflective role of the researcher in interpretation, this methodology was suitable for the available data. MS Excel was used for the coding. The final inter-code reliability was strong (Cohen's Kappa = 1) between the codes. To account for missing data in the codes, the researchers also estimated percent agreement where 96 percent of codes and 79.4 percent of subcodes had an agreement.

Content Evaluation

For questions (3) and (4), a smaller subset of the data was examined using traditional qualitative content analysis. With this method, categories were consistently and unambiguously extracted from text data. After reading and rereading the replies of the respondents, convergence of coding was compared, and incongruities were settled through conversations with knowledgeable professors. The remaining replies were then subjected to an iterative application of the codes and categories. After discussions and deliberations, there was a total agreement.

Results

Regarding how teachers and counselors view the school's involvement in students' mental health, four separate themes emerged from the content analysis.

Theme 1: Assistance

This theme focuses on the various ways that schools may help children to safeguard and enhance their mental health. There were found to be three sub-themes:

Utilizing a Holistic Approach to Treatment. Prioritizing pupils' intellectual and emotional needs.

Ensuring Security and Safety. This sub-theme emphasizes the importance of protecting children's welfare and wellbeing in a school setting.

Advocating for Students' Rights. Supporting students' mental health by speaking out for them.

Theme 2: Frontline

This sub-theme emphasizes on how both educators and counselors consider themselves to be on the frontlines and are frequently the pupils' initial point of contact and admission into mental health care facilities. Four underlying motifs were found:

Detection and Evaluation of Mental Health Issues. This relates to the initial evaluation and detection of pupils' mental health issues.

Prevention. Preventative measures are needed in schools since they are as crucial as therapy itself.

Providing Interventions. This sub-theme describes the interventions or programmes that schools offer to kids directly and are geared toward helping them learn how to manage their mental health issues.

Referral. Being on the frontlines makes it clear that referring people to the right mental health doctors and appropriate mental health programs was the key.

Theme 3: Cooperation

Teachers and counselors highlighted in unison how schools are a small part of a larger network that involves various organizations to support students' mental health. Instances encompassed families, the larger community, and external mental health organizations. To improve collaboration, it is important to communicate about adolescents' mental health issues, the services and programs that are on offer, and the additional supports that are readily accessible. Teachers and counselors stressed the value of being able to communicate with parents freely and honestly about their kid's psychological wellbeing.

Theme 4: Education

The provision of psychoeducation to pupils to raise their understanding of mental health issues and readily available therapies and services was of prominent focus. Two related sub-themes were found:

Education on Psychological Wellbeing. Increased student awareness about the signs and symptoms of mental illness, elements that lead to or worsen mental health, and techniques for caring for their mental health as well as that of others, are essential.

Help-seeking. It is the responsibility of the school to apprise kids about how to obtain support and the many kinds of help that are available in school.

Results of Content Analysis

Programs, services, and school initiatives were the three primary categories used to aggregate questions on the methods educators and counselors utilized to address the mental health problems of students. Programs are described as independent interventions that teachers or counselors might choose to carry out on a private or institutional level. School initiatives are described as any plan created by personnel for a particular purpose inside the educational setting.

Discussion

Thematic analysis uncovered four overarching themes that illustrate how teachers and counselors view schools' contributions to students' mental health as multifaceted, intricate, and integrated into a wider system. These themes also imply that schools frequently choose a reactive rather than a proactive and systematic approach to mental health, adopting various techniques to meet the urgent needs of students and the constraints of the situation. Findings provide light on what teachers and counselors believe they should be doing to help students' mental health as well as how other important stakeholders should assist them. Schools should oversee the assistance offered, spot emerging issues, deliver mental health interventions (i.e., treatment and prevention), make referrals, work with other professionals, and offer mental health education.

Recommendations

The findings of this study suggest that policy makers should distinctly define the duties of those school employees that are participating in mental health promotion, prevention, and intervention as part of wider organizational reforms. Reduced job ambiguity will improve accountability. Schools would be able to identify people in need early and give help either directly or through referral if an online service were made available as part of a continuous, regular evaluation, which might eventually lessen the pressure on teachers to identify those in need. Additionally, this would make it easier to administer medicines. For those with mild to moderate symptoms, tailored online programs could be made available, freeing up counselors to handle the more severe cases. If counselors' total workload was lighter, they would have more time to adopt preventative measures.

References

Alam, A. (2022a). Mapping a Sustainable Future Through Conceptualization of Transformative Learning Framework, Education for Sustainable Development, Critical Reflection, and Responsible Citizenship: An Exploration of Pedagogies for Twenty-First Century Learning. ECS Transactions, 107(1), 9827.

Alam, A. (2022b). Investigating Sustainable Education and Positive Psychology Interventions in Schools Towards Achievement of Sustainable Happiness and Wellbeing for 21st Century Pedagogy and Curriculum. ECS Transactions, 107(1), 19481.

Alam, A. (2022c). Social Robots in Education for Long-Term Human-Robot Interaction: Socially Supportive Behaviour of Robotic Tutor for Creating Robo-Tangible Learning Environment in a Guided Discovery Learning Interaction. ECS Transactions, 107(1), 12389. Retrieved from https://doi.org/10.1149/10701.12389ecst

Alam, A. (2022d). Positive Psychology Goes to School: Conceptualizing Students' Happiness in 21st Century Schools While 'Minding the Mind!' Are We There Yet? Evidence-Backed, School-Based Positive Psychology Interventions. ECS Transactions, 107(1), 11199.

Emotional Intelligence Manages Sustainable Development for an Organization - the Contribution of Psychological Well-Being

S. Sujitha[a] *and Dr. S. Vasantha*[b]

[a,b]Vels Institute of Science, Technology & Advanced Studies (VISTAS), Chennai
E-mail: [a]sujithasatish2007@gmail.com, [b]vasantha.sms@velsuniv.ac.in

Abstract

In this fiercely competitive world, it is an uncommon occurrence for marketers to promote their goods to sustain in society for long-term economic prosperity. Several assumptions essential for the success of the organization are identified, which include the need to incorporate Emotional Intelligence and Psychological Well-Being. But the actual sustainability issues have gotten worse during the same time, when our individuals, and organization, lagged in their communal efforts. Innovation and scaling up efforts of an organization can enrich an understanding of what is required for success. Hence, the purpose of the study is Emotional Intelligence manages sustainable development for an organization along with Psychological Well-Being. The research literature findings suggest that Emotional Intelligence is a crucial factor in promoting and maintaining good relationships within the organization and also for the success of the organization in the market

Keywords: Emotional Intelligence, Employees, Organization, Psychological well-being, Sustainability

Introduction

The Brundtland Report developed the terminology "sustainable development" in the 1980s, defining it as "progress that fulfills the requirements of the current without compromising the capacity of future generations to satisfy their individual requirements" (Mowforth & Munt, 2015). The study's definition of sustainable organizational development is the ongoing, brief growth, and achievement of the goals of the organization (AlAqeel, 2012). Instead of an environmental setting, it aims at generically fostering development in relation to organizations in regard to organizations. To establish sustainable development in terms of the economy, society, and the environment, various policy sectors, including finance, trading, power, farming, and manufacturing, are developed as part of the sustainable development process (Shyle, 2018). People with emotional intelligence have a comparative benefit for an organization, emotional intelligence has emerged as one of the most crucial determinants for success in enterprises. A person with emotional intelligence performs better and fosters an environment where others can do the same, developing leaders at all levels. Employees and managers who have good emotional intelligence can promote a healthy and effective workplace. They work to maintain positive relationships with others by attempting to comprehend things from their points of view. They are more trustworthy, more able to adjust to changing needs, and more capable of acting correctly under pressure. All of these elements contribute to organizational effectiveness.

Organizations must give special consideration to human issues when effectively managing innovation within the workplace. Understanding the intricate psychological connections that take place between people and technology is important as how dependent we are on it in every aspect of life. Emotional intelligence which is the capacity to effectively detect, acknowledge, and control one's emotions for both their own well-being and the well-being of those around them, is an important aspect that can be related to the performance of the organization under such circumstances of Performance, Control progress and Settle disputes. (Nanayakkara et al., 2017) Emotional Intelligence is sustained to enhance employees' performance in an organization both theoretically and empirically. Emotional intelligence has an impact on service quality, which is a basic aspect of customer satisfaction, engagement, commitment, and business success. The key to maximizing organizational performance is to increase the link between a company's services and customer needs, satisfaction, & retention in the service industry. However, when individuals' emotional intelligence fosters coherence between them and their clients, overall organizational performance improves. In the highly stressful workplace of today, when people are motivated to increase efficiency, implement change, and resolve conflicts, Emotional Intelligence is increasingly being acknowledged as a crucial factor of professional success. The study focused on Emotional Intelligence which determines the significance of human resources through certain HR procedures. In the study, secondary data were employed. Information was acquired from books and academic journals, relating to Emotional Intelligence with psychological well-being

A Background of the Study

As new technology emerges and spreads globally, firms that are still using traditional methods are currently encountering issues (Exiron, 2018). Systematic research attention to sustainability and impact management in the context of NBT started to become noticeable with the emergence of the global sustainability agenda in the last decades of the 20th century. An employer-employee relationship is a long-term cohesive bond that arises out of mutual benefit. The stronger the bond, the stronger the propensity of the individual's commitment toward the organization. The psychological contract theory (P.C.) frequently explains this link. Chris Argyris first proposed this idea in 1960, later by Levinson in 1962. It is an unsaid commitment, agreement, or perception of mutual obligation in an employee-employer relationship. The transference theory from psychoanalysis is used in Levinson's thesis to describe how employees see the employer as a substitute parent. As P.C. is more a state of mind that obligates both employee and employer in an employment relationship to be committed, dedicated, and responsible towards each other. This state of mind within an employee often restrains him from making hasty decisions and enables him to nurture a long-term relationship with the employer.

Human Resource Management is a strategic method for managing personnel well and raising employee performance in firms (HRM). It has been a crucial component of current companies over the years. By differentiating them from their rivals and implementing cutting-edge techniques like recruiting, screening, compensation practices, retention strategies, performance evaluation, and so on, HRM aids businesses in achieving their objectives, missions, and goals. Relationships are a crucial human need, both in personal and professional contexts, and they play a significant role in fostering and supporting people's health and well-being (Vincent Egan et al., 2005) Since relationships are essential to one's own well-being (Emma and Dianne, 2008), strengthening strong relationships can also promote well-being at the workplace by encouraging a healthy and sustainable development of an organization. (Di Fabio, Annamaria (2017). Therefore, in this paradigm, sustainability and sustainable development call for boosting the Psychological well-being of employees and organizations through the strengthening of healthy relations and the development of a good account in a normal organizational environment. The development of wholesome and long-lasting organizations increasingly relies on the reservoir and basis of emotional intelligence. The Bar-On model of trait emotional intelligence offers a conventional view of trait emotional intelligence, understanding it as the perspective of someone's social and emotional talents that control a person's interaction with oneself and others and enable them to meet environmental pressures.

The results of the current UN event on Emotional Intelligence and the Sustainability Development Goals and the analysis of literature, both show that the cognitive/emotional and relational capacities that make up the EI concept can facilitate the growth of transformative skills. Both are crucial in promoting persistent change, such as resolving disputes, fostering peace, and creating diverse and just organizations and society. All in all, they have to do with a person's capacity to be self-aware as well as those of others, to distinguish between various emotions and assign them proper labels, to consider one's thoughts after using feelings, emotions and behavior, and to control emotions to challenge and adapt to various surroundings to effectively achieve fair and equal and just priorities.

Thus, one of the effective methods to significantly affect sustainability results is through the ability to critically evaluate one's generalizations, and cognitions leads to perhaps embracing new concepts. Emotional Intelligence can thus have an impact on this ability. EI affects interrelatedness in this context, which refers to how we connect to ourselves, to one another, to the environment, and to the future. This determines how we comprehend and address sustainability. This is because sustainability ultimately revolves around our relationships, especially those with ourselves, the things we hold, and how we perceive ourselves in relation to the rest of the world. Thus, altering how we interact in these relationships is essential to accomplishing the Sustainability Development Goals, especially Sustainability Development Goals 16, which serve as the cornerstone for all other Global Goals. This holds for all organizations, including corporations, political bodies, and educational institutions, not just specific agencies

The Objective of the Study

This article will show how emotional intelligence will impact productivity and also explore how emotional intelligence will manage Sustainable Development for an Organization with the contribution of psychological well-being.

Literature Review

The study states that the primary prevention approach is used to examine how the psychology of sustainability and sustainable development contribute to organizational well-being. It deals with sustainability in terms of enhancing the standard of living for every human being as well as the ecological, economic, and social environments. The psychology of sustainability and sustainable development is seen as a key preventative viewpoint that can promote well-being in organizations at all many levels, from the individual to the group to the organization to inter-organizational processes, according to Di Fabio (2017).

The study aims to identify the effect of Emotional Intelligence on the Psychological Well-Being of employees in 70 branches of Mehr Eqtesad Bank in Tehran, Iran. The researcher collected 300 samples using random sampling, from which the 270 data were scrutinized and used for analysis. Structural Equation Modeling (SEM) analysis was done and the results show an association between Psychological Well-Being & Emotional Intelligence among employees in Iran. (Ahmadi et.al., 2014).

One of the most important elements of a healthy business is emotional intelligence. Without it, there is a higher likelihood that undesirable phenomena like organizational stress and emotional attrition may manifest. However, even if there are just minor gains in productivity, emotional intelligence has a real positive impact on the employees of the company.

Working in an atmosphere with high Emotional Intelligence is one thing; working in one with low EI is quite another. While an environment devoid of Emotional Intelligence presents a barrier to sustainability, the former provides the ideal conditions for the growth of sustainable businesses according to Catalin pirvu (2020).

Research Methodology

In the study, secondary data were employed. Information was acquired from books and academic journals, relating to Emotional Intelligence with psychological well-being along with Sustainability development goals.

Sustainable Development for the Futuristic Organization

Emotional intelligence has been incorporated both implicitly and explicitly into many sustainable organizations, concepts, and strategies for organizational leadership as a result of there is numerous studies on Emotional Intelligence, leadership, organizational change, and ethics/values. The Full Spectrum Method, created by Monica Sharma and used in a number of UN Programmes and projects supporting sustainable development, is one example of a transformational leadership approach that uses it as a key tenet. A high level of emotional intelligence is now required for sustained success in the age of globalization because "Talents" work virtually as well as across other value systems along with Mental health.

Leaders must demonstrate a high level of Emotional Intelligence (EI) in their day-to-day execution if they want to see their firm succeed in the long run.

1. Evaluating the teams' readiness by analyzing their competencies and inclination
2. Directing rather than commanding the conversation when the team members are expressing their opinions.
3. Acquiring the ability to direct the leadership authority at hand in a constructive and gratifying manner
4. Removing their egotistical and stubborn attitudes, adapting to the circumstance, particularly to dealing with the complex scenario
5. Set a good example for others to follow, and you'll have a much greater chance of navigating difficult situations.
6. Understanding that there is no "one size fits all" approach to situational leadership.

Discussion and Conclusion

A high level of emotional intelligence is now required for sustained success in the age of globalization because "Talents" work practically as well as across other value systems. The spirit of the crew could be destroyed by a high-powered boss, which would cause the team members to perform poorly and become disengaged. This would ultimately occur in declining productivity. The two factors of emotional intelligence most strongly correlated with PRM in the degree of aspects were well-being and sociability, demonstrating the significance of self-management in interpersonal interactions and relationship formation. While sociability has to do with how well we get along with other people, well-being is the perception of having many strong good qualities. These results provide additional evidence in favor of the idea that the trait of Emotional intelligence can be viewed as an optimistic personal difference that is in additional linked to fostering workplace well-being through strengthening and managing positive interactions, developing respect and relationality, and ultimately resulting in healthy and sustainable organizations and work environments. The research's conclusions can help management create plans that will be successful and raise job satisfaction. Organizations can take note of these results and plan training sessions and learning opportunities for staff members to raise the level of their emotional intelligence as a result of the study's emphasis on the critical workers' role in emotional intelligence in relation to corporate ethics.

Today's businesses must overcome a wide range of obstacles if they are to achieve long-term success. Numerous studies have shown that an individual's ability to perceive emotion affects how well they handle interpersonal situations at work and organizational politics. An organization should be "People & Financial Driven" in order to achieve commercial sustainability and success. Even for leaders with outstanding expertise and understanding, EQ is a critical component of success. Promotion to a leadership position is given without verification; no formal rules or competency models are adhered to. It simply occurs based on the decision-gut maker's instinct, perceived value, or relationship. There is no denying that if the leaders are not adhering to the set principles and instead employing other emotive techniques of differentiating; leaders who lack emotional intelligence require prompt intervention. It can be difficult, but not impossible, to accept the shift in perspective required to meet the requirements and demands of a multigenerational workforce in a variety of contexts. Challenges can be overcome by persistently working at enhancing one's emotional intelligence in a step-by-step fashion.

References

Ahmadi, S. A. A., Azar, H. K., Sarchoghaei, M. N., and Nagahi, M. (2014). Relationship between emotional intelligence and psychological well-being. International Journal of Research in Organizational Behavior and Human Resource Management, 2(1), 123–144.

AlAqeel, A.A. (2012) Factors Influencing the Sustainable Development of Organizations, University of Gloucestershire.

Catalin pirvu (2020) Emotional intelligence – a catalyst for sustainability in modern business. Theoretical and Empirical Researches in Urban Management, 15(4) (November 2020), pp. 60–69 (10 pages).

Di Fabio, Annamaria (2017). The Psychology of Sustainability and Sustainable Development for Well-Being in Organizations. Frontiers in Psychology, 8, 1534

Gallagher, Emma N., and Vella-Brodrick, Dianne A. (2008). Social support and emotional intelligence as predictors of subjective well-being. , 44(7), 1551–1561.

Erixon, F. (2018). The Economic Benefits of Globalization for Business and Consumers: https://ecipe.org/publications/the-economic-benefits-of globalization for-business-and-consumers/ (Accessed 18.2.2019).

Mowforth, M., and Munt, I. (2015) Tourism and Sustainability: Development, Globalisation and New Tourism in the Third World, Routledge, UK.

Nanayakkara, S. M., Wickramasinghe, V., and Samarasinghe, G. D. (2017). Emotional intelligence, technology strategy and firm's non-financial performance, 2017 Moratuwa Engineering Research Conference (MERCon), 2017, pp. 467–472, doi: 10.1109/MERCon.2017.7980529.

Shyle, I. (2018). Awareness of individuals and businesses in Albania for sustainable development, European Journal of Multidisciplinary Studies, 3(1), pp.46–54

Egan, Vincent, Austin, Elizabeth J., Saklofske, Donald H. (2005). Personality, well-being and health correlates of trait emotional intelligence, 38(3), 547–558.

The Enactment of Social Sustainable Goals in IT Organizations

B. Vidya Sri[a] *and Dr. S. Vasantha*[b]

[a]Vels Institute of Science, Technology and Advanced studies (VISTAS), Chennai, India
[b]Vels Institute of Science, Technology and Advanced studies (VISTAS), Chennai, India
E-mail: [a]vidyasri.3093@gmail.com; [b]vasantha.sms@velsuniv.ac.in

Abstract

The United National Global compact focused on the organization's relationship with its employees. The element of social sustainability is managing the externalities that depend on people, both positively and negatively. When employees feel appreciated and comfortable in their work environment, they are more likely to be productive. This equality promotes long-term economic prosperity. This paper focused on the economic pillars of social goals like Industry, Infrastructure, and economic growth. This study focuses on the factors that paved the way for achieving sustainable development goals in the workplace post covid. The importance of training and the integration of social sustainability in organizational decision-making are the central points of this study. Sustainable management has evolved into a value creation that enables companies to operate more profitably and develop effectively. The study provides a detailed understanding of the strategies used in the IT sector to achieve sustainable development goals.

Keywords: Covid, Employees, Environment, IT Organization, Strategy, Wellbeing

Introduction

The ability of a business to remain relevant and competitive in today's fast-paced environment has become a crucial mission. To be viable, every company must concentrate on its numerous segments. Sustainable development goals (SDG) are now a key topic of discussion in organizations during all phases of policy creation and decision-making. Hence, these economic goals must follow community and environmental practices. Businesses can now use SDG to explore possibilities for solid growth while upholding their collective responsibilities. Various factors make SDG valuable to IT companies. The essential components include reducing risk, ensuring growth, realigning operations, and encouraging new capital for investment.

For any type of Organization knowing the risk before it occurs is essential. To protect the organizations from hazards and to ensure workplace security. Developing a sustainability approach eliminates the risks in IT Industries. The company must have a futuristic perspective on the environment if it wants to succeed with its strategies. These businesses are moving towards corporate sustainability when they encounter growth.

Realigning organizational environments and implementing strategies in action will enhance the performance of an enterprise. Furthermore, it will build a humanistic culture that could be operated to achieve the goal on all levels. Since it will have evolving processes in the long term, it must be well planned before realigning.

Not every organization in the current business climate will have the same quantity of funds to realign the business independently. Some agencies offer financial support to companies that want to shift from their present position and assist them toward sustainability change.

To protect future generations, all businesses, regardless of size, must adopt these sustaining aims. Sustaining these Goals can help us to ensure that the firm survives in this rapidly changing economy.

Methodology

The research paper examines the sustainable growth plans and objectives of the IT industry. The secondary information sources used to gather the data for this conceptually based study included books, journals, articles, and websites.

SDGs in IT Organizations

Meeting employee requirements is the only way for any business to prosper. A successful business strategy must be implemented to satisfy its employees. Organizations seek their policy planners where they may invest, implement innovations, and target ideas that will help their business growth decisions while drafting long-term policies. Many businesses encourage to diversify into new industries by the United Nations' objectives. SDG's central goal is to transform the company for the benefit of its clients and workers.

This article used TREX Company Inc. as an example to discover more about SDG in IT firms. It is a business that produces goods from recycled materials. They have enjoyed a strong track record for sustainable growth in 2021. Their report emphasized how, as a cluster, they have been striving towards "creating a brighter tomorrow together." They highlighted a few key moments, such as:

Making investments to promote sustainability and lessen ecological consequences. By revising their safety procedures completely and doubling the size of the Environmental, Health, and Safety team (EH & S), they have been taking priority on employee career development and safety. The company has also expanded its recruiting teams to increase placement diversity around the globe to alleviate poverty in a particular zone, which contributes to lowering employee inequality. The organization did establish sustainability even higher on its list of priorities.

Not just this organization, but other IT companies also set standards and introduced policies to guarantee that organizational programs and initiatives are handled sustainably. Engaging the majority of the workforce to provide innovative thoughts and ideas is another difficulty in sustainability. As a result, every stakeholder will have a chance to contribute to the organization's success while fostering innovation. Economic growth can be attained by adopting regulations that support ethical hiring practices.

Operational considerations should come first for any IT organization when it is new to sustainability. The business must therefore keep its consumers and customers updated and informed about its policies and procedures. Here, we can examine a few situations of how SDGs are employed in businesses.

A T-shirt manufacturer in North Carolina seeks maximum transparency in its supply chain. The company's CEO emphasized that sustainability is not just applicable to certain types of businesses or commodities. To guarantee visibility in the manufacturing unit, they have invited customers to view their premises if they are eager to take an assembly tour of the design company. If supported, these types of operations can contribute to the growth of industries.

The jewelry industry is another example. The Kimberley procedure is used by the majority of diamond business manufacturers. The Kimberley Process is a certification standard that governs the diamond industry which assures that neither compelled slavery nor gemstones from environmentally harmful restricted areas are used in their manufacturing. Sustainable development objectives of companies using climate growth methods here give birth to healthy growth.

A tiny proportion of IT companies have started to promote energy and water conservation among their staff members. These businesses will demonstrate their commitment to resource sustainability when they begin to do so on a large scale.

Infosys, a leader in the development of the next generation of digital services, won the Horse for Sources award (HFS) in the sustainability and ecosystem innovation categories in June 2022. It has been given this recognition, due to its global leadership in environmental, social, and governance (ESG) and its capacity to satisfy customer needs. Infosys was chosen as the champion out of 200 IT companies because of its constant and favorable influence on community members, society, and ecology.

Infosys was applauded for mitigating all 17 Goals of Sustainable Development (SDGs) by integrating its tools, technologies, and firm ecosystem. Infosys will be the organization to establish a norm on sustainability, according to the head of HFS research, and it will dominate the IT industry for the next 50 years. It has also taken several steps to rebuild biodiversity in the areas where they function as a result of this status.

Every company which respects environmental factors and which worries about the hazards that occur in the environment will attain its sustainability position.

These companies not only consider their profitability or status in the community but also look for more future aspects of sustainability.

Technology makes it possible to create innovative products and services with lower power, material, and water consumption as well as reduced operational wastage, which can simultaneously improve organizational excellence and ecological sustainability.

The fundamental skills required to develop, create, and ensure sustainable systems and applications are outlined in the Principles of Green Software Design which push the firms to grow in the direction of SDG.

Even Accenture has decided to go carbon neutrality by 2025. They have made it a point to consume less water and have aimed to reprocess all of their electronic waste. As previously said, every IT organization is directing its operations toward these sustainable goals.

Measures to Implement SDG

It is the responsibility of national and local government agencies to promote private sector participation in the global effort to achieve the SDGs. By doing this, businesses can reinvent their tactics for achieving their objectives.

Several articles have provided tips for achieving these objectives. Organizations can take a few actions to achieve these SGD goals, including:

1. Joy boxes should be made available to all organizations. Employees can contribute a sum of money to poverty eradication anytime they feel inspired and appreciative. When workers are aware of these kinds of initiatives, they are more likely to be optimistic, and their optimism will bring happiness into the lives of many individuals.
2. Numerous businesses have already started offering one-on-one wellness sessions to employees to assure their general well-being. The implementation of counseling services is still in its early stages. Even during times of covid and after that, organizations began to offer a few extra hours of wellness classes like yoga, meditation, and health awareness tests. Employees who experience stress at work will also experience adverse effects on their well-being. Large corporations and new ventures have created stress-reduction initiatives, and even their policies have been revised. As a result, employees will engage better.
3. Organizations must be conscious that they do not discriminate based on gender when hiring new employees. If progress toward the workplace, and gender equality is not made, our economy and environment will suffer greatly.

   When the same person gets praised and recognized, it also causes emotional stress and workplace harassment. It also causes emotional stress and workplace harassment when the exact individual is praised and recognized. Sustainable development in the workplace will result from assuring equal remuneration for all positions and encouraging participation from both genders equally.

4. Businesses can actively look for candidates from various backgrounds and provide an ethical work environment. Counseling sessions can be conducted to help people to recognize their skills, and working parents can use hybrid working arrangements to create favorable working conditions.
5. Innovation was crucial for both IT businesses and other industrial sectors. Organizations are urged to use eco-friendly measures and build water-saving facilities on-premises rather than acquiring a new place to set up the industry and create a new environment.

   We may achieve these sustainable aims by employing solar energy power plants, conserving water, recycling commodities, developing rain harvesting systems, and planting green and sustainable gardens in industrial regions.
6. In the current environment, many IT firms use hybrid working conditions, which aids us in combating climate change. Employers can cut costs and save money by allowing employees to work from home. Additionally, it reduces travel, which will help to reduce carbon emissions. The reduction of carbon footprints on the part of major IT organizations contributes to a global shift in the climatic factor.
7. To lessen the number of cans, tins, and plastics that companies discard, enable employees to use recyclable pet bottles rather than single-use ones. This will shorten the product's life in the water.
8. To set concrete goals that aid in the fulfillment of SDGs, organizations should collaborate or join forces with other entities. All of these are implementable gradually in businesses. These minor adjustments will have a larger impact on future endeavors.

After agreeing to practice sustainability, a company's next step is to notify its local and international stakeholders about its policy. It facilitates businesses' disclosure of sustainability data, which enhances business performance (Guidry and Patten, 2010).

Doing business in Asia focuses primarily on sustainability. Implementing sustainability might help people to determine whether the company is actively contributing to society. For maintaining a healthy environment for future generations, other corporations' competitive strategies must align with these sustainability aims (Wu et al., 2014).

Every company must nurture activities to build relationships and reputation to achieve sustainable development goals. By generating revenue for the business, this SDG will aid in resolving global issues when a company wants to establish a new service or when it is about to introduce a new product.

If the SDG is not implemented, the company feared losing the chance to compete in the massive targeted market. Therefore, it is preferable to make sure that the SDG's stakeholders and company companions grasp the strategic plan. To attain sustainability, every organization must cultivate interactions and reputation-building initiatives.

The GRI - Global reporting initiative Standards, which are the most broadly accepted social sustainability set of standards, now include revised guidance to assist businesses in integrating sustainability goals.

They have introduced a few tools that assist businesses in focusing on how they are currently contributing and how they might effectively contribute going forward to meet these goals. This evolution is shown as a bridge between the SDG and the GRI. This set of criteria aligns with the 17 SDGs and makes it simpler to review, track, and check progress.

Future Scope and Conclusions

It is believed that substantially larger scopes are necessary to achieve these sustainability goals. In the same way, creating specific boundaries enables us to incorporate current businesses and technologies.

The future scope of these SDGs can be built by solving the operational challenges caused by global warming, which aids us in raising the potential for sustainability in the IT sector.

It also contributes to the transformation of how we live and work by moving towards a low-carbon society, which entails lowering energy consumption and carbon emissions, as well as the toxins produced during manufacturing and by implementing renewable and alternative energy sources in commercial operations.

References

Galpin, T., and Hebard, J. (2018). Strategic management and sustainability. In Business Strategies for Sustainability (pp. 163–178). Routledge. https://doi.org/10.4324/9780429458859-10

Guidry, R. P., and Patten, D. M. (2010). Market reactions to the first-time issuance of corporate sustainability reports: evidence that quality matters. Sustainability Accounting, Management and Policy Journal, 1, 33–50.

Heras-Saizarbitoria, I., Urbieta, L., and Boiral, O. (2022). Organizations' engagement with sustainable development goals: From cherry-picking to SDG-washing? Corporate Social Responsibility and Environmental Management, 29(2), 316–328.

Ordonez-Ponce, E., Clarke, A., and MacDonald, A. (2021). Business contributions to the sustainable development goals through community sustainability partnerships. Sustainability Accounting, Management and Policy Journal, 12(6), 1239–1267.

Pedersen, C. S. (2018). The Sustainable Development Goals (SDGs) are a Great Gift to Business! Procedia CIRP, 69, 21–24. https://doi.org/10.1016/j.procir.2018.01.003

Pradhan, P., Costa, L., Rybski, D., Lucht, W., and Kropp, J. P. (2017). A Systematic Study of Sustainable Development Goal (SDG) Interactions. Earth's Future, 5(11), 1169–1179. https://doi.org/10.1002/2017EF000632

Wu, T., Jim Wu, Y., Chen, Y. J., and Goh, M. (2014). Aligning supply chain strategy with corporate environmental strategy: a contingency approach, International Journal of Production Economics, 147, 220–229.

Role of Higher Education in Achieving the Sustainble Development Goals (SDGs)

Yasmeen Bano[a] *and S. Vasantha*[b]

[a]School of Management Studies, Sathyabama Institute of Science and Technology (SIST), Chennai, India,

[b]School of Management Studies, Vels Institute of Science, Technology & Advanced Studies (VISTAS), Chennai, India

E-mail: [a]yasmeenbano.soms@sathyabama.ac.in; [b]vasantha.sms@velsuniv.ac.in

Abstract

This study presented the important and substantive role of higher education and its relationship to achieving SDGs. The purpose of the study is to identify the role of higher education in achieving sustainable development goals. The study identified three major aspects of higher education (Economic growth, quality education, and Innovation) which can lead to attaining the sustainable development goal. This quantitative study focuses on the factors for achieving the SDGs through higher education. Data were gathered for the study from both primary and secondary sources. Structured questionnaires were used to gather primary data with information regarding sustainable development in higher education. The study concluded that higher education provides Economic growth, quality education, and innovation for attaining sustainable development goals. The study's findings apply to an institutional level. HEIs may consider SDGs as a critical issue to be included in their strategic and policy documents.

Keywords: Academicians, economic growth, Innovation, Higher education (HE), quality education, SDGs.

Introduction

The Agenda for Sustainable Development, 2030 was adopted in September 2015 by members of the UN (United Nations), highlighting the significant role that higher education plays in achieving the SDGs and emphasizing the need to support and develop HEIs' capacity to prepare students for green work and other activities.

The 2030 Sustainable Development Goal Agenda depends on the success of higher education institutions (HEIs). HEIs are officially addressed under SDG 4 on quality education, it has an impact on all of the goals. HEIs aid in the development of the social, environmental, and economic aspects. Because of their central role in networks of partners from the public sector, the private sector, and the nonprofit sector, they are one of the most significant incubators of concepts and solutions to global issues and have a great deal of potential to have a beneficial impact. "Sustainable development" promotes the welfare of both people and the environment (UNESCO, 2016). Importantly, unlike the Millennial Development Goals of 2000, which had poor countries as their primary target, Agenda 2030 has a global scope. It lays out an ethical and aspirational vision of a better future that depends on collaboration and interdependence.

Goal 4 of the 17 SDGs focuses on high-quality education, which binds nations to "Ensure inclusive and equitable quality education and promote lifelong learning opportunities for everyone," which are the focus of the document. Goal 4 is the only SDG that is focused on universal education. The global education community's 2030 aims and objectives are outlined in Goal 4 from the viewpoint of lifelong learning. The term "lifelong learning" refers to all learning activities carried out throughout an individual's life to develop skills, Knowledge, and abilities from the perspective of personal, professional, and societal (UIL, 2015).

The quantitative data used in this study to measure the importance of HE in achieving sustainable development goals was measured, and the results were compatible with one another. It identified the three major aspects of SDGs such as Economic growth, quality education, and innovation. The study presents an in-depth understanding of the role of HE in sustainable development goals.

Objectives

- To study the role of HE in achieving the SDGs
- To examine the aspects of HE in achieving the SDGs

Literature Review

Aarts et al. (2020) and McCowan (2019) highlight one of the goals of SDG, SDG 4 which asks for equal access to postsecondary education, especially higher education, as part of lifelong learning opportunities for everyone. Campuses have identified actions to address economic and social issues, including investing endowments in line with sustainability principles, forming community collaborations, and committing to provide all students with an affordable education (Barlett and Chase, 2004; Rowe, 2007; Chankseliani and McCowan, 2021). The dependence on financial help is sustained by the current international finance models for HE in low-income countries. Instead, the help ought to go into increasing these nations' capability for higher education. Heleta and Moodie (2017) The paper emphasize the significance of continuous initiatives to strengthen higher education capacity in the Global South, as well as the value of academic freedom, Institutional autonomy, and local institutional innovation for the full achievement of higher education's developmental role.

Methodology

The study emphasized aspects of higher education which include Economic growth, quality education, and Innovation. Structured questionnaire has been adopted for data collection and to examine the perception of Academicians. There was a total of 32 questions included which is measured on 5 points Likert scale. For collecting primary data, questionnaire has been distributed to 210 academicians from different colleges and universities by conducting a purposive sampling technique. Using the rule of thumb, the minimal sample size was calculated to be (32*5=160) respondents. According to Gail (1994), the minimal number of observations must be 4 to 5 times greater than the total number of statements used to assess the construct of the study. The research design used in this study is descriptive.

SPSS is used for measuring the reliability which demonstrates the consistency of the questionnaire. Reliability (Cronbach's alpha) of Economic growth, quality education, and Innovation are 0.856, 0.881, and 0.861 respectively. Overall reliability of the questionnaire is 0.895. The association between the independent factors (economic growth, high-quality education, and innovation) and the dependent variable has been determined using equation modeling (Achieving SDG).

Data Analysis and Results

Hypothesis

H1: Economic growth is positively associated with achieving the SDGs.

H2: Quality education is positively associated with achieving the SDGs

H3: Innovation is positively associated with achieving the SDGs

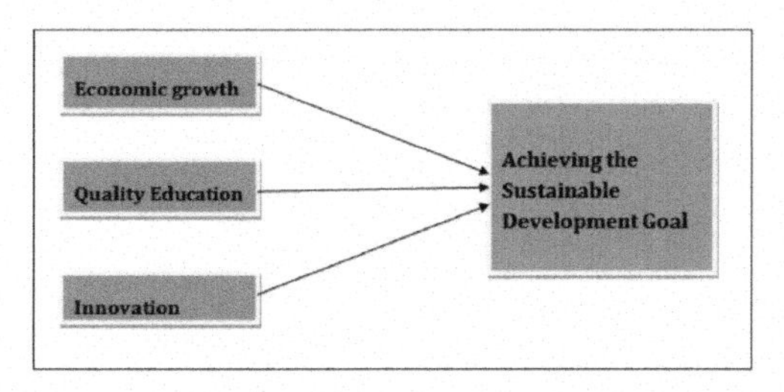

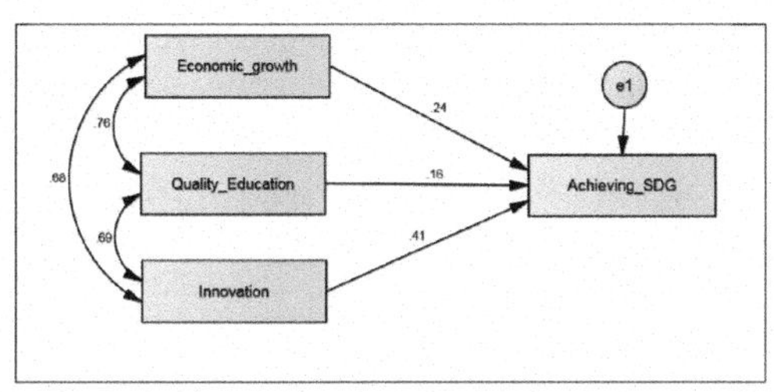

Fig. 1. Proposed Conceptual framework on aspects of HE in achieving the SDGs

Fig. 2. Structural Equation Model (SEM) for the Aspects of HE in achieving SDGs

The Variables used in the conceptual framework are:

I. **Observed, endogenous variables:** Economic growth, quality education, and Innovation on Achieving SDG.

II. **Unobserved, exogenous variables:** e1

The path analysis model is drawn for testing endogenous and exogenous variables using AMOS.

Table 1 explains the causal relationship between Economic growth, quality education, and Innovation in Achieving SDGs.

Economic growth has a positive effect on Achieving SDGs. Economic growth has an unstandardized coefficient value of 0.228, indicating a favorable impact. The Economic growth will increase by 0.228 times for every unit increase in Achieving SDGs. The unstandardized coefficient value was determined to be significant at the 1% level of significance since the p-value was less than 0.05. This study has proved that Economic growth is positively associated with achieving the SDGs. Hence, H1 of this study is accepted.

Table 1. Variables used in Structural Equation Model

Dependent Variables		Independent Variables	Unstandardized Coefficient	S.E.	Standardized Coefficient	T Value	P-Value	Result of Hypothesis
Achieving SDG	<---	Economic growth	.243	.075	.237	3.056	.002	H1 is supported
Achieving SDG	<---	Quality education	.161	.077	.163	2.084	.037	H2 is supported
Achieving SDG	<---	Innovation	.415	.071	.407	5.822	.000	H3 is supported

Quality education has a positive effect on Achieving SDGs. Quality education has an unstandardized coefficient value of 0.161, indicating a positive effect. The Quality of education will increase by 0.161 times for every unit increase in Achieving SDGs. The unstandardized coefficient value was determined to be significant at the 1% level of significance since the p-value was less than 0.05. H2 has been accepted because the direct relationship between Quality education and achieving the SDGs is significant in the study.

Innovation also has a direct positive effect on Achieving SDGs. The unstandardized coefficient value of Innovation is 0.415. The estimated positive indication suggests that positive effect of Innovation will increase by 0.415 times for every unit increase in Achieving SDGs. The unstandardized coefficient value was determined to be significant at the 1% level of significance since the p-value was less than 0.05. H3 is also accepted and indicated the direct relationship between Innovation and achieving the SDGs is significant to the study.

To address the issues of sustainable development, higher education is crucial. However, the industry is capable of far more than just providing sophisticated education. It can develop and unearth groundbreaking research, and related services to communities. Being merely named in the global targets is insufficient. Governments, international organizations, and universities must collaborate for the sustainable development agenda to be fully implemented.

Implications

Students may prefer to adopt SDG achievement (Economic growth, quality education, innovation) as their personal career choice in the future. SDG considerations must be taken as a requirement during internal quality assurance procedures. This study supports lifelong learning possibilities for students in HE and promotes comprehensive, equitable, high-quality education.

Conclusion

This study provides details on higher education and achievement of SDGs. Universities developed their understanding of sustainability in all related academic fields, research, inter-university collaboration, and partnerships with governments, as well as the honorable duty of HE to work for a sustainable future. inclusive formal and informal education, promotion, and curricular integration to encourage SDG learning are ideas for enhancing the implementation of education for sustainable development (ESD).

References

Aarts, H., Greijn, H., Mohamedbhai, G., and Jowi, J. O. (2020). The SDGs and African higher education.

Barlett, P., and Chase, G. (2004). Sustainability on campus: Stories and strategies for change. Cambridge, MA: MIT Press.

Chankseliani, M., and McCowan, T. (2021). Higher education and the sustainable development goals. Higher Education, 81(1), 1–8.

Gail, M. H. (1994). Sample size estimation when time-to-event is the primary endpoint. Drug information journal, 28(3), 865–877.

Heleta, S., and Tohiera, M. (2017). SDGs and higher education – Leaving many behind. University World News. Retrieved from http://www.universityworldnews.com/article.php?story=20170427064053237

McCowan, T. (2019). Higher Education for and beyond the Sustainable Development Goals. Cham, Switzerland: Palgrave Macmillan.

McCowan, T., and Unterhalter, E. (eds.), (2021). Education and development: An introduction (pp. 275–293). London: Bloomsbury.

UIL (2015). Conceptions and realities of lifelong learning (Background paper for Global Education Monitoring Report 2016). Paris, France: UNESCO.

UNESCO. (2016a). Education for people and planet: Creating sustainable futures for all (Global Education Monitoring Report 2016). Paris: Author.

"Happiness Engineering": Acceptance and Commitment Therapy for University Students' Classroom Engagement, Mental Health, and Psychological Flexibility

Ashraf Alam[*,a] *and Atasi Mohanty*[a]

[a]Rekhi Centre of Excellence for the Science of Happiness, Indian Institute of Technology Kharagpur, India
E-mail: [*]ashraf_alam@kgpian.iitkgp.ac.in

Abstract

In the current investigation, the researchers have used a multisite randomized controlled trial employing Acceptance and Commitment Therapy (ACT) to see whether ACT-based interventions help university students improve their classroom engagement, mental health, and psychological flexibility. Students from four engineering institutions were recruited and allocation was done randomly to one of two groups: wait-list control (n = 68) or intervention (n = 68). This intervention developed by the researchers was named "Happiness Engineering." University pupils in the intervention group, over one month, actively participated in four 2.5-hours long workshop sessions and completed activities like meditative practices and observation grids. From MANCOVAs and ANCOVAs, it was unveiled that students in the intervention group demonstrated more psychological flexibility than those in the control group. Students also reported higher levels of classroom engagement, contentment, happiness, and well-being, as well as fewer symptoms of depression, despair, anxiety, nervousness, hopelessness, and stress.

Keywords: University Students, Classroom Engagement, Mental Health, Psychological Flexibility, Acceptance and Commitment Therapy, Multisite Randomized Controlled Trial, Pedagogy, Curriculum

Introduction

The ACT's transdiagnostic character makes it especially suitable for settings at colleges and universities. Despite the possibility of specific mental health illnesses in students, many of them suffer from stress, interpersonal conflict, and prejudice, for which there is no known diagnosis (Alam, 2022a).

The Present Investigation

By conducting a randomized controlled trial in four different universities, investigating the effect of an ACT intervention on classroom engagement — a factor that has, to our knowledge, never been studied before is measuring psychological flexibility to capture all six ACT processes. The current study thus aims to advance previous research. More specifically, it aimed to investigate how university students' psychological openness, mental health, and classroom involvement were affected by the four ACT-based workshops. Four mental health indicators (stress, anxiety, depression, and well-being) were employed as primary outcomes, while psychological flexibility was used as a process outcome. Engagement in the classroom served as a supplementary outcome variable.

The following are the study's hypotheses

After the workshops, students in the intervention group will report more psychological flexibility than students in the control group.

After the workshops, students in the intervention group will report fewer signs of stress, anxiety, and depression and a better level of well-being than students in the control group.

When the workshops are over, more intervention group students than control group students will report better classroom engagement.

Method

Participants

During the spring semester of 2021 and the autumn semester of 2022, 136 students (n = 136) took part in the study. For data collection, two universities from Delhi and two from Bengaluru were selected (Dayananda Sagar University, Bengaluru, 27.6%; Christ University, Bengaluru, 21.4%; Delhi Technological University, 30.9%; and Indian Institute of Technology Delhi, 20.1%). Pupils were primarily female (69.3%) and enrolled in full-time studies (92.4 percent). The average age of the group was 26.384 (standard deviation: 3.19), and 81 percent were from BE/B. Tech programs.

A minority of the learners were pursuing a doctorate (8.9 percent), while a quarter (24.7%) of the students were enrolled in master's programs. Nearly a third of the students (32.1%) were from Uttar Pradesh, while the remainder were from Karnataka (11.2%), Delhi (17.6%), Bihar (21.9%), and other states across the nation (17.2 percent). Table 1 details the pupils' characteristics. Students did not receive any payment or other forms of reward for attending the workshops.

Table 1. Participant Characteristics

Groups	**Age M (SD)**	**Sex** (%)		**Degree** (%)			**Regime** (%)		**Sessions** (%)		**Sites** (%)			
		Male	Female	Bachelor	Master	Doctorate	Part-time	Full-tine	Fall	Winter				
Intervention (n = 68)	31.98 (9.98)	24.7	75.8	55.3	33.9	12.2	27.5	72.9	49	51	26.7	33.9	14.9	21.9
Control (n = 68)	31.34 (8.29)	31.1	69.9	63.7	27.3	8	31	69	54.7	45.3	42.9	27.3	10.1	21.9

Intervention

Students took part in a researcher-developed intervention. The researchers have given this intervention the name "Happiness Engineering". The intervention was marketed to students as a method to complete their coursework well while leading a fulfilling life. It was made up of four 2.5-hour sessions that were given to groups of 9 to 14 students over four weeks.

The first workshop's main themes were committed action processes and ideals. To integrate their personal beliefs into realistic objectives and activities, students needed assistance in identifying what was most important to them in most of their life domains (such as school and family). They were urged to make little lifestyle adjustments that would bring them closer to their beliefs.

The second workshop focused on cognitive diffusion and acceptance. To determine if they were attempting to manage or avoid these inner experiences, students were asked to examine challenging ideas, feelings, or sensations. They were then prompted to consider the long-term viability of experience avoidance tactics and urged to switch to acceptance as a substitute.

Participants were further assisted in overcoming challenging ideas and were advised to keep their beliefs at the forefront of their minds despite these concerns. In the third workshop, students studied several meditation techniques that would help them develop mindfulness and connect with a secure, ongoing sense of self from which they may observe and accept shifting events. The fourth workshop's objective was to consolidate all of the knowledge acquired so far and make sure that it was applied to the student's daily lives.

The intervention includes a variety of ACT activities, including behavioral activation techniques and "the Polk and Schoendorff matrix or the Bull's Eye activity" (Alam, 2022b). The trainer's guide and the participant's guide served as its foundation. The first provides a thorough explanation of how each workshop should be conducted. This manual was created to standardize the workshops and make sure they are given consistently from one semester to the next across institutions. At the start of the first workshop, the students were given the participant's handbook. It includes activities that may be completed at home or through live seminars. Additionally, throughout the intervention, participants can download the overheads from the workshop, practice guided meditations at home, or discover recommendations for extra reading on the website. Two research associates, one with professional expertise in psychology and the other with a formal background in education and learning sciences, led the Happiness Engineering sessions (a male and a female). They were ignorant of the study's hypotheses; however, they were familiar with ACT. They conducted the workshops with two groups: one under the direction of the researcher and the other under the direction of a group counseling specialist who was not affiliated with the present investigation. This process was put in place to make sure that the workshop leaders followed the ACT guiding principles, diligently.

Research Approach

A multisite randomized controlled trial was utilized in each semester (at each study site) to examine the hypotheses. Students in the intervention group attended Happiness Engineering workshops held at their institutions over a month and were required to complete tasks at home in between sessions. In the control group, learners were offered the Happiness Engineering sessions after being put on a waiting list for four weeks.

Procedure

Counseling services' mailing lists from the four institutions listed above, advertising on several social networking sites (Facebook, LinkedIn, Twitter, Quora, and Instagram), and posters posted throughout the campuses were used to find participants. Students who agreed to participate in the study completed the pre-intervention survey before the conclusion of this lesson. The researcher then used a computer program (Alam, 2022c) to randomly allocate them to the intervention or the wait-list control group. Students in the intervention group began their workshops one week later. All students took the post-intervention survey the same week; those in the intervention group completed it at the conclusion of their fourth workshop, while those in the wait-list control group did so online.

Measures

Process and outcome factors were examined both before and after the intervention. Table 2 lists Cronbach's alpha values for each scale.

Table 2. Randomization's effects on post-program student psychological adaptability, mental health, and school engagement

Outcomes Measures			Intervention Group				Control Group									
	Pre	Post	Pre		Post		Pre		Post		MANCOVA		ANCOVA			
	α	α	*M*	*SD*	*M*	*SD*	*M*	*SD*	*M*	*SD*	*F*	*df*	*F*	*df*	*d (itt)*	*d (pp)*
Psychological Flexibility											13.43	3.137				
FFMQ	.84	.89	3.11	.33	3.11	.29	3.21	.37	3.11	.28			5.69	1.213	.31	.51
MEAQ	.83	.86	3.59	.64	3.89	.71	3.67	.65	3.67	.57			27.75	1.213	.59	.58
Values questionnaire	.91	.92	4.15	.97	4.49	.79	4.19	.91	4.19	1.12			16.87	1.213	.51	.42
Mental Health											7.72	4.142				
MSP-9	.81	.90	3.31	.46	2.88	.39	3.42	.35	3.32	.48			23.18	1.213	(-).49	(-).49
WBMMS	.92	.96	3.20	.52	3.51	.48	3.21	.67	3.19	.61			28.84	1.213	.59	.79
GAD-7	.83	.90	17.11	484	13.92	4.39	17.74	4.71	17.32	4.87			21.39	1.213	(-).51	(-).38
PHQ-9	.84	.84	21.01	579	17.29	4.76	19.98	6.31	19.23	6.17			13.85	1.213	(-).46	(-).48
School Engagement																
AES	.92	.91	4.39	.91	4.71	.88	4.41	0.89	4.29	.89			14.49	1.213	.42	.47

Process Controls

A short version of the Five Facet Mindfulness Questionnaire (FFMQ), the Multidimensional Experiential Avoidance Questionnaire (MEAQ), and a values questionnaire created specifically for this study were used to measure psychological flexibility. These scales were designed to measure all six ACT processes. Four psychological flexibility processes — contact with the present moment, cognitive diffusion, self as context, and acceptance—were measured using the FFMQ short version (24 questions).

The Kaiser-Meyer-Olkin value (.79) and Bartlett's test of sphericity, X2 (41) = 559,39, p.001, indicating that the data were appropriate for this kind of study. Two factors with eigenvalues larger than one were found using EFA, and the scree plot showed a distinct break after the third component. Researchers employed "web-based parallel analysis (WPA) with the Monte Carlo simulation" to assess this two-factor approach. Similar findings were obtained from this WPA, which indicated the existence of two factors that together accounted for 66.2% of the variance (Factor 1 = 49.6%, Eigenvalue = 4.88; Factor 2 = 13.9%, Eigenvalue = 1.51). Following that, these components were isolated using oblimin rotation. The two variables have a .57 correlation.

Main outcome indicators

Four factors—stress, psychological well-being, anxiety, and depression—were used to evaluate mental health. Using the Psychological Stress Measure, stress was assessed (PSM-9). Using Well-Being Manifestations Measure Scale (WBMMS), psychological well-being was assessed. The General Anxiety Disorder Questionnaire (GAD-7) and the Patient Health Questionnaire (PHQ-9) were used, respectively, to assess anxiety and depression.

Alternative Outcome Measure

The Academic Engagement Scale (AES), which has 14 questions and three subscales (perseverance, passion for studies, and positive and negative aspects of university), was used to evaluate classroom engagement.

Attrition

Using analyses of covariance (ANCOVA) adjusting for these variables at baseline, the power analysis was created to evaluate the mean change across groups in psychological flexibility (H1), mental health indicators (H2), and school involvement (H3) during the intervention period. An alpha of 0.06, a power of 0.79, and a big effect size (f = 0.39) were used in the study. A total sample size of 79 was sought. It was planned to enroll at least 117 individuals in the trial, assuming that ACT treatments would be effective with an attrition rate of 23%.

Analysis of Data

The effects of randomization on mean levels of all outcome measures at post-intervention were studied by the researchers using multivariate analysis of covariance (MANCOVA) and univariate analysis of covariance (ANCOVA) models, adjusting for these measures at baseline. The Carlson and Schmidt approach was used to compute the effect sizes and report them using Cohen's d.

Results

Initial Analysis

The Kolmogorov-Smirnov and Shapiro-Wilk tests were used to determine whether the data were normal. The Box's M test was used to determine if variance-covariance matrices were homogeneous. No outliers or assumption violations were found. The expectation-maximization (EM) technique was used to impute the missing data at both measurement periods.

Before randomization, pupils were evaluated at baseline (pre-intervention), as was already indicated. First, chi-square statistics and cross-tabulations with adjusted standardized residuals were used to assess how well the groups (intervention vs. control) could be compared across each research site's variables for gender, educational attainment, study style, and semester. The findings demonstrated that with regard to these socio-demographic factors, the groups were not substantially different in any of the sites. After that, analysis of variance (ANOVA) was conducted to see whether student ages varied by group, and the study locations revealed no such variation, $F(1, 141) = 1.17, p = 0.28$.

Discussion

Students who participated in the intervention group outperformed those in the control group on all psychological flexibility measures. Compared to the control group, students who participated in the 'Happiness Engineering' workshops also reported greater post-program well-being and fewer signs of stress, anxiety, and depression. The intervention had a favorable effect on academic engagement as well, which raises the possibility that ACT may lessen dropout rates.

References

Alam, A. (2022a). Mapping a Sustainable Future Through Conceptualization of Transformative Learning Framework, Education for Sustainable Development, Critical Reflection, and Responsible Citizenship: An Exploration of Pedagogies for Twenty-First Century Learning. ECS Transactions, 107(1), 9827.

Alam, A. (2022b). Investigating Sustainable Education and Positive Psychology Interventions in Schools Towards Achievement of Sustainable Happiness and Wellbeing for 21st Century Pedagogy and Curriculum. ECS Transactions, 107(1), 19481.

Alam, A. (2022c). Social Robots in Education for Long-Term Human-Robot Interaction: Socially Supportive Behaviour of Robotic Tutor for Creating Robo-Tangible Learning Environment in a Guided Discovery Learning Interaction. ECS Transactions, 107(1), 12389.

Work-Life Balance and Its Challenges for Medical Professionals in the Health Care Sector

Sandhya Shelar[a] *and Anita Khatke*[b]

[a]Savitribai Phule Pune University, Pune, India
[b]Savitribai Phule Pune University, Pune, India
E-mail: [a]Sandhyashelar021@gmail.com, [b]Anitakhatke7@gmail.com

Abstract

Today it is very difficult to achieve work-life balance in our life. Every employee is facing a dilemma to handle the ideal work-life balance in their personal and professional life. There are various challenges for medical professionals like global competition, personal life/family values and an aging workforce, changing technology, handling shift duty and family at a time. Attrition rate and employee engagement are the present challenges that exacerbate work-life balance of medical professionals. Therefore there is a need to study in detail about medical professionals. The factors influencing work-life balance of healthcare sector and the challenges to maintaining the work-life balance to improve work performance is been studied in this paper. Primary and secondary data are used. To analyze the data Descriptive statistics have been used. Responses have been collected from medical staff of hospitals.

Keywords:- Work-Life Balance (WLB), emotional well-being, Family satisfaction, work satisfaction

Introduction

Medical professionals are facing an ethical dilemma when it comes to work-life balance and feels stressed at work. This will have an adverse effect on the level of satisfaction of the patients and their dependents. Due to rapid changes in technology, medical professionals need to always update their knowledge and learn new techniques on a regular basis. As updating knowledge and handling shift duty and family at a time is a challenging task for medical professionals.

There is an increasing trend of burnout (losing interest in the profession) among medical professionals due to various reasons. Due to low work-life balance, it's becoming challenging to maintain good employee engagement among employees who work in Hospital Industries. Due to changes in the management and advancement of technology, it is very challenging to retain employees. Overall performance of employees going through any such problems either at the workplace or in the family will be affected.

This research study will focus more on the challenges to handle the work-life balance and how medical professional employees are coping with it. There are many challenges due to an imbalance in personal and professional life which affects hospitals' attrition rate and employee engagement. Especially the healthcare industry which provides emergency care depends on its human resource. Healthcare sector needs to ensure that the employees should not leave their job frequently and there should be good engagement among the employees for which they should bring some changes in the workplace policy. Apart from employee engagement, there are some more unique challenges which are as follows.

a) Managing a shrinking pool of healthcare devices
b) Increasing demands of the patient and employers are very difficult to deal with.
c) Performance and responsibility of work handled by them.
d) Uncongenial working schedules
e) Managing compromised sleep

The above challenges are very important for the hospital in terms of employees work-life balance which needs to be verified in terms of WLB.

What Is Work-Life Balance?

Work is the effort that puts an individual physically and mentally to complete any task. It is an activity that we do regularly and consistently and which brings results in monetary terms. Life is the most important aspect of existence that processes, acts, reacts, evaluates, and evolves through day-by-day growth. Many physical and conscious developments happen continuously. Life refers to the state of existence which is characterized by various drivers and various desires which bring causes of stress in life and have an impact on the work a person undertakes.

Work life balance is considered to be important for business and academics. The research study focuses mainly on assessing the impact of work life balance and challenges towards it. In a broad sense, work-life balance is a very big concept where an individual employee gives either more preference to work or family. The challenges like globalization; downsizing and flexible working patterns have a great impact on every organization. Due to modernization, employees are struggling to manage their work and family responsibilities.

Why Is Work-Life Balance Important?

Today work-life Balance is an important subject in everyone's life. It is an important aspect of a healthy work environment. Maintaining work-life balance helps reduce stress and helps prevent burnout in the workplace. When people have ample time and support for their personal life as well as their career, they are satisfied. It focuses on their well-being and mental health as they have more time on their hands. This translates to better relationships with their employers and fellow employees, as well as boosts in productivity and performance. Employees' efficiency and productivity will also improve. In today's economy, employers need to focus more on time and effort in retention policies as it is difficult to attract and retain top talent in the healthcare sector. It's critical to build a reputation for supporting and encouraging work-life balance.

What is the Healthcare Industry?

The Healthcare industry in India comprises hospitals, medical devices, clinical traits, outsourcing, telemedicine, medical tourism, health insurance, and medical equipment. The healthcare industry is an aggregation and integration of sectors within the economic system that provides goods and services to treat patients with utmost care. The industry is growing at tremendous pace owing to its strengthening coverage, services and increasing expenditure by public as well as private players.

What are the Hospitals?

Hospital is an institution providing medical and surgical treatment and nursing care for sick or injured people. Currently, hospitals are largely staffed by professional physicians, surgeons, nurses, and allied health practitioners, whereas in the past, this work was usually performed by members of founding religious orders or by volunteers.

There are various types of hospitals. The most well-known form of the hospital is the General Hospital which carries an emergency department to handle urgent health issues such as fire and accident victims as well as medical emergencies.

Types of Hospitals

There are different types of Hospitals that are typically subsidized by the Government, for-profit or non-profit health agencies and health insurance providers or different charities, trusts or donations which largely depend upon the functionality, size, location, ownership, and specializations. Basically, the hospital can be classified into one of three groups which are as:-

a) Nonprofit hospital
b) Publicly Owned Hospitals
c) For –Profit Hospitals

Hospitals are differentiated according to functions and special care and services which they provide. The infrastructure is very important which includes not only the different machines, medical devices, lab, and technical machines but also the beds and rooms availability. As the patient should feel more comfortable and relaxed with this. Mostly the beds are made after a client requires specific treatment and when there are no other people in the room. There are different types of bed in hospitals with various names:-

a) Simple Beds:- Open Bed Closed Bed
b) Occupied Bed
c) Special Beds: Operation Bed Cardiac Bed Blanket Bed Fracture Bed

Who are the Medical Professionals?

Medical professional means any person licensed or certified to provide health care services to natural persons, including but not limited to a physician, dentist, nurse, chiropractor, optometrist, physical or occupational therapist, social worker, clinical dietitian, clinical psychologist, licensed professional counselor, licensed marriage and family therapist, pharmacist or speech therapist.

Medical professional means any medically qualified doctor including but not limited to anesthesiologists, radiologists, pathologists, surgeons, cardiologists, general practitioners, and obstetricians, Medical professional also extends to include students doctors, medical interns, dentists, and midwives, etc.

Work-Life Balance of Medical Professionals and Variables Considered for Work-Life Balance

The following are the phrases that were used while studying the challenges and work-life balance of the employees who are working in Health care sectors. Following scales were used to measure the work-life balance.

a) Feel frustrated and stressed due to working conditions
b) How frequently you skipped meals
c) Healthy and properly balanced diet
d) Had difficulty in sleeping
e) Slept less than 5 hours at night
f) Do you work without any breaks?
g) Quality time with family
h) Attend family functions and get together
i) Technology brings frustration
j) Arrived home late due to work
k) Every day go happily to work

Above all responses include the options like No never, Sometimes, Less than 1 day, occasionally, not applicable, a moderate amount of time, etc. The analysis was purely dependent upon the scores and the responses received from the individual employees working in the hospitals

Health Care Sectors and Their Challenges Towards Wellbeing of Employees.

Every employee in the Health care sector is facing problems towards the work that is working hours, roles in the job, night shift, and so on. This also creates problems with the employee's ability to take care of his health; the employee is constantly juggling to ensure a balance between the care he gives to his patients and the care he gives to himself and his family. As per various doctors and consultants in infectious diseases, there is no limit to what one wants to do to help patients get better, health care professionals must ensure that they remain healthy too. Medical professionals always try to achieve this fine balance between personal and professional commitments which is always a challenging task for them, particularly for those who have families and dependents. This often brings stress to the employee's life.

Literature Review

D. B. Dhas, P. Karthikeyan (Dec. 2015) - International Journal of Research in Humanities and Social Studies Vol-2 Issue This article (Work-

Life Balance Challenges and Solutions) provides human resource professionals with a historical perspective, data, and possible solutions for organizations and employees alike to work/ life balance. Three factors: global competition, personal lives/family values, and an aging workforce present challenges that exacerbate work/life balance.

S Sharma and J Parmar (January 2016) "Work-Life Balance of Medical Professionals in Government Hospitals of Himachal Pradesh" This study focused on assessing the impact of work-life balance determined by work-family conflict and family work conflict on the well-being of individuals employed in the private sector in India. Well-being was measured by levels of family satisfaction, work satisfaction, and psychological distress. The research tried to find out the effect of stress at job and its effect on their health—mental as well as physical on job and overall work-life balance.

Work-Life Balance in the Health Care Sector (2017) K Shivakumar & V Pujar KLS Institute of Management Education and Research, Belagavi, Karnataka, India published in Amity Journal of Healthcare Management 1(2), (45–54) ©2016 ADMAA. In this research paper, the researcher found that it is difficult to establish the right balance between work and family in the Healthcare sector. This paper studied in detail the various parameters which affect work-life balance and also suggests some practices which could aim at reducing work-life conflict.

Researcher I Gautam and S Jain (2018) Doon University, Deharadun reveals in this study with research title "A study of work-life balance: Challenges and Solution" which is purely on a literature basis and this study focused on assessing the impact of work-life balance determined by work-family conflict and family work conflict on the wellbeing of individuals employed in the private sector in India. Researchers further conclude that well-being was measured by levels of family satisfaction, work satisfaction, and psychological distress.

The research study of R. Mohan and S. Aveline found in their research that the lifestyle of employees has been changing due to rapid change in emerging trends and their likings specifically software Industry people were more prone towards fashion and therefore the work-life balance get impacted. The work process also sometimes standing hours and strict supervision or deadlines of the projects for which the employee always compromised their personal life or health with work or work pressure.

Study Gap in the Research

The previous study conducted was for the Banking sectors, IT sectors, Pharma Industries, Police department, Education, Transportation Industries or other service sectors likewise very few research conducted on Hospital Industries were more focused on Medical professionals, not Nurses. In India, very few research was done on the work-life balance of Nurses not on primary data basis so researcher found this was a big gap in research for further study.

Objectives of the Study

1. To study the relationship between demographic variables and other factors which affect the work-life balance of medical professionals.
2. To study the influence of work-life balance on the family life of the employees
3. To study the challenges faced by employees in the Health sector concerning maintaining work-life balance

Research Methodology

Research Design:- A research design is very important as it provides guidance and support to our research study. The research design used for this study is Descriptive research design used for the analysis. For this, the structured questions are prepared and get it filled from the employees.

Sampling:- The sampling covered the employees from the Health Care sector in Pune and Mumbai. Primary data was collected through structured questionnaires from the different private and public hospital staff and other medical units.

Data Collection

Researcher had more focused on the primary data for the authentic source of responses. Therefore the primary data was collected through Google questionnaires and sometimes from personal interactions. Due to the pandemic and covid problem, the emergency department was not entertained to the researcher but the researcher managed to get the questionnaire form filled out by the medical professionals. There were 120 questions that were distributed in the hospitals. Almost targeted more than 500 employees or medical staff out of which only app. 400 responses were collected out of which few were not answered properly so not considered. The secondary data were collected form past research papers., journals, magazines, books, and newspapers.

Data Analytics

Hospital-Wise Distribution of Respondents

From the above table 1 and the calculation which had found that most of the patients and patient parents preferred Government hospital compared to other hospitals as it shows 37 % of respondents had responded. Semi Government hospitals also provide services and customers or patients prefer the semi Govt. hospital also compared to private and hospitals under trust.

Table 2. Hospital wise Distribution of Respondents

S. No.	Type of Hospital	No. of Respondents	Percentage
1	Government Hospital	163	37
2	Private Hospital	97	22
3	Semi-Government	124	28
4	Hospitals under Trust	58	13

Demographic Variables

Demographic variables were used to analyze the demographic profile of the respondent in the approximately 74% of responses received.

Demographic variables state the frequency and percentage of the respondents for each of the descriptive words such as gender, age, education, experience, occupation of employees, and marital status, which is used to analyze the demographic profile of the respondents. In table 2 we can clearly note around 55% percent respondents are male and 45% of respondents are female, where maximum number of respondents around 70% fall in the age group of 30–40 years. The sample is almost equally distributed among married i.e. 40% and 60% of the population is unmarried. It can also be seen that 40% of the respondents are graduates and 15% of the respondents are post-graduates and hardly ten percent of them have a higher degree as well.

Most of the respondents, like 20% of the respondents have a total working experience of 5–10 years. It was also found that the experience of more than 10 years to 15 years and 15 years it 20 years are almost the same. That is 30% respectively and experience above 20% for the experience above 20 years. It has been found that the monthly income of the majority of respondents is below rupees five lakh p.a. Through the research it has also been observed that Doctors, Nurses, and brothers are more compared to other designations and it is the same that is 25% which is comparatively more with other designation.

Table 3. Demographic Variables

Demographic Variable	Frequency (N=442)	Percentage (%)
Gender		
Male	243	55
Female	199	45
Age (years)		
22–30	88	20
30–35	142	32
35–40	168	38
Above 40 years	44	10
Married Status		
Married	177	40
Unmarried	265	60

Education Qualification		
Intermediate	110	25
Graduate	176	40
Pharma Diploma holder	45	10
Postgraduate	66	15
Highest Degree	45	10
Designation in Hospitals		
Doctor	110	25
Surgeon	67	15
Lab Technician	22	5
Physiotherapist	89	20
Nurses and Brothers	110	25
Helper	44	10
Income in (Rs)		
Rs. 15000–20000	35	8
Rs. 20000–25000	44	10
Rs.25000–30000	88	20
Rs. 30000–35000	111	25
Rs. 35000–40000	124	28
Above Rs. 40,000/-	18	4
Experience in Years		
5 yrs. to 10Yrs	89	20
10yrs. To 15 yrs.	132	30
15 yrs. To 20 yrs.	132	30
Above 20 years	89	20

Table 4. Work-Life Balance and its challenges for Medical Professionals towards employee engagement factors

S. No.	Work-Life Balance	Frequency (N=442)	Percentage (%)	Mean
1	Work Stress	270	61	186
2	Job Satisfaction	172	39	
3	Attitude	320	72	
4	Job commitment	185	42	
5	Competency	170	39	
6	Target Achievement	220	50	
7	Career Development	120	27	
8	Rate of absenteeism	150	34	
9	Health Issues	260	59	
10	Family and personal time issues	130	29	
11	Shift duties	280	63	
12	Grievances if any	70	16	
13	Excess workload	150	34	
14	Long working hours	160	36	
15	Uncooperative supervisor	90	20	
16	Technology adopted in hospital	310	70	
17	Culture of working place	110	25	

Findings

a) There is a relationship between the work-life balance of medical professionals and their job engagements.

b) Due to stress, the mistakes ratio increased and it has been found that the stress level of medical professionals is comparatively high.

c) There is a huge difference between the work-life balance of married and unmarried medical professionals.

d) Medical professional employees always have health issues as compared to other sectors.

e) Employees always need to compromise with their personal and professional life because of the job.

f) The employees working in health care sector are more prone to hospital work as there is no option of working from home for medical professionals.

Conclusion

As lifestyle is changing rapidly health problems or diseases also increased due to which there is always a great demand for the healthcare sector and healthcare products. In respect of the increasing demand for hospitals and awareness among the people, medical professionals are always in high demand.

With the research study, it has been found that the work-life balance of medical professionals is comparatively low. Researcher concludes that if the proper work policy is maintained in the hospital industry then the employee can easily manage their emotional well-being.

References

Dhas, B., and Karthikeyan, P. (2015). Work life balance challenges and solutions. IJRHSS, 2(12), 10.

Gautam, I., and Jain, S. (2018). A study of work life balance: Challenges and Solution, 1(24).

Sharma, S. and Parmar, J. (2016). Work life balance of medical professionals in Government Hospitals of Himachal Pradesh, (0) (0).

Shivakumar, K., and Pujar, V. (2016). Work Life Balance in the Health Care Sector, IPAJM, 45(2).

Aveline, S., and Kumar, R. (2017). Employee Engagement and Effects of Work-Life Balance in Software Industries in Chennai. IJPAM, 116(22), 459–466 ISSN: 1311(2017).

Website

1. www.ijrhss.org/pdf/v2-i12/2.pdf
2. http://indusedu.org/pdfs/IJREISS/IJREISS_2439_90357.pdf
3. https://www.researchgate.net/publication/311364902_
4. https://www.researchgate.net/publication/333456881
5. https://www.researchgate.net/publication/311353885
6. https://acadpubl.eu/jsi/2017–116-13–22/articles/22/39.pdf

Political Representation of Aesop's Beast Fables in Augustan Age

Sanaa Parween,[*,a] *Dr. Jayatee Bhattacharya,*[b] *and Dr. Simi Malhotra*[c]

[a]Amity University of Noida, India
[b]Amity University of Noida, India
[c]Jamia Milia Islamia of New Delhi, India
E-mail: [*]pseudoliterarian@gmail.com

Abstract

Social and political life of seventeenth-early eighteenth century England witnessed both preservation and transition of Aesop's fable. Augustan fabulists took a special interest in using Aesop's beasts to pass on a political message throughout the period. These fables were an essential part of cultural battlefield that accommodated conflicts as well as negotiations after the Revolution of 1688. This paper aims at highlighting the political representation of Aesop's fables and the social unrest that inspired some of the famous collections of Aesop's fables with their reflections. The two collections taken into account for this paper are Roger L'Estrange's and Samuel Croxall's fables with their interpretation.

Keywords: Fables, Aesop, Politics, Beast Fable, Augustan Age

Introduction

Fables have been a source of rich and ripe wisdom that provide insight into human nature through the beast. The structure of a fable is focused, simple and crisp despite that there is complex nature of the underlying context that keeps the genre in popularity. (Newbigging, 1895)

Aesop is a familiar name for readers across ages, it is instantly followed by the mention of his fables, popular for their wit and satire. It is still a topic of debate whether he existed or is he just a fragment of imagination, but the fables that are attributed to him are widely read and enjoyed. Even though these fables are collected, told, and retold across the world, there is the underlying factor of context and cultural background that impacts the interpretation.

Joseph Addison describes the valuable status of fables, he mentions them as "the first Pieces of Wit" that appeared in ancient times and continue to instruct even in the modern-day world. This assertion furthers the acknowledgment of the status of fables as a literary genre, the distinguished narrative structure that allows delivering a moral across ages without causing conflict. The freedom that comes with the rhetoric of this genre allows the fabulist to mold the tale and convey the desired message through it. (Addison et al., 2014)

Fable translations are an essential part of fable scholarship. The highlights of the contemporary events along with the commentary by writers of the compilations give a fair chance to understand the historical perspective and influence on the fable (Daniel, 1982).

The dawn of political changes in seventeenth-eighteenth-century England brought about several literary developments that stemmed from the sharp divide in opinions and programs. The staunch sides of Whigs and Tories, the former stood for supremacy and personal freedom whereas Tory party was all about Divine Rights and hierarchy.

It was during the late seventeenth century and early eighteenth century that English fable began to pick up pace again. It was the political history of this period as well as an increase in literacy rate that encouraged the readership of the fable. It was also during this period that many collections of fables came into existence. Among some of the famous and widely read collections was John Ogliby's *Fables of aEsop, Paraphras'd in Verse* published in 1651. His verse fables were in iambic couplets, aimed at delivering the beast fable or animal tales for their didactic efficacy. (Lewis, 1996) I was followed by Roger L'Estrange's collection known as *Fables of aEsop and Other Eminent Mythologists: With Morals and Reflexions* published in 1692.

Aesopic collections factor in the inequitable power relationships that are depicted through the master-slave relationship, humans replaced by animals where a weaker animal, a prey in the food chain is in the opposite of a stronger, cunning beast much above in the food chain. The natural order of beasts offers a close understanding of power in the fables. (Acheson, 2009)

Augustan period witnessed remarkable changes after the Restoration period, this was in the closing years of William III and the succession of Queen Anne. The political opinions and taking sides in the tug-of-war between Whigs and Tories mattered the most. Each party attempted to bribe and bag authors to propagate their ideology through literary works. This involvement of politicians to influence the literary genre went both ways, one with the politicians seeking literary figures to hire, and second, the literary figures looking for a place in the hierarchical ladder for themselves. As a result, the authors showed a staunch political bias.

Periodicals and coffee houses helped boost the spirit of readership as well as fueled intellectual needs. It gave way to several famous clubs that accommodated intellectuals, politicians, writers, and thinkers and the clubs also acted as a milestone in the development of prose literature. Along with the development of intellectual readership, this age saw a decline in drama but the readership of several periodicals and prose was on the rise.

Political culture of late seventeenth and early eighteenth century was the dominant factors in shaping many literary works, this was a direct strategy to promote literary patronage. To raise the value of parties, the use of literature by Whig and Tory parties yielded several biased authors who wrote for financial gains. The literary works were politically charged with vocabulary and interpretation that rounded up the influence of writings on the readers to gain more followers for a certain party.

Neoclassical defenders of ancient restraint, however, clashed head-on with the agents of aggressive modernity in discussions of Aesop's fables at the time. The "ancients" used fables to counteract the impolite effusions of modern writing, while the unrepentant "moderns" appropriated Aesop to justify, if not sanctify, their excesses. The aim of collecting and translating Aesop's fable was not limited to the literary readership, the underlying agenda of influencing minds through the interpretation of these fables was a crucial part of the process. The powerful materialism and preciseness of fable allowed the compiler to add personal touches to the reflections and understanding (Lewis, 1996).

The examination of the politics of fable during eighteenth-century England largely involves Aesop and the treatment of his fables before the popularity of the genre. The prose verses of fables are universally taught in schools for their didactic nature, this order of instructions was the same for Augustan England as well. To understand the development of beast fables as means of political tools in the hands of writers, the analysis of a common fable from two major collections follows.

Samuel Croxall published his *Fables of aEsop and Others* in 1722 which was a rejoinder to the contribution made thirty years prior by Sir Roger L'Estrange known as *Fables of aEsop and Other Eminent Mythologists: With Morals and Reflexions*, published in London 1692. (Patterson, et al. 1987) Both collections were politically driven and the philosophy behind the interpretation of the fable was influenced by the propaganda of the respective parties. L'Estrange political philosophy was influenced by the belief that society depends on hierarchy that is divinely chosen.

In order to highlight the distinguished use of Aesop's fable in political expression, this paper will explore two fables from each collection. The aim of this is to analyze the interpretation given by each fabulist and relate to the political influence behind it.

In the preface of his collection of fables, L'Estrange mentions the "Dark and Doubtful" parts of Aesop's fable composition that came down to us that it is impossible to distinguish the genuine ones from the copied. The motive behind it can be seen as removing the criteria of interpretation with the motive of including these "reflexions" in the pedagogy.

The fable under this discussion is *A Wolf and a Crane* in which a wolf seeks the help of the Crane when a bone gets stuck in his throat. The Crane agrees to help only on the condition that the wolf will reward him in return. But after the bone is removed, it is time when the Crane demands its reward and the Wolf asks him to be grateful that the Crane got his head back from the throat of the Wolf, and it is rewarding enough (L'Estrange, 1669).

For this the moral and reflexion that L'Estrange condemns the demand of the Crane to expect a reward for a good deed that it has done. Along with this, he points out the symbolism of bone as calamity that might befall someone, and the moral is that one should not be ungrateful for the perseverance that it receives. The conscience of a cause should be enough as a reward is the crucial message of this fable as per the understanding of L'Estrange.

Samuel Croxall disagreed with the version and agenda of L'Estrange and decided that it was time to replace the version with a new one. He was opposed to the idea that children should go through such a massive volume of morality without a scope of freedom and liberty spared.

In his collection, the fable, *The Wolf and the Crane* narrate the same fable with the same characters and plot but with a different application to it. (Croxall, 1863)

Croxall in his application of the fable elaborates that this fable is not an example of ingratitude, rather it is cautionary advice to the reader. The advice is that one should be careful in choosing whom to extend the courtesy of good deeds. This is followed by stressing the fact that one should not serve someone with no honor, as it will result in the suffering of the helper.

Croxall showed how to use fables for their satirical quality and political points and by doing so was successful in preserving the ancient oral tale in written structure (Blackham, 1985).

This contrast in interpretation of the same fables by two different fabulists highlights different approaches based on influence. Individual reading of these fables might procure interpretations that is entirely different from the above. But knowing the literary and cultural history of the periods when these fables were compiled and interpreted, it shapes up to be an interesting observation.

The recurring fact that the writers were bribed to write in favor of the party and often influenced to join the party divides the genuineness of fable translations. The irony of the ordeal points to the reflection of power and politics in the narrative discourse.

Conclusion: The political representation of Aesop's beast fable by writers of late seventeenth century and early eighteenth century was influenced by finance and power more than the literary expression and creative outlet. These collections of fable with their interpretations were aimed at pedagogical indoctrination as well as to popularize English fables.

The journey of fable translations goes through the eye of political unrest and gets influenced by it. It is the same with several other European translations where the reader can find traces of history trailing between the lines.

References

Newbigging, T. (1895). Fables and fabulists, ancient and modern. Frederick A. Stokes Company. https://archive.org/details/cu31924026944870/page/n13/mode/2up?q=fable+and+fabulists

Addison, J., and Steele, R. (2014). The Spectator, Volume 1 Eighteenth-Century Periodical Essays: Volume 1 (H. Morley, Ed.). The Project Gutenberg eBook. https://www.gutenberg.org/files/9334/9334-h/9334-h.htm (Original work published 1891).

Daniel, S. H. (1982). Political and Philosophical Uses of Fables in Eighteenth-Century England. The Eighteenth Century, 23(2), 151–171. http://www.jstor.org/stable/41467265

Lewis, J. E. (1996). The English fable: Aesop and literary culture, 1651–1740. Cambridge University Press.

Acheson, K. (2009). The Picture of Nature: Seventeenth-Century English Aesop's Fables. Journal for Early Modern Cultural Studies, 9(2), 25–50. https://doi.org/10.1353/jem.0.0032

Patterson, A. (1987). Fables of Power.

Sharpe, K., and Zwicker, S. N. (Eds.), Politics of Discourse: The Literature and History of Seventeenth-Century England. University of California Press.

L'estrange, R. (1669). Fables, of Æesop And Other Eminent Mythologists: With Morals and Reflections. John Gray and co.

Croxall, S. (1863). Fables of Aesop and others: Translated into English with Instructive Application. Boston: T.O.H.P. Burnham.

Blackham, H. J. (1985). The fable as literature. In Internet Archive. London; Dover, N. H.: Athlone Press. https://archive.org/details/fableasliteratur00blac/page/n7/mode/2up

Miro Application of Web Whiteboard for Sustainable Development in Teaching and Learning Research

B. R. Aravind[a]

[a]Assistant Professor in English, Kalasalingam Academy of Research & Education, Tamilnadu, India
E-mail: [a]aravind.abur@gmail.com

Abstract

This research paper aims to identify the effectiveness of the Miro App in teaching and learning research methodology for researchers in humanities. Also, to find the instructional differences among researchers and whether the usability of the Miro App results in an enhanced involvement in the teaching-learning process. Fifty research scholars from humanities have participated in the study. Synchronous Online Flipped Learning Approach was employed in the study. The teaching effectiveness scale was administrated to collect the required data from both the control and experimental groups to study the impact of the Miro App and Non-miro App instructional methods. The research findings revealed that the Miro App method of teaching was beneficial to the researchers for better and more specific understanding of the concepts. Additionally, the teaching effectiveness scale results reassured that the researcher in the experimental group has rated a high level of usability in effectiveness for the Miro App teaching method. All the test results were analyzed using the SPSS software package. Particularly, paired t-test value comparison was used to compare the results of control and experimental group participants.

Keywords: Miro App, Web Whiteboard, Synchronous Online Flipped Learning Approach, Teaching and Learning Research

Introduction

The health crisis brought over by COVID-19 has pushed colleges all over the world to shift away from conventional educational methods and toward digital or blended instructional methods. To execute a learning approach that satisfies modern criteria, the higher education sector needs continuous advancement and refinement of methods of education, ways of providing instruction, and effective teaching.

The competition among conventional and information and communication technology (ICT)-based instructional methods in academia is now very intense. The usage of ICT in education is actively increasing as a result of recent educational developments. As the major goal of instruction is to keep students interested in the content, any field of learning must fully implement numerous ICT tools alongside traditional teaching and learning techniques in this situation.

Using Miro, developers can upload and communicate in locations to brainstorm and make presentations. Miro is indeed a digital or app-based interactive learning platform. Miro interactive whiteboards are beautiful because of how adaptable they are. Particularly with Miro, one could collaborate and work, exhibit it, and thereafter complete the process all within the application. Users could zoom around within the spaces' infinite canvases in order to arrange information. Sections or sheets are made by putting "frames" on the board. The organizational priorities combine elements of whiteboard, lecture, and tool functionality. Users can build custom designs directly from the homepage, but to save time, the website also provides a large number of pre-made layouts.

Similarly, the Miro application web whiteboard is used as a teaching aid and is suitable for instructional teaching and learning in all disciplines. We can broadly predict from Ng and Beatrice et al. (2022) study that the teaching method has shifted from online to hybrid learning which includes a variety of technologies that simplifies the teachers and students. The authors have explored aiding organic chemistry with the Miro web whiteboard in hybrid learning. The study encounters that using the Miro platform enhances both social and academic learning. On the flip side, Jones and Megeney (2022) have done a case study on how digital pedagogy has been established during the pandemic period. It highlights the usage of iPads for incorporating teaching mathematics in higher education using the Miro application web whiteboard. With the students of Middlesex University, the author discusses the use of virtual whiteboard apps that is useful for problem-solving in maths. These ripples set off to influence the usability of the Miro application on the web whiteboard for research scholars' acquisition in teaching research methodology from the humanities.

Research Objectives

- To identify the effectiveness of Miro Application in teaching and learning research methodology for researchers in humanities.
- To find the instructional differences among the researchers' involvement in the usability of the Miro App in the teaching-learning process.

Research Questions

Are there any differences between the Miro Application group participants and the Non-Miro application group participants?

What is the perception of researchers towards the instructional method for teaching and learning research methodology?

Significance of the Study

The present study highlights the usability of the Miro application for teaching and learning research methodology. When it comes to teaching the research methodology, which is considered to be vast and more theoretical makes it less arduous for the research scholars by using the Miro application for the learning process. The perks of using the Miro application research scholars believe that they can have hands-on sessions lively throughout the teaching-learning process it's like a parallel and simultaneous process. Hence, it is effective for research scholars in humanities to have practical knowledge and to use the methodology learned effectively.

Research Methodology

The Synchronous Online Flipped Learning Approach (SOFLA) was implemented for the study. This approach has eight steps for the learning approach these principles are aligned to emerge with online instruction. Marshall (2020) illustrated in the study that Marshall and Rodriguez Buitrago (2017) developed this SOFLA framework, which gives us a clear view of Online flipped learning. This approach enables participation in real-time class sessions as they happen in front of a blackboard in Miro it takes place through a virtual whiteboard. Nevertheless, the use of online platforms for the teaching-learning process has become more trending during the pandemic period where many interactive sessions in between the classes are considered to be the optimal choice for many institutions admits Shim's (2022) study. After the COVID-19 pandemic forced all the teaching modes to swift to online instruction that's where the idea of synchronous online flipped learning was an ice-breaker for many such theoretical classes. The 50 researchers were tested using the synchronous online flipped classroom approach along with the Miro app.

Research Design

The research plan of the study is exploratory research where the participants are divided into two groups control and experimental. The control group participants are Non-Miro app researchers and the experimental group is the participants exposed to the Miro app. Both groups are divided into two sessions before and after. During the after session, the control and experimental group undergoes progress testing with the teaching effective scale. Next to the progress testing, both the groups take up paired t-test value comparison is analyzed between the participants in the SPSS tool to find the target value. The exploratory study is implemented to investigate the research questions that are statistically proven using SPSS software.

Reliability and Validity

The reliability test is validated using Cronbach Alpha. A value of 0.702 will indicate a high level of internal consistency for the scale. The Cronbach alpha value of all the items mentioned indicates good reliability so that the items are retained for the main study. The validity procedure was put forth using a questionnaire that was examined under expert validation for content evaluation.

Results and Discussion

The findings of the study are to examine the differences between the Miro App participants and Non-Miro participants and the effectiveness of using the Miro application. The participants are divided into before and after sessions. In the before session, the participants are divided into experimental and control groups. In the differences are mentioned between the control and the experimental group, where the blue bar represents the control group which are non-miro app participants and the orange bar indicates the experimental group which is Miro app participants respectively. In this session currently, the groups are divided to know the participants equally participating. In the next step where both the groups take post-session. Each group has 25 research scholars as participants, in the group and the groups are tested with 50 questions each carrying 2 marks and which calculates 100 total marks for each participant. The control group was not exposed to the Miro application but the experimental group was exposed to the Miro application. The research scholars had effective lecture sessions in the Miro application since they were able to participate actively in hands-on practice because the app allows the participants to edit or ask for doubts and access it simultaneously with the instructor.

The participants were given 50 questions, as a result, to find the effectiveness of using instructional teaching with and without Miro application and understand the theories of research methodology. The experimental group results were evident to prove the effectiveness of teaching research methodology through the Miro platform for the research scholars. The overall mean average is showcased in Table 1 that the Miro application participants experimental group had a higher mean value of 64.88 than that of non-Miro application participants which had a value of 46.16 as shown in Table 1. Hence from the results, it is evident that there is a significant difference in the average value of the experimental group, and as mentioned in RQ 1 it is proven that the differences between the groups are high in the overall average mean value.

Table 1. Overall Average between the Groups

	Differences in Miro and Non-Miro App Group Participants	
	Control Group	Experimental Group
	Mean	Mean
Overall Average	46.16	64.88

Synchronous Online Flipped Learning Approach (SOFLA) With Miro Web Whiteboard

As discussed in the research methodology, the SOFLA mirrors online flipped learning that works outside the classroom that replicates the crucial part of flipped learning instruction (Marshall and Kostka.2020). The method of flipped learning when implemented makes the lecture interactive and dynamic enables learning since, the participants or research scholars are also able to participate as they can do in real-time class sessions and also would be lively and interact with the instructors as well, as stated in the study by Marshall and Kostka (2020). With Miro web whiteboard, the synchronous online flipped learning matches apt for the study as it blends both the application with web whiteboard and the flipped instruction way of teaching compared to the traditional instructional method. Guihua (2020) reveals in the study that SOFLA allows a simultaneous interaction between the instructors and the participants to break the stereotypical way of teaching online through watching courseware or video lectures. During the process, the instructors have the option to flip the class into a lecture one or turn it into a discussion class where the participants are allowed to discuss, share and interact directly with the teachers (Guihua, 2020). In the Miro web whiteboard, the instructors have the access to supervise the participants' or scholars' learning process and to investigate their learning effect. The execution of the Miro web whiteboard along with SOFLA is analyzed with the view to upshot the effectiveness and to know the overall perceptions of the researchers' on instructional teaching and learning.

From the data, the researcher's perception of instructional teaching and learning is determined using a questionnaire that was validated and tested by an expert in this field of teaching. The perceptions are divided into six sections course content and planning, communication and interaction, instructional materials, worksheets and assignments, learning outcomes, and overall evaluation of the course. The course content and planning are evaluated by the research methodology course content that was taught on the Miro application web whiteboard. The planning was further discussed by the teacher using a flowchart to know the blueprint of the course. Secondly, communication and interaction were based on the way of communicating with the participants effectively and on lively interaction during the course. Next, the instructional materials play a vital role because sharing through the Miro application web had more effective results than the traditional method of circulating the materials because the instructors can create a group where they upload the materials that are accessible to the participants. The aftermath of the instructional materials is to test them by assigning worksheets and assignments to participants. Once the worksheets and assignments are assigned, the next process will be to find out the learning outcomes of the participants. Therefore, we can analyze the overall evaluation of the course using the instructional teaching and learning method of Miro application. Hence, it is an exploratory study we tend to analyze the research questions that have been proven statistically using the SPSS tool.

Table 2. Results of Researchers' Overall Perception of Instructional Teaching and Learning

	The Researchers' Perception of Teaching and Learning	
	Control Group	Experimental Group
	Mean	Mean
Course Content and Planning	2.62	3.13
Communication and Interaction	3.16	3.41
Instructional Materials	2.82	3.34
Worksheets and Assignments	2.66	3.68
Learning Outcomes	2.84	3.26
Overall Evaluation of the Course	2.97	3.47
Overall Perception	2.845	3.381

Table 2 illustrates the researchers' overall mean value in the perception of instructional teaching and learning between the control and experimental group. Table 2 depicts the mean values using the SPSS tool for six perceptions as discussed previously between the control and experiment groups. The course content and planning mean value shows positive results as 3.13 from the control group value of 2.62. The perception of Communication and interaction outcasts the mean value of 3.41 that of non-Miro participants was 3.16. The mean difference in instructional materials was slightly higher than the control group value of 2.82 and the users of the Miro application value was 3.34. The researcher's perception in the section of worksheets and assignments after the usability of the Miro application had an extreme hike in the mean value of 3.68 than that of non-users who had 2.66 which had a 0.10 rapid rise in this perceptive. The learning outcomes showed some equivalent changes in the average score between the groups 2.84 and 3.26 respectively. The overall evaluation of the course shows an upsurge in the mean value of the control group and experimental group from 2.97 to 3.47 respectively. The overall perception average of the experimental group is 3.381 significantly higher than the control group's perceptions of 2.845 under the six sections. Thus, using the Miro application is found effective than teaching traditional method. The research concepts in the research methodology are made to understand using instructional learning using the Miro app web whiteboard. The researchers in the experimental group are found to be motivated and interested in acquiring research elements and tend to be confident while using the Miro app presented on the web whiteboard. However, it is found that web-based instructional learning is the best tool for the teaching and learning process.

Conclusion

The study presented and discussed can be summed up as experimental group participants seem to focus their attention more on research elements and show interest in increasing their research skills. The participants should be encouraged by the instructors to be independent in their learning process which helps with retention for a long time. The findings indicate that researchers' average scores match our research questions. The findings of RQ1 and RQ2 show that the control group participants had low mean and the scholars were unable to use their instructional method. Furthermore, it emphasizes participants to expose web-based tools through various activities. The instructors or the teachers should design activities that encourage the participants to be active in participation by employing their instructional teaching and learning methods through practice. Concerning the perceptions, strong significant differences are observed between the control and the experimental group. The RQ2 outcasts the teaching effectiveness scale of learners, because the mean value for the experimental groups' perceptions is significantly higher than the control group's perceptions in all sections. The mentioned instructional teaching and learning methods are assumed to provide a concrete understanding and address the research gap in the literature regarding the use of the Miro app on web whiteboards could influence researchers' acquisition. This study provides supportive statistically significant evidence and data for an operative tool like a virtual web whiteboard in assisting teaching and learning acquisition. Ultimately, researchers' familiarity with web-tools activities has a positive impact on the learning process to enhance and increase research skills. With respect to all these aspects, along with the findings, it could be evident to conclude that the usability Miro app on the web whiteboard is more resourceful for researchers to enhance their learning.

References

Aravind, B. R., and Rajasekaran, V. (2021). Exploring Dysphasia Learners' Vocabulary Acquisition through the Cognitive Theory of Multimedia Learning: An Experimental Study. International Journal of Emerging Technologies in Learning (IJET), 16(12), pp. 263–275. http://dx.doi.org/10.3991/ijet.v16i12.22173

Domínguez-Lloria, S., Fernández-Aguayo, S., Marín-Marín, J. A., and Alvariñas-Villaverde, M. Effectiveness of a Collaborative Platform for the Mastery of Competencies in the Distance Learning Modality during COVID-19. Sustainability, 13, 11, 5854 (2021), DOI: https://doi.org/10.3390/su13115854

Guihua Ma, (2020). The Effectiveness of Synchronous Online Flipped Learning in College EFL Reading Course During the COVID-19 Epidemic https://doi.org/10.21203/rs.3.rs-84578/v1

Jayakumar, P., Suman Rajest, S., and Aravind, B. R. (2022). An Empirical Study on the Effectiveness of Online Teaching and Learning Outcomes with Regard to LSRW Skills in COVID-19 Pandemic. In Technologies, Artificial Intelligence and the Future of Learning Post-COVID-19 (pp. 483–499). Springer, Cham. https://doi.org/10.1007/978–3-030–93921-2_27

Jones, M. M., Megeney, A., and Sharples, N. Engaging with Maths Online-teaching mathematics collaboratively and inclusively through a pandemic and beyond. MSOR Connections, 20, 1 (2022), DOI: https://doi.org/10.21100/msor.v20i1.1322

Marshall, H. W., and Kostka, I. (2020). Fostering Teaching Presence through the Synchronous Online Flipped Learning Approach. Tesl-Ej, 24(2), n. 2.

Ng, B. J. M., Han, J. Y., Kim, Y., Togo, K. A., Chew, J. Y., Lam, Y., and Fung, F. M. (2021). Supporting Social and Learning Presence in the Revised Community of Inquiry Framework for Hybrid Learning. Journal of Chemical Education, 99(2), 708–714. Paper presented at the Performance Management Association Conference. Dunedin, New Zealand.

Shim, E., and Inti, S. (2022). Effectiveness of the Synchronous Online Flipped Classroom on Students' Learning During the COVID-19 Pandemic. EpiC Series in Built Environment, 3, 670–678.

A Study of Female Identity and Marital Discord in the Selected Works of Anita Desai

Sukhman Randhawa,[a] *Jayatee Bhattacharya,*[b] *and Dhananjay Singh*[c]

[a]Amity University, Noida, India
[b]Amity University, Noida, India
[c]Jawaharlal Nehru University, New Delhi, India
E-mail: sukhman470@gmail.com

Abstract

The article aims to examine the functioning of marital relationships and the problems associated with interpersonal relationships to study the subsequent repercussions in the lives of the selected characters of *Cry the Peacock* and *Where Shall We Go This Summer?* It attempts to examine how the operation of personal relationships and societal expectations work together to shape the identity of an individual within the conceptual framework of societal trends, norms, and gazes of men which results in their subsequent disillusionment. Further, the research deals with the journey of neurotic mental state to the ultimate redemption of the protagonists through their frustrations and baffled desires, which leads to marital discord.

The overall objective of the article is to outline the causes of the plight of female characters, the expectations, and roles imposed on them, the aspects affecting the marital relationship, and a journey to self-actualization and self-redemption.

Keywords: Feminine Identity, Gendered Roles, Inter-personal Relationships, Institution of Marriage, Societal Expectations, Suppression.

Representation of Indian Women

Anita Desai has very efficiently depicted the psychological and physical concerns of Indian women. Her work has evoked an enthusiastic response from critics within and outside India for the extraordinary representation of her fictional characters and the haunting exploration of their psyche. It shows her interest in an existential dilemma of her characters, as she depicts the alienation and isolation of her characters from society and their hollow and negative approach toward life.

The major focus of Desai remains on depicting the fully awakened Indian women, who fight to search for their identity. She develops a character that struggles with societal expectations and moves from a sense of alienation toward self-realization. Each of her characters is sketched as isolated, fragile, struggling in the search for her identity and solitude of her happiness. The characters of Anita Desai are not drawn superfluously; rather they are made of blood and flesh, which possess certain strengths and weaknesses.

The portrayal of Desai's protagonists is an attempt to redefine the personal and social relationship between man and woman, sexually and non-sexually through the concept of Kate Millet in *Sexual Politics*. She believes that for centuries, the relationship between men and women has been of dominance and sub-ordinance. It pictures the Indian women being emotionally dependent on their male relationships like on their father, brother or husband but they also gather the strength to say the word of their inner thoughts. As Shantha reports:

Anita Desai's novels constitute together the documentation through fiction, of radical female resistance against a patriarchally defined concept of normality. She finds the link between female duality, myth, and psychosis intriguing; each heroine is seen as searching for, finding, and absorbing or annihilating the double that represents the socially impermissible aspects of her femininity (Krishnaswamy, 1984).

The representation of *Cry the Peacock's* protagonist is completely different from Indian societal expectations. Usually, Indian women consider their husbands as Godly figures and never disrespect their duties towards them or their families. In this novel, Desai has portrayed Maya in a different light, she attempts to highlight Maya's mental trauma caused due to psychological longing. She like other Indian women expected a handsome and caring husband but to her disappointment, he was completely the opposite.

Anita Desai reveals the condition of women by depicting women being deprived of the freedom to choose their husbands. Gautama is portrayed as a man whose, "understanding was scant, love was meager" (Desai, 1980). Maya tries to please her husband to get a single word of appreciation or care but Gautama is devoid of such emotions. Maya confronts, "Is there nothing... Is there nothing in you that should be touched ever so slightly, if I told you, I live my life for you?" (Desai, 1980) Maya craves to express her feelings and desires to him and makes an effort to be more communicative, but eventually, Gautama completely neglects to communicate with her. Gautama stays distant and does not understand her feelings of love and care whereas, Maya longs for involvement and understanding which brings the weird state of estrangement between husband and wife while living under the same roof as strangers.

Dissatisfaction makes Maya an introverted person, who started to comfort her and find solace by staying close to nature or her pet dog Toto. She talks to her dog and treats him like her child because she feels that, "childless women do develop fanatic attachments to their pets, they say." (Desai, 1980) and her dog's death took a serious toll on her mental health, "there were still spaces of darkness in between...Death lurked in those spaces, the darkness spoke of distance, separation, loneliness...I cried to myself—What is the use? I am alone." (Desai, 1980) Maya couldn't keep her head calm after her dog's death and started to live in illusion.

Desai creates her modern protagonists with traditional values but while struggling with societal expectations, protagonists do not compromise their self-respect. Desai beautifully pictures the primitive to the modern era, where women are discriminated against, dominated, and ripped off of their identity. She is greatly concerned about women's bonding with their male partners and voicing out her protagonist's inner thoughts. Her work is, "an effort to discover, to underline and convey the true significance of things." (Sharma, 2013) As, "a woman writer is more concerned with thought, emotion and sensation" (Sharma, 2013).

Similarly, Desai brings out the same theme of the search for identity through her protagonist Sita, in *Where Shall We Go This Summer?* During the primitive age, people gave more importance to the idea of joint families and spending time with their relatives. During that time women were considered as mere unpaid maids who had to adjust but in the modern era, women rejected the concept of self-compromise and came out of those boundaries in search of their identities. Sita got fed up with taking care of her children, family, and the needs of her husband.

Therefore, Sita felt alienated, as she had nobody to talk to and Raman rarely spent any time with her. She felt utterly dominated by her husband, as she had no freedom to make any decisions. Jung believes that "It is in middle age, the second stage of life, that people suddenly wake to a feeling of emptiness and lack of meaning in life" (Fordhan,

1976). Desai profoundly depicts the torments and sufferings of Sita's life as a representation of modern woman's quandary when she feels unwanted. After spending 20 years with her husband, Sita in her forties realize that "She was always waiting...she could not inwardly accept that this was all there was to life...leaving her always in this grey, dull-lit, empty shell. I am waiting...although for what, she could not tell" (Desai, 1982).

Thus Sita felt that her life was confined to four walls with few characters, and no happiness but disappointments. She told her decision of aborting the child but Raman discouraged her decision. She finally decides to leave the city and moved to Manori Island, where she believed that people perform abortions in a much better way. However, the disagreement between Raman and Sita leads to a quarrel, where while looking at a couple she realized that she has missed a life of, "Tender, loving...He loved her – but who was he? Her father or husband? Or a lover? I don't know. I watched them...They were like a work of art...they were inhuman, divine." (Desai, 1982) Sita strongly believes that she doesn't want to give birth to her fifth child to which Raman explains, "Not much longer to go now, Sita, it'll soon be over. You are doing a blunder." (Desai, 1982) Apparently, Sita's decisions were not acknowledged and she was ill treated by her husband. She feels frustrated being stuck in a melancholic life of doing household chores. Her interpretation of city life is totally opposite to her expectation, she believes that city life is wicked for her kids' upbringing, therefore in anger, she says, "I am trying to escape from the madness here, escape to a place where it might be possible to be sane again." (Desai, 1982)

Sita considers the island as her only escape from reality because she felt so unwanted in Raman's family. Due to her suffering and mental trauma she decides to live in the illusionary world, where she is liked and respected, unlike in the city, where she was disrespected, exploited, and cheated. Soon later, when her children couldn't cope with village life, she reconsidered her decision and came back to the city with her husband and children for the sake of her children's future. In the end, she accepts reality and compromises her identity.

In this respect, Desai reflects the conditioning of Indian women as frail and submissive who is always dependent on men. However, when they try to break the cages of traditional social norms, Sita who defies the stature of being docile was tagged as "mad" and Maya who defies being a quiet woman was tagged as "insane." Both the characters in their rebellion tried to escape reality. Maya went into her imaginative world, where she connected more with nature and her dog, and on the other hand, Sita tried to escape city life and went to Manori Island for peace. Nonetheless, both had to return to reality because as portrayed by Desai, a woman in Indian culture cannot abandon her husband. A woman is conditioned to be compliant with the ways of her husband and to stay happy within the four walls of home.

Desai does not present a harmonious environment for females. She depicts the disturbed family environment, where all the females are suffering in their conditions. Since childhood, a girl is taught to be selfless and self-sacrificing which results in torment and suffering. In this process, women feel a sense of identity crisis, lost individuality, and endless suffering.

Marital Discord

Anita Desai skillfully represents the theme of marital disharmony in her novels. She believes that the issue is majorly concerned with a lack of understanding, communication, and trust, along with faithlessness, insincerity, and a soul-grinding process of compromise. Desai in a conversation with Jasbir Jain explains, "All human relationships are inadequate. Basically everyone is solitary... involvement in human relations in this world invariably leads to disaster" (Jain, 1987). Desai attempts to explore intricate family relationships and meaningless marriages to unravel the socio-familial bond of upper-middle-class Indian women. She reveals the plight of women being victimized under traditional norms whereas men enjoy the freedom to make decisions, and pave their lives and women remain passive and powerless in the process. Her main concern is to project the psychic state of her protagonists.

Kate Millet through her observation remarks on the role of the family, as an agent of the patriarchy. She states, "Patriarchy's chief institution is the family. It is both a mirror and a link of the larger society...the family not only encourages its members to adjust and conform... Women have ruled through the family alone..." (Millet, 1969) Desai seems to challenge the power structure of familial and social relationships. She subtly argues that the plight of women is directly caused by power imbalance. In *Cry The Peacock,* Desai represents Gautama's family as made up of bricks rather than hearts. As they never spoke of love and care, rather "they spoke—of discussions in parliament, of cases of bribery and corruption... They had innumerable subjects to speak on...of political scandal and intellectual dissent" (Desai, 1980).

In addition, when Maya suggests her desire to go to the mountains, Gautama completely neglects it and her emotions. She feels isolated like, "a body without a heart, a heart without a body." (Desai, 1980) In her deep sense of sorrow, she feels that "All order is gone out of my life... Thoughts come, incidents occur, then they are scattered, and disappear...Those are no longer my eyes, nor this my mouth...Strangers surround me (Desai, 1980).

Desai delineates a loveless bond of marriage where Maya's emotions are defeated by Gautama's philosophical gibberish. Due to this scenario, Maya undergoes a mental breakdown, "There were countless nights when I had been tortured by a humiliating sense of neglect, of loneliness, of desperation that would not have existed had I not loved him so, had he not meant so much" (Desai, 1980). Sharma effectively explains the disharmony, "Engrossed in his busy schedule, Gautama continues to ignore Maya's needs remaining callously immune even to her physical desires. This is how Maya usually suffers the agony of her unfulfilled desires." (Kumar, 2000)

Maya never understood the meaning behind Gautama's philosophy. His indifference adds to Maya's insanity. She was afraid of death and the shadow of Albino kept haunting her throughout her life, "Torture, guilt, dread, imprisonment these were the four walls of my private hell, one that no one could survive in long. Death was certain" (Desai, 1980). Since every human deal with their problems in their way, considering Indian beliefs, customs, and traditions, Indian women are shy and sensitive by virtue, therefore, they fall prey to sentiments. In her agony, Maya's insanity prevails over her senses, fogs her logic and she pushes Gautama over the parapet and he falls to death. Her giggle and laugh explain her freedom, but it was short-lived. She in her insanity jumps from the balcony and meets the same fate.

Similarly in *Where Shall We Go This Summer,* Desai has developed a character, Sita, who is suffering from rejection and living a hopeless married life in the city. Roy, in *The Aesthetics of An (Un)willing Immigrant: Bharati Mukherjee's Days and Nights in Calcutta and Jasmine.*, comments that Anita puts forward the real picture of misunderstandings between husband and wife. He believes that the cause of disharmony is bad temperaments and communication gaps. Since both the characters, Sita and Raman are sketched as opposites; like Sita deeply rooted in her illusionary world and Raman in cruel reality therefore their married life is a mere compromise.

The representation of Desai's protagonists is mere showpieces at social gatherings as men come back from work to eat and sleep. Desai is against this dry vegetable existence, which her protagonists rebel against. Desai pictures Sita as a victim of maladjustment because Raman keeps himself busy in his business and doesn't care much about his family and children. He assumes that Sita would take care of them. Sita like other Indian women had great expectations from her marriage but to her disappointment, she feels exploited, exhausted, and illuminated. She feels like a machine in the city, whose work is to give birth. In this process, she lost her identity and she decides to move to Manori island but Raman suppresses her feelings and she feels like a person who has "lost all feminine, all maternal belief in childbirth, all faith in it and began to fear for it as yet one more act of violence and murder in a world that had more of them in it than she could take" (Desai, 1982).

For Sita life has become, "unthinkable that anything should happen- for happening were always violent" (Desai, 1982) this makes Sita say that her fifth child "would not be born..." (Desai, 1982) because she finds the violence in a metropolitan city as unacceptable. R. K. Srivastava rightly elucidates Sita's plight as "The incident in which large numbers of crows assault and kill an eagle becomes symbolic of Sita's plight amid violence prevalent society. By giving birth to a child, she would only contribute to the violence of the world" (Srivastava) For Sita, her mental world is more real because she realizes that her "own life is like a shadow, absolutely flat and uncolored" (Desai, 1982) when she witnesses the devotional love of the Muslim couple and realizes what her life lacks.

Soon Sita accepts the stark reality, as what cannot be cured must be endured. Therefore, when Raman comes to meet her in Manori she accepts to go back to city life for the sake of her children. Sita reveals her unhappiness in her married life when asked about it by the couple but Raman accuses her of being inhumane and his remarks show his dominance when he states, "Any woman – anyone would think you are inhuman. You have lived comfortably, always in my house. You have not had any worries" (Desai, 1982). This accentuates Sita's alienation and she realizes "what a farce marriage was, all human relationships were" (Desai, 1982).

Ramesh K. Srivastava reveals:

"At Manori when Sita and Raman meet after a long time...to her shock, she comes to know that Raman had nothing to say...Yet Raman...wins Sita back...But Sita's coming back to Bombay is not a gesture indicating her realization of the existence of love and understanding between them, but rather...that there is no magic left in the island capable of sustaining her in need. Hence it is in a state of disheartening helplessness due to a wish to compromise with her husband." (35)

Through her protagonists, Desai outlines the women's revolt against self-righteous, self-satisfied, egoistic, and conceited men in marital life. Like many other Indian women, Sita fought for her identity but settled and sacrificed her desires and expectations for the sake of her children's future and happiness.

Desai represents both her Indian protagonists as victims of unsatisfied marriages because of their husbands and their familial bonds. On one side, Maya is deprived of sympathy and admiration therefore she violently reacts to her husband's rejections as it makes her feel worthless. Therefore, in her anger and aggression towards her husband, she falls prey to an act of self-destruction. Whereas on the other side, Sita realizes that there is no escape from her troubles therefore, she comes to terms with her life by negating any expectation of miracles or possible solutions and emerges as more mature and wise person in the process. The failure of protagonists in an attempt to secure their marriage is not considered as an individual's failure but as a society's failure.

Conclusion

The novels of Desai have clearly explored the daily injustices and exploitation of women in an androcentric society. Desai set forth the desire of Indian women to be acknowledged as a person, capable of feelings and intelligence. She discourages the step of her protagonists towards exile because she doesn't believe that the exile from society will solve any problem rather, the major focus should be on the effort to exist in society and maintain their individuality than feeling the lack of belongingness.

Desai depicts that in patriarchal society, a woman is expected to be an ideal wife, mother, and caretaker with the attributes of submissiveness, tolerance, and self-sacrifice. In Indian culture, a girl since childhood is embedded with the idea of patience, self-abnegation, and acceptance of lower status than men accredited with the traits of being shy, pure, faithful, and gentle as a wife and loving and caring as a mother.

Desai reveals her interest in the intricacies of human relationships. For her, fiction is the mode of expression to outline the baffling nature of relationships. For Desai fidelity, faithfulness and mutual support are the foundation of Indian marriages. According to Hindu view, fidelity till death is the highest law of marriage to be followed.

In the works of Desai, women pass through intense mental turmoil and achieve their integrated selves. The modern woman struggles for the right outcome as they respect themselves and do not wish to be crowded or intimidated by anyone. They have the courage to raise their voice and rebel against any kind of humiliation or oppression. Even so, in the current fictional framework, the protagonists of Desai, Sita, and Maya succumb to the pressures of socio-familial expectations. Their self-development, therefore, was impossible thematically. They were incapacitated not only by Indian traditions and social expectations but also by their constitutions.

References

Krishnaswamy, S. (1984). The Woman in Indian Fiction in English. 1950–80. APH. Publishing.

Desai, A. (1980). Cry the Peacock. New Delhi: Orient Paperback.

Desai, A. (1982). Where Shall We Go This Summer? Orient paperbacks.

Sharma, M. (2013). Marital Discord in Anita Desai's Novels. Global Journal of Human Social Science Research.

Fordhan, F. (1976). An Introduction To Jung's Psychology. Hannondsworth: Penguin.

Jain, J. (1987). Stairs to the Attic: The Novels of Anita Desai. Jaipur: Printwell Publishers.

Millet, K. (1969). Sexual Politics. New York: Avon.

Kumar, S. (2000). The Reverse Patterns Of Journey in Anita Desai's Cry the Peacock and Where Shall We Go This Summer? Critical Essays on Anita Desai's Fiction. (ed), Jaydipsingh Dodiya. Pub., IVY, Publishing House, New Delhi.

Srivastava, R. K. Anita Desai at work: An Interview.

Role of Digital Competency in Sustainable Quality Education

G. Bhuvaneswari,[*,a] Rashmi Rekha Borah,[b] Binu Sahayam D,[c] Manali Karmakar,[d] Moon Moon Hussain[e]

[a,b,c,d]School of Social Sciences and Languages, Vellore Institute of Technology, Chennai, India; B. S. Abdur Rahman Crescent Institute of Science and Technology, Chennai, India

E-mail: [*]bhuvaneswari.sb@gmail.com

Abstract

Since younger generations are constantly changing, their needs and circumstances have changed, and they now require a high level of digital competence. For this reason, teachers must learn how to respond to student's needs in a way that is safe, didactic, and educational. To accomplish this, it is necessary for teachers to carry out the teaching-learning process of students and to encourage the acquisition of critical competencies in learners. To use ICT in the classroom critically, dynamically, and creatively, instructors need a certain set of skills, competencies, knowledge, and attitudes, which is what is meant by "digital competency." This paper aims at finding out the gap between maintaining sustainable quality education and adaptability of teachers in perceiving digital literacy.

Keywords: digital classroom, digital competency, language teaching, and teacher's professional development, sustainable quality education

Introduction

This review investigated the amount of digital proficiency that teachers needed to effectively teach engineering undergraduates. This investigation is being conducted through a survey. There were 85 teachers from an engineering institution who took part, representing a variety of academic fields. A 14-question survey on "Digital Competency" was utilized. Using the Cronbach-Alpha method, the instrument's reliability was assessed, and a reliability coefficient of 0.75 was found. After the data were analyzed, the study questions were addressed using the mean and standard deviation. The results showed that there were several variables influencing competency, including age, gender, and place of residence (rural/urban). The findings also included information on teachers' motivation to learn about and equip themselves with digital competency.

Literature Review

Children and adults now play, access information, communicate, and learn in entirely new ways thanks to the digital revolution. Children in preschool are already comfortable with digital devices. (Caena et al., 2019). Although many teachers were raised in environments that primarily rely on digital tools, this does not imply that they are digitally competent. (Li and Ranieri, 2010). Utilizing the advantages of digital technology as well as addressing their drawbacks constitutes digital competency (Napal-Fraile et al., 2018). Educators have to create conditions and chances for in-depth learning encounters that might reveal and enhance students' capacities. Teachers are expected to be catalysts for meaningful learning rather than merely facilitators, (Bhuvaneswari et al., 2022) using their creativity to select from a variety of tactics that can be combined and tailored to the needs of each student and setting. (Raghul et al., 2021). Teachers must include technology in their learning-teaching processes given the growing importance of technology in business life and career (Tondeur et al., 2017). Deep learning is unleashed by technology when it is carefully combined with the other essential elements of the new pedagogies (Jayakumar et al., 2022).

Sustainability Education for Sustainable Development (ESD), which has been described as UNESCO, in 2014, is a term that is frequently used to refer to education. Every person can get the knowledge, skills, attitudes, and values (Binu Sahayam et al., 2022) essential to sculpt a sustainable future through education for sustainable development.

Purpose of the Study

The purpose of this study is to identify the teachers' digital competency, and factors affecting learning skills and identify the willingness of teachers in transforming into a digital environment.

Research Questions

1. How depth is the awareness of digital literacy among the teachers to maintain sustainable quality education?
2. Do the Digital skills possessed by teachers differ based on gender, age, and place of residence they belong to?
3. What are the factors affecting the teachers in getting adapted to digital transformation?

Methodology

Research Design

A survey with a descriptive format was used for this study. The research was done for the faculty members who instruct engineering students. The professors at an engineering institution come from several disciplines. As both engineering and non-engineering courses, such as English, Psychology, and Social Sciences, are taught in engineering institutions, this study was carried out to provide a diverse analysis of the lecturers.

Participants

These study targets teachers of various disciplines of knowledge were considered for the study as digital tools are used by faculty across the disciplines. Totally 85 teachers were asked to fill out the questionnaire and results are considered based on their age and place of residence also. Convenience sampling technique was used to select the sample for the study.

Research Instrument

A 14-question survey on "Digital Competency" was utilized. Using the Cronbach-Alpha method, the instrument's reliability was assessed, and a reliability coefficient of 0.75 was found. After the data were analyzed, the study questions were addressed using the mean and standard deviation. The answers were analyzed on a 5-point scale of Agree, Strongly agree, Neutral, Disagree, and Strongly disagree.

Reliability Test

Since the Cronbach-Alpha approach is most appropriate for polytomous scored items, it was used to estimate the instrument's internal consistency. The device had a reliability index of 0.75, which indicates its reliability.

Administration of the Questionnaire and Data analysis

The Questionnaire was administered to the faculty via google forms and results obtained from the forms were collected and analyzed using mean, standard deviation, and percentage analysis.

Results and Findings

Among 85 teachers, 44 were females and 41 were males. The questionnaire was answered by the teachers who were between the age group 25–45. The average age of the teachers is 35.

Table 1. Digital Literacy of the teachers

Questionnaire	Responses from Women		Responses from Men	
	Agree	Disagree	Agree	Disagree
I use my phone to check my mails	37	7	33	8
I use online search engines to find information	36	8	31	10
I have an online profile on a social media platform	23	21	34	7
I have a blog of my own.	17	27	18	23
I have created data using online tools	24	20	19	22

Table 2. Factors affecting Digital Tools – Gender, Age, and Place of Residence

Questionnaire	Agree	Disagree
Internet is accessible easily from my home	39	35
I reside in a city	36	32
Digital platform is only for youngsters	5	6
I use smartphones to teach in the class	8	15

Table 3. Percentage Analysis

Categories	Average Percentage – Women	Average Percentage - Men
Digital Literacy of the Teachers is accessible easily from my home	27.4	27
Factors affecting using Digital tools	22	22

The percentage analysis shows that there is no difference in possessing digital literacy between men and women as the average percentage of women is 27.4 and the average percentage of men is 27. Similarly, the factors affecting using digital tools also do not have major variation in percentage which means that there is no significant difference here too.

Table 4. Adaptability

Questionnaire	Agree	Disagree	Neutral	Strongly Disagree	Strongly Agree
I am ready to try new methodology in order to reach my students' level	36	10	9	8	22
I am able to shift to new technology without difficulty	27	9	15	18	16
I explore new digital tools	41	11	4	4	25
I like chalk and talk method rather than digital tools	18	8	9	31	19
I don't mind learning technology from my students	52	5	8	13	9

The percentage analysis shows the result as follows; The responses both agree and strongly agree are considered to decide the adaptability of teachers and here any other factor such as age or gender is not taken into consideration. While 68 % are ready to adopt a new methodology "I am ready to try new methodology to reach my students" level, 50% of the teachers do not find difficulty as a part of adaptation process. 77% of the teachers are curious to learn and "explore new tools," 37% of the teachers like the earlier chalk-and-talk methodology rather than digital tools. This number is quite low when compared to others. 71% of the teachers are ready to learn from future generations as well.

Conclusion

Assuring environmental protection and conservation, advancing social equity, and fostering economic sustainability are all goals of education for sustainable development (ESD), which encourages the development of the knowledge, skills, understanding, values, and behaviors necessary to create a sustainable world.

This study aimed at finding the digital literacy of the teachers, factors affecting learning digital tools, and adaptability of teachers in accepting the digital environment of education. Adaptability of the teachers is high as per the result which aligns with the behavior that is necessary to maintain sustainability in any field.

References

Bhuvaneswari, G., Borah, R. R., and Hussain, M. M. (2022). Willingness to Communicate in Face-To-Face and Online Language Classroom and the Future of Learning. In: Hamdan, A., Hassanien, A. E., Mescon, T., Alareeni, B. (eds) Technologies, Artificial Intelligence and the Future of Learning Post-COVID-19. Studies in Computational Intelligence, Springer, Cham. 1019, 237–253. https://doi.org/10.1007/978-3-030-93921-2_14

Binu Sahayam, D., Bhuvaneswari, G., Bhuvaneswari, S., and Thirumagal Rajam, A. (2022). Stress-Coping Strategy in Handling Online Classes by Educators During COVID-19 Lockdown. In: Hamdan, A., Hassanien, A. E., Mescon, T., Alareeni, B. (eds). Technologies, Artificial Intelligence and the Future of Learning Post-COVID-19. Studies in Computational Intelligence, Springer, Cham. L, 1019: 51–65, https://doi.org/10.1007/978-3-030-93921-2_4

Caena, F., and Redecker, C. (2019) Aligning teacher competence frameworks to 21st century challenges: The case for the European Digital Competence Framework for Educators (Digcompedu). Eur J Educ, 54, 356–369. https://doi.org/10.1111/ejed.12345

Jayakumar, P., Suman Rajest, S., and Aravind, B. R. (2022). An Empirical Study on the Effectiveness of Online Teaching and Learning Outcomes with Regard to LSRW Skills in COVID-19 Pandemic. In Technologies, Artificial Intelligence and the Future of Learning Post-COVID-19, 1019, 483–499 Springer, Cham.

Li, Y., and Ranieri, M. (2010). Are "digital natives" really digitally competent?-A study on Chinese teenagers. British Journal of Educational Technology, 41(6), 1029–1042, https://doi.org/10.1111/j.1467-8535.2009.01053.x

Napal-Fraile, M., Peñalva-Vélez, A., and Mendióroz-Lacambra, A. (2018). Development of Digital Competence in Secondary Education Teachers' Training. Education Sciences, 8(3), 104. https://doi.org/10.3390/educsci8030104

Raghul, E., Aravind, B. R., and Rajesh, K. (2021). Difficulties Faced by Special Education Teachers during COVID-19 Pandemic. International Journal of Early Childhood Special Education, 13(2).

Tondeur, J., Aesaert, K., Pynoo, B., Braak, J. V., Fraeyman, N., and Erstad, O. (2017). Developing a validated instrument to measure preservice teachers' ICT competencies: Meeting the demands of the 21st century. British Journal of Educational Technology, 48(2), 462–472. https://doi.org/10.1111/bjet.12380

Digital Infrastructure for SHGs of Tribal Women in Odisha: Means for MSMEs to Achieve SDGs

Gopal Krishna Jena,[a] *Lipika Mohanty,*[b] *Sarbeswar Mohanty,*[c] *and Sukanta Chandra Swain,*[d,*]

[a,b,c,d]School of Humanities, KIIT Deemed to be University, Bhubaneswar, Odisha, India
E-mail: *sukanta_swain@yahoo.com

Abstract

Women in tribal communities can overcome obstacles in life, and they have done so from the dawn of time, but their lack of financial resources makes it difficult for them to progress. It can be eliminated if credit is made readily available to destitute indigenous women. Self Help Groups (SHGs) have emerged as the most successful programs for eradicating poverty and empowering women, particularly among the most vulnerable members of society. Moreover, the adoption of digital infrastructure may turn the situation around as it has been succeeding to do so in many micro, small, and medium enterprises (MSMEs). This study aims to determine if SHGs with digital support help achieve the first three Sustainable Development Goals (SDGs), which include eradicating poverty, no hunger, and promoting good health and well-being. According to qualitative research, digital intervention in SHGs can help with SDG fulfillment.

Keywords: Well-being, SHGs, Tribal Women, Digital Infrastructure, Need Analysis

Introduction

In tribal society, women constitute a major working force and supplement the livelihood of the family. But their contribution is ignored because of their custom and habits. They are virtually invisible in the economic sphere. That is why; over the past years participation and empowerment of women have become the catchwords in the rural development program of our country. When it comes to participatory development and women's empowerment, "women self-help groups (WSHGs)" be the most effective technique since 1992. WSHGs improve the socioeconomic condition of the tribal women and to meet the condition of the tribal women folk and to meet the challenges in future years and fulfill our planned goals as well as regional income distribution. It helped to reduce the poverty and that has helped in accelerating the rate of growth of income, output, employment, and lively hood in the tribal region.

Review of Literature

SHGs play an important role in providing credit, training, thrift, and enhancing the decision-making process of tribal women. Empowering rural women both Government and non Government organizations play a very important role (Rao, 2016). The provision of microcredit to members of self-help groups can empower rural women. During post-SHG period income, saving, and employment increased (Sarania, 2015). Through SHGs and information technology, tribal women can support the development of the local economy and promote tribal culture. They will become more powerful and the neighborhood will benefit as well (Jena et al., 2022). SHGs are crucial tools that empower rural women for self-determination, community development, and nation-building (Sharma, 2013).

During the post SHGs situation, the income and employment of the tribal women increase tremendously, and SHGs in concert a significant task in the enhancement of women (Reji, 2013). The involvement of the rural poor in SHGs has a significant contribution to their social empowerment, which includes improving their self-confidence, altering their position of ancestors and humanity, contact skill, and other observable beneficial change (Puhazendi, 2001). SHGs play a vital role in the development of tribal women through all-around development of the socioeconomic condition, participation in the decision-making process and capacity-building measures, livelihood, and changing the level of awareness in every work of their life (Lenka et al., 2015). Women are empowered through SHGs by various financial activities such as savings, borrowings, budgeting, and rotating funds. Women are thereby becoming self-reliant and self-dependent because of various motivational programs and schemes organized by SHGS (Rajendran et al., 2013). Through digitization, India's extensive cultural history may be maintained. However, in light of the constantly evolving nature of technology, the preservation procedure must be flexible (Mohanty et al., 2022). SHGs are becoming a vehicle for development. Thus it is pertinent to study the financial problem of SHGs facing difficulties in getting loans; hindrances in the approval of funds, inadequate operational expenses, Inadequate cooperation within groups, refund of loans by SHG members, etc. (Das, 2012).

Research Gap

Based on reviewing existing literature, the research gaps identified are as mentioned below.

- ✓ There is paucity of systematic studies on SHGs in the empowerment of Tribal communities.
- ✓ Although some studies are there on empowerment in Tribal Districts, no concise study is there on the empowerment of tribal communities, particularly women, in less Tribal dominated Districts like Nayagarh District.
- ✓ Studies are not there to unfold whether digital infrastructure can change the fate of tribal women by way of achieving efficiency in a better way.

Objectives

The objectives of this study are to;

- ✓ ascertain whether SHGs contribute to the well-being of the tribal women, and
- ✓ explore whether access to digital infrastructure by SHGs can be the facilitator for catering to well-being in a better way.

Methodology

Since the Study is to assess the SHGs in tribal women empowerment, selecting a majorly Tribal District or a completely Non-Tribal District is not apt. Selecting a District with less Tribal population is just to study the Tribal population, particularly Tribal women. Nayagarh is such a District. The data collection techniques employed in this study are multi-stage sampling and purposeful sampling. Purposive sampling is used to choose the district, but multi-stage sampling is used to choose the blocks, GPs, villages, SHGs, and SHG members. There are eight blocks in Nayagarh District. Out of this, two Blocks - Daspalla and Nuagaon are selected because more tribal population is there in those two Blocks.

- Nayagarh District consists of eight Blocks.
- Two Blocks are selected (where more tribal people live)
- From each Block, 20 GP (2 × 20GP = 40 GP)
- From one GP, one village is selected - (40 × 1=40 villages are selected)
- From one village, one SHG - (40 × 1=40 SHGs are selected)
- From one SHG, at least two members are selected, i.e., two members from 20 of the SHGs and three members from 20 SHGs are selected (Total sample size is 100 members)

A qualitative study is done. In-depth interviews, observation methods, and focus group discussions are followed to obtain qualitative data. The relevant secondary data are collected from periodicals, books, journals, newspapers, websites, Block officer, Non-Government Organizations and DRDA office.

SHGs - An Initiative Towards Women's Empowerment and Well-being

The SHGs is a small group of people of same community coming together and making contribution small amounts of saving in a common pull and this money will be provided to the needy person with a very low rate of interest and every member have a bank account. The bank also provides financial help to the SHGs. The main objective of SHGs is to achieve economic development among the rural poor by way of; enabling the economic progress of members by engaging in income-generating activities, encouraging the members in productive activities, promoting the habit of saving, helping to reduce poverty, motivating members establishing small enterprises, helping self-confidence among the members, and taking collective decision by the members.

SHGs have a high potential to augment the well-being of the community involved if they are handled properly. However, it is pertinent to know whether the SHGs of the study area have contributed to the well-being of tribal women. Well-being is a very composite concept having multiple parameters. However, for tribal women, the most important parameters are the monthly expenditure amount and the education of their children. Thus, two parameters have been considered in this study. These are; a) The respondents' monthly expenses both before and after joining their respective SHGs and b) Concern for Children's education before and after joining SHGs. It was revealed that, except for one respondent, all respondents' monthly spending had significantly increased since joining their respective SHGs. It demonstrates that with the help of SHGs, the income and spending of the members of the SHGs have increased. The only woman's ad hoc decrease in monthly spending after joining the SHG versus before joining the SHG was caused by her rationalization of several expenditure sectors.

Moreover, it is discovered that the majority of the respondents have grown more concerned with their children's education. They had no interest in their children's education before joining the SHGs. Thus SHGs have, in turn, enhanced the quality of life for their stakeholders.

Need for Digital Infrastructure for Efficiency of SHGs

Digital infrastructure is a collection of resources, including Internet, Broadband, Mobile Networks, Communications Satellites, Data Centers, End-User Devices, Internet of Things, Applications "application programming interface (API)" integration, and the Cloud, that call for the utilization of data and computerized tools, systems, processes, and methodologies.

The principal objective of digital infrastructure the programme, which includes e-marketing, Mahila e-hat, expansion of Telecom network, e-services, Broad band facility for proper implementation of welfare, education, and health program, was transfer of digital technology to the rural areas and the benefits can be reached in the doorsteps of the poor people of the country. Benefits of digital infrastructure are; 1) With the help of digital technology women's self-help groups can sell their products and get the appropriate price for their product, 2) Computerization of land records, 3) Direct Benefit Transfer DBT to the beneficiaries, and 4) Banking facility at the doorsteps.

Out of 100 respondents, only 22 women have access to mobile phones for voice communication and only four women have smart mobile phones both for voice communication and internet data usage. However, those who have smartphones, simply use them occasionally as they don't put data pack regularly and they don't know much about the usage of smartphones constructively. No respondent has the access to a computer or laptop. Thus, for day-to-day functioning, all the SHGs depend on transmitting messages through physical messengers. In the process, they kill enough productive time and wait long for anything to happen. Although the respondents didn't feel the need for digital infrastructure like smartphones and computers/laptops with internet connectivity for their efficiency, as they were given the situation of a facilitator that can reach out messages at the destination at their fingertips, they expressed their happiness. Getting the raw materials for their cottage industries and selling their output to the appropriate customers at reasonable prices are the main concerns of every SHG that the respondents belong to. As they came to know about the fact that all these things can be done easily by way of digital infrastructure, they felt the need for such devices. However, they have some reservations regarding the adoption of such devices out of apprehensions. Thus, it is essential to make them access all possible digital infrastructures and know the effective usage of such infrastructure by way of training them aptly.

Perception of SHG Members on Digital Infrastructure as a Means to Achieve SDGs

Only 20% of the members analyzed have access to digital infrastructure and, as a result, are aware of its benefits, according to demographic primary data acquired from 100 respondents, or the tribal women who are members of several SHGs. However, it was important to ascertain whether each member had an opinion regarding the use of digital infrastructure. In order to determine the SHG members' perceptions, a few variables have been identified and those are;

- Using digital infrastructure by self or someone else for SHG functioning is fascinating and worthwhile (VAR1),
- It is exciting and worthwhile to use a smartphone to conduct an online search for information on how SHGs can operate more effectively. (VAR2),
- Using various apps on a smartphone, whether for oneself or others, to augment the performance of SHGs is interesting and valuable. (VAR3),
- It's intriguing and worthwhile to hear success stories of SHGs or MSMEs on a smart mobile phone, either for oneself or another. (VAR4),
- It's exciting and worthwhile to upload photos of the SHG activities using a smartphone, whether it's done by you or someone else. (VAR5),
- Using a smartphone to record and post films of SHG activities, whether by oneself or someone else, is interesting and worthwhile. (VAR6),
- It's interesting and worthwhile to use a smart mobile phone to get directions or use GPS to communicate with other SHGs and government offices (VAR7),
- Reading (or listening) social media postings, particularly posts related to SHGs or MSMEs by self or someone else is fascinating and worthwhile (VAR8),
- Watching TV shows, movies, etc. pertaining to SHGs/MSMEs/SDGS on a modern gadget is fascinating and worthwhile (VAR9),
- Utilizing social media to exchange SHG inputs and outputs will help your team work together more effectively and contribute to the SDGs (VAR10),
- Use digital infrastructure to promote the activities (including products/services produced) of an SHG will be beneficial from the point of view of effectiveness(VAR11),
- The SHG's performance will improve thanks to the use of digital infrastructure, which will also assure well-being and make it easier to fulfill SDGs (VAR12).

The independent variables are VAR1, VAR2,..., and VAR11, and the dependent variable is VAR12.

To ascertain if digital infrastructure can assist the SHGs in achieving the SDGs, 100 SHG members in the Nayagarh district of the state of Odisha provided information on the selected variables. The respondents' levels of agreement with the statements were calculated using a 5-point Likert scale. The levels of agreement are transformed to standard points as follows: 1 for Strongly disagree, 2 for Disagree, 3 for Neither agree nor disagree, 4 for Agree, and 5 for Strongly agree.

Test of Reliability

Since the Cronbach's Alpha value is 0.945, the variables selected and the data gathered on those variables have succeeded in the reliability test (as presented in Table 1 and 2)

Table 1. Summary of Respondent Processing

		N	%
Cases	Valid	100	100.0
	Excluded[a]	0	.0
	Total	100	100.0
a. List-wise omission hinged on all variables in the process.			

Source: Output from SPSS

Table 2. Statistics for Reliability

Cronbach's Alpha	N of Items
.945	12

Source: Output from SPSS

ANOVA for Hypothesis Testing

To ascertain whether the employment of digital infrastructure in SHG operations has the potential to attain SDGs, ANOVA and regression analysis are used.

Null Hypothesis: Use of digital infrastructure in the operation of the SHGs does not have the perspective to achieve SDGs.

ANOVA was calculated to test the null hypothesis, and the results are depicted in Table 3.

Table 3. ANOVA[a]

Model		Sum of Squares	df	Mean Square	F	Sig.
1	Regression	78.772	11	7.161	362.653	.000[b]
	Residual	1.738	88	.020		
	Total	80.510	99			
a. Dependent Variable: VAR12						
b. Predictors: (Constant), VAR11, VAR4, VAR7, VAR8, VAR1, VAR6, VAR10, VAR5, VAR3, VAR9, VAR2						

Source: Output from SPSS

The p-value, which is displayed in the last column of Table 3 is 0.000, and the level of significance (i.e.), is commonly deemed to be 0.05. The null hypothesis is rejected since the p-value (0.000) is less than the level of significance (0.05). It is crucial to conclude that SHGs may be able to achieve SDGs by operating on digital infrastructure. Even though the majority of SHGs do not use the digital infrastructure to promote their activities that assist fulfill the SDGs, it is apparent that technology has a role to play in this regard based on the interest and perspectives of the SHG members.

Identification of Influential Variables Using Regression Analysis

The use of digital infrastructure will improve SHG performance, ensure well-being, and make it easier to achieve SDGs, therefore identifying the independent variables that have a big impact on the dependent variable is important. Tables 4 and 5 show the results of the regression analysis and the model description and coefficients.

Table 4. Summary of the Model

Model	R	RSquare	RSquare (Adjusted)	Std. Error (of theEstimate)
1	.989[a]	.978	.976	.14052
a. Predictors: (Constant), VAR11, VAR4, VAR7, VAR8,VAR1, VAR6, VAR10, VAR5, VAR3, VAR9, VAR2				

Source: SPSS output

In regression model, a higher R square number denotes a good fit for the model. The model fits well because R Square is 0.978 (Table 4), meaning that independent factors account for 97% of the variation in the dependent variable. The number of independent variables used to predict the dependent variable is taken into consideration by the Adjusted R-square. We can use it to check whether introducing new variables to the model genuinely improves its fit. When there are several variables in the regression model, it is preferable to utilize an adjusted R-square. If the adjusted R Square value rises with more independent variables, then more independent variables improve the model. The adjusted R square in our model is 0.976, which is extremely near to the upper extreme value. As a result, the number of independent variables in this model is appropriate, and it does not need any additional independent variables.

Table 5. Coefficients[a]

Model		Coefficient (Unstandardized)		Coefficient (Standardized)	t	Sig.
		B	Std. Error	Beta		
1	Constant	–.116	.177		–.658	.513
	VAR1	.082	.025	.078	3.256	.002
	VAR2	.505	.076	.534	6.640	.000
	VAR3	.024	.051	.019	.467	.641
	VAR4	.025	.021	.025	1.210	.229
	VAR5	.004	.028	.005	.156	.876
	VAR6	.006	.031	.005	.201	.841
	VAR7	.087	.033	.058	2.662	.009
	VAR8	.008	.029	.008	.286	.776
	VAR9	–.010	.031	–.011	–.328	.744
	VAR10	–.074	.038	–.064	–1.958	.053
	VAR11	.397	.053	.395	7.513	.000
a. Dep. Var.: VAR12						

Source: Output from SPSS

The "p" values for four of the 11 independent variables VAR1 (Using digital infrastructure by self or someone else for SHG functioning is fascinating and worthwhile), VAR2 (It is exciting and worthwhile to use a smartphone to conduct an online search for information on how SHGs can operate more effectively), VAR7 (It is interesting and worthwhile to use a smart mobile phone to get directions or use GPS to communicate with other SHGs and government offices) and VAR11 (Use digital infrastructure to promote the activities, including products/services produced, of your SHG that will enhance the performance) are less than the "α' value. The dependent variable is thus considerably influenced by these four independent variables and the Regression Model is; VAR12 = –0.116 + 0.082 VAR1 + 0.505 VAR2 + 0.087 VAR7 + 0.397 VAR11.

Standardized regression coefficients (Beta) have been taken into consideration for identifying the explanatory variable which has the greatest impact on the dependent variable. A variable with a higher β value has a stronger influence on the dependent variable. Beta Values for four significant variables are VAR1 = 0.078, VAR2 = 0.534, VAR7 = 0.058, and VAR11= 0.395, as shown in Table 5. The variable VAR11 has the highest β value. It indicates that this variable has the most influence on the dependent variable.

Conclusion

There is no second thought on the inference that SHGs play a great role in the well-being of the tribal women studied and act as a facilitator in achieving SDGs. The credit of augmented earnings expenditure and cautiousness of children's education is attributed to the SHGs. However, the contribution of SHGs towards the well-being of the tribal women would have been much better, had the members been made accessible to digital infrastructure for the functioning of the SHGs. Thus, there is a need for embedded digital infrastructure in the functioning of SHGs studied. Mere making the digital infrastructure available to tribal women won't serve the purpose. Besides making those accessible, proper training for tribal women needs to be catered to for effective usage of those digital infrastructures. Findings of this study will help bureaucratic agencies, NGOs in the concerned areas, researchers, civil societies, and political bodies to shift their focus towards digital infrastructure that facilitates bettering the well-being of the tribal women and their households.

Acknowledgments

As a token of acknowledgment, the researchers are obliged for the cooperation of the respondents without which the article would not have been possible.

References

Rao, A. (2016). Role of Micro Finance in the Empowerment of tribal women, International Journal of multi disciplinary advanced Research Trends, 3(1), 213–214.

Sarania, R. (2015). Impact of SHGs on Economic Empowerment of Women in Assam, International Research Journal of Interdisciplinary and Multi-Disciplinary Studies (IRJIMS), 1(1), 148–159.

Jena, G., and Swain, S. C. (2022). Need for Technology Intervention in Functioning of SHGs Run by Tribal Women to Promote Tribal Culture, ECS Transactions, 107(1), 11957.

Sharma, M. K. (2013). A case study on Socio-Economic condition of Self Help Group Members in Golaghat District of Assam, International Journal of Innovative Research and development, 2(4), 186–195.

Reji, Dr. (2013). Economic Empowerment of women through SHGs in Kerala, International Journal of marketing, financial services and management research, 2(3), 2–4.

Puhazhendhi, V., and Satyasai, K. J. S (2001). Economic and social empowerment of Rural poor through SHGs, Indian Journal of Agricultural Economic, 56(3). July–September 2011, 450.

Lenka, C., and Mohanta, Y. (2015). Empowerment of Women through participation in Self Help Groups – A Study in tribal areas internet .J. Home Sci.Ext and com. Manage, 2(2), 126–131.

Rajendran, S. M., William, A. T., and Raja, V. D. (2013). Micro Finance and Empowerment of women Through SHGs in Kanyakumari District. Indian Streams of Research Journal, 1–11.

Mohanty, L., and Swain, S. C. (2022). Use of Digital Technologies by the MSMEs to Preserve Cultural Heritage of India and Achieve Sustainable Development Goals, ECS Transactions, 107(1), 2022, 14343.

Das, S. K. (2012). Ground Realities of Self Help Groups – Bank Linkage Programme. An Empirical Analysis, International Journal of Research in Social Sciences, 2(2), 464–479.

Fetishism: Paradoxing the Narratives of Sustainable Development Goals

Manali Karmakar, Rashmi Rekha Borah, D. Binu Sahayam, G. Bhuvaneswari*

School of Social Sciences & Languages. Vellore Institute of Technology, Chennai, India
E-mail: *manali.karmakar@vit.ac.in

Abstract

Good Health and Well-Being, and Responsible Production and Consumption are enlisted as the third and twelfth goals of the Sustainable Development Goals (SDGs) project. However, this paper aims to argue that amidst the exponential growth of online shopping, the mentioned goals are standing in paradox to each other. This paper would specifically concentrate on offering a review of the existing research conducted during the ongoing COVID-19 pandemic era. The paper underlies how the sustainability development goals about responsible production and consumption, and the protection of the mental health and well-being of the citizens are being in constant threat. The paper argues that the threat is triggered due to the spike in digital shopping culture. The paper through the review of the existing research would like to draw attention to how the thoughtless buying behavior of the consumer has a toll on their mental health as well as the environment.

Keywords: Online Shopping, Digital Culture, Fetishism, Consumerism, Mental Health, Environmental threats

Introduction

Sustainable Development Goals (SDGs) may be explained as an outline designed by United Nations General Assembly to sensitize mankind about the rising social problems that can be strategically resolved to create a sustainable society for the present as well as the future generation. There are 17 goals enlisted under the SDGs project. This paper will be focusing on goals number three and twelve, i.e., Good Health and Well-Being, and Responsible Consumption and Production. The paper argues that amidst the exponential growth of online shopping, the mentioned goals are standing in paradox to each other. Global e-commerce has undergone a sea change during the COVID-19 pandemic. The paper argues that the government's promotion of digital shopping to combat the spread of COVID-19, has not merely changed the behavior of the consumer who is psychologically tuned to an instant shopping culture, it has escalated the existing environmental and mental health issues. The spike in digital shopping has a negative impact, on the environment witnessing a depletion in natural resources as well as the mental health and well-being of the consumers who are succumbing to a fetish culture leading to a range of mental disorders such as compulsive buying disorder. Environment protection experts, psychologists, and psychiatrists have shown concern regarding the imbalance visible in the environment, as well as in the mental health of consumers due to their thoughtless consumption of resources via digital mode. This paper aims to offer a quick overview of the expert narratives to offer food for thought to gauge the current status of sustainable goals number three and twelve in the context of online fetish culture.

Online Shopping and Compulsive Buying Disorder (CBD)

The term fetishism has its root in anthropology where it refers to a sacred or symbolic object that embodies supernatural power. Karl Marx has extended the concept of fetishism into the domain of consumerism to explain how in the capitalist culture, consumers fail to appreciate the producers' physical, emotional, and intellectual investment of labor that went into the production of the entity. Fetishism may further be defined as a "perceptual disorder" where we turn unsympathetic towards the contribution of the producers producing a commodity. The commodity is perceived as an active agent defining and deciding the taste and desire of the consumers. It will not be an exaggeration to state that fetishism is reincarnated in the current culture of digital shopping. Choubassi et al., argue that "Contemporary consumer culture is forever steered towards a capitalist society of commodity fetishism, with globalization and the new technologies of E-commerce and digital advertising, the notion of commodity fetishism that Karl Marx talked about in 1867 (Marx, 1992) is now taking a new higher dimension of domination" (pp. 1–2). The purchased commodities may appear to be an extension of consumers' personalities. The purchased commodities are conceived to be symbols of the agentic status of the consumers, however, studies have highlighted the illusory factors underlying the assumed empowered status of the consumers who gradually succumbs to their insurmountable purchasing desires, leading to mental disorders such as compulsive buying disorder (CBD). Compulsive buying disorder may be defined as ungovernable buying behavior that leads to distress and financial impairment. In other words, a person with compulsive buying disorder suffers from an irresistible desire to buy products without taking into consideration family requirement and financial status. The term was first explained clinically by Emil Kraepelin in the year 1915 who describes the condition as oniomania or buying mania. Swan-Kremeier et al. (2005) referring to Kraepelin's study explicate the scientific explanation of the disorder thus:

> Kraepelin's original description was later expanded by Bleuler (1924), who commented on the uncontrollable and impulsive nature of the symptoms. Recent decades have brought a reemergence of interest in describing, defining, and classifying compulsive buyers in both consumer behavior and psychiatric literature (p. 185).

Anas et al. (2022) in their study have argued that the digital shopping culture during the pandemic has led to a shift in consumer behaviors who to fight isolation and loneliness have taken shelter in the online platform to circumvent the social and emotional crises imposed by the sudden lockdown. The study identifies an emerging pattern in the consumers' ways of impulsively buying products, a behavior. Verma and Naveen (2021) state that CBD has to be studied through a renewed perspective to understand how the pandemic has turned into a catalyst for the disorder. COVID-19 is a traumatic event and normatively studies have focussed on how work-from-home (WFH) culture has triggered a range of physical and psychological ailments such as orthopedic pain, depression, and sleeping disorder. In the domain of consumer studies, the focus is laid on understanding the financial setbacks and their impact on the standard of the consumer's lifestyle. However, limited attention is paid to understanding the interrelationship between the instant accessibility of contactless shopping and CBD. Fear and shopping convenience are conceived to be important variables for studying CBD during pandemics. Wang et al. (2021) study the interrelationship between the COVID-19 pandemic, sense of control, impulsive buying, and moderate thinking. Moderate thinking is a Chinese philosophy and cultural value. but it has its root in modern Western cultural values. Moderate thinking is primarily governed by three prime concepts. Firstly, plurality indicates the ability to gauge a particular phenomenon through multiple views. Secondly, integration.

That indicates the ability to integrate the internal and external information pertaining to the phenomenon, and thirdly, harmony. Choosing a middle point after taking into consideration the larger picture and its effect on the phenomenon. Moderate thinking may be conceived as a key concept that may help regulate the unruly behavior of the consumer in a fetish culture. Wang et al. study identify that during the pandemic "the effect of the COVID-19 pandemic on impulse buying is weakened in consumers with high moderate thinking, and the effect of the COVID-19 pandemic on impulse buying is enhanced for consumers with low moderate thinking." Naghavi et al. (2021) also supports the existing narrative on compulsive buying disorder stating, "the rapid growth of e-commerce marketplaces amid the pandemic has aggravated the risk of shopping addiction, especially with digitally savvy societies like Indonesia, where the use of online shopping platforms such as Shopee, Tokopedia, and Bukalapak has become a trend among youth" (para. 3). The ongoing research narratives emphasize how how the sustainability development goals pertaining to responsible production and consumption, and the protection of the mental health and wellbeing of the citizens are being in constant threat due to the spike in digital shopping culture that was assumed to introduce an order of sustainability, equality, democracy, and private happiness.

Impact of Compulsive Buying Behavior on the Environment

United States Census Bureau (2022) reports that consumer behavior changes during the pandemic have brought immense change in the landscape of e-commerce. As per the recent 2020 U.S. Annual Retail Trade Survey (ARTS) e-commerce "increased by $244.2 billion or 43% in 2020, the first year of the pandemic, rising from $571.2 billion in 2019 to $815.4 billion in 2020." However, the tremendous growth has increased the concern of environmental experts. World Economic Forum diagrammatical represents the impact of online shopping on the environment thus:

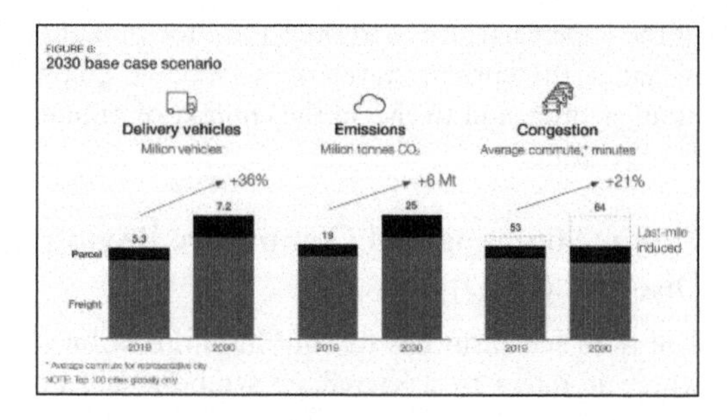

Fig. 2. Online shopping and its environmental impact including carbon emissions. Image: World Economic Forum

Online shopping culture is driven by the promise of reasonable price, speed, and convenience which has changed the psychological tuning of the consumers who are habituated to an instant shopping culture. In an article Igini (2021) argues that the promise of instant satiation of consumers' desires exerts an immense load on global logistics support. Impulsive buying of commodities has led to the extraction of natural resources more in comparison to the actual requirement, thus the consumers' irresponsible consumption of the products fuelled by the online shopping trend has problematized the existing strategies designed to achieve the goal of sustainability. Igini explains:

"Amazon, the world's leading online marketplace in terms of traffic, has found the perfect formula to satisfy all three of them, by training its customers into believing that free and fast shipping is something they should have, thus forcing smaller retailers to adopt the same strategy in order to keep up with giant competitors...In September 2021, several industry groups representing over 65 million transport workers wrote an open letter to heads of state at the United Nations General Assembly, warning that, if this trend continues to grow at the same rate, global transport systems are going to collapse." (para. 2)

Studies have also brought to light that product packaging contributes to a large extent to CO_2 emissions. It pollutes the ecosystem, and also adds an enormous amount of waste to our landfills. The present study contests the notion the normative notion of e-commerce as eco-friendly. Studies have urged consumers to behave responsibly and reduce their reliance on online shopping.

Conclusion

The review of the literature pertaining to consumer behavior, mental health, and the environment brought to light the interconnectivity underlying the sustainability goal. The article demonstrates how COVID-19 as a natural catastrophe triggers anxiety and panic among consumers leading to a mental health crisis that has a direct impact on the retail landscape leading to environmental devolution. The review paper enables researchers in the area of sustainable development to cognize the 17 SDGs as an ecosystem that is interdependent on each of the enlisted goals for its sustainability. From the review of the literature, it is identified that CBD is a new phenomenon and the medical world is yet to recognize it as a disorder that can be included in the Diagnostic Statistical Manual (DSM 5). In addition to this, interconnectivity between e-commerce and the environment as an area of research is evolving. Hence, this literature review encourages more interdisciplinary work to understand the issues of CBD amidst the pandemic crisis and its entanglement with digital shopping and the environment.

References

Anas, M., Khan, M. N., Rahman, O., and Uddin, S. M. F. (2022). Why consumers behaved impulsively during COVID-19 pandemic? South Asian Journal of Marketing, 3(1), 7–20. https://doi.org/10.1108/sajm-03–2021-0040

Choubassi, H., Sharara, S., and Khayat, S. (2019). E-commerce and Commodity Fetishism Violence in New Media Marketing. In: Jallouli, R., Bach Tobji, M., Bélisle, D., Mellouli, S., Abdallah, F., Osman, I. (eds) Digital Economy. Emerging Technologies and Business Innovation. ICDEC 2019. Lecture Notes in Business Information Processing, vol 358. Springer, Cham. https://doi.org/10.1007/978–3-030–30874-2_15

Igini, M. (2021, December 20). The Truth About Online Shopping and its Environmental Impact. Earth.Org. Retrieved July 23, 2022, from https://earth.org/online-shopping-and-its-environmental-impact/

Marx, K., and Engels, F. (2021). Wage Labour and Capital/Value Price and Profit (Paperback (Combined) ed.). Intl Pub Co Inc.

Naghavi, N., Waheed, H., Allen, K., and Pahlevansharif, S. (2021, October 15). How COVID-19 has increased the risk of compulsive buying behavior and online shopping addiction among young consumers. The Conversation. https://theconversation.com/how-covid-19-has-increased-the-risk-of-compulsive-buying-behavior-and-online-shopping-addiction-among-young-consumers-161018

Swan-Kremeier, L. A., Mitchell, J. E., and Faber, R. J. (2005). Compulsive Buying: A Disorder of Compulsivity or Impulsivity?. In: Abramowitz, J. S., Houts, A. C. (eds). Concepts and Controversies in Obsessive-Compulsive Disorder. Series in Anxiety and Related Disorders. Springer, Boston, MA. https://doi.org/10.1007/0–387-23370–9_10

U.S. Census Bureau. (2022, April 26). Annual Retail Trade Survey Shows Impact of Online Shopping on Retail Sales During COVID-19 Pandemic. Census. Gov. Retrieved July 23, 2022, from https://www.census.gov/library/stories/2022/04/ecommerce-sales-surged-during-pandemic.html

Verma, M., and Naveen, B. R. (2021). COVID-19 impact on buying behaviour. Vikalpa. 46(1):27–40.

Wang, S., Liu, Y., Du, Y., and Wang, X. (2021). Effect of the COVID-19 Pandemic on Consumers' Impulse Buying: The Moderating Role of Moderate Thinking. International journal of environmental research and public health, 18(21): 11116. https://doi.org/10.3390/ijerph182111116

Whiting, K. (2020, February 9). Is online shopping bad for the environment? World Economic Forum. Retrieved August 1, 2022, from https://www.weforum.org/agenda/2020/01/carbon-emissions-online-shopping-solutions/

A Systematic Review Study on the Quality of Life Associated with Depression Among the Elderly in India

S. Indhumathi[*,a] *and D. Binu Sahayam*[b]

[a]Research Scholar, School of Social Science and Languages, VIT Chennai, Tamilnadu, India
[b]Assistant Professor, School of Social Science and Languages, VIT Chennai, Tamilnadu, India
E-mail: [*]indhumathi.s2021a@vitstudent.ac.in

Abstract

Depression is seen as a significant impediment to quality of life. To document the current state of research, this paper evaluates the body of literature on the quality of life of the elderly related to depression. Through the search, 15 papers were found, bulk of which was on depression that lowers quality of life and minority on other relevant concerns. Research on relationships between depression and elderly people's quality of life as well as therapeutic options for depression are mostly lacking and should be given higher priority. More study is required to increase conceptual knowledge of depression severity within the nation's cultural setting. To address the multiple enacted, perceived, internalized, and layered forms of depression, context-specific therapies are required. The evaluation advises creating an agenda focused on the elderly quality of life and interventions to support the government's objective of enhancing the elderly quality of life and well-being.

Keywords: Depression, Quality of life, Elderly, Treatment programs

Introduction

Older adults are those who are 60 years of age or older (United Nation, 2019). India is in the midst of a demographic change that will eventually lead to population aging (Ahuja, 2014). According to the Longitudinal Ageing Study of India, 319 million Indians would be in their 60s or older by 2050, constituting roughly one-fifth (19.4%) of the population. (The Hindu, January 7, 2021). In 1980, people aged 60 and over made up just 5.9% of the country's population. India's two states with the most elderly people are Tamil Nadu and Kerala. In this survey, there were more elderly women than elderly men (Times of India, 2021). Maintaining physical health, social engagement, life satisfaction, psychological resources, and personal development are key to successful aging. This study helps us comprehend the significance of QOL for the elderly and the factors related to depression that affect their QOL. Let's examine the definitions of QOL and depression to gain a deeper grasp of the study. The WHO defines the quality of life as "an individual's view of their place in life in relation to their objectives, aspirations, standards, and concerns and within the framework of the culture and value systems they live in" (WHO)."Depression is a prevalent mental illness, consistent sorrow and a lack of interest in formerly fulfilling or joyful activities are its defining traits." Additionally, it may impair appetite and sleep. Depression mostly affects the elderly quality of life and may be brought on by retirement, a lack of social interaction, loneliness, stress, or the death of a loved one (WHO). Both the terms "QOL" and "depression," as well as how they can be considered in relation to the study, are defined with clarity.

Objective of the Study

- To study the QOL associated with depression in people aged 60 & above.
- To recommend some suggestions based on the findings from the review.

Statement of the Problem

Globally, ageing is an unavoidable social issue. With this rapid expansion, age-related issues are affecting an increasing number of senior people. In the aged population, depression is a significant health issue since it can have catastrophic effects, including a tendency toward suicide. The elderly quality of life is impacted. In this review study, how depression and quality of life are related in India based on previous studies.

Need of the Study

Geetha (name changed) from Erode is a 70-year-old widow who comes from a low-income family. She was depressed as a result of her husband's death, and she was neglected by her son's family, causing her quality of life to suffer. She revealed:

> *I've never been concerned about anything in my life. I began to feel lonely and alienated when my husband died. My son has abandoned me, and he was the one who placed me in the elderly home. I miss my family terribly. However, my sons believe I am a burden to them, which causes me to become depressed and unwilling to live.*

The need for the study is to understand how the elderly are viewed and treated in society. Every individual in society will experience this same phase. One in six individuals in the world will be 60 or older by 2030. The number of persons in the globe who are 60 years or older will double by 2050. (WHO, October 4, 2021). Therefore, there is a need to increase awareness of the importance of senior support and care.

Review of Literature

This review study examines the QOL of the elderly in relation to depression in adults 60 years of age and older. The material that is now accessible demonstrates the importance of QOL in elderly people as well as the depression they encounter as they age. The number of people over 65 in the world is still increasing quickly. Currently, 617 million individuals, or 8.5% of the world's population, are 65 or older. By 2050, this number is anticipated to rise to about 17 percent of the global population, according to a recent analysis titled "An ageing planet" (National Institute of Health, 2015). The growing number of aged individuals calls for a deeper level of research on how well the elderly are doing. A researcher must conduct a review of the relevant literature for the study in order to gain a thorough understanding of the topic. Only after conducting a review can a researcher begin to address the issue that will be the focus of the investigation. This chapter provides a summary of the context and circumstances in which the study was done and aids in formulating the researcher's problem. Effective health promotion initiatives are required, having an emphasis on diagnosis and chronic disease prevention according to a study by Dasgupta (2018) that examined the quality of life of the aged in West Bengal. The findings also emphasized the need for geriatric care with counseling and an emphasis on old-age pensions, both of which will help to further improve their quality of life. Another study by Raju and Nidhi (2014) examined depression and quality of life (QOL) among older women in urban India. The study findings suggested that elder abuse has a markedly detrimental effect on these women's quality of life, necessitating an urgent shift in social attitudes in addition to policy and programmatic interventions. Similar to the epidemiological study, a study on the state of the elderly's quality of life in Thirumalizai, Tamil Nadu's urban region, noted that elderly people require assistance and care from family and society (Parasuram, 2021). The level of quality of life was found to be average in elderly people, and the social relationship domain was comparatively lower than the physical, psychological, and environmental domains. Another study titled "Quality of life and its associated factors using WHOQOL-BREF among elderly in urban India" (Kumar et al., 2014). In a study comparing the elderly people's quality of life in nursing homes and in their own houses (Panday et al., 2015), the elderly people living in nursing homes had a higher superior quality of life than senior citizens who reside in families. According to the study of Elderly people's quality of life in Wardha district,Maharashtra (Mudey,2011),those who reside in urban areas reported considerably poorer levels of physical and psychological QOL than those who reside in rural areas. In comparison to the urban population, the elderly in rural areas reported significantly lower QOL in the areas of social and environmental interactions. Older men in northern India's rural elderly population were found to have a higher quality of life (QOL) than older women in the physical, psychological, social, and environmental domains. Using WHOQOL-BREF, it was determined that elderly people in urban Puducherry had an average quality of life and its associated characteristics (Kumar et al., 2014; Raj et al., 2014). A study on the physical health and quality of life satisfaction of elderly people in Varanasi, Uttar Pradesh, found that the vast majority of elderly persons had an average quality of life which has an impact on the elderly's quality of life (QOL).

Methodology

Google Scholar, Pubmed, Scopus, and Springer were searched to identify articles related to the title focusing on four keywords: QOL, elderly, depression, and India using screening tools. This study included the articles published during the years 2005–2020 were included. Review articles, case studies, and those variables non-related to QOL were excluded. Keeping in mind the title of the study 20 articles were carefully examined and finally, 15 were selected. For the purpose of this review, 15 articles were selected based on QOL and depression

Findings

Depression and quality of life

This review study revealed that depression was a common problem among the elderly, which had an impact on their physical and mental health as well as their quality of life.

According to Srinivasan P et al study findings, the elderly experienced low to fair QOL, mild to moderate depression, and a moderate-high negative association between quality of life and depression. Depression in institutionalized seniors was also found to be more prevalent in males. According to a different study that looked at elderly subjects, Losing one's health is equated with a poor quality of life, but having a good quality of life is equated with a larger variety of categories, including activity, money, social life, and family relationships—criteria that differed from subject to subject. As a result,

health appears to be a reliable sign of poor quality of life but a poor predictor of successful aging. The quality of life of older persons can be improved dramatically by treating depression and anxiety. The management of depression and anxiety can significantly enhance the quality of life for older people. Chronic illness and low educational attainment lower elderly people's quality of life and raise their rates of depression. Quality of life is negatively impacted by the intensity of pain and depth of grief. The quality of life for elderly people may increase as a result of efforts to ameliorate these situations.

Compared to elderly individuals living in a family setting, QOL was higher for those who were residing in old-age facilities. In India, depression is very common among the elderly. Data on symptom profiles and the effectiveness of different therapeutic approaches for the treatment of depression in the elderly from India are, however, lacking.

Discussion

There is a need to improve the field of geriatric psychiatry because the senior population is growing throughout Asia and India. Therefore, it is important to comprehend the common illnesses in this population. The material that is currently available and originating from India indicates that depression is substantially more prevalent among the senior population. According to cross-national research, depression among the elderly may be more common in India than in other developing and underdeveloped nations.

According to the results of "A study on QOL between elderly persons living in old age homes and inside family setup by Rishi Pandey et al in 2015," QOL was higher for elderly people living in old age homes compared to elderly people living in family settings based on the depression level.

Barua et al. (2011) evaluated the median prevalence rates of depression in an elderly Indian community and compared them to comparable rates worldwide through an analysis of international literature.

Implication of Social Work Practice

The assistance of social workers is crucial in assisting the aging population. Social workers who work in elderly care have a unique opportunity to connect deeply and meaningfully with elderly people and their families, to change unjust institutions at both the individual and communal levels, and to know that their efforts directly benefit those in need. Here the researcher used the methods of social work practice:

1. Social casework is a primary method of social work. Through social casework, the social worker identifies the elderly problems and needs and the social worker will provide therapy for an elderly client who is suffering from depression and also encourage the elderly to pursue stimulating activities.
2. Using social group work method, the elderly care support groups will motivate the elderly through peer speech and some activities through which they could cope with their problems.
3. Activity-based community organization programs can be conducted for the elderly.
4. Social action plays a very important role in formulating government policies. It can access the availability of Helplines, resource centers, and schemes for the elderly so that they can be able to provide proper and needful assistance that improves their mental health and quality of life.

Suggestions

- The amount of Indian research on geriatric depression is scant. Even though there have been researches evaluating the majority of older people with depression have only utilized various rating measures, despite depression being common among them. Diagnostic tools have only been used in a small number of studies to support the diagnosis of depression. The studies that are accessible are all single-center studies. As a result, numerous research using a two-stage evaluation is required to investigate the prevalence of depression in the elderly. No research has, to date, particularly addressed the prevalence of depression among the elderly and how it impacts the quality of life.
- To research the incidence rates, it is necessary to follow up on a cohort of elderly people. Organize or sensitize different motivation or entertaining programs for the elderly by the intervention process by professional social workers.
- Government in collaboration with NGOs supports funding in initiating spirituality programs, and entertainment shows and makes them busy with activities for improving their Quality of life and also to engage in physical and mental activity.
- Monitor and treat depression by professionals that may lead to an increase in the level of quality of life of elderly.
- Recreational activities like gardening, yoga, bird watching, reading, and spending time with their peer groups, etc can be recommended with the help of their family or through caretakers to keep them engaged to maintain their quality of life.

Conclusion

As a result of this review, it indicates that depression is quite prevalent in aging. Research on the many facets of depression among the elderly in India is seriously lacking. As a result, there is a pressing need to focus on geriatric depression because it has an impact on their quality of life. A long-term study examining several facets of depression is required.

References

National Institute of Health-funded census bureau (March 28, 2016). World older population grows dramatically. Url:https://www.nih.gov/news-events/news-releases/worlds-older-population-grows-dramatically

Dasgupta, A., Pan, T., Paul, B., Bandopadhyay, L., and Mandal, S. (2018). Quality of life of elderly people in a rural area of West Bengal: A community-based study. Med J DY Patil Vidyapeeth, 11(6), 527–531.

Raju, S., and Nidhi, G. (2014). Quality of life of Older women in urban India. (Shodhganga).

Parsuram, G., et al. (2021). An epidemiological study on quality of life among elderly in an urban area of Thirumazhisai, Tamilnadu. Journal of Family medicine and Primary care, 10(6), 2293–2298. doi: 10.4103/jfmpc.jfmpc_1636_20

Kumar Ganesh, S. (2014). Quality of Life (QOL) and Its Associated Factors Using WHOQOL-BREF Among Elderly in Urban Puducherry, India. Journal of Clinical and Diagnostic Research, 8(1), 54–57. doi: 10.7860/JCDR/2014/6996.3917

Panday, R. (2015). A study on quality of life between elderly people living in old age home and within family setup. Open Journal of Psychiatry Allied Science, 6, 127–131. doi: 10.5958/2394-2061.2015.00010.5

Mudey, A., Ambekar, S., Goyal, R. C., Agareka, S., and Wagh, V. V. (2011). Assessment of Quality of Life among Rural and Urban Elderly Population of Wardha District, Maharashtra, India. Ethno medicine, 5(2), 89–93.

Raj, D., et al. (2014). A study on quality of life satisfaction & physical health of elderly people in Varanasi: An urban area of Uttar Pradesh, India. International Journal of Medical Science and Public Health, 3(6). doi:10.5455/ijmsph.2014.140320145

Srinivasan, P. (2015). Elderly: Depression and quality of life. International Journal of Applied Research, 1(13), 538–540.

Barue, A. (2011). Prevalence of depressive disorders in the elderly, Annals of Saudi Medicine, 31(6), 620–624. doi: 10.4103/0256-4947.87100

Influence of Personality and Sector of Employment on Perceived Social Support and Work Family Conflict

W. Cyril Allen Jonathan,[*,a] and K. N. Jayakumar[b]*

[a]Ph.D. Research scholar, Department of Psychology, Periyar University, Salem
[b]Assistant Professor Department of Psychology, Periyar University, Salem
E-mail: *cyrilbg7000@gmail.com

Abstract

Locus of control is associated with one's cognitive-behavioral and emotional character. which emerged in the social learning model of personality. To explore the role of "social support (PSS)" and "work-family conflict (WFC)" across "locus of control (LOC)" and sector of employment (SOE) the present study employed incidental sampling technique to collect data using Google Form. 223 Participants responded to the 20 items Locus of Control Instrument, 10 items Work and Family Conflict Scale, and 12 items Multi-Dimensional Scale of Perceived Social Support scales along with demographic details. Results revealed that locus of control and sector of employment differentiated PSS and WFC. In the LOC, internals had significantly higher PSS and low WFC. On the other hand, externals had lesser PSS and higher WFC. The LOC and SOE had a positive influence on PSS and a negative influence on WFC. The results and implications of this study are discussed in the main article.

Keywords: locus of control, social support, work-family conflict, employees.

Introduction

The Sustainable Development Agenda 2030, emphasizes the significance of family farming and family welfare. Further, the agenda also makes a number of inferred connections to families, family members, and family management. For instance, in formulations like promoting collective responsibility within the family and home as suitable for the nation, promoting healthy lifestyles, and promoting well-being for everyone of all ages. This becomes evident that a range of direct and indirect family policies and practices would be useful for the accomplishment of several "Sustainable Development Goals (SDG)" and targets. All SDGs and targets are interlinked resulting in the complexity of understanding, addressing, and solving problems. Family diversity, collaboration, and partnerships with contributors, as well as how to involve families in the development and implementation phase, are additional significant considerations in the design and implementation of family policies.

Work-family balance can be perceived as a conflict in which encounters in work and family duties are mutually incompatible. However, a well-understood work-family balance is the balance acquired because one can properly connect the requirements of productive employment with that of the family. Despite recent requests for a more balanced approach and recognition of good connections between work and family, the issue remains dominated by work and conflict perspectives. In addition, concerns about work-family balance remain in the social, economic, and demographic spheres. Work-family balance influences psychological, and physical health, mental health, dietary and lifestyle habits, child and adolescent health, employers, and health systems (UN DESA, 2016).

Dimensions of work-family conflict that can be defined as distinct but related concepts have various antecedents and outcomes. These include work-to-family conflict (WFC) and family-to-work conflict (FWC), which is when family responsibilities negatively affect work. High levels of work-family conflict and vice-versa are linked to poor results for the employees personally, their families, and the organizations whom they work for. These observations have increased the focus on developing interventions and identifying organizational and regulatory considerations that assist in reducing employees' perceptions of work-family conflict (Haslam, 2012).

Social support is the emotional and physical support that friends, family, and co-workers provide. It is the emotional and practical support a person feels and accessible to them when they feel like they need it (Zimet, 1988).

Social support is the presence of supportive interactions and the enhancement of the quality of those connections. When people receive social support from their families and places of employment, they are better able to maintain a healthy work-life balance and live in peace and harmony. Similarly, emotional support and mentorship from coworkers and supervisors combine to provide social support (Oguegbe et al., 2021).

Locus of Control was defined by Rotter as "the degree to which persons expect that reinforcement or an outcome of their behavior is contingent on their behavior or personal characteristics versus the degree to which persons expect that the reinforcement or outcome is a function of chance, luck, or fate, is under the control of powerful others, or is simply unpredictable." Rotter (1966) concluded that individuals differ along a range of internal and external beliefs. Individuals who have an internal viewpoint are more likely to think that their actions will determine the course of their behavior. Those who adopt an external belief are more likely to hold an alternative viewpoint. He also stated that "if a person perceives a reinforcement as contingent upon his behavior, then the occurrence of either a positive or negative reinforcement will strengthen or weaken the potential for that behavior to recur in the same or similar situation." As a result, a person's behavior is much less likely to change when they do not see a reinforcement being depended on it.

Research Method and Tools

To explore the role of perceived social support (PSS) and work and family conflict (WFC) across locus of control (LOC) and sector of employment (SOE) was used. The present study employed the Incidental sampling technique and 223 data were collected using Google Forms.

1. **The Locus of Control instrument (Clawson, 2008).** This 20- item scale has two-dimension i.e., internal locus of control and external locus of control. The responses were scored in a True / False dichotomous response.
2. **The Work and Family Conflict Scale (Haslam, 2012).** This 11-item scale has three dimensions viz. work-to-family conflict (1–5), family-to-work conflict (6–10), and work-life balance (1 item). The responses were scored on a 7-point rating scale i.e., 1- very strongly disagree, 2- strongly disagree, 3- mildly disagree, 4- neutral, 5- mildly agree, 6- strongly agree and 7- very strongly agree. The investigator has re-established the reliability Cronbach $\alpha = 0.88$ and this scale has adequate validity.
3. **Multi-dimensional Scale of Perceived Social Support (Zimet, 1988).** This 12-item scale has three dimensions viz. family (3, 4, 8, and 11), as well as friends (6, 7, 9, and 12) and confidence in the actions of peers (1, 2, 5, and 10). The responses were scored on a 7-point rating scale i.e., 1- very strongly disagree, 2- strongly disagree, 3- mildly disagree, 4- neutral, 5- mildly agree, 6- strongly agree and 7- very strongly agree. The investigator has re-established the reliability Cronbach $\alpha = 0.88$ and the scale has adequate validity.

Along with assessment tools, socio-demographic details like age, gender, order of birth, family type, marital status, and socio-economic status were recorded.

Table 1. Locus of control and organizational culture among employees in the workplace

Variable	LOC	N	Mean	Sd	t-value
PSS	External	167	65.40	12.98	−3.28*
	Internal	56	71.59	9.60	
WFC	External	167	43.51	14.40	4.79*
	Internal	56	33.34	11.51	

Note. **p<0.05. NS= Not Significant*

It is evident from Table 1 that the Mean and Standard Deviation of the two groups based on the locus of control on perceived social support are M=65.40 and 71.59; SD=12.98 and 9.60 respectively. The Mean and SD for an internal and external locus of control on work and family conflict are M=43.51 and 33.34; SD=14.40 and 11.51 respectively. This indicates that internal and external locus of control differ significantly on perceived social support (t= −3.28, p>.01), and work and family conflict (t= 4.79, p>.01).

Table 2. Work sector and organizational culture among employees at the workplace

Variable	SOE	N	Mean	Sd	t-value
PSS	Edu	101	64.79	12.67	−2.37*
	IT	122	68.74	12.10	
WFC	Edu	101	44.57	15.12	3.49*
	IT	122	37.97	13.11	

Note. **p < 0.05. NS= Not Significant*

It is noticed from Table 2 that the Mean and Standard Deviation of the two groups based on the sector of employment i.e., education and IT sector on perceived social support are M = 64.79 and 68.74; SD = 12.67 and 12.10. The Mean and SD for education and IT sector employees on work and family conflict are M = 44.57 and 37.97; SD=15.12 and 13.11 respectively. This indicates that internal and external locus of control differ significantly in perceived social support (t= −2.37, p>.05), and work and family conflict (t= 3.49, p>.01).

Table 3. Influence of sector of employment and locus of control on work and family conflict among employees at the workplace

Predictor Variable	Work and Family Conflict			
	B	*SE*	*β*	*t*
Sector of employment	−3.48	2.03	−0.12	−1.71
Locus of control	−8.48	2.33	−0.25	−3.64*
	F = 13.06			
	R = 0.32			
	R^2 = 0.10			

Note. **p<0.05. NS= Not Significant*

Table 3 shows the multiple regression analysis for the influence of sector of employment and locus of control (independent variable) on work and family conflict (dependent variable). The findings indicated that sector of employment (β = 0.12, t= −1.71) and locus of control (β = −0.25, t= −3.64, p < 0.05), significantly predicted work and family conflict (R^2 = 0.10; F = 13.06; p < 0.01) and together they accounted for 10 percent of the variance.

Table 4 shows the multiple regression analysis for the influence of the sector of employment and locus of control (independent variable) on perceived social support (dependent variable). The findings indicated that sector of employment (β = 0.08, t = 1.11) and locus of control (β = 0.18, t= −2.50, p < 0.05), significantly predicted work and family conflict (R^2 = 0.05; F = 6.10; p < 0.01) and together they accounted for 5 percent of the variance.

Table 4. Influence of sector of employment and locus of control on perceived social support among employees at the workplace

Predictor Variable	Perceived Social Support			
	B	*SE*	β	*t*
Sector of employment	2.02	1.81	0.08	1.11
Locus of control	5.21	2.08	0.18	2.50*
	F = 6.10			
	R = 0.22			
	R^2 = 0.05			

Note. **p<0.05. NS= Not Significant*

Discussion

The current study adds to an existing body of research by extending our understanding of the comparison of locus of control and sector of employment on perceived social support and work-family conflict among employees. Given the precise findings, the current study provides a better understanding of how locus of control (internal and external) and sector of employment (education and IT) show differences in perceived social support and work and family conflict, and it expands this knowledge by presenting the concept that this comparison may differ across groups and individuals. Between internal and external locus of control participants on perceived social support and work and family conflict, internals has higher scores in perceived social support, and externals have the higher mean score in work and family conflict (Table 1). Between education and IT sector of employment participants on perceived social support and work and family conflict, IT sector employees have higher scores in perceived social support, and education sector employees have higher mean scores in work and family conflict (Table 2). Regression analysis indicates that among locus of control and sector of employment on work and family conflict, locus of control is significantly influenced by work and family conflict. 10%. of work and family conflict were predicted by locus of control (Table 3). Whereas among locus of control and sector of employment on perceived social support, locus of control is significantly influenced by perceived social support. 5%. of perceived social support was predicted by locus of control (Table 4).

Conclusion

Various studies have demonstrated that numerous factors are critical in determining the social support and work-family conflict of employees in an organization. Organizations must devise effective measures to reduce the difficulties and pressures on them. This is because families are social systems that are vulnerable to changes in an individual's biological, psychological, economic, social, and legal situations. Therefore, the employee's personality and sector of employment may be conflicting, which might lead to difficulties between work and family life. Since this study aims to test the internal and external locus of control as well as education, IT sector of employment as predictors of work-family conflict and social support. We need to rely on the social learning theory. The contribution of locus of control plays a significant effect on employees' work-family attitudes and behaviors in their day-to-day lives. When it comes to their work outcomes, individuals who scored high on the external or internal LOC tend to have different beliefs and behaviors. Employees may not be performing to their full potential due to various levels of LOC. Furthermore, personality has been one of the key conflict indicators that will determine how employees behave under pressure and stressful circumstances.

Recommendation

The institution and management heads must execute human resources programs to strengthen their staff members' expectations and emotional aspects of work-life balance. specifically, flexible time helps employees to adjust and adapt to their work arrangements, which must be encouraged at the workplace. Managers and heads of the institution/organizations have to make every effort to provide employees with the necessary social support they deserve, as this will be affecting and boost their performance and motivation.

The researchers also suggest that organizations should encourage a conductive culture that harnesses employees and managers to understand the need to maintain a good work-life balance for better workplace productivity and a harmonious work environment.

References

Clawson, James G., and Gerry Yemen. (2008). "The locus of control."

Haslam, D., Filus, A., Morawska, A., Sanders, M. R., and Fletcher, R. (2015). Child Psychiatry & Human Development, 46(3), 346–357.

Oguegbe, T. M., Iloke, S. E., Ezisi, J. O., and Ofoma, E. B. (2021). Work Life Balance Within The Private Sector: The Predictive Roles of Social Support and Locus of Control. European Journal of Social Sciences Studies, 6(4).

Rotter, J. B. (1966). Generalized expectancies for internal versus external control of reinforcement. Psychological monographs: General and applied, 80(1).

United Nations Department of Economics And Social Affairs, Division of Social Policy and Development, 2016 family experts group meeting.

Zimet, C. A., Dahlen, T. C., Zimet, G., and Farley, C. C. (1988). Perceived organizational support and extra role performance which leads to which? The Journal of Social Psychology, 149, 119–124.

Zimet, G. D., Dahlem, N. W., Zimet, S. G., and Farley, G. K. (1988). Journal of personality assessment, 52(1), 30–41.

Prophesying the Future Retailing Model of Emerging Markets with Special Focus on India

Pradeep Alex[*,a] *Dr. Arti Agrawal*[a] *and Dr. Danish Hussain*[a]

[a]CHRIST (Deemed-to-be University) Lavasa, Pune, India
E-mail: *pradeep.alex@res.christuniversity.in

Abstract

In recent decades, we have seen paradigm shifts in the retail world. Sellers and customers are drifting to online space due to an increase in digital literacy which empowers customers to take smart decisions. This shift is happening at different speeds and paces depending on the economic stage of the market. In this study, after analyzing the trends of digital adoption in different markets, researchers propose a retail evolution model and conclude that all the markets will reach a maturity stage where online and offline channels co-exist and synergize. Omni-channel capability is the key success factor for any traditional shop to sustain itself in the evolving retail world. Hence, an Omni-channel ecosystem should be developed to offer equal opportunities to retailers irrespective of their size. Ideally, this ecosystem should be managed by the government so that it ensures that customers and small retailers are not exploited.

Keywords: Omni-channel, Ecommerce, Retail

Introduction

Our businesses are facing disruptions due to various factors. In the retail world, the topmost one is digital disruption. Why do we call it a disruption instead of an opportunity? The reason is that the traditional retail community is not equipped to take advantage of digitalization and convert disruption into an opportunity. Here, the government of each country has to come forward in developing digital infrastructure and a strong policy frame to ensure that digitalization is leveraged at all levels, especially at the small business unit. If we take the example of India, the government is owning companies in strategic sectors like petroleum, natural gas, railways, etc. with few exceptions in the recent past. This helped to avoid monopoly in the market in these sectors and ensure that the common man is not affected. However, this is not the case with digital infrastructure or e-commerce.

Digital life is the future. Virtual meetings, online purchases, online courses, and AI-based platforms are part of our daily life. However, no government in developing countries except a few like China take action to ensure equality in this area, on the other hand, it allowed big players to freely operate in their markets. The government of these countries was not focused on setting up a strong foundation (investments and R&D) in digital space through its public sector arms and it made dependency on private players with global expertise. Ideally, government should have provided digital infrastructure to all small time businesses, develop market places and educate small retailers to leverage technology to grow business.

In this context, researchers are trying to forecast the future of the Indian retail industry. With more than 1.3 billion population, a large geographic spread, and different cultures, social habits, and economic stages, India is a country of diversities. These diversities increase the scope of e-commerce. The simple reason is that online business provides all products under the sky that meets the requirements of different consumer segments and gives the flexibility to make the offering customized at a personal level. Secondly, when we think about emerging markets like India, Indonesia or Brazil we need to consider the digital adoption rate. Few decades ago, the government in these countries was taking efforts to increase literacy rates. That was a tedious and troublesome process. But today when we see the rate of digital adoption in these countries we will get surprised. Without any major push, people started using digital gadgets and they are evolving in that process. Thanks to the affordability of these gadgets, internet connectivity, and the excellent user experience.

Today, we are discussing digital adoption. However in countries like India where the ratio of the young population is high, the use of technology is going to reach the next level, so we may eventually discuss digital exploitation.

Review of Literature

E-commerce adoption rate is different by country and category. Online grocery sales in USA are just 3%, whereas it is around 10% ~ 15% in countries like South Korea and UK. In USA e-commerce penetration for footwear is 20% and consumer electronics is 40%. One reason for the lower adoption of E-commerce in sectors like grocery is that even now, consumers depend on written lists, aid planning, etc. (Caine and Paratore, 2019). If online platforms are bridging such gaps in the purchase journey they will be able to increase the adoption rate heavily. Another challenge faced by virtual retailers is that some customers are not willing to use their credit/debit cards due to trust issues (Chopra, 2016). But the recent Covid-19 situation increased the adoption rate of plastic money and digital payments and it helped online retailers to overcome the challenge of consumer hesitation towards online transactions. As online retailing solves customer pain points it will register higher growth rates. There is a strong synergy between small local shops and online retailers and with the support of online retailers, mom and Pop shops can increase the value they offer to local customers and become more profitable (Chopra, 2016).

In the case of chain stores in the high-value goods category, the cost of maintaining the inventory at a different location increases investment requirements. However, online sellers can maximize choices for consumers while maintaining a lean inventory (Chopra, 2016). When traditional retailers integrate themselves with online it allows them to enhance business.

The sales happening at the retail level are determined by three factors: the market environment, intertype competition, and the marketing mix (Ingene and Takahashi, 2016). Normally traditional retailers do not think about creating a market environment, they operate in an environment that already exists. With collective effort and a digital ecosystem, retailers will be able to create a new market environment that can provide additional growth opportunities. Faster technology adoption by customers, rapid adoption of the digital medium by brands, the convenience offered by online retailers and reduced growth of large physical stores due to increased property prices, and higher cost of operation accelerated the growth of online retailing (Sinha, Gokhale, and Rawal, 2015). By developing online retail capabilities, physical stores can be more competitive because it helps to expand the categories they sell, avail the benefit of economies of scale, and operate on lean inventory. By leveraging already existing assets and boosting certain capabilities and resources, mom-and-pop retailers would be able to utilize gaps in e-commerce and use opportunities arising from them, thus gaining a competitive edge over online players (Agnihotri, 2015).

Price–role orientations vary between nations, however, some nation-independent factors exist. Price consciousness does not vary between two nations for low-price functional items, such as bath soap, biscuits, etc (Zielke and Komor, 2015). By developing online sales capabilities, traditional shops can increase the level of business, enhance cost competitiveness and thus offer competitive prices to customers.

The Retail Evolution Model

Please take an assessment of the time our youngsters spend on mobile phones, tablets or laptops. While working, traveling, or even when they communicate face to face with others, these gadgets are attracting their attention and time. If you add simple logic to this fact, one thing is clear - tomorrow's market is a digital market. It can be a fully digital marketplace or a digitally aided market.

Building Omni Channel Capability is a must for the survival of traditional retail shops in future, researchers developed a retail evolution model that can give insight to retailers on the next steps they should consider. The proposed retail evolution model is given below:

All Category	**Future Situation** **Evolved Consumer:** For the consumer there is no difference between E-commerce and Traditional shops, it is all about convenience and smart purchases. An ecosystem that addresses all service issues is in place.	
Non-perishable/ Do it by yourself Category	Consumer mind set: **Prone** Ecommerce penetration at moderate level, High dependency on traditional shops	Consumer mind set: **Convenience seeker** Ecommerce penetration at High level
Perishable goods/ Items that needs service support	Consumer mind set: **Folkloric** Ecommerce penetration at low level; Very high dependency on traditional shops	Consumer mind set: **Investigator** Ecommerce penetration at moderate level, there is dependency on traditional shops
	GDP: High contribution from agriculture & manufacturing	GDP: High contribution from the service sector

Fig.1: Retail evolution model developed by researchers

In emerging markets, ecommerce mind set can be described as Prone and Folkloric. Prone represents a customer's willingness to buy online if there is no after-sales service required and the goods are non-perishable, whereas folkloric represents the consumer mindset to go with traditional retailers whom they know because they believe that the traditional retailer can offer the best service. That means in the case of products that require service, the e-commerce penetration will be low until the online sellers address all the service needs successfully. In the case of developed countries, the consumer mindset is more favorable to e-commerce and they are ready to try e-commerce even in product categories that need after-sales service. Depending on the category the retailer deals with and the economic stage of the country, retailers should think about e-commerce adoption to ensure that they are surviving and growing.

The retail evolution model is developed after reviewing case studies from different sectors and markets. Based on the insights, researchers identified a few actions for the retailers in emerging markets which are given below:

Bridge the convenience gap: Convenience is one of the major reasons of consumer preference towards e-commerce. Brick-and-mortar shops need to develop Omni-channel capability to deliver service quality equal to e-commerce.

Create and manage marketplaces: Creating and managing an online platform or marketspace is the key to develop Omni- channel capability. Researchers call this online platform an ecosystem and suggest government or local bodies develop and control such marketplaces.

Collective power to offer the best deals: Attractive price offer is another reason towards e-commerce. Online players can provide the best deals because of their scale. The unbounded market of e-commerce gives them the power to play in large volumes and transfer the benefits of economies of scale to consumers. Brick-and-mortar shops should come together on a single digital platform to get the benefits of scale and compete with large e-commerce players. Governments in emerging markets should develop and host such large digital platforms for traditional retailers to make sure that the large business community of retailers survives. To ensure better income distribution, the survival of these shops is very important.

Implement digital tools: Traditional retailers should develop digital capabilities to enhance customer experience. Also, they should focus on data analytics to increase the efficiency of their business. The online platform should collect relevant data, derive insights and actionable points, and make it available to retailers.

Be organized: Scattered traditional retailers do not have the power to invest and increase their business scope. Hence the ecosystem should help unorganized retailers to come together and work in an organized way for their collective benefit.

Leverage network: Home delivery may not be a feasible option in far-fledged locations and rural markets. Here comes the importance of pick-up locations. Retail stores can act as pick-up locations or aggregators. It can reduce the delivery cost.

Collective bargaining: When retailers come together to a common ecosystem it will help retailers to have better bargaining with the manufacturers.

Influence socio-cultural elements: In emerging markets, joint family structures have prominence. This family system influence purchase behavior. Due to strong bonding and interdependency between family members, shopping is a group activity. Hence the trend of people moving out of their homes as a group to do purchases is normal. Retailers should equip their store infrastructure to cater to the needs of such consumer groups.

Conclusion

E-commerce is growing exponentially. However, its growth rate varies by category and economic stage of the country. Today e-commerce penetration may be low in certain markets or categories which need service. But the situation will evolve and reach a maturity stage where all the barriers to e-commerce adoption will be removed and customers will be comfortable with the e-commerce for all category products. Thus, traditional shops should develop Omni-channel capability so that they can leverage online to grow their business. The Omni-Channel ecosystem is capable of covering different stakeholders like customers, suppliers, etc. and the ecosystem should help the physical stores to reduce their operating costs by following lean inventory, faster rotation of capital, space-saving, etc. An E-commerce ecosystem can help physical stores to reduce the cost of their operation and expand their business to add categories and generate additional revenue and profit. Researchers give importance to the existence of small retail shops especially in emerging markets because their existence ensures better equality in income distribution and restricts monopoly. Ideally, the government should take initiative to develop an Omni-channel ecosystem which should help small retailers, particularly the Kirana shops to come together, collaborate, and offer end-to-end solutions to the customers. This will be the key factor for the survival of small retailers.

Acknowledgments

We would like to acknowledge and give our thanks to our colleagues and friends for guiding us to complete this work.

References

Agnihotri, A. (2015). Can brick-and-mortar retailers successfully become multichannel retailers? Journal of Marketing Channels. 22(1), 62–73.

Caine, S., and Paratore, M. (2019). Omnichannel grocer is open for business–and ready to grow. Bain & Company.

Chopra, S. (2016). How omni-channel can be the future of retailing. Decision. 43(2), 135–144.

Ingene, C. A., and Takahashi, I. (2016). The evolution of Japanese retailing: 1991–2007. In: Campbell, C., Ma, J. (eds). Looking forward, looking back: Drawing on the past to shape the future of marketing. Developments in Marketing Science: Proceedings of the Academy of Marketing Science. Springer, Cham.

Sinha, P. K., Gokhale, S., and Rawal, S. (2015). Online retailing paired with Kirana—A formidable combination for emerging markets. Customer Needs and Solutions. 2(4), 317–324.

Zielke, S., and Komor, M. (2015). Cross-national differences in price–role orientation and their impact on retail markets. Journal of the Academy of Marketing Science. 43(2), 159–180.

Family Conflict and Rivalry in *The Shipwrecked Prince* and *King Lear*: A Comparative Study

Beenish Mir,[a,*] *Dr. Jayatee Bhattacharya,*[b] *and Dr. Gazala Gayas*[c]

[a,b]Amity Institute of English Studies and Research, Amity University, Noida, India.
[c]Cluster University, Srinagar, J&K, India.
E-mail: [*]mirbeenish021@gmail.com; jbhattacharya@amity.edu; gazalagayas@gmail.com

Abstract

The present paper stands on the proposition that family conflict and rivalry are ubiquitous. However, conflict and rivalry need to be resolved for the smooth functioning of society according to the theory of functionalism. Literature plays a vital role to help resolve family conflicts. It portrays the universal nature of conflict, its repercussions, and the ultimate need to resolve it to live in peace since family conflict is normal; it's the repair that matters. This paper reflects the universality of knowledge shared in different texts and contexts by analyzing and comparing *The Shipwrecked Prince*, a Kashmiri folktale that reverberated in the valley of Kashmir since times immemorial, and *King Lear*, Shakespeare's 17th C tragedy. Family Conflict Theory forms the basis of understanding as content analysis of the two stories is undertaken to portray the generality of human societies and human nature through the lens of literature.

Keywords: Conflict theory, Family, Functionalism, Kashmiri folktale, Rivalry, Social stability, *The Shipwrecked Prince, King Lear*

Introduction

Sociologists, particularly the functionalists, view the coming together of those components of society to form a system that is disparate yet dependent on each other. Similar to how multiple organs in a physical body execute heterogeneous tasks to sustain the whole, various institutions within a society each carry out distinct functions to keep that society functioning. Family is a basic and essential building block of a society. As structural functionalists posit, there must be a family unit to ensure an individual has access to food, shelter, and money (functional prerequisites). Every family member has a very distinct job to perform in the family as a whole. The important tasks of teaching the young the aspects of social behavior and tending to the emotional needs of its members are carried out by families, and the performance of these functions imparts stability to a family which in turn leads to financial security and formulates a harmonious society (Haralambos and Holborn, 2013). However, where there is family, there is conflict. Every family member proposes his or her interests and concerns, resulting in a range of interests that compete with one another, turning the family into a battleground of conflicting interests. The myriad of conflicting interests culminates in aspects like selfish behavior, envy, problems related to finance and multiple opinions which end in family conflicts. People are inspired to represent these conflicts time and again through different medium – folklore being one of them.

Kashmiri folktale is an idiosyncratic genre of Kashmiri folklore that is by one way or another associated with the real world despite the presence of heavenly and inexplicable components. The mysterious things develop as per a particular rationale. In folktales there is a co-existence of miracle and reality which strikes us enigmatic as Lutz Rohrich believes, "...the folktale's magical events are not isolated; kings, craftsmen, and farmers, the mother, the stepmother, and siblings – all people exist in reality" (Rohrich and Tokofsky, 1991, p. 66). These people share certain relationships as are shared by people existing in the real world. One of the relationships is the one that exists within a family and where there is family, there is conflict and rivalry. Conflict of family and rivalry between family members is a recurrent theme seen in various folktales. A common feature found in these tales is children neglected by their parents and a feud between family members (Scherf, 2009, p. 81). Family relationships, particularly those between children and their parents, are society's essential component that begins at birth and culminates in death. A flux in this unit leads to a flux in the overall stability of the society according to functionalists who consider the family as one of the constituent elements of a society vital for its stability.

The Shipwrecked Prince (Knowles, 1893), and *King Lear* (Shakespeare, 1877), are the two source materials compared to understand how family conflict is represented and resolved. As Shakespeare's play *King Lear* was influenced by Geoffrey of Monmouth, who claimed that a collection of Welsh or Armorican folktales that Archdeacon Walter had given to him was the source of his ideas, one can say *King Lear*'s plot is based on a folktale. The purpose of this comparison and the hypothesis thereby is that family and kin relationships illustrate the universality of knowledge. This study aims to help readers comprehend the place of literature in society and social relations, given that both texts share similarities in terms of conflict and its resolution.

Conflict Theory and Family

Functionalists view the family as a micro-society wherein every member contributes to the family's survival and continuation by performing a multiplicity of vital functions. The family exhibits among its members, much like any other social unit, the same opposing interests seen in other social groups in society, which leads to conflict.

Karl Marx's political thought, which held that conflict has always been a crucial component of social existence throughout human history, was where conflict theory was first explicitly conceived. Marx also contended that the stages of a person's existence in society inevitably result in revolution. Several scholars have analyzed family from the lens of conflict theory. For instance, according to Andersen and Taylor (2007), power dynamics inside the family serve as a reflection of socioeconomic disparities in society at large. Another faction of scholars expresses the view that conflict culminates in growth and sustenance, and that growth promotes cooperation both within the family and in society. After a conflict, when we negotiate, our interests briefly align. However, conflict invariably returns, necessitating ongoing negotiation to promote progress and understanding; as a corollary, conflict is dialectic in nature (Klein and White, 2014). According to conflict theory, friction and conflict result in fostering possibilities of discussion culminating in resolution, thereby significantly improving the relationships between family members. This dialectic may eventually result in harmony, and members of the family may learn more about one another. This dialectic is illustrated in literature as well, although to the best of the researchers' current knowledge, minimal to no substantial work has been done in this area. To fill the gap, this paper examines the two literary works under discussion via the prism of family conflict theory.

Discussion

The Shipwrecked Prince

This Kashmiri folktale was collected by J. H. Knowles belonging to the Church Missionary Society posted at Srinagar, Kashmir, who narrated it first in *The English Antiquary* before including it in his collection of Kashmiri folktales titled *Folk tales of Kashmir* which was published in 1893. Knowles himself mentioned that a Brahman had related this story to him after having heard it from a Mohamedan.

The Shipwrecked Prince follows the story of the relationship between a father and his son, a father-in-law and his son-in-law, and between brothers-in-law, and narrates the adventures of the youngest of four sons. To test the intelligence of his sons, the king enquires about their views regarding his hold on the kingdom. Three of them respond by claiming that it is their father's good fortune that he owns the kingdom and exercises wise leadership over it. The youngest son does not agree with his brothers and says with an impudence that the king is powerful because of his (the youngest son's) good fortune. It is unexpected for him to respond in such a prideful manner. The king is unpleased with his reaction and orders the lad to depart from the palace. The king soon regrets his decision and orders the prince to return, but his emissaries are unable to convince him. He starts an unusual profession, marries three more females, and sustains three shipwrecks in the time span. He lands himself in the palace of his father-in-law, a king, where his brothers-in-law envy him for the love and appreciation he receives from the king. Due to their jealousy, they want to take credit from the prince for slaughtering a leopard, a jackal, and a bear but are defeated by the prince. After learning that the country of his father has been overrun by foreigners, and the king along with the royal family has been taken prisoner, the prince marches out with his army to combat the conquerors. He succeeds in defeating the adversaries and restoring his father's rule over the country. By doing so, he convinces the elderly monarch to acknowledge that his son's claim which had forced him to abandon his palace, his country, and his family, was, in fact, legitimate. The king admits how he misunderstood his vanity and haughtiness for that of his son, for which he had to pay a price. The king passes over his throne to his youngest son and the rest of their lives pass in peace.

King Lear

King of Britain, Lear, after ruling for sixty years without his son who could be his heir, decides to split his realm among his three daughters Goneril, Regan, and Cordelia, and marry them to appropriate men. To see which of them is worthy of the crowning glory of his realm, he asks them who loves him the most. Goneril claims that she loves him more than her soul by calling upon the Promised Land as evidence. Regan swears on oath that she cherishes her father beyond all other beings. His youngest daughter Cordelia, who is dearest to him, responds that she has always loved him as a father and that anyone pretending otherwise must be hiding her true feelings behind a facade of flattery. She then regrets how her sisters have tricked their father. Dissatisfied with her response, the king forbids Cordelia from having any part of the kingdom. He gives his other two daughters one half of the kingdom, marrying them to the Duke of Cornwall and Duke of Albania, and orders to transfer the remaining portion to them after his death. Cordelia gets married to the Frankish monarch who accepts her without a dowry. After some time, the husbands of the older daughters rise and take Lear's kingdom from him. It is followed by Lear's quarrels with his daughter Goneril about the number of his retainers, his flight to Regan and his quarrel with her, and ultimately his return to Goneril, who does not agree to receive him back unless he dismisses all his attendants. These events prompt him to board a ship for Gaul and seek Cordelia. She is moved by his plight and provides him with a retinue. Her husband gathers an army and invades Britain with King Lear being restored to the throne of the entire kingdom. The old monarch rules for up to two years before dying, and Cordelia replaces him on his death

Comparison

Family serves as the center of focus of the two stories, with all of the activities revolving around family members and their competing interests. Through the course of their narratives, both families battle to remain unified. In both cases, opposition or disagreement between the members acts as the backdrop for family conflict and sibling rivalry. Another context for rivalry and conflict is motivated by avarice and self-interest.

Conflict in the family begins in the form of communication between the family members. As Azcona (2009) states "Conflict in a family group is the resultant of the dynamic forces and energies that flow in different directions among the members, through verbal and non-verbal communication." (p. 4), in *The Shipwrecked Prince,* the conflict between the king and his youngest son, the prince, commences through verbal communication when the prince does not say as was expected from his father. Similarly in *King Lear*, Lear and Cordelia have a conflict of opinion after dialogues between them, as Lear does not believe Cordelia's love for him was genuine and the conflict with his other two daughters, Regan and Goneril also resulted from the communication that took place between them regarding inheritance. Hence, communication or in other words, the interaction between the family members becomes the locus of their conflict.

Research on parent-child conflict began in the late 1990s. The conflict between parents and children is normal and is part of family life. It is the resolution of that conflict that keeps the family healthy and stable. Literature can provide family therapy too by portraying how the conflict between parents and children occurs and resolution can be attained. In the families of the two stories, a conflict arises between a father and one of his children, who certainly speaks in a way not expected by the father. This leads to the father expelling his child from the family and the inheritance. One of the reasons that can be cited for such behavior of the father is the inferiority complex of the father and his thirst for always staying in power in the family. Similar to how a father is the leader of a family, in the stories he is also a monarch who is the head of the state or a country. Therefore, apart from the external conflict, there is an ongoing internal conflict between the head of the family and the head of the state. In other words, there exists a contradiction and conflict between kingship and fatherhood in the two stories. According to Chang (2010), "Lear demands obedience from his daughters, and no defiance is allowed" (p. 3).

Similarly, in *The Shipwrecked Prince*, the king could not accept his youngest son's unexpected defiance while stooping to the flattering behavior of his other sons. Lear and the father of the prince do not recognize that they had been addicted to a kingly habit for a long time and as such, they treated their children as their subordinates whose responsibility was to please their kings through adulation. Both fathers gradually realize what they had done and how they had been blinded by their arrogance and pride. Lear begged for Cordelia's forgiveness, and while doing so, he collapsed and died. Lear transitioned from believing that Cordelia had failed him to understanding that it was actually he who had done so. He betrayed her by not acknowledging the sincerity and purity of her love for him. Hence, a shift in Lear's attitude and level of trust is seen towards the end. Similar to this, the father of the shipwrecked prince conceded that he had mistaken his own hubris and arrogance for that of his son, for which he had to face consequences.

The envious behavior and greed for power and inheritance, both of which are part of the deadly sins, prompted another conflict in the stories i.e., destructive conflict. In *The Shipwrecked Prince*, the brothers-in-law could not bear the unconditional love and affection that the king, their father, was showering upon his son-in-law, the prince of the story. This resulted in a feeling that arises when someone desires something that another person has, whether it be a material item, a status symbol, or imagined success, termed as envy, one of the seven deadly sins. They took all the opportunities to downplay the prince before the king, like in horse-riding and hunting. Their rivalry did not bear anything except for shame as their true color was revealed. As such, they did not get anything from the inheritance of their father, who was the father-in-law of the prince. In *King Lear* too, we see how Lear's two eldest daughters, their husbands, and assuredly Edmund - Lear's son - suffer from greed, the most notable aspect of human nature. Greed, another of the seven deadly sins leads them to not get anything in the end as they had wished for. Greed only resulted in rivalry among the family members as the daughters turned against their father and also against each other, and eventually ended in death.

Conclusion

The objective of this research was to examine how literary texts represent conflict, one of the key elements that define all human societies.

Since healthy families support social order and economic stability, conflict and rivalry must be managed if society is to function effectively. The findings of this paper reveal that the universality of conflict, its ramifications, and the final necessity of reconciliation to live in peace is manifested via literature. Family conflict theory was the basis of understanding the two texts and it was comprehended that family conflict and rivalry between its members are common in both *The Shipwrecked Prince* and *King Lear*. Literature helps to instigate the need for conflict resolution within families as through these tales, it can be concluded that conflict is normal; it is the repair that matters. This paper leads to an understanding of the functions folktales perform in society, as well as the positioning of Kashmiri folktales and, in turn, Indian folk literature within global circuits of receipt and repurposing.

References

Andersen, M. L, and H. F. Taylor. (2007). Sociology: Understanding a Diverse Society. Belmont, CA: Wadsworth/Thomson Learning.

Azcona, M. C. (2009). Resolution of Family Conflicts through Literature, Peace, Literature, and Art, Vol. II.

Chang, Wan-Yi. (2010, July). Conflict between Kingship and Fatherhood in William Shakespeare's King Lear. https:/web.ntpu.edu.tw/~shueng/King_Lear.pdf

Haralambos and Holborn, (2013). Sociology Themes and Perspectives, Eighth Edition, Collins. ISBN-10: 0007597479.

Klein, D. M., and J. M. White. (2014). Family Theories: An Introduction. SAGE Publications, New York.

Knowles, J. H. (1893). The Shipwrecked Prince. Folk tales of Kashmir. London: Kegan Paul Trench, Trubner, pp. 355–392

Rohrich, L., and P. Tokofsky. (1991). Folktales and Reality, Indiana University Press, Bloomington.

Scherf, Walter (2009). Family Conflicts and Emancipation in Fairy Tales. Children's Literature. 3. 77–93. 10.1353/chl.0.0399.

Shakespeare, William. (1877). King Lear. Oxford: Clarendon Press.

The Subtle Warnings Signs of Suicidal Thought and Behaviour Exhibited by Hannah Baker in "*Thirteen Reasons Why*" by Jay Asher

Mahoor Zahid,[a] *Dr. Deepali Sharma,*[b] *and Dr. Varsha Gupta*[c]

[a]Research Scholar, Amity Institute of English Studies and Research, Amity University, Noida
[b]Assistant Professor, Amity Institute of English Studies and Research, Amity University, Noida
[c]Associate Professor, Rajdhani College, Delhi University
E-mail: *mahoor.zahid.mz@gmail.com; dsharma6@amity.edu; varshaguptarc@gmail.com

Abstract

Defined as self-inflicted death, suicide has been a significant issue around the world. It is a powerful and emotional behavior, and a variety of professionals not limited to those of mental health, are intrigued by it. It is one of the serious causes of mortality in the world. The emotional factor makes it an interesting topic to discuss in literature. Because of increasing interest in the concept, this study investigates *Thirteen Reasons Why*, a novel in which the main character shows self-destructive behavior. The character Hannah Baker is analyzed, to understand the psychological factors responsible for her suicide. Descriptive qualitative methodology using the theory of Psychoanalysis and defense mechanism by Sigmund Freud and Durkheim's Theory of Suicide is used to analyze the reasons. The study also deals with how the final act of the main character could have been prevented and how her situation was not handled well.

Keywords: *Thirteen Reasons Why*, Suicide, Psychological Factors, Durkheim's Theory of Suicides

Introduction

I hope you're ready because I'm about to tell you the story of my life. More specifically, why my life ended (Asher, 2007). Addressed by a wide range of disciplines, suicide is an emotionally powerful behavior. Suicide doesn't have a universal definition however it can be defined as an intentional self-inflicted death. The concept of suicide is subjective when it is used in everyday conversation but the idea is more complex than "self-inflicted death." The founder of modern suicidology, Edwin S. Shneidman, opposed the idea of medicalizing suicide. According to him, it was more of a human condition. He explained suicide as

"Suicide is not a disease (although there are those who think so); it is not, in the view of the most detached observers, and immorality (although, as noted below, it has often been so treated in Western and other cultures); and, finally, it is unlikely that anyone theory will ever explain phenomena as varied and as complicated as human self-destructive behaviors. In general, it is probably accurate to say that suicide always involves an individual's tortured and tunneled logic in a state of inner-felt, intolerable emotion" (Shneidman and Leenaars, 2010).

Suicide has always been an intriguing concept thus making scholars, philosophers, and sociologists lecture and explain the intricate motives behind it. Novels, especially young adult novels are consistently feared and questioned whenever suicide is dealt with in the novel. The idea of an adolescent as the center of the conflict in such records is least entertained and usually concludes in the banning of books. With the de-medicalization of the concept by Edwin S. Shneidman, the concept can be examined in terms of human behavior. Many elements can be responsible for why an individual wants to end their life. Suicide refers to a person's deliberate wish to die and the actions taken to follow that wish. Originally written in 2007, the novel *Thirteen Reasons Why* by Jay Asher, is a story of a teenager, Hannah Baker, and her reasons to kill herself. We track a boy named Clay Jensen, who received a box of seven cassettes that Hannah documented when she was alive. These tapes were sent to twelve people and they had to listen to the reason after listening to all the tapes, they had to move on to the next one on the list. Throughout the novel, we get two different narratives, one of Clay Jenson, his part of the journey he takes to uncover the reason behind Hannah's death, and the second narrative is of Hannah Baker and her thirteen reasons why. The novel is about a girl who commits suicide because of the rumors spread about her, the bullying she encountered, and the sexual abuse.

In this research, the researcher attempts to study and focus on the potential warning signs of suicidal behavior evinced by the protagonist in *Thirteen Reasons Why* by Jay Asher.

Method

In this research, a qualitative method of investigation is used by the writer. Studying and understanding the meaning that organizations provide to a social or human issue is possible using the qualitative method. The research process includes new questions and techniques. The core issue dealt with in this research is the analysis of the subtle signs of suicidal tendencies exhibited by Hannah Baker throughout the novel. The primary source of data for this study is the novel *Thirteen Reasons Why* by Jay Asher, which was released in 2009. Books, journals, and websites pertinent to this research serve as secondary data. Throughout the study, the writer collected the data by reading the novel rapidly, summarising it after reading and taking notes, and then classifying the data. Thus, data display, data reduction, and the study's conclusion are the approaches used in the research.

Discussion

The researcher delivers the following discussion:

Subtle signs of Suicidal Behavior are exhibited by Hannah Baker throughout the novel.

Literature paves the course to explore one of the taboos one of them being suicide. *Thirteen Reasons Why* by Jay Asher is one of the best-selling novels that explore the topic of suicide and its depiction of it among teenagers. Hannah Baker, a younger teenager who depicts the sign of suicide when things turn out to be a tragedy for her from the start of middle school, eventually takes an inevitable step and kills herself. The writer examines and investigates the subtle signs of suicidal tendencies displayed by Hannah throughout the novel. The conclusion of the study is anticipated to present fragments of details on the acknowledgment of the subtle signs of self-destructive behavior and suicide prevention.

The warning signs of suicidal behavior exhibited by an individual are

- A previous suicide attempt
- Psychological problems
- Sexual abuse

Lack of support system, strained interactions with parents or friends, and experience social isolation (Cash and Bridge, 2009).

As discussed above, the signs of suicidal behavior, Hannah Baker exhibited the following subtle signs of self-destructive behavior: Depression followed by hopelessness, Escape, help-seeking, sexual abuse, hostile school environment.

Depressed

Hannah Baker was a melancholic sort of an individual, throughout her story, her thoughts are sketched as despairing. She might not be depressed at the opening of the novel but as the story unfolds we see her becoming depressed and ultimately ending her life. It is hard to identify a happy point in her life in the narrative from the first tape to the last. Disappointment, anxiety, pain, and sadness are significant elements of her life.

"Depression is an emotional state characterized by feelings of melancholy and dread, a sense of worthlessness and guilt, separation from others, lack of sleep, food, and sexual drive, as well as a loss of interest and enjoyment in everyday activities" (Kring et al., 2018).

Hannah Baker began displaying early signs of depression which people around Hannah might not have noticed. In addition, depression is a common disorder characterized by "depressed mood, loss of interest or pleasure, feelings of guilt or low self-worth, disturbed sleep, low energy, and poor concentration" (Bhowmik et al., 2012).

People have different ways to handle pain and these sign might vary depending upon an individual. And many signs might indicate that a person is depressed.

Signs of depression

Feelings depressed or unhappy

Loss of enjoyment or interest in routine activities

Fatigue, exhaustion, and a loss of vitality can make even the smallest chores feel laborious.

Feelings of guilt or worthlessness, obsessing over past mistakes, or blaming yourself when things don"t go as planned (Bhowmik et al., 2012).

Whenever it comes to depression and its symptoms, each person is impacted differently, therefore symptoms differ from person to person.

Reason for depression

A specific reason for depression is unknown, Some depression runs in the family, and because of biological vulnerability, the depression can be inherited. A sequence of outside factors, such as a significant loss, challenging relationship, financial issue, or other unfavorable change in life patterns, might be responsible for depression. The illness can occasionally be caused by a confluence of genetic, psychological, and environmental factors. As a side effect, taking some drugs can cause depression. In particular, drugs used to treat conditions including cancer, seizures, severe pain, and contraception can cause depression (Bhowmik et al., 2012).

Depression is worldwide and a lot more people than anticipated are suffering, the reason might be different but the disease renders the same pain and isolation. Hannah Baker was no different, a teenager who moved to a new town and wanted to start fresh. She wanted to be her person and not let any prejudice affect her impression. But life had other plans for her.

Hannah Baker did not decide to take her life on the spur of the moment, but rather as a result of repeated incidents of bullying, sadness, and hopelessness, as well as a lack of self-expression. In her cassettes, Hannah Baker discusses chaos theory, which describes a pattern that appears random or complex but always has a beginning from which the entire process begins, as well as the butterfly effect. Like a butterfly's wings flapping, it signals the beginning of a hurricane. Furthermore,

"As each tape is shown, we notice the events have a snowball effect on Hannah's life drastically changing everything. There is a lack of understanding of these core concepts of the show and the need for this show. One can never really understand the psychological trauma a person goes through before committing suicide, some dots are left unconnected. We never have all the motives, either the person is unwilling to share every bit of detail or they are successful in their concluding act, suicide (Johnson, 1987). *Thirteen Reasons Why* makes the association between each factor that leads to one girl's suicide making people more aware of the effect their action has on someone" (Martin, 2019).

In analyzing the course of the subtle signs of suicide, depression remains one of them in the novel, the writer will focus on the stressful life events as the main cause of Hannah's depression in the novel.

Hannah Baker was bullied, sexually harassed and cyber-bullied throughout high school, she started feeling betrayed and started to isolate herself.

"Betrayal. It's one of the worst feelings. I know you didn't mean to let me down. Most of you listening probably had no idea what you were doing—what you were truly doing" (Asher, 2007).

The things Hannah wanted to achieve gradually and slowly she began to lose touch with her passion. She desired to achieve something in herself, she wanted to go to college. She liked poetry, it was her escape, her therapy "But it's always cheaper than a therapist. I did that for a while. Poetry, not a therapist" (Asher, 2007).

With all the life events going on at school, she slowly started to feel depressed and hopelessness started to pave its way in her life. She started to isolate herself. Isolation and loneliness are fundamental points of depression and depression when not treated lead to hopelessness and drastic steps. It's not self-isolation or suicide that kills a person, lack of understanding and empathy are equally accountable for it. All the events taking place in Hannah's life were making her association with society weaker. She started to shut herself. Individuals who tend to shut themselves, when feeling hurt and ignored tend to commit suicide. This is known as egoistic suicide, according to Durkheim, which is when an individual becomes "detached from society" (Durkheim et al., 1966).

Hannah Baker was encountering the same, her social interaction and bond were failing thus making her take a drastic step toward suicide.

Escape

Escapism was a defense mechanism that Hannah used for her depression. Her problem was testing her beyond her limits, she started to have suicidal thoughts. She even confronted her trying to figure out ways to kill herself. Least painful way to end her misery. "So I've decided on the least painful way possible. Pills (Asher, 2007) She couldn't face her problems anymore, the more she tried to help herself the more lost she felt. The best way she decide to end her problems was to commit suicide. The signs of suicide include verbal signs, according to the study, people who think about suicide, express hopelessness through their words, Occasionally express a desire to die, explain feelings of entrapment, and express how agonizing pain is (Villines, 2019).

Hannah was displaying all the signs of suicide, she was contemplating methods to end her life. Not only that she did communicate about her pain being unbearable to her school counselor who just neglected her and did nothing to aid her. Hannah needed help and that's why she for one last time went to her school counselor to talk and seek help. She talked about how she wanted everything to stop "I need everything to stop" (Asher, 2007). Despite her cry for assistance, no one understood her, she went home and escaped the suffering and pain she was feeling. Not only verbal but behavioral changes were also portrayed by Hannah.

Hannah's signs were subtle but not unnoticed, from the moment she cut her hair to when she quit her job at Cresmount, she was calling for help but nobody noticed, not her friends, her teachers, and not even her family. Hannah was losing hold, she was not able to keep up with the events occurring in her life, and the best way to terminate the pain was to end her life. With her school life being a mess and parents being too involved with their business and no one to talk to she finally chose to let go of her pain forever and killed herself but swallowing pills.

Change in Appearance

Change in appearance is one of the common signs exhibited by the person, attitude or behavior changes along with the sudden change in physical appearance are the signs to look for as signs for help. Hannah started to feel alone and started to shut people out, she started to ignore people and even quit her job at the Crestmont. Hannah desired to get noticed, she wanted people to know that things were not moving right. So she decided to chop her hair. She wanted her parents to notice the new hairstyle and converse about it but it didn't happen. Her parents who loved her were so engaged with their business that they skipped the sign. They were getting distant and Hannah had no one to talk to, it wasn't that her parents ignored her but they were spending less time together. Low self-esteem, negative behavior, hopelessness, and as well as problems controlling emotions and maintaining interpersonal relationships, which lead to her psychological disorder and further increased her self-destructive tendencies. This was one of the signs exhibited by Hannah regarding her deteriorating mental health.

Conclusion

Based on the analysis of the novel this paper presented the subtle signs of suicide that were exhibited by the main protagonist Hannah Baker throughout the novel.

Hannah Baker had a self-destructive personality that was evident in her actions. She was engaging in destructive behavior that was replicated in her day-to-day interactions. The subtle behaviors she exhibited throughout were examined in this paper. These behaviors included giving others the impression that she intended to pass away, harboring negative thoughts about herself, failing to act to address problems, committing social suicide, and self-sacrifice. Neglecting the indications can have disastrous consequences, thus appropriate diagnosis and behavior improvement are crucial parts of prevention. The symptoms and the behavioral shifts were highlighted in this study.

Furthermore, data on the subtle signs are examined using Durkheim's Theory of Suicide. The study concludes that if Hannah Baker's symptoms had not been ignored or taken for granted, someone could have been able to help her.

References

Asher, Jay (2007). Thirteen Reasons Why, 10, 13, 176.

Shneidman, Edwin S., and Leenaars, Antoon (2010). Suicide. Suicidology Online, 8.

Martin, Jerusha (2019). The Butterfly Effect: Analysis of the Snowball Effect in Thirteen Reasons Why. Literary Endeavour, 101.

Cash, Scottye J., and Bridge, Jeffrey A. (2009). Epidemiology of youth suicide and suicidal behavior. Current opinion in pediatrics, 21(5), 613–619. doi:10.1097/MOP.0b013e32833063e1

Kring, A. M., Johnson, S. L., Davidson, G., and Neale, J. (2012). Abnormal Psychology Twelfth Edition-DSM 5 Update. In Wiley https://doi.org/10.1037/031838

Bhowmik, Debjit, Kumar, S., Paswan, Shravan, and Dutta, A. S. (2012). Depression-symptoms, causes, medications and therapies. Pharma Innov. 1. 32–45.

Durkheim, Émile, John A. Spaulding, and George Simpson (1966). Suicide: A Study in Sociology.

Villines, Zawn (2019). Medically reviewed by Timothy J. Legg, How to help a teen with depression.

Demographic Variables and Job Satisfaction Among College Lecturers

K. Savitha[a] *and J. Venkatachalam*[b]

[a]Ph.D. Research Scholar. [b]Professor

[a,b]Dept of Psychology, Periyar University, Salem - 636011

E-mail: *savitha29k@gmail.com; chalam66@yahoo.com

Abstract

Job satisfaction of individuals will have an effect on any organization, be it higher productivity or better performance. Job satisfaction of the lecturers will ensure better education for the students and thereby the growth of the Education Institutions. Thus, the current study aims at exploring the gender differences in Job Satisfaction of lecturers. Data were collected from 149 lecturers, through a purposive sampling method by online mode. *Job satisfaction scale* by Dr. *Amar Singh* and Dr. T. R. Sharma (1999) was included for measuring Job satisfaction dimensions. The results revealed no significant gender differences in job satisfaction. In this context, a greater understanding of educators' Job Satisfaction could facilitate the development of more effective policy practices that would increase not only the level of educators' satisfaction, commitment, and morale but also improve the performance of the Educational Institution system.

Keywords: Job Satisfaction, College Lecturers, Gender

Introduction

A job is not just a primary source of revenue; it is also a crucial aspect of living. Each worker loses a significant portion of their day at work, which also affects their social standing. Given for many people, employment plays a crucial role in their lives, and job satisfaction is significant to overall wellbeing. If workers are happy, they will work more efficiently and deliver better-than-average results, which will increase earnings. Employee satisfaction also increases the likelihood that they will be inventive, and creative, and make discoveries that help any organization advance and improve through time and in response to shifting market conditions. On the national agenda, improving educational performance is a top priority, with educators and policymakers putting special emphasis on accountability, curriculum reform, teacher quality, school choice, and other related issues. A successful system is built on a strong teaching faculty. An educational institution's main requirement is to attract and keep high-caliber teachers (Sharma and Jyoti, 2006).

Teacher plays a significant role in students' life. At various levels, students receive a variety of knowledge and skills that will help them lead fulfilling lives. As a result, a teacher can effectively convey the idea of sustainable living to all of his or her students. In just a few decades, the planet of today won't look the same. Although the future doesn't look good, education can help students be both ready for change and equipped to stop more harm. Students will be equipped to survive in an uncharted future if teachers, academics, and parents educate them now. More significantly, students of today might grow up to be future academics, scientists, and activists who will defend our sustainable development. Special strength of teachers is alumnae association, the student who is influenced by these teachers may implement sustainable development in their business or implement it in their lifestyle. The present study also had questions related to sustainable development in Job satisfaction scale. For example - "Do you feel that some job or profession in some way adds to the economy and development of the nation?" Regarding job satisfaction of lecturers, teacher who is satisfied with their job, he or she alone can think of sustainable development beyond the syllabuses if they are really satisfied with their job, teacher who loves their job and is satisfied with their job can only think of positive change in students not only on students' academic but also making students in the contribution to economic development, conservation of energy, resources etc.

Understanding the components involved in the development of quality teachers is necessary. One of those essential components is job satisfaction in which the important element influencing institutional dynamics is teacher job satisfaction, which is typically regarded as the main dependent variable used to measure how well an organization's human resources are performing. For an effective educational system, it is crucial to comprehend the aspects that influence teachers' job satisfaction. Teachers who report feeling satisfied with their jobs tend to be more committed (Reyes and Shin, 1995), enthusiastic (Chen, 2007), and inclined to have good attitudes toward their daily tasks and the effort they put forth when dealing with students (Caprara, Barbaranelli, Borgogni, and Steca, 2003). The main objective of the proposed research has been designed to investigate gender differences in job satisfaction among college lecturers to fill the vacuum in the research and add more to the literature.

Participants

The participants were 149 lecturers from different Colleges, Southern India. Majority of the sample are male (55%) lecturers.

Measures

Job satisfaction was measured by using an instrument by Singh, A., and Sharma, T. R. (1999), a specific attempt is made to pinpoint the underlying elements of job satisfaction. This instrument assesses both job intrinsic and job extrinsic aspects, consisting of 30 statements to be responded on 5-point rating scale.

Job Satisfaction of Lecturers-Gender Wise Comparison

Simple t-test was performed to examine the difference between the genders. As can be seen from Table, there is no significant difference between the male and female samples with regards to their job satisfaction, where the obtained t value for job intrinsic, extrinsic, and total job satisfaction scores are 0.317, 0.786, and 0.318 respectively. The finding reveals that, that there is no significant gender difference in job satisfaction. However, this finding is not surprising because even within identical national contexts, studies on teacher gender and job satisfaction are typically characterized by a high degree of inconsistency. As a result, while one study of the job satisfaction of English teachers found no obvious gender differences (Crossman and Harris, 2006), further a study by Oshagbemi, T. (2000) also reveals no so significant differences, which is similar to the results of the present study, where "t" values are not significant. Hence the hypothesis is not accepted. Zafarullah Sahito and Vaisanen (2020) suggested that no significant difference was found in the levels of job satisfaction between male and female teachers.

In addition, Poppleton and Riseborough (1990) in their study reported that women were more satisfied with their teaching positions. In a similar way relationship between teacher gender and job satisfaction are far from consistent, pointing to either higher levels of job satisfaction for women (Liu and Ramsey, 2008; Ma and MacMillan, 1999) or men (Liu and Ramsey, 2008). The reasons for gender differences in job satisfaction would be: First, female employees have lower expectations than male employees but still feel satisfied; second, because of social pressures, it is more likely that females will not express their dissatisfaction.

Conclusion

This study adds to our understanding of the differences between men and women lecturers in job satisfaction who teach in colleges. There are two key reasons why the connection between gender and teacher job satisfaction is seen to be significant – First considering that teaching is a common profession choice for women in many nations, it is important to look into how gender affects job satisfaction and determine the causes of gender differences in job satisfaction. Secondly, the career goals, future advancement, and professional growth of female teachers are likely to be influenced by how satisfied they are with their work. Finding offer insightful knowledge about the issue because it is crucial to provide all advantages and a favorable working environment in educational institutions. Decision-makers must acknowledge the connection between teacher job satisfaction and educational excellence. A key focus of efficient policy-making should be how to enhance teachers' working conditions. Unfortunately, the importance of teachers' job satisfaction is sometimes overlooked while creating educational policies. For the sake of teachers themselves as well as students and institutions, the ministry of education or private business should priorities maximizing teacher satisfaction and minimizing disaffection.

Table 1. Job Satisfaction of Lecturers-Gender wise comparison

Components of Job Satisfaction	Male		Females		"t" Value
	M_1	SD_1	M_2	SD_2	
Job Intrinsic	31.88	6.794	32.22	6.473	.317 NS
Job Extrinsic	38.62	10.533	39.06	9.145	.786 NS
Job Satisfaction Total	70.50	15.755	71.28	14.251	.318 NS

References

Amarasena, T. S. M., Ajward, A. R., and Haque, A. K. M. A. (2015). The effects of demographic factors on job satisfaction of university faculty members in Sri Lanka. International Journal of Academic Research and Reflection, 3(4), 89–106.

Caprara, G. V., Barbaranelli, C., Borgogni, L., and Steca, P. (2003). Efficacy beliefs as determinants of teachers' job satisfaction. Journal of educational psychology, 95(4), 821. DOI:https://psycnet.apa.org/doi/10.1037/0022-0663.95.4.821

Crossman, A., and Harris, P. (2006). Job satisfaction of secondary school teachers. Educational Management Administration & Leadership, 34(1), 29–46. doi: https://doi.org/10.1177%2F1741143206059538

Kapa, R., and Gimbert, B. (2018). Job satisfaction, school rule enforcement, and teacher victimization. School Effectiveness and School Improvement, 29(1), 150–168. doi: https://doi.org/10.1080/09243453.2017.1395747

Oshagbemi, T. (2000). Gender differences in the job satisfaction of university teachers. Women in review Management. doi: https://doi.org/10.1108/09649420010378133

Poppleton, P., and Riseborough, G. (1990). A profession in transition: Educational policy and secondary school teaching in England in the 1980s. Comparative Education, 26(2–3), 211–226. doi: https://doi.org/10.1080/0305006900260205

Reyes, P., and Shin, H. S. (1995). Teacher commitment and job satisfaction: A causal analysis. Journal of school leadership, 5(1), 22–39. doi: https://doi.org/10.1177%2F105268469500500102

Sharma, R. D., and Jyoti, J. (2009). Job satisfaction of university teachers: An empirical study. Journal of Services Research, 9(2).

Tadesse, B., and Muriithi, G. (2017). The influence of employee demographic factors on job satisfaction: A case study of Segen Construction Company, Eritrea. African Journal of Business Management, 11(21), 608–618. doi: 10.5897/AJBM2017.8403

Zembylas, M., and Papanastasiou, E. (2004). Job satisfaction among school teachers in Cyprus. Journal of Educational Administration, 42(3), 357–374. doi: https://doi.org/10.1108/09578230410534676

Sustainable Crisis: Psychoanalytical Reading of Populism and Trauma in Select War Narrative

Elizabath Jansy[*,a] *and Shobana Mathews*[b]

[a]Christ Deemed to be University, Bangalore, India
[b]Christ Deemed to be University, Bangalore, India
E-mail: *elizabath.jansy@res.christuniversity.in

Abstract

Sustainable Development has become a ubiquitous discourse in contemporary times. As it revolves around the three pillars, namely economy, environment, and society, it intends to meet all-inclusive rudiments of the present and future unvaryingly. The role of children in achieving sustainable goals is profound. But children are subjected to traumatic events that scar their mental, emotional and physical growth. A study on the tactics of the incipient populists has become a need of the hour. Though populism would refer to a political movement that appeals to the ordinary people and stands for the people, citizens have to be perspicacious about the evolving maneuvers of the populists who threaten a nation's egalitarianism system. This paper exposes the representation of trauma among children elicited by populists, curtailing the global vision of achieving sustainability, in the narrative Beasts of No Nation by Uzodinma Iweala. Employing the method of textual analysis, the study uses trauma theory to substantiate the arguments. The paper will specifically discuss the fragmented identity and dissociated psyche of children.

Keywords: Populism, Childhood Trauma, Sustainable Crisis, Dissociation.

Introduction

Populism has emerged as a topic of poignant discussion in recent times. While populism remains to be a "contested concept," (Mudde and Kaltwasser, 2017) a recent approach considers populism, as a political approach engaged by an explicit type of leader, who ponders power as well as pursues to head based on direct and without any intervention or support from their supporters (Mudde and Kaltwasser, 2017). While such a leader can also claim to be the one who stands for the goodwill of common people, citizens have to be perspicacious about the evolving maneuvers of the populists who pose a threat to a nation's egalitarian system. Literary Narratives have represented populists and populism for ages. While populism impacts society universally, a study on the detrimental effect of populism on children cannot be trivialized as it disturbs the whole Sustainable Achievement Goals. History testifies to the agony of children who are subjected to atrocious activities like abduction, thrashings, rape, bullying, racism, insult, hurt, agony, and abandonment. UNICEF's 2020 Annual Report underscores how 2020 was a year like no other. School closures increased vulnerability to abuse, mental health strains and loss of access to vital services have hurt children deeply (UNICEF, 2021). In addition to this, a study about trauma caused to children, as a result of populism as well as populist leaders is a need of the hour to establish a healthy sustainable society and environment across the Globe.

Theorizing Trauma: Trauma Novel enunciates profound loss, intense pain, and extreme fear of the individual and society. The literary trauma theory enunciated by Kali Tal, and critics such as Cathy Caruth, ruminates the responses to a traumatic experience as a trait that is an intrinsic or inborn feature of traumatic memory and experience (Balaev, 2008). They add that it distorts the psyche as well as divides the sense of identity in the traumatized self. Caruth defines trauma as that which occurs after the real experience of the traumatic event as "trauma cannot be discerned in the simple or real traumatic occurrence in a victim's past, but rather in the way it was specifically not known at the first occurrence, return to disturb and haunt the traumatized survivor later on (Caruth, 1996). The paper argues this representation of trauma in the belatedness of the traumatic event by the child protagonist Agu. The paper also exposes the intergenerational transmission of trauma, dissociation of psyche, and the identity of the traumatized child victim as well as argues the representation of intense trauma experienced by a child, instigated by a populist who deprives all the happiness the child deserves, as represented in the narrative, *Beasts of No Nation*.

Though the Sierra Leone Civil War has a lot of similarity to the war portrayed in the narrative, the historical background of the war is not set in any particular country nor does the narrative refers to any specific war but voices out for the many wars that took place in Africa, in which children were psychologically manipulated and were recruited as child soldiers. The leader of the rebel group manipulates the children and proves to be a tactful populist. Beasts of No Nation is centered in an unnamed African Country, where the protagonist, Agu, a young boy of eleven years old, tells the story of a traumatic and brutal life to which he was exposed. Agu after the death of his father in the war encounters a group of rebel fighters, led by a self-claimed commandant, an otherwise populist, who elects him to be a soldier and gradually forces the child soldiers to kill, loot, lie, rape, and murder. They confront adverse poverty, hunger, illness, and mass murders as well as abuse. Toward the end, his troop kills the commandant. Agu finally ends up in a refugee camp where he tries to come in terms with all the traumatic events he had encountered. It's a story of a helpless child wrecked and torn by the violence he was forced to be a part of, initiated by a populist.

Onyekachi Eni, Chukwu Romanus Nwoma, and Chukwuka Ogbu Nwachukwu "recontextualize" (Eni et al., 2020) the normal conjecture of a child's victimization, in the mainstream discourse, using the literary narratives, Ahmadou Kourouma's *Allah is Not Obliged* and Uzodinma Iweala's *Beasts of No Nation*, in the article, "The Child Soldier as a Mercenary: An Interpretive Recontextualization of Ahmadou Kourouma's Allah is Not Obliged and Uzodinma Iweala's *Beasts of No Nation*." The paper argues the conscious will of children to be child soldiers and contends the normal victimization of the child soldiers, in the mainstream discourse as forced soldiers, as well as their development employing Bronfenbrenner's bioethical systems theory and the methodology used is interpretative textual analysis.

Eleni Coundouriotis compares the contemporary African Narratives to the literature that thrived during the colonial and post-colonial periods and argues the lack of historicity (Coundouriotis, 2010)most of them novels, and argues that the recent texts (those published since the mid-1990s and the blind political conditioning, that is seen in the contemporary narratives, in the article titled, "The Child Soldier Narrative and the Problem of Arrested Historicization," using select Novels of Ahmadou Kourouma and Emmanuel Dongala, using the theory of decolonization by Frantz Fanon. Obi Nwakanma, in the article titled, "Metonymie Eruptions: Igbo Novelists, the Narrative of the Nation, and New Developments in the Contemporary Nigerian Novel," argues the efficacy of Igbo Novelists (Nwakanma, 2008) and the influence of their writings in the contemporary Nigerian Literature. The author exposes the significant role of Nigerian literature in building a sense of National belonging among them. The hegemony maintained by the elite post-colonials is articulated using textual analysis in the narratives *Arrows of Rain, Half of a Yellow Sun, Subernia's day* et al.

Among the various studies conducted on childhood victimization in the Narrative, *Beasts of No Nation,* there is a lacuna found in studying the trauma of the victimized child soldier because of the Populist, who is exclusively responsible for the mental, emotional, psychic, and physical deleteriousness of the child. This paper argues the representation of a populist leader who through his tactics traps an innocent child into inhuman atrocities that fracture the whole progress of the child, in the narrative, *Beasts of No Nation* by Uzodinma Iweala. The paper also argues how the narrative represents the collective trauma represented by the individual child protagonist, Agu who expatriates the historic and intergenerational trauma the child is a victim of.

Cry of the helpless - Populism and the Traumatized Childhood: The trauma incited by populist governance need not be necessarily evolved by a liberal democracy alone. A populist can evolve on a dictatorial self-choice too, depending on the command one might exert over the select community. "A reasonable definition will then result: Populism can be either inclusive or exclusive, carried out from "below" or "above" and forced" (Hartleb, 2017). Iweala uncovers the trauma elicited by such a populist on children through the narrative *Beasts of No Nation.* This force by which the commandant becomes a populist for his exigencies, under the label of, for the good of people and children, turns to be the most biased and atrocious leader who evokes fear and traumatizes the child Protagonist, Agu and his fellow friends, using his tactics. The mannerism and ways in which the commandant presents himself create a false impression of a leader in the minds of innocent children. Agu states, a man is coming down from the truck. He is looking like the leader. All the other soldiers are following each movement he is making. He is moving slowly like an important person to make sure that everybody looking at him is knowing he is chief (Iweala, 2005).

Iweala magnificently represents how a false populist uses his tactics in becoming a leader traumatizing a whole group of innocent, helpless children. Having heard and seen soldiers before the war began, Agu seems to be influenced to be a soldier. After losing his father in war and being trapped in the hands of the evolving populist who is the commandant, Agu agrees, while asked, "Do you want to be soldier?" (Iweala, 2005) without understanding the trauma he is going to endure nearly. The commandant lures Agu and states, "If you are staying with me, I will be taking care of you and we will be fighting the enemy that is taking your father" (Iweala, 2005). Believing the commandant and his ornamented words, Agu becomes a Soldier for the commandant and his troop. Agu is taught to do everything that would deeply traumatize him from looting, beating, stealing, to killing and abusing.

The children were used as puppets for the selfish motives and needs of the commandant and his troop. Later the children are taught to take drugs but manipulated stating that it is the gun juice for active working. Though Agu dislikes it initially, he becomes addicted to it in time. Agu is profoundly traumatized as the commandant sooner sodomizes the innocent child who is unable to even recognize what is happening but just feels intense pain. Agu states, "Each time I am going to Commandant, I am feeling that I should not go in because I am knowing what he is wanting to do to me" (Iweala, 2005). He heartbreakingly inflicts mental, emotional, and physical trauma on Agu but tactfully justifies it before the child, as he states, "it is what a commanding officer is supposed to be doing to his troop" (Iweala, 2005). Agu feels depressed that he states, "I was feeling I can never be smiling again" (Iweala, 2005). As the contemporary trauma theory states, "Trauma creates a speechless fright" (Balaev, 2008). Agu is unable to resist or defend himself so he keeps quiet in midst of extreme pain, due to fear, though he wishes to defend himself. This speechlessness of trauma is represented in the fiction as Agu charts, "Agu, he is saying. Let me tell you something. I am not wanting to know anything he is telling me (Iweala, 2005). But I am saying Sah! Yes, Sah!" (Iweala, 2005). Agu is later taught to abuse women and children, wounding his sense of identity. He quotes, "I am too young to be knowing about these thing" (Iweala, 2005) and this sense of doubt puts him into a dillemma of guilt and uncertainty.

Plea of the Bruised Child Psyche: Traumatic experience, as Freud indicates suggestively, is an experience that is not fully assimilated as it occurs (Caruth, 1996) but is comprehended and intensely felt in its "belatedness" or its memory. Agu comprehends the trauma he is encountering, not in its very occurrence but later in its remembrance. He states, "I am thinking of home. How many time am I thinking of my home when we are in the bush?" (Iweala, 2005). The memory of separation hurts him relentlessly. He recollects the moments he bid his mother and sister goodbye for the last time. The traumatic memory of his father being shot dead in front of his eyes haunts him often. He longs for the past days of joy but realizes he can no longer be the old happy child and states, "All we are knowing is that before the war we are children and now we are not" (Iweala, 2005). Though the very moment of trauma experience does not affect him, the latent memory of the shocking experience traumatizes Agu.

Agu is no longer able to follow the commands as the multiple trauma he encounters every day, wounds his psche. He is lost and broken that the narrative represents the dissociated psche of the traumatized child. Dissociation is a break in how your mind handles information. You may feel disconnected from your thoughts, feelings, memories, and surroundings. It can affect your sense of identity (Wiginton, 2021). Dissociation involves the temporary alteration or separation of normally-integrated mental processes in conscious awareness (Butler and Palesh, 2004). Iweala represents the dissociated psyche of the child as Agu states, "I am walking with my hand stretching out in front of me because I am trying to catch all of those thoughts that is floating around me so I can make sure no part of me is missing" (Iweala, 2005). Agu no longer can be at peace nor sense that the extreme trauma he has encountered even deprives him of his sleep and sense of his self. He states, "Nothing is the same anymore. I am not being able to be sleeping at all when it is time to sleep. Each time I am lying down my head, some voice inside of me is shouting so I cannot even be closing my eye. I am fearing that I am not knowing myself anymore" (Iweala, 2005).

The child is mentally, emotionally, and psychologically traumatized and he loses his sense of identity and consciousness. He doubts his existence and strives between his past and present paradoxical ethics. Agu states, "We are almost one hundred and twenty of us standing at attention" (Iweala, 2005). Thus, the narrative exposes the plight of groups of children who are collectively traumatized, beaten, forced, sodomized, and assaulted by a self-claimed populist.

Transferred Travail: Intergenerational transmission of Trauma: The intergenerational theory of trauma offers a paradigm to examine the representation of trauma in literature. Iweala delineates the intergenerational transmission of historic trauma through the narrative. The theory of transhistorical trauma "indicates that a massive trauma experienced by a group in the historical past can be experienced by an individual living centuries later who shares a similar attribute of the historical group" (Balaev, 2008). Agu's father who is already a survivor of a similar war does not defend nor react to the atrocities meted over him or his generation, but simply remains silent. While it was commanded to leave the place where they resided during the onset of war, Agu quotes, "I am wanting to open my mouth and scream, but my mother and my father are keeping quiet so I am keeping quiet also" (Iweala, 2005).

This trait of being quiet and silent when threatened is intergenerationally passed and acquired by Agu when he is later traumatized and abused by the commandant. Though he wanted to react and defend himself he remains quiet just like his father. It is evident while Agu is sodomized by the commandant and he exclaims, "I don't want to be a soldier at all. But I am not saying any of this. I am not saying anything at all" (Iweala, 2005). While Agu is abused by the commandant, he stays quiet without defending just like his father who stayed quiet midst of the war. The author represents how this defenselessness seems to be intergenerationally transferred to Agu. Agu is traumatized that he loses the stability of his mind and charts, "sometimes I am telling her, I am hearing bullet and scream in my ear and I am wanting to be dying so I am never hearing it again" (Iweala, 2005), while he finally reaches the rehabilitation center and speaks to Amy, who tries to help him and other children who were victims of war. Responding to Amy, Agu converses that he had seen worse events that cannot be even talked about and adds, he wishes not to express it as he wants to be happy. He tells Amy, "I am wanting to be happy in this life because of everything I am seeing. When I am saying all of this, she is just looking at me and I am seeing water in her eye" (Iweala, 2005).

Conclusion

The narrative painfully communicates the violence that was incited on the innocent children, by a self-claimed populist for his own tactful and selfish need thereby curtailing the endeavor towards achieving global sustainability. While mature men and women are themselves prone to fall for the tactics of the tactful populists, the narrative exposes the atrociousness that a populist is capable of exhilarating among children. Children, who simply abide by what is taught, have to be provided with responsible leaders who are capable to mould them into healthy individuals. They do deserve a joyful childhood free from all prejudice, plight, pain, abuse, and trauma. The narrative serves to be an eye opener to shield children from all abductions as well as understand the tactics of any evolving tactful populists thus achieving sustainability in its fullness. Let it not be forgotten, "What you do to child matters, and they might never forget" (Morrison, 2015).

References

Balaev, M. (2008). Trends in Literary Trauma Theory: An Interdisciplinary Critical Journal, 41(2), 149–166.

Butler, L. D., and Palesh, O. (2004). Spellbound: Dissociation in the movies. Journal of Trauma and Dissociation. 5(2), 61–87.

Caruth, C. (1996). Unclaimed Experience Trauma, Narrative and History. John Hopkins University Press.

Coundouriotis, E. (2010). The child soldier narrative and the problem of arrested historicization. Journal of Human Rights. 9(2), 191–206.

Eni, O., et al., (2020). The Child Soldier as a Mercenary: An Interpretive Recontextualization of Ahmadou Kourouma's Allah is Not Obliged and Uzodinma Iweala's Beasts of No Nation. Critique - Studies in Contemporary Fiction.

Iweala, U. (2005). Beasts of No Nation. In John Murray Publishers.

Mudde, C., and Kaltwasser, C. R. (2017). Populsim. A very short introduction. Populism: A Very Short Intoduction.

Nwakanma, O. (2008). Metonymie eruptions: Igbo novelists, the narrative of the nation, and new developments in the contemporary Nigerian novel. Research in African Literatures. 39(2), 1–14.

UNICEF. (2021). UNICEF Annual Report 2020 |UNICEF.https://www.unicef.org/reports/unicef-annual-report-2020

Impact of Problematic Internet Use on Psychological Well-Being, Hyperventilation and Chronic Fatigue Syndrome Among Youth

S. James Robert[*,a] *and S. Kadhiravan*[b]

[a]Research Scholar, Department of Psychology, Periyar University, Salem, India
[b]Professor and Head, Department of Psychology, Periyar University, Salem, India
E-mail: *jamesrobert463@gmail.com

Abstract

The prevalence of internet use increased worldwide, especially among youth. People use the internet for various reasons such as education, business, reading news, shopping, and so on. Although the internet provides a variety of benefits, it also has significant negative impacts. This study was conducted to explore the impact of problematic internet use on psychological well-being, hyperventilation, and chronic fatigue syndrome among youth. 448 college students from different colleges in Salem city were selected through stratified random sampling and the data was collected with Internet usage scale by Robinson and Mukundan (2016); General health questionnaire - 28 by Goldberg (1978); Nijmegen questionnaire by Van Doorn et al. (1982); and Chalder fatigue scale by Chalder (1993). Results revealed that the problematic internet use of youth had a significant positive association with somatic symptoms, anxiety, social dysfunction, and severe depression of psychological well-being as well as with hyperventilation and chronic fatigue syndrome.

Keywords: Problematic internet use, Psychological well-being, Hyperventilation, Chronic fatigue syndrome, Youth

Introduction

The use of the internet is prevalent in daily life and is expanding rapidly across the globe and became an effective system in eliminating human geographical barriers. It has redefined communication and resulted in an exponential rise in human interaction, without the limitations of time and geography. People use the internet at all phases of their lives, particularly young people in college for the preparation of assignments, presentations, internship papers, and reports. The usage of the internet is also indispensable for finding out the most recent information. Da Cruz and Alvarez reported that 95% of college students were estimated to be using the internet (Da Cruz and Alvarez, 2015). Problematic internet use (PIU) among youth has emerged as a result of the expansion of high-speed internet access and the availability of digital gadgets in India over the years. Although the benefits of the internet help individuals succeed academically, it also hinders their performance and leads to behavioral addictions. Excessive usage in web surfing, social media, pornography, freelancing, and video games have been reported to be characterized by anxiety, agitation, headache, backache, weight gain or loss, blurred vision, dry eyes, sleep disturbance, depression, poor academic performance and carpal tunnel syndrome (Uddin et al., 2016). Uddin et al. reported that excessive internet usage is a newly emerging mental health and social issue among youth causing neurological complications, psychological disturbances and social problems (Uddin et al., 2016). A person's life is completely taken over by these multifaceted compulsive, cognitive, and behavioral symptoms.

Problematic Internet Use and Psychological Wellbeing

The burden of health issues is increasing, with significant effects on global health and economic consequences due to poor control over PIU. Poorolajal et al. reported that about half of all mental health problems begin at the age of 14 and three-quarters by mid-20s (Poorolajal et al., 2017). Among young people, college students make up a sizable portion who struggle with social isolation, depression, and anxiety. Even though for the majority of young people, college can be an enjoyable time, it can also be accompanied by significant stresses that may increase the risk of mental illness, such as lengthy study hours, abnormal sleep habits, academic pressures, and so on that would result in poor psychological wellbeing. Students who use the internet excessively tend to struggle to focus on their academics and other duties, leading to poor psychological well-being.

Problematic Internet Use and Hyperventilation

Researchers reported that improper use of internet such as frequency, duration, and incorrect positioning will affect individuals' health, especially muscles and nerves also cause various physical problems such as pain in the neck and shoulders (Nyi Sukmasari, 2015). The situation of aggravation caused by smartphone use in a darkened environment may make the eyes work harder than usual, which, if done for a prolonged amount of time, will make the eyes quickly tired and result in headaches or neck pain. Hafiz and Khairul reported that PIU makes students lazy and weak causing excessive pulse to use internet even more (Hafizd and Khairul, 2014). Additionally, it causes the body's physiological processes to malfunction and causes stress, depression, and insomnia.

Problematic Internet Use and Chronic Fatigue Syndrome

Every individual has encountered fatigue at some point in their lives, which is a subjective feeling that is challenging to distinguish from common experiences of fatigue, sleeplessness, overstrain, or exercise depletion. Chronic fatigue syndrome is characterized by fatigue, unrefreshing sleep, exacerbation of symptoms following exertion, and cognitive impairment which often result in drastically reduced levels of daily functioning (Mateo et al., 2020). PIU has a notable detrimental effect on physical health, resulting in symptoms like headaches, musculoskeletal pain, and fatigue. Students who spend too much time online get unmotivated to learn, are unable to focus, have trouble sleeping, and experience tingling, muscle stiffness, and disorientation.

Need for the Study

The Internet has made the world smaller by allowing everyone access to information and fostering connections with individuals all over the world. It has drawn many people to spend more time online, which could have negative psychological effects and social repercussions. PIU can lead to physical health issues such as dry eyes; carpal tunnel syndrome; repetitive motion injuries; pain in wrist, neck, back, and shoulder; migraine headaches; numbness, and pain in fingers (Park et al., 2013) which are the symptoms of hyperventilation. Students are not an exception to the adverse effects of PIU such as physical inactivity, lack of adequate sleep, severe depression, loneliness, and social anxiety that could also result in chronic fatigue syndrome. If these problems are untreated, they could result in severe impairment of students' functioning at college and influence their subsequent development (Rasmussen et al., 2020). Hence, it is imperative to explore the impact of PIU on psychological well-being, hyperventilation, and chronic fatigue syndrome among youth.

Materials and Methods

A normative survey was used in order to determine the impact of PIU on psychological well-being, hyperventilation, and chronic fatigue syndrome among youth.

Tools Used

Internet usage scale by Robinson and Mukundan, 2016; General health questionnaire - 28 by Goldberg, 1978; Nijmegen questionnaire by Van Doorn et al. (1982); and Chalder fatigue scale by Chalder, 1993 were used to collect the data.

Results

Table 1. Correlation between PIU and psychological well-being among youth

PWB/ PIU	SS	A	Social Dys.	Severe Dep.
Mailing / Texting	0.23*	0.23*	0.22*	0.22*
SNS / Apps	0.24*	0.33*	0.16*	0.27*
Entertain	0.19*	0.21*	0.14*	0.23*
Sexual sites	0.22*	0.28*	0.23*	0.31*
IoB	0.29*	0.31*	0.20*	0.32*
Overall PIU	0.32*	0.37*	0.25*	0.37*

Note: *p < 0.05; PWB- Psychological Wellbeing; SS-Somatic symptoms; A- Anxiety; Dys- Dysfunction; Dep-Depression; SNS- Social Networking Sites; IoB- Influence on Behaviour

From Table 1, it is noticed that all the dimensions of PIU and psychological wellbeing were positively correlated.

Table 2. Correlation between PIU and hyperventilation among youth

PIU dimensions	Hyperventilation
Mailing / Texting	0.109*
SNS / Apps	0.206*
Entertain	0.140*
Sexual sites	0.178*
IoB	0.213*
Overall PIU	0.235*

Note: *p < 0.05; IoB- Influence on Behaviour

It is observed from Table 2 that mailing & texting, SNS / Apps, entertainment, sexual sites, and influence on the behavior of PIU had a significant positive association with hyperventilation of youth.

Table 3. Correlation between PIU and chronic fatigue syndrome among youth

PIU dimensions	Chronic fatigue syndrome
Mailing / Texting	0.208*
SNS / Apps	0.191*
Entertain	0.171*
Sexual sites	0.199*
IoB	0.169*
Overall PIU	0.246*

Note: *p < 0.05; IoB- Influence on Behavior

From Table 3, it is noted that all the dimensions of PIU had a significant positive association with chronic fatigue syndrome, indicating that students are experiencing a significantly high level of chronic fatigue syndrome due to poor control in their use of internet.

Table 4. Differences in PIU by gender, family type, and area of living

Category	M	SD	t
Boys	120.33	22.28	11.77*
Girls	98.09	17.65	
Nuclear	106.68	22.63	2.19*
Joint	111.82	23.35	
Rural	107.84	22.78	0.74^{NS}
Urban	109.47	22.81	

P<0.05; NS-Non-significance; M-Mean; SD-Standard Deviation

Table 4 demonstrates the statistically significant difference in PIU by gender and family type, however, there was no significant difference according to residential area. Boys' PIU was noticeably greater than girls' (120.33 vs. 98.09). PIU risk was higher for students from joint families than for students from nuclear families (111.82 vs. 106.68).

Discussion

The present study highlights the impact of PIU on psychological well-being, hyperventilation, and chronic fatigue syndrome among youth. PIU and psychological well-being had a significant positive association with each other (Table 1). Problematic internet users are prone to hold maladaptive cognitions that would result in anxiety, severe depression, and social dysfunction. Studies also revealed a direct relationship between PIU and psychological problems such as depression, loneliness, and social isolation (Berte et al., 2019). Social networking sites provide several potentials for misunderstandings, poorly managed expectations, and the exaggeration of maladaptive traits, which can make people feel more alone. The present finding is consistent with the findings of Berte et al. posited that psychopathological disorders such as depression, loneliness, and social anxiety were necessary elements in the etiology of pathological internet use (Berte et al., 2013).

PIU was significantly and positively related to hyperventilation (Table 2). People who spend too much time online playing games and using social networking sites report having trouble sleeping, feeling lightheaded, and paying attention. As a result, they may develop insomnia, daytime fatigue, and headache. Long-term use of social media or playing online games adds to a person"s stress load, which can lead to problematic behavior like exhaustion, decreased attention, and increased sensitivity that causes pain, irritability, and illogical thoughts. It can also cause a chronic headache, loss of interest in activities, hallucinogenic perceptions, loss of appetite, and immune system dysfunction (Table 3). Hence, the positive relationship between PIU and chronic fatigue syndrome is quite logical.

Furthermore, youth differed significantly in PIU based on their demographic categories such as gender, type of family, and area of residence (Table 4). Girls had a lower level of PIU than do boys with a given sample size in this study. This can be a result of the fact that girls frequently receive more parental supervision than boys, which might help keep them from using the internet for extended periods. Youth from joint families had a higher level of PIU than youth from nuclear families. Traditional Indian households are evolving in terms of lifestyle. No matter where they live, everyone, especially young people, has easy access to the most recent technologies. Hence, it is reasonable that youth did not have any significant difference in PIU based on their area of residence.

Conclusion

The findings of the study revealed that students exposed to PIU were at risk of developing poor psychological well-being. Parents have to guide them at home to limit themselves from excessive use of the internet, and the same could also be done at educational institutions by the teachers and counselors for better psychological well-being.

Findings also indicated that students had symptoms of hyperventilation and chronic fatigue syndrome due to PIU. Students may be encouraged for regular exercises such as walking, running, and bicycling to keep engaged. Further, it is witnessed that PIU had a significant influence on hyperventilation and chronic fatigue syndrome through the mediation of anxiety and severe depression. Yoga and meditation could help students to tackle anxiety and depression. It may also help students to develop healthy internet usage habits. Together, these would prevent them from hyperventilation and chronic fatigue syndrome.

Moreover, problematic internet users need to be diverted to focus on other hobbies and activities that do not require online access. Teachers may encourage the students to turn off their phones during classes as a form of a digital break to avoid using the internet unnecessarily. Considering the increased access to the internet among youth, the authorities and parents should implement appropriate strategies to decrease the harmful effects of PIU on their health. Studying the underlying reasons for PIU and prevention of the same is essential to safeguard the youth from this problem.

Acknowledgments: This work was supported by University Grants Commission (UGC) Government of India (NET-JRF).

References

Da Cruz, J. D. A., and Alvarez, T. (2015). Small Islands, Big Problems: Cybersecurity in the Caribbean Realm. Small Wars Insur. 2(7), 44am.

Uddin, M. S., Al Mamun, A., Iqbal, M. A., Nasrullah, M., Asaduzzaman, M., Sarwar, M. S., and Amran, M. S. (2016). Internet addiction disorder and its pathogenicity to psychological distress and depression among university students: A cross-sectional pilot study in Bangladesh. Psy. 7(08), 1126.

Wetterneck, C. T., Burgess, A. J., Short, M. B., Smith, A. H., and Cervantes, M. E. (2012). The Role of Sexual Compulsivity, Impulsivity, and Experiential Avoidance in Internet Pornography Use. Psychol. Rec. 62, 3–17.

Poorolajal, J., Ghaleiha, A., Darvishi, N., Daryaei, S., and Panahi, S. (2017). The prevalence of psychiatric distress and associated risk factors among college students using GHQ-28 questionnaire. Iran. J. Public Health. 46(7), 957.

Nyi Sukmasari, Radian. (2014). Sering Pakai Handphone ? Hati – hati Terkena Text Neck. http://health.detik.com/read/2014/04/24/170104/2564766/763.

Hafizd, Khairul. (2014). artikel pengaruh-Smartphone-bagi-remaja.html. http://khairul1992.blogspot.com/2014/04.

Mateo, L. J., Chu, L., Stevens, S., Stevens, J., Snell, C. R., Davenport, T., and VanNess, J. M. (2020). Post-exertional symptoms distinguish myalgic encephalomyelitis/chronic fatigue syndrome subjects from healthy controls. Work (Preprint), 1–11.

Park, S., Hong, K. E., Park, E. J., Ha, K. S., and Yoo, H. J. (2013). The association between problematic internet use and depression, suicidal ideation and bipolar disorder symptoms in Korean adolescents. Aust N Z J Psychiatry. 47(12 Pt 1), 153–159.

Rasmussen, E. E., Punyanunt-Carter, N., LaFreniere, J. R., Norman, M. S., and Kimball, T. G. (2020). The serially mediated relationship between emerging adults' social media use and mental well-being. Comput. Hum. Behav. 102, 206–213.

Berte, D. Z., Mahamid, F. A., and Affouneh, S. (2019). Internet Addiction and Perceived Self-Efficacy Among University Students. Int J Ment Health Addict, 1–15.

Psychological Distress Among IT Sector Employees During COVID-19 Pandemic in India

S. Aamira Zackiya[*,a] *and J. Venkatachalam*[b]

[a,b]Periyar University, Salem, India
E-mail: [*]aamira.vlr@gmail.com

Abstract

The COVID-19 pandemic is the most drastic public health crisis, inflicting significant psychological and physical distress all around the globe. People's living and working circumstances have changed drastically since the nationwide lockdown was imposed owing to the COVID-19 outbreak. This sudden transition has caused significant impact in the mental health of the employees, inflicting distress. Hence, the present study aims to investigate the psychological distress among the employees working in the IT sector during the second wave of the COVID-19 pandemic in India. In this study, General Health Questionnaire short-form (GHQ-12) was used to assess the psychological distress of the study sample through simple random sampling technique. Participants (N=156) responded to the GHQ-12, and the demographic data were also collected. Results revealed that about 11% of the sample are suffering from severe psychological distress whereas 60% with moderate psychological distress. The overall GHQ score showed a significant positive relationship with all three dimensions of GHQ-12 namely anxiety and depression, social dysfunction, and loss of confidence. The findings also show that there is a significant gender difference in the overall psychological distress, as well in the anxiety and depression dimension. Specific psychological interventions can be provided in order to alleviate the psychological distress of the employees.

Keywords: COVID-19, Employees, IT Sector, Psychological Distress

Introduction

The World Health Organization (WHO) proclaimed COVID-19 a pandemic on March 11, 2020, owing to its vigorous transmission throughout the world (Giorgi et al., 2020). The rapid transmission of this virus is through interpersonal encounters such as shaking hands, touching contaminated surfaces (fomites) and objects, and then exposing it to the eyes, nose, or mouth according to the Centers for Disease Control and Prevention (2020). Due to the lack of a vaccine, confining and controlling this virus posed a significant challenge for the entire world (Williams et al., 2020). Several measures have been implemented to curb its transmission, including travel restrictions and aircraft cancellations from and to countries with high levels of contamination, as well as quarantining infected persons (Giorgi et al., 2020). Eventually, a lockdown was enforced, which includes the closure of academic institutions, professional bodies, consulting firms, organizations, and public venues such as restaurants, movies, museums, etc.

This pandemic period has elevated anxiety and depression rates, which were relatively high during periods of excessive mortality and stringent confinement restrictions (Kumar and Nayar, 2021). These additional restrictions, such as self-isolation and quarantine, have had an impact on people's normal routines, and livelihoods, and several studies have reported various psychological consequences like poor sleep quality, stress, cyberchondria (Zackiya and Venkatachalam, 2022) anxiety, depression, poor well-being, post-traumatic stress disorder, and so on (Li et al., 2020).

COVID-19 and Workplace

Undeniably, COVID-19 is an unconventional crisis for workplaces. This pandemic has affected the global economies and societies massively which has resulted in furloughs and layoffs as per the World Economic Forum (2020). Many employees were laid off as numerous firms and industries were forced to shut down, while other essential employees faced a tremendous increase in workload and job strain. The risk factors for poor mental health have risen, such as financial insecurity, unemployment, and fear, while protective factors such as social connections, employment, educational participation, physical activity, daily routine, and access to health care, have declined. Employees might get affected negatively due to the physical distancing measures even though it is effective in reducing the transmission of COVID-19 (Williams et al., 2020). Several findings imply that working conditions have worsened due to which employees are more prone to suffer from mental health issues such as stress, depression, and anxiety. From previous research, two major issues of the workplace have been identified due to COVID-19. They are as follows:

Loss of Employment

One of the most prevalent stressors reported by research is the fear of losing one's job and one's income (Koh and Goh, 2020). Millions of people have lost their employment as a result of the recession. People who lose their jobs during an economic downturn are more likely to suffer from poor mental health (Drydakis, 2015) and they struggle for food, shelter, and a living, leading to depression, suicide, and self-harm (Kumar and Nayar, 2021).

Work from Home

Due to the enforcement of the lockdown, many employees were forced to mandatory work from home (WFH). Both positive and negative effects of working from home are identified. WFH's positive effects include work-life balance, flexibility, time-saving, quality time, and comfort, while negative effects include multitasking, decreased work motivation, increased costs, distraction, and limited communication (Mustajab et al., 2020). In addition, employees must balance working from home with other obligations such as homeschooling, childcare, and other caregiving responsibilities. These additional obligations, together with the constraints of prolonged remote working and the demands of achieving productivity levels, can exacerbate stress and mental health issues among the employees.

Methodology

Aim of the Study

The present study aims to assess the psychological distress among the employees working in the IT sector in India during the second wave of the COVID-19 pandemic.

The research objectives are:

- To assess the level of psychological distress among IT employees during the COVID-19 pandemic.
- To find out the gender difference in the psychological distress among IT employees during the COVID-19 pandemic.
- To assess whether specific demographic factors have a substantial impact on psychological distress among IT employees.

Participants

This study was carried out among 156 employees working in the IT sector in the southern region of India. The sample comprised 65 males and 91 females. The age range of the participants is 23 - 45 years of age (M_{age} = 27.53, SD = 3.90).

Procedure

Descriptive survey method was adopted, along with simple random sample technique. This research was carried out using an online survey using Google forms with standardized questionnaires, during the second wave of COVID-19 in India. The survey link was shared through emails. Participation was voluntary and all the participants provided online consent.

Measures

General Health Questionnaire (GHQ-12): This is the short-form version of the original GHQ-60, developed by Goldberg and Williams (1988). It is a screening tool for determining the level of psychological distress a person has encountered in the last few weeks. It has 12 items and three dimensions: anxiety and depression, social dysfunction, and loss of confidence. On a four-point scale, respondents assessed their experience with the symptoms in each item.

Demographic Information: Demographic information regarding participants' age, gender, area of living, marital status, number of hours spent on social media, and whether infected with COVID-19 was collected.

Statistical Analysis

The data were analyzed for descriptive and inferential statistics using the SPSS (Statistical Package for the Social Sciences) software to test the hypotheses.

Results

Table 1 shows the demographic characteristics of the research sample. In this study, the female 58% (n = 91) respondents are higher than the male 42% (n = 65). While 61% (n = 95) of the sample lives in a nuclear family setup, 39% (n = 61) are of joint family. 40% (n = 62) of the sample are living in the rural area, urban (n=58) 37% and semi-urban 23% (n=36). The distribution of the sample based on marital status showed that about 74% (n = 116) of the sample were married, and 26% (n = 40) of the sample were unmarried. About 50% (n = 78) of the sample spent less than two hours per day using social media, 29% (n = 45) of the sample spent two to five, and 21% (n = 33) of the sample spent more than five hours per day, further, about 46% of the sample have reported that they have infected with COVID-19.

Table 1. Demographic Characteristics of the Study Sample

Characteristics	Category	Frequency	Percentage
Gender	Male	65	42
	Female	91	58
Type of family	Nuclear	95	61
	Joint	61	39
Area of living	Rural	62	40
	Urban	58	37
	Semi-Urban	36	23
Marital status	Unmarried	40	26
	Married	116	74
Social Media Usage	< 2 hours	78	50
	2-5 hours	45	29
	> 5 hours	33	21
No. of people infected with COVID-19		73	46

Note. n = 156

Table 2 shows the inter-correlation of GHQ dimensions. All three dimensions of GHQ namely anxiety and depression, social dysfunction, and loss of confidence (r = 0.85, 0.86, and 0.78; p< 0.01) showed a significant positive relationship with the overall GHQ score. Further, anxiety and depression showed a significant positive relationship with the other two dimensions, namely social dysfunction and loss of confidence (r = 0.53 and 0.56; p < 0.01). Similarly, social dysfunction dimension showed a significant positive relationship with loss of confidence (r = 0.57, p < 0.01).

Table 2. Mean, Standard Deviation, and Inter-Correlation of GHQ Dimensions

Scale	M	SD	1	2	3	4
1. Anxiety and Depression	4.50	2.93	—			
2. Social Dysfunction	6.33	3.02	0.53**	—		
3. Loss of Confidence	1.87	1.64	0.56**	0.57**	—	
4. GHQ total	12.71	6.41	0.85**	0.86**	0.78**	—

Note: GHQ - General Health Questionnaire;
** $p < 0.01$, * $p < 0.05$. (2-tailed)

Table 3 shows the level of psychological distress among the IT sector employees. During the COVID-19 pandemic, low psychological distress is manifested in 22% (n=34) of the employees, moderate psychological distress is manifested in 67% (n=105) and approximately 11% (n=17) of the employees suffer from severe psychological distress.

Table 3. Level of Psychological Distress among IT Employees

Level	Frequency	Percentage
Low Psychological Distress	34	22
Moderate Psychological Distress	105	67
Severe Psychological Distress	17	11

Note. n = 156

Table 4 presents the gender difference in the psychological distress of the employees. It was evident that there is a significant gender difference in the anxiety and depression dimension and the overall GHQ score. Females were found to have higher scores in the anxiety and depression dimension and in the overall GHQ score (t = 2.18, 2.14; $p<0.05$) compared to males.

Table 4. Gender Difference in the Psychological Distress among IT Employees

Variables	Gender	M	SD	t
Anxiety and Depression	Male	3.90	2.71	2.18*
	Female	4.93	3.01	
Social Dysfunction	Male	5.83	2.65	1.79
	Female	6.70	3.22	
Loss of Confidence	Male	1.69	1.61	1.15
	Female	2.00	1.66	
GHQ total	Male	11.43	5.60	2.14*
	Female	13.63	6.82	

Note: GHQ-General Health Questionnaire; * $p < 0.05$. (2-tailed)

The investigator attempted to know whether there is any significant prediction of socio-demographic variables on the psychological distress of the IT sector employees during the COVID-19 pandemic and the results are presented in Table 5. Multiple regression analysis was carried out to examine the prediction of psychological distress. Gender and social media usage showed a significant influence on the psychological distress of IT employees during the COVID-19 pandemic. Further, this analysis revealed that 11% of the variance is influenced by these predictor variables over the psychological distress among IT employees (F=5.98, R=0.33, R^2 = 0.11; $p < 0.05$).

Table 5. Multiple Regression Analysis Predicting Psychological Distress among IT Employees

Dependent Variable	Predictor Variables	B	SE	β	t
GHQ	Gender	2.24	1.01	0.17	2.22*
	Type of Family	1.18	1.08	0.91	1.09
	Area of living	0.78	0.63	0.10	1.23
	Marital status	-1.65	1.27	-0.11	-1.29
	Social media usage	1.61	0.60	0.20	2.68*
R = 0.33 R^2 = 0.11 F = 5.98*					

Note: GHQ - General Health Questionnaire;* $p < 0.05$; (2- tailed)

Discussion

The present study was carried out to assess the psychological distress among IT sector employees during the second wave of the COVID-19 pandemic. It was apparent that the infection of the coronavirus is widespread in the nation with 46% of the study sample being infected which was consistent with the results of Ranjan et al. (2020). In this current study, 67% of the employees are affected with moderate psychological distress and 11% of the employees suffer from severe psychological distress. One interesting finding is that about 21% of the employees spent more than five hours per day using social media which can be attributed to compensating for their interpersonal needs during isolation. Aleman and Sommer (2020) reported that individuals living alone were also recognized as a high-risk group during a pandemic when physical distancing is implemented. There is a significant gender difference in psychological distress and also in anxiety and depression dimensions. Female employees tend to experience higher psychological distress than males. This could be due to the additional obligations demanded by society, where females have to take care of household chores, their children's homeschooling, and other caregiving obligations amid their work-from-home tasks. A similar conclusion was reported in recent research where female employees report anxiety and depression at rates up to twice as high as males with financial difficulties (Li et al., 2020).

A lack of adequate psychological support from the employer can be a risk factor for poor mental health. Especially, among younger people with higher educational backgrounds, job insecurity, lengthy periods of isolation, and future uncertainty exacerbate psychological distress (Giorgi et al., 2020). To avoid psychological distress, companies must pay greater attention to the vulnerable population such as the elderly, women, and migrant workers (Qiu et al., 2020). Various organizational and work-related interventions, such as improving workplace infrastructures, implementing correct and shared anti-contagion measures, and implementing resilience training programs, can all help alleviate this circumstance.

Conclusion

The present study found that a significant number of IT sector employees suffer from moderate to severe psychological distress leading to poor mental health during the pandemic. The findings of the study, suggest the importance of designing and implementing mental health support programs and services for an employee by organizations. These support services should be made available to all the employees of the company even during the phase of working from home to mitigate their psychological distress, potential anxiety, stress, and mental health challenges which ultimately increase their productivity levels.

References

Drydakis, N. (2015). The effect of unemployment on self-reported health and mental health in Greece from 2008 to 2013: a longitudinal study before and during the financial crisis. Soc. Sci. Med. 128, 43–51.

Giorgi, G., Lecca, L. I., Alessio, F., Finstad, G. L., Bondanini, G., Lulli, L. G., Arcangeli, G., and Mucci, N. (2020). COVID-19-related mental health effects in the workplace: a narrative review. Int. J. Environ. Res. Public Health. 17(21), 7857.

Koh, D., and Goh, H. P. (2020). Occupational health responses to COVID-19: What lessons can we learn from SARS? J Occup Health. 62(1), e12128.

Kumar, A., and Nayar, K. R. (2021). COVID-19 and its mental health consequences. J Ment Health. 30(1), 1–2.

Li, Z., Ge, J., Yang, M., Feng, J., Qiao, M., Jiang, R., and Yang, C. (2020). Vicarious traumatization in the general public, members, and non-members of medical teams aiding in

COVID-19 control. Brain Behav. Immun. 88, 916–919.

Mustajab, D., Bauw, A., Rasyid, A., Irawan, A., Akbar, M. A., and Hamid, M. A. (2020). Working from home phenomenon as an effort to prevent COVID-19 attacks and its impacts on work productivity. TIJAB. 4(1), 13–21.

Qiu, J., Shen, B., Zhao, M., Wang, Z., Xie, B., and Xu, Y. (2020). A nationwide survey of psychological distress among Chinese people in the COVID-19 epidemic: implications and policy recommendations. Gen Psychiatr. 33(2), e100213.

Williams, S. N., Armitage, C. J., Tampe, T., and Dienes, K. (2020). Public perceptions and experiences of social distancing and social isolation during the COVID-19 pandemic: A UK-based focus group study. BMJ Open. 10(7), e039334.

Zackiya, S. A., and Venkatachalam, J. (2022). Sleep Quality in Relation with Cyberchondria among Young Adults during COVID-19 Pandemic. ECS Trans. 107(1), 14943–14957.

Triangulation Study on LGBTQ Inclusion with Sustainable Development Goal 10 using Twitter Data and Topic Modelling

V. P. Ganesh Aravindh, Dr. V. B. Kirubanand, and Mohammed Fairos Mirza

Christ University, Bangalore
E-mail: ganesh.aravindh@res.christuniversity.in; kirubanand.vb@christuniversity.in; mohammed.mirza@cs.christuniversity.in

Abstract

Our study deals with sentiment analysis in implementing SDG 10 through an illustration that addresses the problems of the LGBT community. The LGBT community includes lesbians, gays, bisexuals, transgender people, and other queer individuals. Using a full archive API search from Twitter, 1.6 million tweets were extracted across five years about LGBTQ communities and used for this research. It uses Machine Learning (ML) and Natural Language Processing (NLP) to perform the standard text preprocessing and identify sentiment in tweets. Textblob calculates sentiments from tweets. Latent Dirichlet Allocation (LDA) topic modeling to identify hidden topics from a collection of tweets that helps to find issues faced by the LGBT community; This model achieved a state-of-the-art performance since no other similar methods in this domain have been carried out which in turn would help government bodies reduce inequalities and implement SDG 10.

Keywords: NLP, Sentiment Analysis, Topic Modelling, LDA, SDG 10, LGBTQ

Introduction

Every human, regardless of sexual orientation or gender identity, deserves equal access to opportunities and services, and their safety should be ensured. Young people who identify as LGBT, on the other hand, Lesbian, Gay, Bisexual, Transgender, Intersex, or Questioning (LGBTIQ+) are among the most marginalized and socially excluded members Because of their gender, they are particularly vulnerable to stigma and discrimination. actual or perceived sexual orientation and gender identity. Stigma and discrimination can negatively impact the rights of LGBTIQ+ youth. In addition to their fundamental right to live free from violence and discrimination, bullying can lead to an increased likelihood of avoiding school and low personal and academic self-esteem. The United Nations decided to develop a set of global goals to end poverty and inequality by 2030, and equality groups pushed for the rights and needs of lesbian, gay, bi, and trans people to be taken into account. The Sustainable Development Goals (SDGs) were agreed upon and signed by 193 governments in 2015 on the premise that they apply to everyone, everywhere, and will "leave no one behind."

The basis for the inclusion of LGBTI persons within the SDGs is captured by the design of the goals and the formal international treaties and resolutions they are built on. Moving to the goals themselves, SDG 10 focuses on reducing inequalities within and among countries. Target 10.2 states, "by 2030, empower and promote the social economic and political inclusion of all, irrespective of age, sex...or another status," a classification commonly interpreted as including LGBTI persons.

Hence this paper deals with the triangulation study of Inequalities faced by the LGBTQ community with the Sustainable Development Goal 10 integrating Data Science Modelling techniquTor to analyze the opinion of people and the issues faced by LGBTQ communities; social media was considered as the point of the data source. Twitter data was extracted, and the sentiment of the tweets was classified as Positive, Negative, and Neutral with Vader (Valence Aware Dictionary for Sentiment Reasoning model). Most importantly, we have built a Latent Dirichlet Allocation (LDA) with Grid search to perform Topic modeling, which thereby helps in identifying the general topics.

Related Work

The word sentiment has three meanings: Opinion layer, Emotion layer, and Emoticon layer. Sentiment analysis is the field of research that analyses individuals' thoughts, feelings, perceptions, behaviors, and emotions of individuals towards things like such as goods, products, political parties, people, problems, incidents, issues, and their attributes. Sentiment Analyses are carried out at levels like the Document level, Sentence level, and Aspect level. These sentiment analyses are conducted to identify the best sentiment out of the statement for analytic purposes.

Also, topic modeling is another important method of identifying the emotion or sentiment toward a group of people. Topic Models are algorithms for discovering main themes in large and unstructured documents. Topic modeling helps us to identify essential topics from large-scale archived data. LDA is used to determine the significant issues from the corpus. LDA is one of the most popular probability topic models and has proved to have better results when compared with others.

Grid search is one technique that enables the model to be robust by tuning the optimum values in the hyperparameters. Enabling grid search will improve the model performance at the time of execution. Generally, grid search takes more time to compile as it identifies the best parameters.

Need for Study

The equality and non-discrimination guarantee provided by international human rights law applies to all people, regardless of sex, sexual orientation and gender identity, or "other status." In any of our human rights treaties, there is no fine print, no hidden exemption clause that might allow a State to guarantee full rights to some but withhold them from others purely based on sexual orientation and gender identity. Hence identifying the inequalities faced by the LBTQ communities was essential, and therefore we have utilized,

The Vader model used Sentiment Analysis to analyze text and determine the sentiment expressed (positive, negative, or neutral) "(Mandloi and Patel, 2020)." Further, an LDA model was used to achieve Topic modeling to recognize the general topics.

Objectives

Primary Objective

Identifying the most prevailing topics using Latent Dirichlet Allocation (LDA) to classify and categorize the text in a document and the number of words per topic and perform the Grid search algorithm to identify the optimal parameters for the problem to automate the trial-and-error method

Secondary Objective

Categorizing the opinions expressed in the tweets into Positive, Negative, and Neutral with Vader.

Materials and Methods

Data Description

Social media mining from Twitter was utilized to obtain the opinion of people and the issues faced by LGBTQ communities. Twitter Academic Research access has been acquired to download the data. More than 13 million tweets across five years, i.e., 2017 to 2021, have been collected. The data collection was carried out using hashtag search. The search hashtags related to LGBTQ communities were extracted individually and pre-processed.

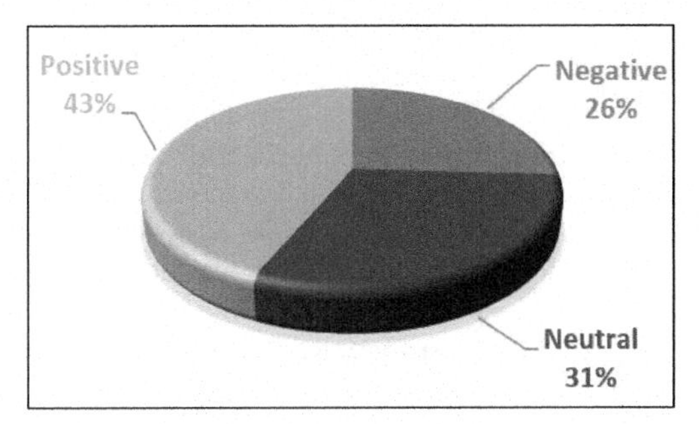

Fig. 1. Classification of downloaded tweets

Data Pre-processing

The majority of Twitter data is highly unstructured. Typos, slang, and grammar errors are possible. These cleaning steps are applied to the documents to generate structured data.

Convert to lowercase letters: Because text analysis is case-sensitive, conversion to lowercase letters is required. The frequency of each letter is counted in the probabilistic model, so "Text" and "text" are treated as different words if case conversion is not used. All of the notes have been converted to lowercase.

Remove @user and links: Twitter has special rules regarding reserved symbols and links that could confuse later analysis. The @ character is used when the user mentions another user to read this tweet (Ikoro et al., 2020).

Remove punctuations and digits: This is a common step in many text mining techniques. Punctuation in sentences improves readability for humans, but a machine cannot distinguish between punctuation and digits. Because text analysis is not concerned with numbers, punctuation is removed. In most cases, numerical digits do not affect the meaning of the text.

Remove stopwords: Stopwords refer to words that usually have no analytic value, such as "a," "and," "the" etc. These words improve human readability while complicating analysis. Depending on the circumstances, comments can be added to the list of stop words.

Remove extra white spaces: The earlier pre-processing steps can generate additional white spaces. Removing the different areas is necessary for text cleansing.

Stem Documents: Topic modeling to rìognize the general topics. Some words in the text have the same meaning but in different forms. Stemming is the process of eliminating affixes from words to convert the terms into their base form; for example, stemming "run," "runs," and "running" into "run."

Model Framework

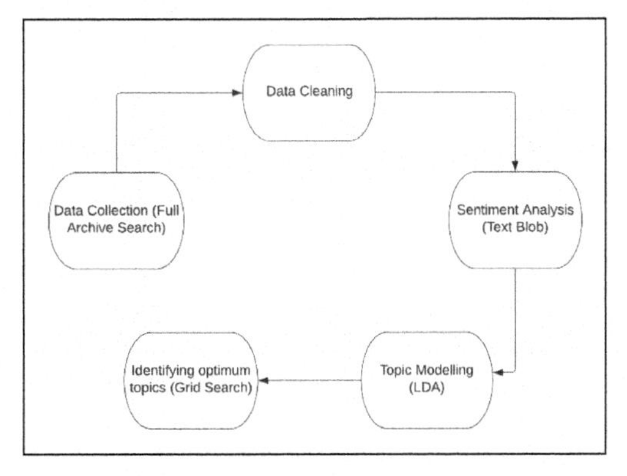

Fig. 2. Block Diagram – significant steps involved in the methodology

Text Mining: Text mining, also called text data mining, is the approach to obtaining high-quality text information. It is used to explore and process extensive unstructured data to extract high-value helpful information to provide insights into a specific scenario like sentiment analysis. Text mining relies on various advanced techniques drawn from statistics, machine learning, and linguistics. Typical text mining includes Text Grouping, Concept Extraction, Sentiment Analysis, and Summaries. This study will use statistical and machine learning techniques to mine meaningful information and explore topic mining.

Topic Model: Topic modeling recognizes words from topics in a document or data corpus. This is useful because extracting words from many tweets takes much longer and is much more complex than removing them from issues within the document.

Topic models can help with organizing, classification, collaborative filtering, and information retrieval. Topic models produce interpretable, semantically coherent topics, which can be investigated by listing the most likely terms for each case (Korencic et al., 2021). Furthermore because they allocate words to subjects based on the context of the tweets, topic models are ideally adapted to provide synonymy (many words with similar meanings) and polysemy (words with numerous meanings). In addition to computing the subjects covered in a collection of tweets, topic models generate a set of individual probabilities that every given tweet in the group is about any of the calculated topics.

Tokenization: Preprocessing input text means transforming the data into a predictable and analyzable format. It decomposes a stream of tweet data into words, terms, sentences, symbols, or other meaningful parts known as tokens.

Lemmatization: Lemmatization generally refers to getting things right through vocabulary and morphological analysis of words, usually with attempts to remove refraction suffixes and prefixes, remove only inflected endings and return a word's base or dictionary form, known as a lemma. Lemmatization is responsible for grouping various inflectional words with the same root form. Example: "communities" become "community," "Meeting" becomes "Meet," "Better," and "Best" becomes "Good." This advantage is that it can reduce the number of unique words in the dictionary.

Count Vectorizer: The count vectorizer is one of the most straightforward and satisfying techniques. It counts the number of times a word appears in the document and uses that value as a weight. Count vectorizer technology provides text document tokenization and builds a lexical of terms. Count vectorizer returns the integer to count words.

Latent Dirichlet Allocation (LDA): LDA is a popular unsupervised clustering technique for text analysis. It is a topic modeling technique in which words are represented as topics and documents as collections of these word topics.

Sampling Topics: LDA will place documents according to the document topics in the space. For example, in our case, with issues relating to LGBTQ (Inequality, Love, Community), LDA will arrange documents into a triangle, with the corners representing the topics. Some documents might have several issues; an example is a document between inequality and gender. For instance, a violence paper can relate to inequality and gender. This kind of distribution is called Dirichlet distribution and is controlled by parameter $\boldsymbol{\alpha}$ "(Yang and Zhang, 2018)'

Sample Words: The second Dirichlet distribution is defined with the parameter $\boldsymbol{\beta}$, which maps the topics in the word space. For instance, consider a few words such as harassment, marriage, group, or similar.

Instead of documenting the parameter $\boldsymbol{\beta}$, place the topic into the space. For example, the issue of love is closer to the word marriage than the word harassment. The multinomial distribution of words for this topic might consist of 73% marriage, 17% group, and 10% harassment. Similarly, for the issue of inequality, the multinomial distribution of words can be defined as 55% harassment, 35% marriage, and 10% group.

The whole above process is mathematically defined as

$$P(W,Z,\theta,\phi;\alpha,\beta)=i=1MP(\theta j;\alpha)i=1KP(\phi;\beta)$$
$$t=1NP(Zj,t|\theta j)P(Wj,t|\phi zj,t),$$

α and β define Dirichlet distributions, θ and ϕ determine multinomial distributions, Z is the vector with topics of all words in all documents, W is the vector with all terms in all copies, M number of records, K number of issues, and N number of words. The entire training process or maximizing probability can be performed using Gibbs sampling. The overall goal is to create each document and word as monochromatic as possible.

Results and Discussion

This analysis gives us a brief understanding of the collected data. LDA topic modeling is applied to analyze the topics hidden in these tweets.

Based on the grid search we ran, it is understood that the learning decay of 0.5 and no. of cases K = 10 is shown as the optimum number of topics in each sentiment category (Positive, Negative, Neutral). The top 10 frequent words in each topic for each sentiment are in Table 1. The results lead to some observations about people's reactions to the LGBTQ Community.

In Topic 1 and Topic 2, the words like "community," "black," "fight," and "hate" shows that people have discussed rising awareness about the community and how they can overcome it showing the positive sentiment from these topics. The words "bully," "abuse," and "crime" in Topics 5 and 6 refer to the violence and abuse the LGBTQ community facing over the years.

Table 1. Top 15 frequent words in top 2 topics for each sentiment

Sentiment	Positive		Neutral		Negative	
Topics	**Topic 1**	**Topic 2**	**Topic 3**	**Topic 4**	**Topic 5**	**Topic 6**
Word 1	people	anti	lgbtq	lgbtq	gay	abuse
Word 2	lgbtq	marriage	trump	gay	force	hate
Word 3	woman	force	protection	video	sex	lgbtq
Word 4	black	condemn	transgender	youtube	bully	gay
Word 5	fight	bully	order	history	anti	crime
Word 6	community	transgender	law	day	fuck	child
Word 7	amp	campaign	worker	new	male	sexual
Word 8	group	lgbtq	federal	check	abuse	man
Word 9	link	male	bathroom	late	transgender	community
Word 10	minority	junta	state	news	tran	amp
Word 11	human	beat	marriage	book	man	assault
Word 12	kill	member	say	music	beat	group
Word 13	hate	abuse	transgend	month	condemn	victim
Word 14	just	law	pass	follow	junta	parent

Topic 3 and Topic 4 show that people are very much involved in supporting and fighting against the inequalities faced by the community; the words "protection," "law," and "marriage" point out and show how people all over the world have a different view on this community. The topic distribution in our model is perfect, as shown in Fig. 3, where the model could distinguish the unique topics properly without overlapping them.

Conclusion

From the research, we can infer that the triangulation study of SDG 10, LGBTQ inclusion with Machine learning, has given desired results as we can identify essential topics across each category of the tweets.

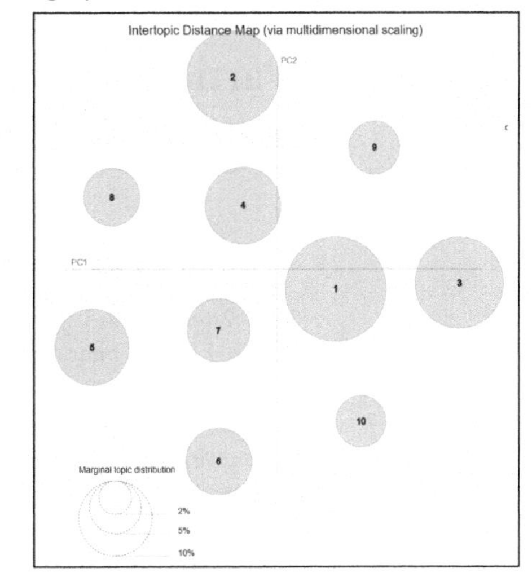

Fig. 3. LDA Topic Distribution

Thus, the research model enables us to identify significant issues /concerns of the LGBTQ community. Topic modeling shows a substantial difference between the topics being discussed, as the intertropical distance map shows that each topic is unique to the cluster.

References

https://plan-international.org/srhr/lgbtiq-inclusion/

Mandloi, L., and Patel, R. (2020). Twitter sentiments analysis using machine learning methods. 2020 International Conference for Emerging Technology, INCET 2020, 1–5.

Ikoro, V., Sharmina, M., Malik, K., and Batista-navarro, R. (2020). 2020 7th International Conference on Social Network Analysis, Management and Security, SNAMS 2020. 2020 7th International Conference on Social Network Analysis, Management and Security, SNAMS 2020, 95–98.

Korencic, D., Ristov, S., Repar, J., and Snajder, J. (2021). A Topic Coverage Approach to Evaluation of Topic Models. IEEE Access, 9, 123280–123312.

Yang, S., and Zhang, H. (2018). Text Mining of Twitter Data Using a Latent Dirichlet Allocation Topic Model and Sentiment Analysis. International Journal of Computer and Information Engineering, 12(7), 525–525

Community Participation in Public Space Planning and Management: Cases of Indian Cities

Bharati Mohapatra[*,a]

[a]Veer Surendra Sai University of Technology, Burla, India
E-mail: [*]bharati_mohapatra@yahoo.com

Abstract

Participation of the community in public space planning and management can be a way to sustainably improve the spatial and social environment of urban areas and develop some social capital in the neighborhood. Quality of place is enhanced by encounters and activities taking place, that generate interaction of people and place, rather than being planned according to stereotype proposals that are not place-specific and even difficult to manage and maintain. Community-based practices and strategies enable decision-makers and planners to encompass people's needs. For spaces to have social, civic, and economic value, they need to be planned and managed from the viewpoint of the community by incorporating people participation approaches. This paper discusses different cases of community participation schemes implemented for developing and managing public spaces in Indian cities. The operational complexities involved in their implementation are brought out in the paper, and improvement measures are suggested.

Keywords: Community Participation, Public Space, Planning, Management

Introduction

Public space is an important developmental objective within the framework of the 2030 UN sustainable development goal 11 and target 7: to provide safe, inclusive, and universal access to green and public spaces, for all sections of the community (UN-Habitat, 2017). Planners and researchers working in city environment, especially public space recognize the fact that the quality of experience one derives from these spaces are dependent on the spatial characteristics of the space, which can be ensured not only through sensitive planning and design but also by devising appropriate strategies for sustainable management of this public realm. *Change in policy,* by adopting an alternative management approach and *change in perception,* primarily among decision-makers, make it possible for Urban Local Bodies (ULB) to build flexible and adaptable management systems and processes and create accessibility to a wider spectrum of people and resources for implementing their initiatives. Co-management offers several new strategies that allow land and resource managers to be more responsive to social concerns. Civic bodies and planning authorities in India have also been trying to adopt alternative management schemes to improve the quality of the urban environment, to meet the challenges of resource mobilization and timely action.

Literature Review

The quality of public spaces and their design is widely researched and discussed in academics and policy making (Hartmann and Jehling, 2018; Muller, 2019, whereas, the effective management and maintenance of public spaces require more debated and alternative methods need to be explored (Duivenvoorden, et al., 2021). Managing public spaces in cities is important in the face of other major challenges ahead, like climate change adaptations, urban mobility, and changing energy demands (Maring and Blauw, 2018). With urban areas undergoing a major transformation, financial resource allocation for the management of public spaces under the fragmented civic bodies involving various departments is crucial. Management of public spaces requires location-specific strategies.

People, public spaces, and the built environment are mutually influenced by each other. People use public spaces and they make the environment livelier which positively influences the surrounding built environment and provides a sense of security and cohesiveness to the community. These interactions between people and spaces, as well as community participation in planning these environments, helps to build more people-friendly public spaces and cities. According to Froud "Better community involvement, or "social participation," is the process of using appropriate methods and tools to explore, negotiate, and very importantly respond appropriately to the question of who gets what, when and how." It has become a socially embedded practice to plan and act on issues related to the living environment and actively engage with the public, instead of only fulfilling abstract interests (Van Herzele, 2004).

The provision of parks in urban neighborhoods is a community development measure and requires the involvement of the residents. Community development initiatives can be identified by: 1. *Ethical Partnership;* facilitating interactions between local communities and public agencies. 2. *Consultation and involvement of the Community;* creating a more representative and participative environment and ensuring that the strategic vision of the public body or agency is mapped with that of the shared interest of the public. 3. *Community leadership;* local authorities play a key role in stimulating bottom-up development through the collaboration for capacity building of the communities (Rogers et al., 1999).

Studies on community participation in urban development projects can be broadly grouped into two categories. The first deliberate about the conceptual standing of community participation as a viable alternative for environmental management. It initiates a dialogue about the extent of validity and essence of community participation, the effectual advantages from the stakeholders, the matter of practicality, the responsibilities and roles in collaborative management, etc. They are often enumerated in a prescriptive tone, like, prescribing the required methods of participation, the stages of community involvement, and the processes involved in putting them to practice (Lepofsky and Fraser, 2002; Lane and Mcdonald, 2005). The second category of literature contains case studies of environmental management projects. The role of community involvement in various environmental programs and different models and methods adopted in projects for the provision, management, and maintenance of environmental resources have been discussed. So much has been written and studied in the field, but many of them have failed in practice, without adequate documentation of the participatory process in the implementation and maintenance of landscape projects.

Residents develop a sense of responsibility and commitment to the project due to their irreplaceable involvement in it. Residents, being the first link in the chain, have experience with the underlying problems of the neighborhood and, consequently, protect the best interests of locals by fostering an in-depth knowledge of local priorities. Contrarily, agencies harbor preconceived ideas when providing funding for redevelopment (Forrest and Kearns, 1999); and the level of community involvement and ownership of strategies are credited for the long-term success and sustainability of change (Carley, 1998).

Community Participation in City Environment Improvement in India

In India, urban planning and management require a multidisciplinary approach for an equitable share of resources to all citizens and to make cities sustainable. The 74th Constitution Amendment Act (CAA) of India aimed to provide local councils with more authoritative control over numerous administrative and financial functions.

In addition, The Good Urban Governance Campaign (2001) was launched by the Indian government in collaboration with the UNCHS, whereby The involvement of local government, civil society, service providers, and other stakeholders is anticipated in interactive discussions. This Campaign envisages improvement in the quality of life of people in cities by reinventing them as inclusive cities.

With urban areas in India accounting for 31.14% of the population (Census, 2011), the demand for improved quality of life has resulted in a strain on physical and social infrastructure. Due to this, the demand for an urban service provider that is effective, efficient, and cost-effective has risen. Basic urban services must be provided, maintained, and upgraded in areas under the jurisdiction of the Urban Local Bodies (ULB), which include Municipal Corporations and Urban Development Authorities. However, due to financial constraints, these ULBs face an acute shortage of resources and capacity on account of their inability to effectively use their revenue-raising power. As a result of which it has become difficult even to maintain the existing social infrastructure at satisfactory levels. This has led to national-level initiatives like the Atal Mission for Rejuvenation and Urban Transformation (AMRUT) and "Jawaharlal Nehru National Urban Renewal Mission" (JNNURM). JNNURM strives to encourage cities to work for the betterment and welfare of the existing service levels in a financially sustainable manner. It has stressed good governance in the municipal context that is based on two principles: Transparency and Civic Engagement and Capacity building measures. The first principle of civic engagement highlights the participatory mechanism, which should be so structured that they have legal standing and administrative power. AMRUT launched in 2015 was also envisaged for urban revival to improve the quality of life by establishing urban infrastructures that can ensure urban transformation. State Annual Action Plans are proposed by the States for implementation of the projects. As per the urban profile of the cities, the ULBs are allocated funds for the execution of different projects. The AMRUT mission guidelines give discretion to the states over how to prioritize the use of inter-ULB funds by setting certain recommendations, wherein there is a guideline for consultation with the elected representatives of the cities. This is how the so-called connection with the community is made through stakeholder consultation with local elected representatives.

In order to improve their urban environments, Indian cities have adopted development strategies that are centered on state action, private sector effort, or direct user participation (in some cases). These models have worked out for certain advantages: stable institutional setup and non-profit measures of the state-cantered actions; the ability to invest freely and more responsiveness to demanding changes in the market-cantered private sector initiatives; and the sensitivity to user needs and public engagement potential of the user-centered approach.

Cases of Community Participation Schemes

The "Bhagidari" scheme, which means a collaborative partnership that is sharing form of partnership, which emphasizes the broader concept of governance rather than an instrument of direct service delivery, was created by the Delhi Government in India. Better service delivery, clean and green Delhi, partnership, and participation of citizens in governance are the goals of the Bhagidari concept. It is a mechanism that facilitates the citizens of Delhi to have interactions with public body officials which enables them to be directly involved in the governance through their representative units like the Resident's Welfare Association (RWA), Market trader's associations, cooperatives, federation, etc. (Maitra, 2005). Bhagidari shifts the system from complete dependence on government officials to decentralizing power and sharing the role of governance. Consequently, this shift sparked a conversation about sharing the responsibilities instead of passing the responsibilities on. Different departments and agencies of the State government, local bodies, and citizen groups are active partners in this scheme. The scheme has become operational in four phases: Planning process, Decentralisation, Empowerment, and Institutionalisation. The activities to be decentralized are decided and concerned officers are identified to monitor and document the process. The RWAs are given the responsibility to carry out development and maintenance activities. The system of rewarding achievers has increased participation. In the third phase, legalizing the RWAs' status to undertake maintenance activities was emphasized. The institutional phase has commenced in January 2004, but there are many issues to be resolved regarding the constitutional standing of the RWAs and unclear political hierarchy that hinders the accountability of Local Bodies and State Departments in the participation process with the local associations like RWA and Merchant Trader's Association (MTA).

For implementing the solutions and strategies brought out in the *Bhagidari* workshops a hierarchical functioning pattern is devised where public utility departments and Deputy Commissioner (Revenue) monitor the implementation of the scheme. Area officers are assigned to collaborate with their area RWAs and the district officers of the public utilities supervise and coordinate the working of these Area officers. Due to this regular interaction, citizen associations and Residential Welfare Associations have come out as dependable and potential partners in shaping policies. The Delhi government scrutinizes the data, ground realities, feedback, opinions, and suggestions and then identifies the "common ground factor" and then finalizes the policy changes. The Delhi Government gives financial and institutional support by creating the "Bhagidari Cell" in the Chief Minister's office and allocating funds for conducting workshops, paying consultants to conduct workshops, and for documentation.

To provide increased green space cover and an enhanced quality of life for Ahmedabad residents, the Ahmedabad Municipal Corporation (AMC) and the Ahmedabad Urban Development Authority (AUDA) have collaborated with the leading corporate house and cooperative sector, to take on the development and maintenance of several parks throughout the city. Ahmedabad has a long history of a breed of entrepreneurs from the powerful mercantile, artisan corporative, and guilds, who have invested in the city as well as participated in its development and management. The business community dominated the local political scene and took interest in civil life during 1950-65 when most of the mayors of the municipal corporation were textile mill owners. But, with changing political scenario the mercantile community gradually lost interest in the quality of civic life. But, with the advent of the 21st century, entrepreneurs have begun to develop a keen interest in the development of community infrastructure, thus manifesting projects like the revival and maintenance of public gardens and urban forestry. There are several positive instances where networking among the NGOs and integration of the community efforts at grass root level has brought advantages to the poor. Self-management capacity of some organizations has proved to be useful in the education and health sectors. However, the dearth of networking among private and public sectors, NGOs, and community-based organizations (CBO) participating in the development process, is persistently evident.

In India, there are numerous instances of public-private partnerships (PPPs), which are typically associated with efforts to privatize the public or government sector. Likewise, The Bangalore Agenda Task Force (BATF), is dedicated to making Bangalore a global city by harboring development projects, modeling cutting-edge city infrastructure, and improving the internal capacity of city corporations and civic agencies Although BATF has been successful in integrating citizen involvement into its agenda, the strategy still needs to take into account, the community sentiments and priorities.

In most of the community participatory schemes operating in India, it is seen that the RWAs have not shown sustained interest in taking care of the recreational facilities, local environment, and monuments, firstly due to lack of expertise in the field, secondly, they are interested more to solve community development issues and thirdly the status of these associations is not yet legalized (Maitra, 2005). Greening the neighborhood, preventing encroachment, and controlling traffic through the residential colony are a few of the issues raised that have an impact on physical planning related to residential areas. This concept involves three broad steps; problems and consensus building, implementation mechanism, and monitoring of implementation. The participation schemes are institutionalized; however, the process structure remains inflexible and ambiguous as per specific location and project. The methods of participation and selection criteria and the role of stakeholders in the process are unclear in most cases.

Operational Complexities

Community-based programs and initiatives for developing public spaces and facilities have to deal with a multitude of networking to ensure coordination and appropriate connection between the stakeholders. Both small-group dynamics and large-group dynamics are employed for arriving at a consensus on solutions in different participatory programs. Bhagidari scheme employed both approaches to planning its actions. Technical limitations were noticed in small group dynamics with 20-30 people as it took a long to create sufficient momentum to facilitate change in the city. Secondly, small groups tend to disperse and fail to sustain the effort, and thirdly, there is less representation of multiple stakeholders from decision-making and implementing agencies. Large group dynamics proved to make the process sustainable as it brings the whole system into one platform by bringing the administrators, implementing agencies, and internal and external stakeholders together to take into account maximum views for an inclusive decision-making process. However, accountability to perform, effective action, and monitoring become difficult when coordinating among a large group becomes difficult. Bhagidari scheme has no constitutional basis. People were concerned that even though Bhagidari began to do good work, all the schemes would be abandoned once the government changed after the elections. The majority of RWA criticize the lack of ability and willingness to put things into action. An RWA cannot force people to act according to the rules. Therefore, some power should be given to the RWA to enforce the idea more practically. In most of the participatory schemes, it is found that there is a lack of networking among NGOs, and private and public sectors including community-based organizations (CBO) participating in the development process. An inherent tension between traditional top-down policy processes and bottom-up development action plans is found to constrain community involvement in public space planning and management. It is also seen that individuals are often more than willing to express their concern for the environment, but this commitment declines when action is requested.

Conclusion

From the literature and case studies, it appears that different viewpoints are held regarding the accomplishment of participation methods. Consequently, opinions on who should participate, in what, and how to vary widely between and among project agencies, politicians, and residents. Several factors have been identified as reasons that explain why support for participation in environmental policymaking varies among the population. These factors include race, political efficacy, information, age, education, and gender as reported in the literature (Johnson, 2009). While a number of factors have been suggested as being important for explaining environmental attitudes, public uncertainty about policy and trust in government and technology are crucial for explaining why citizens identify themselves as environmentalists, but still do not become active in supporting environmental action.

Public willingness to participate depends on factors such as existing controversy; emotions attached to the place, and perceived impact on the community identity. Policymakers' willingness to involve themselves depends on political interests, prior experience with public participation processes, and on their trust in the facilitators of the public participation process. Public willingness to participate depends on factors such as existing controversy; emotions attached to the place, and perceived impact on the community identity. Policymakers' willingness to involve themselves depends on political interests, prior experience with public participation processes, and their trust in the facilitators of the public participation process.

"Public-Private Partnership" in management makes the planning and management process more democratic and user inclusive in many developing countries. For the third-world government, as they cope with resource cutbacks, recession, and organizational adjustment policies, it has become important to identify realistic strategies taking into account the local situation and understand the implications of empowerment for both local communities and implementing agencies. Limits of participation need to be recognized and the involvement of government agencies should be identified to make the process inclusive of state support. A study of different community participation methods and management practices reveals that there is a need for effective management of public spaces and a more integrated strategic action plan to avoid the ineffectiveness of fragmented decisions lacking common consensus practice of managing public spaces and the systematic literature study reveal that first, there is a need in society for effectively managing public space and a demand for a more integral-strategic approach to avoid ineffectiveness of fragmentation, and second, that there is a knowledge gap in academia addressing such management of public space.

References

Carley, M. (1998). Findings—Towards a Long-term Strategic Approach to Urban Regeneration. York, Joseph Rowntree Foundation.

Eva Duivenvoorden, E., Hartmann, T., Brinkhuijsen, M. and Hesselmans, T. (2021). Managing public space – A blind spot of urban planning and design. Cities, 109, 1–3.

Forrest, R., and Kearns, A. (1999). Joined-Up Places? Social Cohesion and Neighbourhood Regeneration, York, Joseph Rowntree Foundation.

Froud, D. (2017). Talking architecture with strangers: why community engagement in development matters more than ever. Ed. Daisy Froud. Making good–shaping places for people Published by Centre for London, UK.

Hartmann, T., and Jehling, M. (2018). From diversity to justice – Unravelling pluralistic rationalities in urban design. Cities, 91.

Johnson, R. J., and Scicchitano, M. J. (2009). Willing and able: explaining individuals' engagement in the environment. Journal of Environmental Planning and Management, 52(6), 833–846.

Lane, Marcus B., and Mcdonald, Geoff (2005). Community-based Environmental Planning: Operational Dilemmas, Planning Principles, and Possible Remedies. Journal of Environmental Planning and Management, 48(5), 709–731.

Lepofsky, Jonathan and Fraser, James C. (2002). Building Community Citizens: Claiming the Right to Place-making in the City. Urban Studies, 40(1), 127–142.

Maitra, Shipra (2005). Decentralized Governance-Delhi's Experience in Participatory City Management. Nagarlok, 2, 8–18.

Muller, A.-L. (2019). Voices in the city. On the role of arts, artists, and urban space for a just city. Cities, 91, 49–57.

Van Herzele, A. (2004). Local knowledge in action. Valuing non-professional reasoning in the planning process, Journal of Planning Education and Research, 24, 197–212.

Surveying Interest and Engagement in Political Discourse

A. Fredrick Ruban*,[1] and Dr. P. V. Arya[2]

[1,2] *Department of English and Cultural Studies, Christ (Deemed to University), Bengaluru, India*
E-mail: *fredrick.ruban@res.christuniversity.in

Abstract

Political discourse has occupied a significant position in the everyday life as human life is interwoven with politics. As Pericles states, politics takes interest in people even if people do not take interest in people. It is evident from this statement that no human can escape from politics and political discussions. Hence, the present research attempts to detect the interest of students in producing political discourse and explores the engagement of the students with the political discourse. The present research adopts a gender-based approach. The survey has been conducted in Bengaluru collecting responses from around three hundred college students, excluding the students of arts programs as they are already into a curriculum that sensitizes them to the political discourse.

Keywords: Political discourse, Interest, Engagement, Undergraduate students

Introduction

Political discourse is a kind of discourse produced by a professional politician or political institution or public in the form of text or talk within a political context. It is treated as the formal exchange of views and opinions shared by the discourse maker to solve societal problems or to recommend suggestions for societal problems. It also functions to persuade the listeners with concrete arguments. Language contributes immensely to the production of political discourse. As for as political discourse is concerned, both the speaker and audience have paramount importance in popularizing the knowledge produced. It can be disseminated by listeners who are interested in political discourse and who are engaged in producing political discourse. In the modern scenario, college students need to have political knowledge for sustainability, especially in metropolitan cities like Bengaluru. Therefore, the present study aims to detect whether the undergraduate students of Bengaluru are interested in political discourse and if they are engaged in producing it. The objectives of the present research are: to find the interest of Undergraduate students in politics, to study the engagement of the Undergraduate students in political discourse, and to compare the political interest and engagement from a gender perspective

Method and Methodology

The present study is a quantitative research using random sampling method. A questionnaire encompassing seven closed-ended questions was shared among the Undergraduate students of Bengaluru and 336 students volunteered to register their responses. Out of 336 students, 199 are male students and 137 are female students. In the questionnaire, questions number: 1, 3, 5 and 7 aim at detecting the "interest" of the students in the political discourse, and questions number: 2, 4, and 6 aim at determining the engagement of the students with the political discourse. The present research employs a gender-based approach to analyze and interpret the data. It makes a comparative analysis to plot the differences and variations present between the male and female responses.

Interest

The entire data analysis and discussion section of the present paper gets divided into two: "interest" and "engagement." The first part deals with the "interest" of the students in the political discourse and the second part discusses the engagement of the students in the political discourse. To begin with the first part of the discussion, it is necessary to consider the first question of the questionnaire as it aims at detecting the interest of the students. The given question is "Does political discourse interest you?" for which the options were "Yes" and "No." Out of 199 male respondents, 146 male respondents have chosen "Yes." It marks that 43.5% of male students have acknowledged the fact that the political discourse interests them. The remaining 53 male respondents, which is 15.85%, selected "No." They were open to register that political discourse does not interest them. The same question was placed in front of 137 female respondents and 81 female respondents have chosen "Yes" and 56 female respondents have chosen "No." In percentage, 24.1% of female students have recorded that political discourse interests them and 16.7% of female students have agreed with the fact that political discourse is not their cup of tea. The majority of the male students have chosen "Yes" and the majority of the female students have chosen "No" to the given question. On a comparison, it can be concluded that the male students are majorly interested in political discourse and female students are not so.

The connecting question to detect the interest of the students in political discourse is "Do you agree that the knowledge of politics is needed for the college students?" for which the given options were "Yes" and "No." This question emphasized the need for political knowledge among undergraduate students. Out of 199 male respondents, 170 respondents choosing "Yes" as their choice agreed that political knowledge is needed for college students. On the contrary, 29 male respondents chose "No" to say that knowledge of politics is not necessary for college students. It has been found that 50.6% of male respondents have agreed that knowledge of politics is needed and 34.8% of female respondents have accepted the need of political knowledge for undergraduate students. On the contrary, 8.6% of male respondents and 6% of female respondents have stated that knowledge of politics is not needed. Comparatively, the majority of male respondents understood the need for political knowledge therefore they selected the choice "Yes" for the given question.

Knowledge is acquired in classrooms whether it may be in virtual classrooms or physical classrooms, which nobody can deny. Since classrooms are one of the places where the act of imparting knowledge takes place, the present research must collect responses in this aspect. The fifth question of the framed questionnaire focused on the aspect of acquiring political knowledge. The given question is "Do you agree that the college classrooms can discuss political issues even when it is minorly related to the prescribed syllabus?" and the options are "Yes" and "No." It is quite surprising to find that 50.3% of male students have chosen "Yes" and have agreed that college classrooms can discuss political issues even when it is minorly related to the prescribed syllabus. This shows that the male students are more inclined toward listening and producing the political discourse whereas the female students were not showing similar interest. Hence, it is found that 32.4% of female students have agreed to have political discussions inside the classrooms and 8.3% of the female students have said "No" to political discussions inside the classrooms. Comparatively, the male respondents are keener to have political discussions inside the classroom.

Acquiring political knowledge can propel students to become politicians and political analysts. To emphasize this idea, the question: "Do you agree that the college students can get involved in politics?" and the choices were "Yes" and "No." Out of 199 male students, 160 students have chosen to say "Yes" to agree with the idea and

39 students have chosen to say "No" to disagree with the idea. Among the 137 female respondents, 103 students selected "Yes" and agreed with the idea of students can get involved in politics and 34 female students disagreed with this idea. It is found that 47.6% of male respondents agreed that students can get themselves in politics and 11.6% of male respondents disagreed with the idea. Likewise, 30.7% of female respondents agreed that undergraduate students can involve in politics and 10.1% of female respondents disagreed with the proposed idea. It can be concluded that the majority of the male respondents agreed with the idea of undergraduate students involving in politics and the female respondents disagreed with it.

Engagement

In this part of the discussion, the data collected to test the engagement of undergraduate students has been interpreted and analyzed. The questions of this section focus on the students' engagement with political discourse. The question: "Do you make political discussions with your friends" was placed in front of both the male and female respondents with the choice "Yes" and "No." out of 199 male respondents, 142 respondents chose "Yes" and 57 respondents chose "No." likewise, out of 137 female respondents, 80 respondents chose the option "Yes" and 57 respondents chose the option "No." It can be interpreted that 42.3% of male respondents have confessed that they make political discussions with their friends and 17% of male respondents have agreed with the fact that they do not make political discussions with their friends. It can be concluded that the male students are more involved in making political discussions rather than female respondents.

To participate in politics, it is necessary to acquire political knowledge and this knowledge can be acquired through mass media. Watching news channels and reading newspapers certainly help a student to acquire this kind of knowledge. There is a need to detect how the respondents are engaged in reading newspapers and watching news channels to acquire political knowledge. Therefore, the fourth question of the questionnaire focused on this aspect. The question placed in front of the respondents was "Do you agree that college students should read newspapers and watch TV news channels daily to acquire political knowledge?" They were asked to choose either "Yes" or "No" as their choice. Among 199 male respondents, 167 students preferred to say, "Yes" and 33 students chose to say, "No." Likewise, among 137 female students, 120 students preferred "Yes" and 17 students chose "No" for the question. It is transparent that 49.7% of male respondents and 35.7% of female respondents have chosen the option "Yes." It is evident that 9.8% of male respondents and 5.1% of female respondents have chosen "No" to say that reading newspapers and watching news channels are not required to acquire political knowledge. Only a meagre group of male and female respondents have felt that watching news channels and reading newspapers are not necessary. Hence, it can be concluded that it is necessary to read newspapers and watch news channels to acquire political knowledge.

Reading newspapers and watching news channels can spur students to participate in politics. It is an activity that keeps them engaged in acquiring political knowledge and thereby conducting discussions. Acquiring political knowledge helps the students to take part in political commentaries and discussions. To detect this aspect, the present research has framed the question: "Is it required to give space for college students to take part in the live political debates conducted by TV channels? and the choices given are "Yes" and "No." Among 199 male respondents, 163 selected "Yes" and 36 selected "No." Similarly, out of 137 female respondents, 115 respondents have chosen "Yes" and 22 respondents have chosen "No." It is evident that 48.5% of male respondents and 34.2% of female respondents have chosen "Yes," and 10.7% of male respondents and 6.5% of female respondents have selected "No." Comparatively, more male respondents have selected "Yes" and agreed that students must be given space to take part in the political debates held by the TV channels. A meagre group of students have felt that students should not be brought into the debates conducted by the TV channels. The study has found that the undergraduate students have felt the need to engage themselves in political activities.

Conclusion

The present study concludes that a majority group of male and female respondents has agreed that they are interested in political discourse. The summary of the finding has been mentioned in the chart pasted below (chart-).

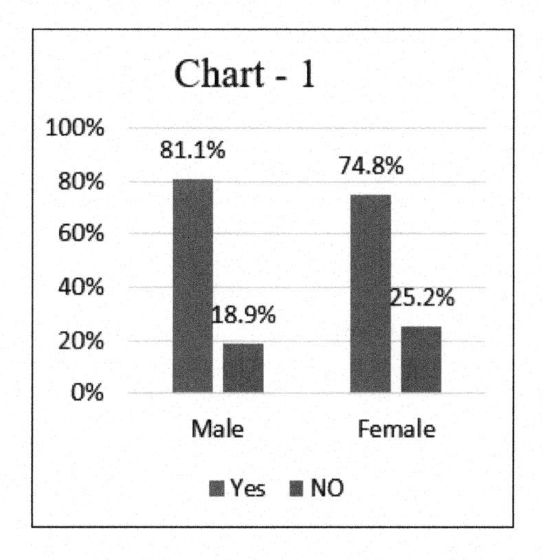

It can be read from the above chart that 81.1% of the male respondents are interested in political discourse and 74.8% of the female respondents are interested in political discourse. This shows that a majority group of students has come forward to say that they are interested in political discourse. They also agree that political knowledge is contributing to their sustainability. It gives them the courage to initiate a conversation with a new peer group and helps them to make more interaction with their friends to strengthen their bonds. It is transparent that 18.9% of male respondents and 25.2% of female respondents have transparently registered that the political discourse does not interest them. It can be concluded that 79% of male respondents have shown their participation in political discourse and 76.6% of female respondents have shown their participation in political discourse. It is evident from the chart that 21% of male respondents and 23.4% of female respondents do not engage with political discourse.

The present study concludes that the majority group of students has agreed that political discourse interests them and a majority group of students has agreed that they engage in producing political discourse. By taking a gender-based approach, it can be concluded that male students are more active in acquiring political knowledge and recreating the political knowledge for sustainability. Comparatively, this group of students are more interested in political discourse and engaged in producing political discourse than the female students.

References

Amaglobeli, G. (2017). Types of Political Discourses and Their Classifications. Journal of Education in Black Sea Region, 3(1), 18–24.

Fairclough, Isabela and Fairclough, Norman (2017). Political Discourse Analysis: A Method for Advanced Students. United States: Routledge.

Griffin, Gabriele. Ed. (2013). Research Methods for English Studies. Edinburgh: Edinburgh UP.

Mills, Sara. (2001). Discourse. Taylor & Francis e-Library.

Pickering, Michael. (2008). Research Methods for Cultural Studies. Edinburgh: Edinburgh UP.

West, Allen B. (1924). Pericles' Political Heirs. Classical Philology. 19(2), 124–46.

Opinion Mining of National Education Policy 2020 to Improve Its Implementation for Women Empowerment

Mausumi Goswami[a] *and Ahini Abraham*[b]

[a,b]CHRIST (Deemed to be University), Bangalore, India
E-mail: *mausumi.goswami@christuniversity.in

Abstract

Opinion Mining is an extremely popular technique to realize the impact of the implementation of any policy. National Education policy 2020 is popularly termed NEP 2020. NEP 2020 is proposed to bridge the gaps in the education system to make it more aligned with sustainable development goals. NEP 2020 promotes education among girls at primary, secondary, and higher education levels. Women empowerment deals with SDG#5. Promoting girls' education through gender equality and women's empowerment is a significant step toward achieving the sustainable development goals SDG 2030 agenda of the United Nations. An artificial intelligence-based framework is discussed in the paper to improve the presence of women through the positive empowerment of women in our society.

Keywords: Opinion Mining, Sentiment Analysis, SDG goals, Artificial intelligence, Deep learning

Introduction

The national education policy was uploaded in August 2020 by MHRD. Women are an integral part of society. To proceed toward sustainable goals it is essential to empower women. NEP 2020 has provided sufficient encouragement to pursue the same. In this Research, Sentiment Analysis is used to identify the opinions of social media users to ascertain the emotions behind the tweets. Sentiment Analysis is performed to identify the possible connection between women's empowerment, places/regions, and overall positive, negative or neutral sentiments. Opinion Mining is an extremely popular technique to realize the impact of the implementation of any policy. National Education policy 2020 is popularly termed NEP 2020. NEP 2020 is proposed to bridge the gaps in the education system to make it more aligned with sustainable development goals, and more employable stakeholders, to promote interdisciplinary study, applications-oriented skill enhancement & outcome-based curriculum, to promote education among girls in primary, secondary, and higher education levels etc. An artificial intelligence-based framework is discussed in the paper to improve the status of women through the positive empowerment of women in our society.

Related Work: Khan et al. (2014) proposed a hybrid classification-based technique built in conjunction with a Twitter opinion mining framework to better evaluate and classify the Twitter feed. Among the polarity classification methods included in the framework were the enhanced emoticon classifier, improved polarity classifier, and SentiWordNet classifier. Rezapour et al. (2017) manually examined the 2016 US election results by annotating the hashtags that were based on a corpus, as well as negation detection testing. The accuracy level increased by 7% as a result of this. Wiegand et al. (2010) presented a research paper on the role of negation in sentiment analysis, as well as various computer approaches for modeling negation. This research focuses on negative word recognition, the scope of negatives, and the limitations and constraints of negative modeling in particular. Without expressly identifying the sentences, Wu et al. (2017) suggested a sentence-level emotional classification technique. Combining two types of weak monitoring with document-level and word-level emotion labels provides an integrated system for learning sentence-level emotion classifiers. Saha et al. (2017) used the text blob method for processing, polarity, and calculation of polarity confidence. The results obtained were validated using Weka with SVM and Naive Bayes algorithms. The rate for Naive Bayes was 65.2%, which was 5.1% higher than the correct answer rate for SVM. The Twitter Streaming API was used by Ibrahim et al. (2015) to collect data on Indonesian presidential election forecasts. The purpose was to use Twitter data to understand public opinion. To do this, the study performed automatic buzzer detection after data collection to remove unwanted tweets, splitting each tweet into multiple sub-tweets and emotionally analyzing them. Based on the political debates and network-based data available on Twitter's timeline, they created a rigorous dataset to identify and politically rank individuals in the 2010 US midterm elections. The characteristics to be analyzed were broadly divided into two categories. (a) Content-based user level. (b) Network level for determining relationships between users. In this work, (Hur et al., 2016) have used three machine learning-based algorithms, such as artificial neural networks, regression trees, and support vector regression, to capture non-linear relationships between box office revenues based on movie review sentiment. Mahalakshmi et al. (2015) showed how to analyze Twitter's emotions using an unsupervised learning approach. SenticNet, SentiWordNet, and Sentis langNet were the three emotional lexicons used to determine the polarity of tweets. In this paper, Wang et al. (2020) proposed an approach for classifying emotions using deep learning models. Using NLP for topic modeling to find key issues related to Covid 19 represented on social media. Classification is performed using the LSTM Recurrent Neural Network (LSTM RNN) model. When studying textual sentiment analysis, continuous relationships between words are important. The research work done on sentiment analysis using deep learning is analyzed with the year of publication and number of citations received. The year 2010 and 2011 saw a huge increase in the work on sentiment analysis but usually on the traditional methods. There was an increase in the work on sentiment analysis from 2014 hence the citations were being split and a lower amount of citations. The most amount of work was noticed in 2017 and 2018. Figure 1 demonstrates the proposed framework to achieve SDG goals through women empowerment and NEP implementation. Figures 2 and 3 show the connection between empowerment and SDG 4,5. Figure 3 emphasizes the various suggested initiatives to promote women's empowerment. The suggested strategies and directives may help a nation to build its women as a strength of the country. These are mentioned only as directives. The suggested measures shall be within the national laws, policies, and guidelines.

Proposed Methodology: The accumulated research is analyzed to introspect the popularity of machine learning applications used for sentiment analysis. A few deep learning applications are found to have had a very huge impact on the scientific community from the year 2013, and 1997. Women empowerment-related research is not reported in this analysis due to the lack of application of deep learning in this area. National Education Policy has included many strategies to improve the status of girls in India in terms of empowering them and encouraging them through various policies floated targeting women empowerment. Figure 1 has included the proposed framework which emphasizes the usage of sentiment analysis to promote woman's empowerment. The steps involved are mentioned in the action. The output indicates tangible output. The outcome represents the non-tangible output.

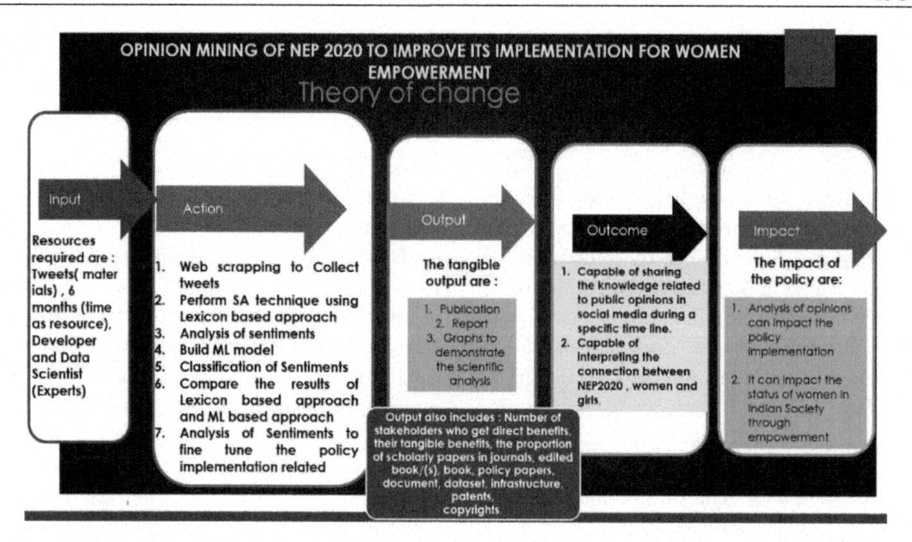

Fig. 1. Proposed framework

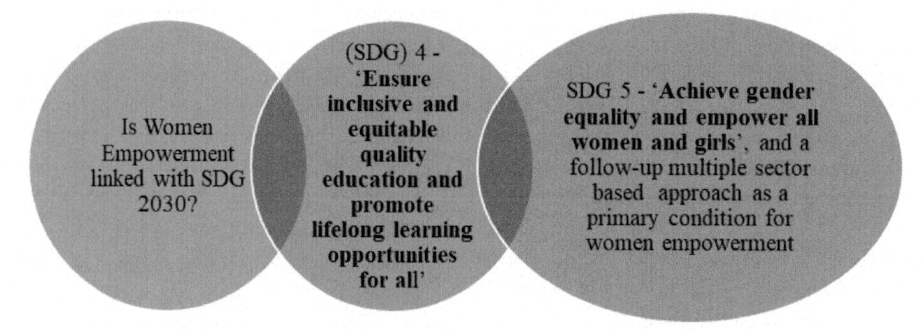

Fig. 2. Connection between Women empowerment and SDG

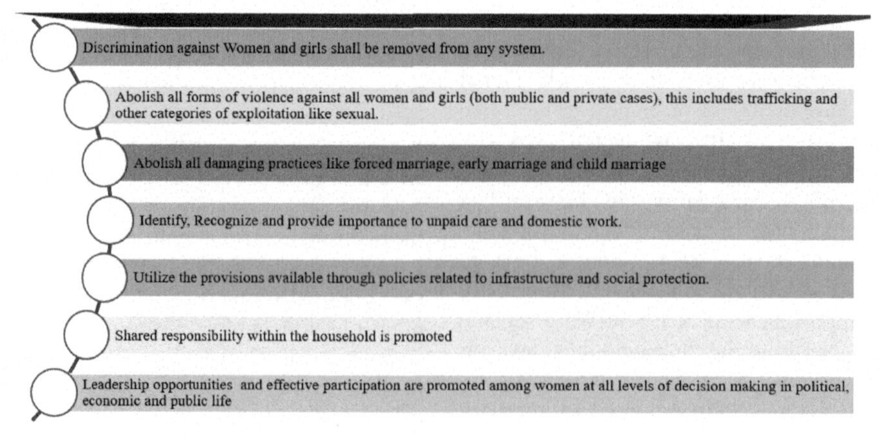

Fig. 3. Women Empowerment and SDG5

Table 1. Years with the corresponding citations

Year	Citations
2017	12
2017	15
2018	23
2017	34
2017	39
2017	40
2016	69
2015	71
2021	96
2015	116
2016	119
2020	150
2018	281
2014	290
2010	329
2014	345
2013	380
2011	685
2018	1065
2017	1083

Table 2. Top Most Years paper with the corresponding citations

Year	Citations
2013	28419
2013	33155
1997	64439

Algorithm: Sentiment Analysis using TextBlob

Input: Text

Output: Polarity and Subjectivity

The steps of Sentiment Analysis using TextBlob Algorithm are:

The collected data is preprocessed. NLTK library TextBlob is called.

The data is now processed using the function sentiment to find the polarity and subjectivity of the sentence. To find subjectivity and polarity it uses individual word scores and takes the average of that. The polarity says whether the sentence is positive or negative with a high score and low score respectively. The subjectivity score with a high value says that the sentence is of high score and the low score is objective.

Conclusion: This paper proposes a framework for women's empowerment through the usage of artificial intelligence. Artificial intelligence has created an impact on all aspects of social life. Social change through empowering women can create an enormous impact on the GDP of the country. This framework will help to improve the tangible and in tangible output of the work. The impact of this work can create further avenues and scope of opportunities to improve the condition of women in society at large. This is a part of ongoing research. Future works will report a few findings of the research through national education policy implementation.

References

Khan, Farhan Hassan, Bashir, Saba, and Qamar, Usman (2014). TOM: Twitter opinion mining framework using hybrid classification scheme. Decision support systems, 57, 245–257.

Rezapour, Rezvaneh, Wang, Lufan, Abdar, Omid, and Diesner, Jana (2017). Identifying the overlap between election result and candidates' ranking based on hashtag-enhanced, lexicon-based sentiment analysis. In 2017 IEEE 11th International Conference on Semantic Computing (ICSC), pp. 93–96. IEEE.

Wiegand, Michael, Balahur, Alexandra, Roth, Benjamin, Klakow, Dietrich, and Montoyo, Andrés (2010). A survey on the role of negation in sentiment analysis. In Proceedings of the workshop on negation and speculation in natural language processing, pp. 60–68.

Wu, Fangzhao, Zhang, Jia, Yuan, Zhigang, Wu, Sixing, Huang, Yongfeng, and Yan, Jun (2017). "Sentence-level sentiment classification with weak supervision. In Proceedings of the 40th international ACM SIGIR conference on research and development in information retrieval, pp. 973–976.

Saha, Shubhodip, Yadav, Jainath, and Ranjan, Prabhat (2017). Proposed approach for sarcasm detection in twitter. Indian Journal of Science and Technology, 10(25), 1–8.

Ibrahim, Mochamad, Abdillah, Omar, Wicaksono, Alfan F., and Adriani, Mirna. Buzzer detection and sentiment analysis for predicting presidential election results in a twitter nation. In 2015 IEEE international conference on data mining workshop (ICDMW), pp. 1348–1353. IEEE.

Conover, Michael D., Gonçalves, Bruno, Ratkiewicz, Jacob, Flammini, Alessandro, and Menczer, Filippo. Predicting the political alignment of twitter users. In 2011 IEEE third international conference on privacy, security, risk and trust and 2011 IEEE third international conference on social computing, pp. 192–199. IEEE, 2011.

Hur, Minhoe, Kang, Pilsung, and Cho, Sungzoon (2016). Box-office forecasting based on sentiments of movie reviews and Independent subspace method. Information Sciences, 372, 608–624.

Pandarachalil, Rafeeque, Sendhilkumar, Selvaraju, and Mahalakshmi, G. S. (2015). Twitter sentiment analysis for large-scale data: an unsupervised approach. Cognitive computation, 7(2), 254–262.

Jelodar, Hamed, Wang, Yongli, Orji, Rita, and Huang, Shucheng (2020). Deep sentiment classification and topic discovery on novel coronavirus or COVID-19 online discussions: NLP using LSTM recurrent neural network approach. IEEE Journal of Biomedical and Health Informatics, 24(10), 2733–2742.

Factors Affecting Entrepreneurship Intention: An Empirical Study with Reference to Indian University Students

*Anshruta M. Rasania[a] and Dr. Namita Kapoor[a]**

[a]Amity School of Economics, Amity University, Noida, India
E-mail: *namita3112@gmail.com

Abstract

This study proposes to examine the implications of multiple factors like Attitude towards Entrepreneurship (ATE), Perceived Behavioural Control (PBC), and Subjective Norm (SN) on entrepreneurial intention (EI) among Indian university students. The study uses feedback from 192 Indian University Students who have self-reported through a structured questionnaire. Partial Least Square-SEM (PLS-SEM) was applied to assess the study's model and the postulated hypothesis. It shows the effects of study variables on Entrepreneurship Intention (EI). Attitude Towards Entrepreneurship (ATE) catalyzes the relationship between Entrepreneurship Education (EE) and EI, while Perceived Behavioural Control (PBC) notably arbitrates the relationship between EE & EI along with SN & EI. India's Education Ministry should consider designing university programs that lead to more influential EE thus promoting entrepreneurship. Subsequent evaluation should consider the impact of Theory of Planned Behaviour (TPB) on entrepreneurial activities in rural areas of India. This study proposes to the Indian Government to make entrepreneurship education a compulsory course in Indian schools to influence youth's attitude toward entrepreneurship.

Keywords: PLS-SEM, Attitude Towards Entrepreneurship, Perceived Behavioural Control, Subjective Norm, Entrepreneurial Intention

Introduction

Stimulating the entrepreneurial spirit is critical for the burgeoning of any economic system. Consideration should be given to entrepreneurship when generating new jobs, expanding existing ones, and driving the economy. Since India is in the midst of a start-up boom, over doubling its unicorns, increasing the country's official tally of unicorns to 54 and displacing the united kingdom to 3rd place ("India displaces the UK to be third top country hosting unicorns." 24th December 2022. The Economic Times), we can perspicuously notice the indispensable role of entrepreneurs and startups in bringing in economic changes and advancements to a country's economy. The following are the objectives of the study:

- To examine the effect of EE and training on students' EI and dreams.
- To identify the impact of other variables like SN, ATE, and PBC on the EI of university undergraduate students in India.
- To assess the mediating effect of ATE and PBC between EE and EI in increasing innovation intent and enthusiasm amongst students.

Theoretical Background

Intentions are still thought of as an accurate predictor of human behavior (Krueger, 2008 (1)). According to (Ajzen, 1991 (2)), an individual's intentions are determined by three antecedents: (a) attitude towards the act, which is the degree to which a person has a favorable or unfavorable appraisal of the behavior in question, (b) subjective norm, which means the degree to which a person perceives social pressure to execute or not execute the behavior, and (c) perceived behavioral control, which is the perceived ease or difficulty of implementing the behavior in question (Armitage and Conner, 2001 (3); Shyti and Paraschiv, 2015 (4)).

- H1: ATE is positively affecting entrepreneurial intention.
- H2: SN is affecting entrepreneurial intention positively.
- H3: PBC is positively affecting EI.
- H4: EE is positively affecting one's ATE.
- H5: EE is positively affecting PBC.
- H6: ATE mediates the relationship between EE and EI.
- H7: PBC mediates the relationship between EE and EI.

The creation of the above-mentioned hypotheses was followed by an exploratory survey. For this purpose, a questionnaire was created and floated to be filled out by the respondents. Responses from 192 individuals were collected.

Review of Literature

TPB was founded by Icek Ajzen (1991). In a nutshell, TPB consists of three elements that predict the formation of intention: (a) attitude toward the behavior, (b) subjective standards and norms, and (c) perceived behavioral control (self-efficacy). It claims that all behavior requires some preparation and that the intention to engage in that behavior can be predicted.

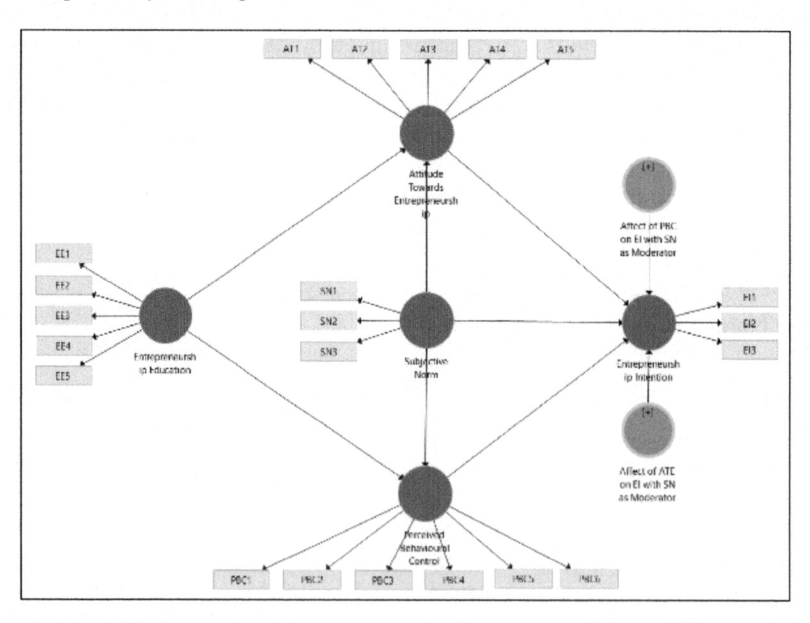

Fig. 1. Smart-PLS Path Model

As per experts, people with distinguishing features or characteristics are more likely to pursue entrepreneurship, while other scholars contend that the necessary knowledge and expertise, the skill of entrepreneurship can be mastered. Personality, parenting, entrepreneurial competence, interest (Rodermund, 2004 (5)), and demographic factors (Nga and Shamuganathan, 2010 (6)) have all been studied to see how they affect entrepreneurship.

An essential aspect of this study is to look into factors that predict EI. Knowing what influences a person's decision to become an entrepreneur will help accelerate business and aid policy research. While research on EI's background is emerging, the decision-making mechanism that promotes entrepreneurial behavior remains an open research problem (Fallah et al., 2018 (7)). EI among students is a novel research area that remains untraversed despite its potential to provide insights into different techniques for establishing new businesses, especially in developing countries (Zreen et al., 2019 (8)).

(Shankar, 2012 (9), as referenced by Basu, 2014 (10)), identifies 6 barriers to EE in India. There is standardization deficiencies, as well as a scarcity of homegrown experience and educated teachers who may serve as mentors. To deal with such issues, Basu (2014) recommends entrepreneurship be taught as a fundamental course.

Research Methodology

In this inquiry, a quantitative technique has been made use of since it aims to discover empirical evidence for assumptions drawn from previous literature. The data and outcomes of the investigation will foretell the relationship between the predicted factors. The questionnaire was written in the English language. It had two sections: an introduction and the main body of the questionnaire which respondents were to rate how much they agreed or disagreed with statements about their psychological characteristics (financial independence, autonomy, achievement orientation, and uncertainty avoidance), non-psychological characteristics (entrepreneurial experience and entrepreneurial knowledge), and environmental conditions (external context). The data was anatomized using SmartPLS to assess the validity (convergent and discriminant) and reliability.

Data Analysis and Discussion

The questionnaire designed had demographic questions about the respondent's gender, age, marital status, parent's employment status, nationality, and family income level.

The questionnaire was based on the Likert scale and the respondents were to rate how much they agreed or disagreed with certain psychological characteristics (financial independence, autonomy, achievement orientation, and uncertainty avoidance), non-psychological characteristics (entrepreneurial experience and entrepreneurial knowledge), and environmental conditions (external context and university environment).

Table 1. Indicator Reliability, Composite Reliability, AVE

Construct Indicators	rho_A	Composite Reliability	AVE
EI	0.7685	0.8580	0.6690
ATE	0.8231	0.8626	0.5576
SN	0.8572	0.9076	0.7662
PBC	0.8243	0.8661	0.5222
EE	0.7844	0.8366	0.5095

Source: Author's Calculation

Table 2. Fornell Larcker Ratio, Cronbach's Alpha

	Cronbach's Alpha	ATE	EE	EI	PBC	SN
ATE	0.8061	0.7467				
EE	0.7611	0.6992	0.7138			
EI	0.7519	0.6115	0.5521	0.8179		
PBC	0.8139	0.5209	0.5349	0.6078	0.7226	
SN	0.8483	0.2701	0.3580	0.3331	0.5076	0.8753

Source: Author's Calculation

Table 3. Heterotrait-Monotrait Ratio (HTMT)

	ATE	EE	EI	PBC	SN
ATE					
EE	0.8388				
EI	0.7488	0.6839			
PBC	0.5972	0.6450	0.7504		
SN	0.3139	0.4450	0.4097	0.6265	

Source: Author's Calculation

Table 4. VIF

	ATE	EE	EI	PBC	SN
ATE			1.3855		
EE	1.1470			1.1470	
PBC			1.7889		
SN	1.1470		1.3900	1.1470	

Source: Author's Calculation

Findings and Conclusions

The primary goal of this research was to discern the factors affecting the entrepreneurship intention of Indian university students. The outcomes of this study show that EE is positively associated with EI amongst Indian university students. Furthermore, ATE was reported to have a full mediation impact on EE and EI, but the mediation impact of ATE is not significant SN and EI. PBC, on the contrary, has a significant mediation impact on both EE and EI as well as SN and EI.

Table 5. Final Analysis

Hypothesis	Statement	T Statistics	P Value	Decision
H1	Attitude towards entrepreneurial behaviour is positively affecting entrepreneurial intention.	5.8208	0.0000	Accepted
H2	Subjective norm is affecting entrepreneurial intension positively.	2.3656	0.0184	Accepted
H3	Perceived Behavioural Control/ Self-Efficacy is positively affecting entrepreneurial intention.	4.4767	0.0000	Accepted
H4	Entrepreneurship education is positively affecting one's attitude towards entrepreneurship.	8.5941	0.0000	Accepted
H5	Entrepreneurship education is positively affecting perceived behavioural control.	17.1232	0.0000	Accepted
H6	ATE mediates the relationship between EE and EI.	5.4058	0.0000	Accepted
H7	PBC mediates the relationship between EE and EI.	3.1712	0.0016	Accepted

Source: Author's Calculation

Suggestions

This study, therefore, presents a specific contribution to the TPB by empirically assisting in the determination of the associations between ATE, EE, and EI. The literature verifies the presence and the influence of entrepreneurship, the principal results of this study, which were based on Indian data, were unique in the sense that they may reflect the perspective of Asia's emerging nations with youth as its main workforce. The study implies that collaboration with organizations and investment in educational R&D is required to inculcate entrepreneurial spirit. The curriculum/ workshops/ guest lectures should be based more focused on the needs and aspirations of students and "one fits all" thinking should be eradicated The study emphasizes the role played by mentors in encouraging and motivating students to build incubators on campus.

References

Krueger, N. F., and Carsrud, A. L. (1993). Entrepreneurial intentions: Applying the theory of planned behaviour. Entrepreneurship & regional development, 5(4), 315-330. https://doi.org/10.1080/08985629300000020

Ajzen, I. (1991). The theory of planned behavior. Organizational behavior and human decision processes, 50(2), 179-211. https://doi.org/10.1016/0749-5978(91)90020-T

Armitage, C. J., and Conner, M. (2001). Efficacy of the theory of planned behaviour: A meta-analytic review. British journal of social psychology, 40(4), 471–499.

Shyti, A.; Paraschiv, C. (2014). Does Entrepreneurial Experience Affect Risk and Ambiguity Attitudes? An Experimental Study. https://doi.org/10.5465/ambpp.2015.79

Schmitt-Rodermund, E. (2004). Pathways to Successful Entrepreneurship: Parenting, Personality, Early Entrepreneurial Competence, and Interests. Journal of Vocational Behavior, 65(3), 498–518. doi: 10.1016/j.jvb.2003.10.007

Koe Hwee Nga, J. and Shamuganathan, G. (2010). The Influence of Personality Traits and Demographic Factors on Social Entrepreneurship Start Up Intentions. Journal of Business Ethics, 95(2), pp.259–282. doi: 10.1007/s10551-009-0358-8

Fallah Haghighi, N., Mahmoudi, M. and Bijani, M. (2018). Barriers to Entrepreneurship Development in Iran's Higher Education: A Qualitative Case Study. Interchange, 49(3), pp. 353–375. https://doi.org/10.1007/s10780-018-9330-9

Aneeqa, Zreen, Muhammad, Farrukh, Nida, Nazar, Rimsha, Khalid (2019). The Role of Internship and Business Incubation Programs in Forming Entrepreneurial Intentions: an Empirical Analysis from Pakistan. Journal of Management and Business. doi: 10.7206/jmba.ce.2450-7814.255

Shankar, R. (2012). Entrepreneurship: Theory and Practice. India: Tata McGraw Hill. ISBN818209269 (ISBN13: 9788182092693)

Basu, R. (2014). Entrepreneurship Education in India: A Critical Assessment and a Proposed Framework. Technology Innovation Management Review, 4(8), pp.5–10. doi:10.22215/timreview/817

Criminogenic Cognition of Juveniles in Conflict with the Law and Use of the Internet with the Victim-Offender Overlap

O. S. Athira,[*,a] *D. V. Nithyanandan,*[b] *and Sarbjit Kaur*[c]

[a]Ph. D. Research scholar, Periyar University, Salem
[b]Assistant Professor, Department of psychology, Periyar university, Salem
[c]Cybercrime investigator, BPR&D, New-Delhi
E-mail: [*]athiraos96@gmail.com

Abstract

Theory and research in social psychology emphasize the significance of particular cognitive processes in promoting and maintaining morally dubious behavior. Children's internet use is frequently portrayed as a sign of society's advancement and a call to digital predators and criminals. The relationship between victimization and the offending of crime and delinquency is known as the victim-offender overlap. This study aimed to explore the cognitive thinking styles and technology use with juvenile victimization and offending overlap in conflict with the law. This study adopted a survey method that is descriptive and associational across Delhi from various observational homes. Around 120 samples. The result shows a significant influence of victim-offender overlap and internet use on criminogenic cognition among juveniles in conflict law. Overall, this study contends that cognitive capacity is a significant criminogenic risk factor with significant ramifications for both the development of structural and multilevel theories of crime as well as corrective solutions.

Keywords: Criminogenic cognition, Internet use, Victim-offender overlap, JUVENILE in conflict with law

Introduction

Children and young adults have been recognized as essential human resources for growth and critical societal change agents. However, it can only happen if they engage in meaningful discussions about the topics most matter to them. Adolescence is a challenging and brief stage of life. Teenagers frequently act impulsively, taking risks without stopping to contemplate the long-term effects of their decisions. Most people gave up committing crimes as a regular part of maturing and growing older. Their delinquency resulted from the interactions between their environment's influence and their own emotional, psychological, cognitive, and brain development (Romer, 2010).

According to the Bureau, 28,830 of the 43,506 crimes against children were perpetrated by juveniles in 2019 that were punishable by the Indian Penal Code (IPC), 1860, and people committed the Special Local Law (SLL) in the age above group. These statistics called attention to the psychosocial problems facing young people who had run afoul of the law in India (NCRB, 2019).

The cognitive model of crime and delinquency carries that criminal behavior is associated with cognitive deficits. Many offenders have had developmental delays in developing several cognitive skills essential to social adaptation. Their cognitive deficits place them at risk for criminal adjustment. Criminogenic cognitions are thinking processes likely to weaken the connection between standards and behavior (Tangney, Stuewig, and Mashek, 2007).

Much information about the connection between victimization and crime is often known as victim-offender overlap. Although crime victims don't always turn into offenders, most offenders have been victims. People who have experienced victimization may develop unfavorable physical, mental, and behavioral traits and some may even commit crimes (Jennings, 2012).

The Internet, a global network of interconnected information systems, has transformed almost every area of human life (Shekhawat and Rathore, 2014). Children typically learn to use computers more quickly and effectively than adults. The majority of them own cell phones, which are essentially portable computers. Young children like to enjoy exploring and trying new things. However, their exploration and experimenting may take them to places prohibited by law and transform them into criminals, sometimes without their knowledge. One effect of this can be seen is shown in the criminal behavior of young people. Most violent scenes for sale are geared toward young people, and many of them can be found online for the lowest prices in the form of video games, music videos, and the like.

Because none of this research evaluated adolescents with overlap between victim and offender and internet use, there is a knowledge gap in this area. Theoretically, this research will contribute to the body of knowledge already available on how information and communication technologies affect juvenile criminality.

The current study aims to determine how victim-offender overlap, internet use, and criminogenic cognition relate to juvenile offenders.

Objectives

To understand the role of demographic factors such as gender and socioeconomic status on criminogenic cognition of juveniles in conflict with the law.

To explore the relationship between criminogenic cognition, internet use, and victimization-offending overlap among juveniles in conflict with the law.

Tools Used:

The following tools have been used to collect data along with demographic data sheet.

I. Criminogenic Cognitions Scale (CCS; Tangney et al., 2006).
II. The Compulsive Internet Use Scale (CIUS; Meerkrek et al., 2011).
III. The Illinois Bully scale (IBS; Espelage & Holt, 2001).

Table 1. *t*-test results comparing males and females on criminogenic cognition.

	G	N	M	SD	T
CC	M	61	54.95	11.90	10.68
	F	59	72.11	3.33	

*Significant at 0.05 level; CC = criminogenic cognition

Table 2. intercorrelation between criminogenic cognition, victim-offender overlap, and internet use

	CC	VOO	IU
CC	-		
VOO	0.917**	-	
IU	0.330**	0.364**	-

*Significant at 0.05 level; CC = criminogenic cognition; VOO = victim-offender overlap; IU = internet use

Table 2 shows that a significant, positive correlation between criminogenic cognition and victim-offender overlap indicated those who showed higher victimization and offending reported high criminogenic cognition, r (118) = 0.917, p = 0.01.

Table 3. effect of victim-offender overlaps on criminogenic cognition.

	B	R^2	F	T
VOO-CC	0.82	0.84	621.1	24.9

*Significant at 0.05 level

From Table 3 it is found that the victim-offender overlap significantly predicted criminogenic cognition, $\beta = 0.82$, $t\,(118) = 24.9$, $p < 0.005$. Victim-offender overlap also explained a significant proportion of variance in criminogenic cognition, $R2 = 0.84$, $F\,(1, 118) = 621.1$, $p < 0.005$.

Major Findings

- Female juveniles in conflict with the law show higher criminogenic cognition than male juveniles in conflict with the law.
- Poor socio-economic status children in conflict law show higher criminogenic cognition than middle socio-economic status juveniles in conflict with the law.
- There is a significant relationship between criminogenic cognition, victim-offender overlap, and internet use among children in conflict with the law.
- There is a significant influence of internet use on criminogenic cognition among juveniles in conflict with the law.
- There is a significant influence of victimization-offending overlap on criminogenic cognition among juveniles in conflict with the law.

Conclusion

This study explores the connection between cognitive ability and delinquent behavior within a sample of juveniles in conflict with the law, drawing on the growing life-course paradigm in criminological theory. Overall, the findings of this study point to the importance of cognitive capacity as a criminogenic risk factor, with significant implications for both correctional interventions and the ongoing development of structural and multilevel theories of crime.

The results prove beyond a shadow of a doubt that post-injury services are rarely provided to violent crime victims who fall under the victim-offender overlap. In many cases, victims are unable to seek assistance because of their ignorance. The study report underlined the necessity for official institutions, including the criminal justice system, healthcare providers, and mental health service providers, to provide victim support. According to the findings, the alignment of those institutions with an adversarial, dichotomized model, typically unsupportive of "bad" offenders even when they are crime victims, may suppress that support.

Practically speaking, this study will aid in the understanding of the need to address the issue of the effects of information and communication technology on juvenile delinquency by the government, social welfare unit, police, teachers, parents, students, social workers, and community development workers. Additionally, it is intended that the study would help these stakeholders in their efforts to develop policies that will lessen the impact of information and communication technology on adolescent delinquency. This study will also motivate groups, both governmental and non-governmental, to plan campaigns, conferences, and seminars aimed at minimizing the impact of information and communication technology on adolescent criminality.

Acknowledgment: The Bureau of Police Research and Development, New Delhi, provided funding for this effort. This research has no funding.

Disclosure of potential conflicts of interest: The authors have none to disclose. The manuscript's contents have been reviewed and approved by each co-author, and there are no competing financial interests to disclose. We attest that the submission is unique and is not already being considered by another publisher.

Ethics approval statement: Ethical clearance was obtained from juvenile justice board 1, New Delhi, to visit the observational homes for data collection. The principles for ethical conduct in research, outlined by the Bureau of police research and development, New Delhi, were applied in all aspects of this study.

References

Baglivio, M. T., Epps, N., Swartz, K., Huq, M. S., and Hardt, N. S. (2014). The prevalence of Adverse Childhood Experiences (ACE) in the lives of juvenile offenders. Journal of Juvenile Justice, 3, 1–23.

Bandura, A. (1990). Selective activation and disengagement of moral control. Journal of social issues, 46(1), 27–46.

Dierkhising, C. B., Ko, S. J., Woods-Jaeger, B., Briggs, E. C., Lee, R., and Pynoos, R. S. (2013). Trauma histories among justice-involved youth: findings from the National Child Traumatic Stress Network. European Journal of psychtraumatology, 4,10.3402/ejpt.v4i0.20274.

Espelage, D. L., and Holt, M. (2001). Bulling and Victimization during early adolescence: Peer influences and psychosocial correlates. Journal of Emotional Abuse, 2, 123–142.

Jennings, W. G., Piquero, A. R., and Reingle, J. M. (2012). On the overlap between victimization and offending: A review of the literature. Aggression and Violent Behavior, 17, 16–26.

Meerkerk, G. J., Van Den Eijnden, R. J., Vermulst, A. A., and Garretsen, H. F. (2009). The Compulsive Internet Use Scale (CIUS): some psychometric properties. Cyberpsychology & behavior: the impact of the Internet, multimedia and virtual reality on behavior and society, 12(1), 1–6. https://doi.org/10.1089/cpb.2008.0181.

Romer, D. (2010). Adolescent risk taking, impulsivity, and brain development: implications for prevention. Developmental psychobiology, 52(3), 263–276 https://doi.org/10.1002/dev.20442.

Shekhawat, D., and Rathore, P. (2014). Internet Usage in College: A Comparison of Users and Non-Users about Self Esteem and Satisfaction with Life. Indian Journal Of Health And Wellbeing, 5(3), 335–340.

Tangney, J. P., Stuewig, J., and Mashek, D. J. (2007). Moral emotions and moral behavior. Annu. Rev. Psychol., 58, 345–372.

Tangney, J. P., Stuewig, J., Mashek, D., and Hastings, M. (2011). Assessing jail inmates' proneness to shame and guilt: Feeling bad about the behavior or the self?. Criminal justice and behavior, 38(7), 710–734.

Testifying Legal Admissibility: Germline and Embryo Editing Focusing on SDG 15-Life on Earth

Dr. Bhupinder Singh[a] *and Arunima Shastri*[b]

[a]Associate Professor, NMIMS Deemed to be University, Chandigarh Campus
[b]Assistant Professor, CHRIST (Deemed to be University) Delhi- NCR
E-mail: talwandibss@gmail.com

Abstract

The expansion of germline and gene editing has sparked a vigorous debate over how severely it can interfere with human DNA throughout the world. the most significant issue that needs to be tackled in the future. creating very specific legal guidelines for who is responsible for harm brought on by gene editing in human embryos and reproductive cells. It is necessary to demonstrate potential legal issues that could occur and that will certainly appear in the future legal sphere. SDG 15-Life on Earth is connected with "Protect, restore and promote sustainable use of terrestrial ecosystems, sustainably manage forests, combat desertification, and halt and reverse land degradation and halt biodiversity loss." Legal problem faced by people is the issue of whether modifying human reproductive cells' and embryos' genes is ethically acceptable. Over the past few years, revelations about genetic engineering and embryo editing have been the focus of discussions and medical research. Certainly, the reason being it enables the precise replacement of a broken gene with a precise gene. Compared to previous methods, this one is significantly faster, easier, cheaper, and safer. Plants, animals, and people can all be genetically modified using this method. discussions on how to protect the biodiversity of future generations by genetic editing of embryos and germ cells. The legal status of human germ cells and embryos should be established in accordance with Indian and international laws governing gene therapy in embryos and embryo editing focusing on SDG 15-Life on Earth.

Keywords: Germline, Modification, Embryo, Life on Earth, Legal Admissibility

Introduction

The expansion of germline and gene editing has sparked a vigorous debate over how severely it can interfere with human DNA throughout the world. The concept of germline creates very specific legal guidelines for who is responsible for harm brought on by gene editing in human embryos and reproductive cells.

Because of the emergence of new technologies and innovations and their potential impact on a wide range of areas, this research paper inspired to conduct a technology foresight process to pursue the achievement of the Sustainable Development Goals (SDGs), which are critical for building a sustainable and equitable future as the objective to enhance people's quality of life, boost prosperity, and safeguard the environment.

SDG 15-Life on Earth is connected with "Protect, restore and promote sustainable use of terrestrial ecosystems, sustainably manage forests, combat desertification, and halt and reverse land degradation and halt biodiversity loss." Legal problems faced by people are the issue of whether or not human reproductive cells' and embryos' genes can be edited legally. Revelations concerning genetic modifications and embryo editing are a topic that has been the focus of discussions and medical research for several years.

Embryo Editing: Perspective Concerning Scrutiny

World is changing faster than ever before. The increasing complexity and fast-paced innovations are occurring in all aspects of life, economically, socially, and environmentally, and they need to be embraced in order to move forward. Advances in technology are not only allowing companies and people to improve quality of life, reduce costs, produce on larger scales and be able to reach much bigger and more sustainable competitive advantages but also increasing the urge to adopt more sustainable practices in companies and to measure their progress towards such goals in recent years. To better direct efforts and make better-informed policy and investment decisions, companies and organizations might adopt technology foresight strategies targeted at examining the longer-term prospects for innovation in science and technology.

Certainly, because it enables the precise replacement of a broken gene with the correct gene. Compared to previous methods, this one is significantly faster, easier, cheaper, and safer. Plants, animals, and people can all be genetically modified using this method. discussions on how to protect the biodiversity of future generations by genetic editing of embryos and germ cells. To address SDG 15—Life on Earth—gene therapy in embryos, germline, embryo editing, and other issues, it is important to clarify the legal position of human germ cells and embryos from international legal regulations to Indian legal standards.

When scientists discovered that they could alter DNA in other organisms using bacterial enzymes that had evolved to protect bacteria from infections, modern molecular biology was born in the 1970s. That scientific advance sparked a lively debate on ethics and safety concerning "recombinant DNA" technology and made clear how crucial openness and transparency are to building public confidence in the scientific community.

The system streamlines genome engineering and makes it accessible to more stakeholders. A recent demonstration of human-germline editing makes these crucial considerations about the proper application of a potent technology more pressing than ever. At this point, one thing is at least certain: we do not yet have adequate knowledge of the potential and boundaries of the new technologies, particularly with regard to engineering heritable mutations.

The field of biology has seen significant upheaval as a result of the quick invention and wide use of simple, affordable, and efficient genome-editing technologies.

It is debatable whether human germline engineering should be used. Some scientists support the technology's rapid development, while others call for its temporary restriction. A total prohibition would be impractical given the general accessibility and simplicity of the technology and would prevent research that could result in future treatments. Instead, it is preferable to reach a solid agreement on a suitable middle ground. Future debates should also cover additional potentially harmful genome editing applications in non-human systems, such as the alteration of insect DNA to "push" specific genes into a population. These discussions should build on this December's gathering.

Concerns of Germline and Embryo Editing

Many important concerns regarding germline and embryo editing are there and some of them are in discussion, One, Safety: The international community of scientists and clinicians must develop common procedures for assessing genome-editing efficiency and off-target implications to make it easier for researchers to compare and assess the outcomes of various experiments for therapeutic relevance. Two, Connectivity: The international conference should serve as a model for more forums where experts enlighten and educate the public about the moral, social, legal, and scientific repercussions of altering the human genome with assistance from the genome-editing and bioethics sectors. Three, policymakers and scientists should coordinate globally to set a shared future course and give clear instructions on what research is and is not ethically acceptable. Four, rules: This partnership should lead to the establishment and application of appropriate oversight to laboratory work that attempts to evaluate the efficacy and specificity of genome-editing technologies in the human germ line. Five, human germline editing should not proceed at this time to create genome-modified individuals, in part due to the unknowable social repercussions.

Testifying Legal Admissibility: Germline and Embryo Editing

Admission of various disparities in worldwide policies and regulations on fragile issues circumscribing using human embryos as research subjects have been able to steadfast the human development, the study claims that in "vitro genome editing in human embryos' remains quintessential in this reference. Heritable alteration is ethically permissible, but only in very narrow situations, the report continues. There aren't any viable alternatives for preventing a serious condition, and there are no efficacious mechanisms to check on the regulation and oversee protocol for creation as well as the guaranteed long-term follow-up, among other things.

Technically speaking, it is clear that despite the stringent guidelines, we are not yet ready to use heritable gene editing for treating human disease as there exists a knowledge void that needs to be filled by rigorous scientific study and technical development to reach efficacy and be vary of any risk. Since the report's publication, questions have been asked about mosaicism and off-target consequences. The former should be solvable shortly given the significant advances in our understanding of the system and its applications being made by scientists all around the world.

The Worldwide Summit's organizers demanded the creation of an ongoing international watchdog to cater to the discussion on the clinical applications of human gene editing. Suggestions have been made by the US National Academies report pertaining to the International Code of Ethics in prospering the idea of human gene editing. The report points out that there needs to be broad and ongoing technical support and extension activities to address the serious challenges raised by promoting the idea of human and non-human primate models; international cooperation can help address the safety and efficacy concerns before heritable gene editing is even considered for clinical trials, for example.

As knowledge develops, ongoing worldwide discussions should continue to evaluate and keep track of how different regulatory systems across jurisdictions are changing to retain certain common principles. To make sure that the diversity of societal, cultural, and political values around the world is viewed as a strength rather than a hindrance to achieving an agreement. Working collaboratively to provide appropriate public engagement tools to assist in deciding on the ethical limits of gene editing should be a significant objective. To promote open science in this critical area, international agreements should also eliminate any restrictions based on nationality, intellectual property, and economic interests.

Nucleases are enzymes that snip DNA at specified sites, allowing for the deletion or rewriting of the genetic material. This process is used in gene editing procedures. Recently, attention has been focused on the very user-friendly CRISPR/Cas9 technique. Currently, somatic, or non-reproductive, cells utilize the technology. For instance, using zinc-finger nucleases, an older gene-editing technique, Sangamo BioSciences of Richmond, California, deleted a gene from white blood cells that produce the receptor that HIV binds to enter the cells.

Under section 53 of the code of civil procedure, 1973 a police officer who reasonably believes that a medical examination of an arrested person will provide evidence as commission of offense can request a medical practitioner to conduct such test on him.

Section 53A of the code of civil procedure, 1973 provides that a police officer who reasonably believes that medical examination of an arrested person who is accused of rape can request a medical practitioner to conduct such test on him if it will provide for evidence as commission of an offense.

DNA Profiling: Deoxyribonucleic acid in short DNA present in cells of every human body, this technology of DNA Testing can be done and used in many ways:

- DNA fingerprinting is used in legal cases as evidence to show parentage or fingerprints left on crime scene.
- It can be used in identifying a dead body that has been damaged badly.
- It also helps in identifying hereditary passed diseases.

To control the usage of DNA Testing the legislation has passed a bill called "*the DNA Based Technology (use and regulation) Bill, 2017*."

Generic medications are those that have the same chemical composition as brand-name medications did when they were still covered by patents. When those patents expire, generic medications are created using the same chemical composition; however, they may vary in terms of manufacturing methods, color, flavor, packaging, etc. Medical Technologies in New Areas is in charge of generic medication.

If forensic human identification techniques result in a successful personal identification, they are successful. However, if the pertinent information and testimony are denied as admissible in court, even the strongest personal identification is meaningless in the prosecution or defense of an accused. The legal guidelines governing the admissibility of expert testimony and evidence are then examined, along with four recognized techniques for identifying people (Identification-establishing techniques (DNA profiling, forensic anthropology, forensic radiography, and forensic odontology), as well as the legal implications of their application in forensic cases, are discussed.

Conclusion

Humanity has changed nature in unexpected ways and at an accelerating rate during the past century. Landscapes are impacted by factors such as population growth, industrialization, urbanization, infrastructure development, and agricultural expansion. These factors reduce the size and quality of the total habitat, which weakens the ecosystem.

Legal problem faced by people is the issue of whether modifying human reproductive cells' and embryos' genes is ethically acceptable. The disclosure of genetic engineering and embryo editing has long been the subject of discussions and medical research. Certainly, because it makes it possible to precisely replace a defective gene with a healthy one. Compared to earlier techniques, this one is a lot simpler, safer, and faster. It is possible to genetically edit humans, animals, and plants. discussions on how to genetically modify germ cells and embryos to preserve biodiversity for future generations. The legal status of human germ cells and embryos should be determined from international legal regulations to Indian legal regulations concentrating on SDG 15-Life on Earth, germline and embryo editing, and gene therapy in embryos.

It seems unrealistic to think that legislative restrictions will make it illegal to change human genes. The idea that the law will guarantee every scenario is a myth. It won't make the world better, but it might make a small difference. This calls for a reconsideration of conventional compensation arrangements in order to include the rights of those whose genomes have been altered. This should be the main focus concerning the law field and specifically personal law. To create such solutions now, it's too soon. The objective of conversation from the law point of view, basically the central point is the concept of accountability concerning the loss, significant for the general public and as a law scholar.

References

Krekora-Zając, D. (2020). Civil liability for damages related to germline and embryo editing against the legal admissibility of gene editing. Palgrave Commun 6, 30. https://doi.org/10.1057/s41599-020-0399-2.

Tetsuya Ishii, and Inigo de Miguel Beriain, (2019). Safety of Germline Genome Editing for GeneticallyRelated "Future" Children as Perceived by Parents, The CRISPR Journal, 2, 370. DOI: 10.1089/crispr.2019.0010.

Cyranoski, D., and Reardon, S. (2015). Embryo editing sparks epic debate. Nature, 520 (7549).

Doudna, J. (2015). Perspective: embryo editing needs scrutiny. Nature, 528 (7580), S6–S6. and stock market returns, 1997–2001. Econ. Geogr. 81(1), 11–29.

Pei, D., Beier, D. W., Levy-Lahad, E., Marchant, G., Rossant, J., Belmonte, J. C. I., and Baltimore, D. (2017). Human embryo editing: opportunities and importance of transnational cooperation. Cell Stem Cell, 21(4), 423–426.

Kaye, D. H. (2010). The double helix and the law of evidence. Harvard University Press.

Baltimore, D., Tatel, D. S., and Mazza, A. M. (2018). Bridging the Science-Law Divide. Daedalus, 147(4), 181–194.

Belikova, K. M. (2020). Some general remarks and legal restrictions in reproductive and therapeutic gene modification in Russia. Systematic Reviews in Pharmacy, 11(4), 431–439.

Knoppers, B. M., and Kleiderman, E. (2019). Heritable genome editing: who speaks for "future" children? The CRISPR Journal, 2(5), 285-292.

Liaw, Y. Q., Turkmendag, I., and Hollingsworth, K. Newcastle Law School, Newcastle University.

Eduardo, D., and Raposo, V. L. (2012). Legal aspects of post-mortem reproduction: A comparative perspective of French, Brazilian and Portuguese legal systems. Med. & L., 31, 181

Exploring Psychological Wellbeing of College Students in Relation to Their Demographic Identity: Predictors and Prevalence

Abhishek Prasad[*,a] *and S. Kadhiravan*[b]

[a]Ph. D. Research Scholar, Department of Psychology Periyar University, Salem, India
[b]Professor, Department of Psychology, Periyar University, Salem, India
E-mail: *abhicmf@gmail.com; kadhir1971@gmail.com

Abstract

The impact of social media use on depression, anxiety, and psychological distress among college students was examined in this study using synthesized data. Data was gathered both online and offline using an incidental sampling method. Ryff's Psychological Well-Being Scales (PWB), a 42-item version scale, was completed by participants (N=241), and demographic data like age, gender, type of institution they attended, birth order, and the number of hours spent were gathered. The findings showed a negative link between psychological well-being and the number of hours spent using social media and the type of institution they attended. The type of university they attended, their birth order, and the number of hours they spent on social media were among the demographic factors that significantly varied. Multiple regression analysis showed 14 percent of the variation in students' psychological wellness was predicted.

Keywords: Psychological Well-being, Social Media, Networking, Psychological Distress

Introduction

The concept of universal human needs and efficient functioning is the foundation of psychological wellness. It relates to reaching one's potential and being fully functional. Realizing one's true nature and reaching one's full potential are both examples of psychological wellness. It has to do with self-actualization and complete functionality (Arslan and Asıcı, 2021). Positive mental states, like happiness or satisfaction, are essentially what psychological well-being refers to. It can also be used to define a person's general functioning and emotional state. It combines having a positive attitude with being able to work well. It is important to remember that being in a good state of health does not mean that everything in your life is perfect or pleasant. It is not necessary for people to always feel happy; the occurrence of painful emotions (such as disappointment, failure, and loss) is a normal part of life. However, being how to control these unpleasant or painful feelings is essential for long-term well-being (Khaute and Malsawmi, 2021).

University students' psychological well-being is an essential component to accomplishing their goals and reaching their potential while going through a significant period of transition (Arslan and Asıcı, 2021). The understanding of psychological health consistently falls behind the understanding of psychological pathology. The disparity is clear in the size of the research (Ryff, 1995). Since children, adolescents, and young people may be particularly vulnerable to the negative psychological effects of the COVID-19 epidemic, there has been a lot of primary research on youth mental health, psychological well-being, and drug use at this time (Zolopa et al., 2022). Numerous studies have employed the 6-dimension conceptualization of psychological well-being as a comprehensive measure of psychological functioning and life outcomes (Tan et al., 2021). Social networking sites (SNS) are widely used in our daily life, which has spurred a growing body of research on the traits of frequent SNS users, with a focus on those who may struggle with self-control and spend a lot of time on SNSs is one of the defining characteristics of SNS addiction. Numerous studies show that spending more time on SNS is linked to less community involvement offline (Muench et al., 2015). In 2021, it's expected that there will be more than three billion active Facebook accounts globally. In the current environment, popular platforms like YouTube, Snapchat, and Instagram are attracting more users, particularly from younger groups (Henzel and Håkansson, 2021).

Measures

Age, gender, the order of their birth, the type of institution they attended, the number of hours they spent using social media, and locality were all noted in addition to the assessment tools.

Ryff's Psychological Well-Being Scales (PWB), a 42 Item Version was used. This scale comprises 42 items and six subscales, including self-acceptance (items 5, 11, 17, 23, 29, 35, and 41), environmental mastery (items 2, 8, 14, 20, 26, 32, and 38), personal progress (items 3, 9, 15, 21, 27, 33, and 39), and positive relationships (items 4, 10, 16, 22, 28, 34, and 40). (item nos. 6, 12, 18, 24, 30, 36, and 42). The replies were scored on a 6-point rating scale, with 1 representing "strongly disagree" and 6 representing "strongly agree." The researcher has confirmed that the scale's validity is sufficient, with a Cronbach alpha of 0.84.

Statistical Analysis

MS Excel was used to screen the data, and the manual was used to score the results. It was done a percentage analysis to determine how the sample was distributed. To restore the scales' dependability, Cronbach alpha was utilized. To examine the relationship between the study variables, Pearson correlation was used. To determine the impact of demographic factors on psychological well-being, multiple linear regression was utilized.

Results

Table 1. "f" value on psychological wellbeing on DV

DV	Demographic Variables	N	Mean	SD
PWB	Government	99	60.54	11.38
	Private	137	58.70	9.48
	Aided	4	43.00	1.00
f-value = 6.00*				

Note: * $p < 0.05$

Based on type of institution, Table 1 displays the psychological wellness ratings (government, private and aided). Government, private, and aided each have a mean score of 60.54, 58.70, and 43.00, respectively. There is a sizable difference in psychological wellness, as indicated by the "f" value (6.00). According to this research, achieving one's life goals to the fullest extent possible is a vital aspect of human performance. Realizing one's goals, as a result, is often seen as a sign of positive personal development and is associated with high levels of general wellbeing (Yang et al., 2021).

Table 2. "f" value on psychological wellbeing based on HSM

DV	Demographic Variables	N	Mean	SD
PWB	2–4 Hours	124	60.50	10.02
	5–7 Hours	87	60.18	11.09
	8–10 Hours	29	50.68	5.43
f-value = 11.93*				

Note: * $p < 0.05$.

Table 2 displays the psychological well-being scores. Government, private, and aided each have a mean score of 60.50, 60.718, and 5.43, respectively. The "F" score (11.93) demonstrates that the number of hours spent on social media has a significant impact on psychological wellness. Numerous theories contend that robust social networks have a positive impact on psychological and physical health. According to the stress-buffering theory, those who have close relationships with others are less likely to become ill from stress (Hogeboom et al., 2010).

Table 3. "f" value on psychological wellbeing and order of birth

DV	Demographic Variables	N	Mean	SD
PWB	First Child	122	60.26	11.72
	Middle Child	53	56.54	8.44
	Last Child	50	61.64	8.95
f-value= 5.17*				

Note: * $p < 0.05$.

First, middle, and last child's respective mean scores are 60.26, 56.54, and 61.64. Considering their birth order, the "f" value of 5.17 indicates a considerable difference. Adler categorized it as the following -siblings are reflected in a large portion of the social media content on the subject: the youngest is independent, fun-loving, and wild, while the middle child is typically forgotten. Firstborns are more likely to be responsible, authoritarian, and under constant pressure to live up to their parents' expectations (Arslan and Asıcı, 2021).

As shown in Table 4, there is a correlation between psychological well-being and demographic factors such as gender, area of residence, field of study, type of institution, order of birth, number of hours spent on social media, and kind of social media used. Statistically significant relationship between demographic factors and general psychological health and among the variables was discovered (r ranges from –0.08 to 0.01). It was discovered that there was no correlation between psychological well-being and the associated socio-demographic factors, specifically the kind of institution (p 0.05, r= 0.15), the amount of time spent on social media (p 0.05, r= -0.23), and the stream of study (p 0.05, r=-0.12).

Table 4. Scale inter-correlation for psychological wellbeing (PWB)

	1	2	3	4	5	6	7
2	0.10						
3	0.13*	0.11					
4	0.06	−0.01	0.01				
5	−0.09	0.00	-0.04	0.00			
6	−0.03	−0.05	0.03	−0.26**	−0.02		
7	−0.01	0.01	0.04	−0.32**	0.24**	0.06	
8	0.01	0.02	−0.12	−0.15*	−0.09	−0.23**	−0.08

Note. $^{*}p<0.05$. $^{**}p<0.01$

(1)=Age (2) = Gender (3) =Locality (4) = Field of research (5) = Institution kind, (6) = Birth order, (7) = the number of hours spent online, (8) = Social media platforms

Table 5 displays the results of multiple regression analysis for the influence of age, gender, place of residence, field of study, type of institution, order of birth, hours spent using various social media platforms, types of platforms used, and psychological health (dependent variable). The results showed that gender (=0.03, t=0.5, p 0.05), area of residence (=0.01, t=0.28, p 0.01), kind of institution (=-0.2, t=-4.05, p 0.01), number of hours spent on social media (=-0.29, t=-40.60, p0.01), and stream of study (=-0.10, t=-1.70, p 0.01) significantly predicted. College students' demographic influences were significant (R2 = 0.14; F = 4.83; p 0.01) and combined them explained 14% of the variance.

Table 5. Impact of demographic factors on PWB

DV	Psychological Wellbeing			
	B	***SE***	***β***	***t***
Age	0.62	2.03	0.02	0.30
GEN	0.86	1.61	0.03	0.53
AOL	0.35	1.25	0.01	0.28
SOS	−2.25	1.32	−0.10	−1.70
TOI	−5.44	1.34	−0.27	−4.05*
OB	−0.72	0.68	−0.06	−1.05
NHS	−4.39	0.95	−0.29	−4.60*
TSM	−4.99	2.42	−0.13	−2.05*
F = 4.83				
R= 0.37				
R^{2}= 0.14				

Discussion

This study reflects an effort to identify factors and the degree of their prevalence. The findings show that media consumption hours, regardless of gender, predicted the impact on psychological well-being. Additionally, social media use did predict that the more time spent on social media, the greater the influence (Arslan and Asıcı, 2021). Demographic factors like study field, place of residence, and birth order were found to have positive associations with one another. Even though the findings of this study are important, more research is required to fully understand how social networks' educational ties function. To better understand the relationship between Internet use and social networking, personality traits, a person's online and offline habits, and longitudinal study designs should be taken into account (Hogeboom et al., 2010).

Conclusion

Significant statistical importance has been shown by social demographic statistics used to assess the prevalence and determinants of their effects on psychological wellness (Yang et al., 2021). Social media use has psychological effects that are related to demographic factors. This study sheds light on the behavior that is becoming more prevalent among college students who have developed a habit of spending a considerable amount of time on social media. The findings of this study may help guide those seeking counseling for those who are more reliant on social media and guidance on how effectively handle such devices for personal usage.

FUNDING: Abhishek Prasad is a recipient of Indian Council of Social Science Research Doctoral Fellowship (ICSSR), Government of India (File No. RFD/2022-23/SC/PSY/53).

References

Arslan, Ü., and Asıcı, E. (2021). The mediating role of solution focused thinking in relation between mindfulness and psychological well-being in university students. Current Psychology, 1-10. https://doi.org/10.1007/s12144-020-01251-9

Henzel, V., and Håkansson, A. (2021). Hooked on virtual social life. Problematic social media use and associations with mental distress and addictive disorders. PLOS ONE, 16(4), e0248406. https://doi.org/10.1371/journal.pone.0248406

Hogeboom, D. L., McDermott, R. J., Perrin, K. M., Osman, H., and Bell-Ellison, B. A. (2010). Internet use and social networking among middle aged and older adults. Educational Gerontology, 36(2), 93–111. https://doi.org/10.1080/03601270903058507

Khaute, G. K., and Malsawmi, H. A study on the psychological well-being of college students. Mizoram Educational Journal, 53.

Muench, F., Hayes, M., Kuerbis, A., and Shao, S. (2015). The independent relationship between trouble controlling Facebook use, time spent on the site and distress. Journal of Behavioral Addictions, 4(3), 163–169. https://doi.org/10.1556/2006.4.2015.013

Ryff, C. D. (1995). Psychological well-being in adult life. Current Directions in Psychological Science, 4(4), 99–104. https://doi.org/10.1111/1467-8721.ep10772395

Tan, Y., Huang, C., Geng, Y., Cheung, S. P., and Zhang, S. (2021). Psychological well-being in Chinese college students during the COVID-19 pandemic: Roles of resilience and environmental stress. Frontiers in Psychology, 12, 671553. https://doi.org/10.3389/fpsyg.2021.671553

Yang, C. C., Holden, S. M., and Ariati, J. (2021 September). Social media and psychological well-being among youth: The multidimensional model of social media use. Clinical Child and Family Psychology Review, 24(3), 631–650. https://doi.org/10.1007/s10567-021-00359

Zolopa, C., Burack, J. A., O'Connor, R. M., Corran, C., Lai, J., Bomifim, E., DeGrace, S., Dumont, J., Larney, S., and Wendt, D. C. (2022). Changes in youth mental health, psychological wellbeing, and substance use during COVID- 19 Pandemic: A rapid Review. Adolescent Research Review, 7(2), 161–177. https://doi.org/10.1007/s70894-022-00185-6

India and Nepal Bridging the Gap with Hydropower Project Enhancing Science and Technological Partnership

Trishna Rai and Pooja Raghav

Madras University, Chennai, India. Vellore Institute of Technology, Chennai, India
E-mail: Trishnarai73@gmail.com

Abstract

Energy diplomacy is a concept intertwined with foreign policy and national security. Irrespective of its self-sustaining ability the demand of every functioning nation is the proper usage of energy resources. This article focuses on building international relations across borders and facilitating nations with economic and energy partnerships. Taking India and Nepal hydro power project as its case study it further focuses on science and technological partnership and the generation of economical and energy diplomacy. Energy diplomacy is ensuring energy security. With massive usage and limited resources renewable energy resources have started to play a pivotal role. The article specifically looks at the integration of energy diplomacy into foreign policy. This paper also provides an analysis of how bilateral energy diplomacy can support the interest of India in Nepal.

Keywords: Energy diplomacy, India and Nepal relations, Hydro-power project

Introduction

Energy diplomacy goes hand in hand with economic and security co-relations among countries. With growing industrialization globalization and ever-increasing consumption, there is always an urgency to fulfill the demands putting a huge amount of pressure on energy resources. The industrial revolution marks the sudden increase in the energy demand. In the last 50 years, there has been a massive increase in energy supply and this curve is only on the rise. This changes the entire functioning dynamics of a nation making energy resources the most valued power and converting it into a national security agenda. The economy and energy have become inseparable.

Energy diplomacy has been integrated into foreign policy and national security mainly through the route of security and economy. In the case of India and Nepal, the collaboration of water diplomacy for the benefit of both nations can be seen as India's usage of hydropower project assistance to strengthen its soft power in Nepal. Foreign policy among countries has energy as a huge driving force. Energy diplomacy entered into national and foreign policy through national security policy and economic policy. Tying national foreign policy, national economic policy, and national security policy into an intertwined partnership. It almost goes hand in hand with each other making sure the smooth functioning of the nation. This makes it almost impossible to measure growth, development, and stability when a factor is deteriorating. Despite becoming the world's most powerful; every country does have a sense of alignment or dependency with other nations irrespective of its supremacy. Complete self-sustenance has become an idealist phenomenon. Especially with globalization, digitalization, and artificial intelligence, it seems almost impossible to be self-reliant. Big powers bind together for their sustenance at the top, and developing nations use diplomacy in aspects of searching the market for commodity, or to pave way for recognition likewise, underdeveloped nations use diplomacy for survival. Different variations and different uses of diplomacy are economic, military or energy. Nations are incapable of abstaining from it. That is where the role of foreign policies plays a significant part.

Objective

1. To study the benefits of energy diplomacy in enhancing bilateral relationships.
2. To understand the changing relationship of Nepal with India.

Methodology

This study aims to understand the deteriorating relationships between India and Nepal and the steps being avoided to bridge the gap between the two countries. The study is compiled through a comprehensive study of secondary data like books, newspapers, and governmental recordings.

India and Nepal

India and Nepal share multiple relationships of culture, religion, kinship, and border. Being a landlocked nation Nepal is surrounded by India on three sides and with the Himalayan ranges in the northern regions, it has a massive dependency on India. One of the significant landmarks of the relationship between the two countries is the signing of the Treaty of Peace and Friendship 1950 which provided an open border with the free movement of people and commodities.

Prime Minister Narendra Modi visited Nepal in May 2018 in order to plaster back the relationship between India and Nepal. Modi along with The Prime minister of Nepal, K. P. Sharma Oli had discussions on working together and enhancing bilateral relations by expanding socio- economic development. They emphasized advancing cooperation on water resources for mutual benefit. They also agreed to enhance bilateral cooperation in the power sector. Furthermore, enhance the social relation between the masses from both countries.

India's Satluj Jal Vidyut Nigam Ltd. (SJVN) has managed to bag the 679 MW Lower Arjun Hydro Power Project after bagging the 900 MV Arun 3 Hydro Electric Project through international bidding, which could be a step towards rebuilding the relationship between the two countries. Prime Minister Narendra Modi of India and Prime Minister K P Sharma Oli of Nepal laid the foundation stone of Arun3 project on the 11th of May, 2018. The ministry of power stated, "India is also endeavoring creation of a common pool for regional power grid and energy market in the neighboring countries for regional peace and improvement of optimum utilization of generation assets." The primary project, ArunIII Hydropower Project at the Arun River in the east of Nepal has an estimated cost of 1.5 billion dollars with 4018.87 million unit's production. The recent venture, Lower Arun Hydroelectric project has an estimated price of 1.3 billion dollars. Both are being constructed on a build-own- operate- transfer (BOOT) agreement in accordance with the SAPDC (SJVN and the Arun III Power Development Company.

The Arun Hydropower Project

During the 2018 visit to Nepal Prime Minister Modi of India and K P Sharma Oli, the Prime Minister of Nepali laid the foundation of the Arun III hydropower project taking a step forward towards building bilateral cooperation in the power sector. This plant is being developed in Sankhuwasabha District in East Nepal with a power generation capacity of 900MW. The project comprises a 70/466-meter gravity dam with a storage capacity of approximately 13.94 million cubic meters. It will include an 11.75km long head race tunnel of 9.5 in diameter, 2 steel pressure shafts, and 4 penstocks. With an estimated length of 192 meters and a width of 10 meters.

The dam will also comprise four underground desalting chambers with a length of 420m, a width of 16m and a height of 24m. The powerhouse of the project will be underground and will be equipped with four vertical Francis turbine units each. The rated capacity of each turbine generator is 225MW. The powerhouse will have a gross head of 308m and a design head of 286.21 m.

The project was 15 % complete by January 2019. Completed works include the construction of a 62m diversion tunnel at Chisopani, Nayabasti of Makalu Rural Municipality 3, while ongoing activities include the construction of an underground powerhouse, measuring 179.5m in length, 22.5m in width and 49.5m in height, along with a transformer house.

The output from the power plant will be transferred to the Nepal-India border through a 300km-long, 400kV DC transmission line, which will be routed along Diding (Nepal) to Dhalkebar (Nepal) towards Muzaffarpur (India).

The second Hydro power project bagged by India is the Lower Arun Hydro power project with an estimated power generation of 697MW. It will be a tail race development therefore water of Arun III Hydropower project will be redirected into the river for Lower Arun Project.

For both India and Nepal, the hydropower project seems to be a way for rebuilding estranged relationships. Irrespective of future political relationships and the ever-changing foreign policies of the two individual sovereign states, something that is common is that cross-border energy trade will be mutually beneficial for both sides. Primarily the economic, environmental, and social aspects are unavoidable on either side. This has created a power trading ecosystem to meet developmental goals and rebuild social and political trust. During 2019, Nepal exported approximately a 35million units of power to the state of Bihar in India and imported close to 2813 million units from India. For the year 2018-2019, Nepal imported 43% of its total power consumption from India. Therefore collaborating on the Hydropower project has a high scope of benefitting the countries.

Apart from the obvious development in the power and energy sectors, the Arun Hydropower project will generate more than 3000 employment opportunities for both Indians and Nepalese. The project brings along the development of the region through the construction of roads, amenities, schools, and hospitals. And finally, the most important political reason is that it helps boost the bilateral relationship between India and Nepal.

Conclusion

To conclude one can estimate the positive outcomes of the project in both India and Nepal. With the changing dynamics in South Asia, the importance of encouraging bilateral relationships with neighboring states is ideal for all nations. A hydropower project boosts the economy, encourages development, and improves bilateral cooperation. Considering the changing scenario of India and Nepal's relationship with China being a major factor this seals another bilateral project withholding India's stand on Nepal.

References

Colgan, J. D. (2014). Oil, domestic politics, and international conflict. Energy Research & Social Science, 1, 198–205.

Rest, M. (2012). Generating power: debates on development around the Nepalese Arun-3 hydropower project. Contemporary South Asia, 20(1), 105–117.

Washakh, R. M. A., Chen, N., Wang, T., Almas, S., Ahmad, S. R., and Rahman, M. (2019). GLOF risk assessment model in the Himalayas: A case study of a hydropower project in the Upper Arun River. Water, 11(9), 1839.

Arun-3 hydroelectric project (900 MW) - SSNR projects. Projects. (2022, April 15). Retrieved August 28, 2022, from https://ssnrprojects.com/projects/arun-3-hydroelectric-project/.

Mea.gov.in. (2022). Retrieved 31 July 2022, from https://www.mea.gov.in/Portal/ForeignRelation/India-Nepal_Bilateral_Brief_Feb_2020.pdf.

SJVN Bags 679 MW Lower Arun Hydro Electric Project in Nepal. Pib.gov.in. (2022). Retrieved 31 July 2022, from https://pib.gov.in/Pressreleaseshare.aspx?PRID=1693606

Systematic Literature Review of Interlinkages between Sustainable Development & Human Development

Neha Biswas[a] *and Namita Kapoor*[*,a]

[a]Amity University, Noida, India
E-mail: [*]namita3112@gmail.com

Abstract

Human Development is essentially promoted by sustainability and is not realistic without sustainability. If human development is related to empowering people to lead educated, healthy, and long and fulfilling lives, then sustainable development stands for making sure that future generations can continue doing the same. Human Development Index (HDI) reflects the level of well-being of a nation as it measures the mean achievement of factors that forecast quality of life. It captures the basic socio–economic dimensions of human development namely, Education, Health, and Income which align well and coincides with the goals of sustainable development. This study aims to establish and review the available literature on the interlinkages between the HDI and the SDGs. The HDI covers only the socio-economic aspects of development, leaving scope for including an environmental aspect like a measure of degradation of natural capital necessary to study the relation between the two approaches to development.

Keywords: Human Development, Sustainable development, HDI, Systematic literature review

Introduction

Human development and Sustainable development are the extensive concepts in the development approach. Sustainable development is a development that stands to cater to the immediate needs without negotiating the capacity of future generations to meet their own needs. There are quite a few links between AGENDA 2030 and Human Development (Opoku, Dogah, and Aluko, 2022). Human development is a multifaceted concept. Human development aims to provide the necessary basic conditions to all individuals and collectively for the betterment of their potential and living a productive and creative life. There was an emerging realization among economists that national income or Gross Domestic Product (GDP), does not include the social or human aspects of development The objective of HDI was to shift the focus of developmental economics to people-centered policies from national income accounting with a measure that could be used to analyze a countries' progress not only in terms of economic or monetary expansion but also in the context of key social outcomes (Batallas Astudillo, 2020). A new approach for the advancement of wellbeing if human was seen in the initial Human Development Report in 1990. The concept of human development stands for the expansion of affluence of human life, rather than of the economy in which they live. The approach is focused on humans and their opportunities and choices. The three parameters used by UN to compute HDI constitute the key aspects of human development. These key aspects are - a long and healthy life (life expectancy), being knowledgeable (mean and expected years of schooling), and having a decent standard of living (income per capita). When these key dimensions are provided initially, the advancement and options to bring improvement in other dimensions of human life will also progress, which lays down the basis to monitor progress on other approaches to development like the SDGs. HDI has been upgraded through the United Nations Development Program's (UNDP) annual reports and has become the single most extensively used measure of human development. The concepts behind HDI cultured the MDGs in 2000, which gave rise to the SDGs in 2015. The MDGs and the UN Millennium Declaration already mirrored the basic concepts of human development – the expansion of human abilities and capabilities by directing to basic deprivations of humans (Antoniades, 2020). In this context, the framework and principles of human development can provide rational contributions to the development agenda of 2030 and the execution of the SDGs Likewise, measures of human development can gain from the SDG indicators as they progress and evolve over the next years Agenda 2030 for Sustainable Development, inclusive of the 17 SDGs, are global goals that took over the MDGs. The SDGs will mold development plans at national levels over the next years. From total elimination of poverty and hunger to being responsive to changes in climate and sustainable practices for our natural resources, agriculture, and food, lies at the core of the agenda 2030.

Many of the SDGs relate directly to the HDI. Practically the measurements of both branches of development, human and sustainable, can be interlinked which raises the need to a growing need to understand the link between HDI and SDGs (P., Costa, Rybski, and Lucht, 2017). The global progression in meeting the SDGs is largely dependent on India's progress. Presently, India is responsible for 20 percent of the SDG gap in 10 of the 17 SDGs globally and greater than 10 percent of the delay in another 6 Goals. This study aims to assess the interlinkages between HDI and SDGs using a case study of India and determine the relationship between both approaches of development.

Methodology

In this paper, a systematic literature review is done to identify the knowledge that exists on the interlinkage between human development and sustainable development. According to studies, a systematic literature review is a transparent, organized, and replicable method of research for the analysis of the existing literature. There are several purposes for conducting a systematic literature review, like to determine gaps within the existing research and propose areas for further research activities or to recommend a framework and to identify recent research scope and potential themes of research (U. Sivarajah, 2017). Thus, this study provides an interdisciplinary overview of the link between the two approaches to development namely, human development and sustainable development.

Review Protocol

The review protocol of this study is based on, the following conditions:

The Scopus database was selected as the source for an authentic list of studies based on the theme of the study.

- To explore the literature, all published journal articles were considered, but to improve standard control, other document or sources such as publications, conference series, books, editorials, and book chapters were chosen to be omitted.
- To improve consistency, the investigation for the studies was bounded to studies published in an 8 years time frame from 2015 (introduction of Sustainable Development Goals) to 2022.
- Because of the interdisciplinary nature of this study, the search involved articles associated with any kind of academic discipline as categorized by the Scopus database relating to the two approaches of development discussed in the study, sustainable and human, that was scripted in the language -English.
- To ensure the suitability of the selected articles, related keywords to the two

approaches of development discussed in the study, sustainable and human & their indicators were considered. All the works whose, abstracts involved words relating to Human Development (i.e., Human development, HDI, Health, Education, Standard of Living) and all the works that involved words directly relating to Sustainable development in their paper titles (i.e., Sustainable Development Goal, Sustainable development, SDGs and 2030 Agenda) were studied.

- Empirical and conceptual articles both were selected.
- Finally, suitability was substantiated by reading and studying the whole article to line up the selected articles with the objective behind this study.

Database Research

Since the review protocol was developed for this study, keywords were first put into the Scopus database following the above conditions. This brought in 16,039 publications being shown. Then, titles, abstracts, and proper scanning of articles were conducted on the assimilated articles according to the developed protocol. This brought in a sample of 149 interdisciplinary studies in the Scopus database. At the final stage of the process, these papers were further investigated. This selection was scanned for empirical and conceptual both, studies using the criteria pointed out in the conditions in the protocol. The final scan led to the final sample of 10 documents.

Interlinkages Between Health and SDGs

There are quite a few links between the approach of Agenda 2030 and human development. Human development captures the key socio-economic dimensions like health, education, and standard of living that directly relate to some of the SDGs like no poverty, good health, education for all, and work and employment (Disli and Koç, 2022). HDI also has indirect links with other SDGs that aim at reducing hunger and promoting peace. So, it can be said that, if a country's HDI value is known, its SDG ranking can be forecasted with precision of 91 percent (Bissio, 2019). So, if the HDI is growing positively in the right direction, it is probable that the SDGs are advancing too, and vice versa. Now we will discuss the interlinkages between each dimension of the index of human development and the SDGs to offer a more comprehensive understanding of the study.

Health and SDGs

Good health and well-being (SDG 3), is one of the 17 goals of Sustainable Development which aims to reduce mortality rates and ensure that people have a level of health that allows them to lead an economically and socially productive life. Health is interlinked with various other SDGs like SDG 1 – No Poverty, SDG 2 – No Hunger, SDG 4 – Quality Education, SDG 5 – Gender Equality, SDG 6 – Clean water and Sanitation, SDG 10 – Reduced Inequalities, SDG 11 – Sustainable cities and communities and SDG 12 – Responsible production and consumption. Good health implies many socioeconomic requirements such as reducing poor education levels, reducing unemployment and insecurity, and improving conditions of life. One of the basic dimensions that HDI measures are health and the parameter for that is Life Expectancy. Life expectancy is used as a proxy of overall population health, and even though health is multi-dimensional, life expectancy is a largely used indicator of health conducted a study on 193 countries and found out that Life Expectancy was highly correlated with GDP per capita and access to electricity, directly interlinking it with the SDGs. The health aspect of HDI is indicated by longevity, and survival factors. Public policymakers and researchers have analyzed other factors that induce influence the health aspect, including insufficient or lack of basic sanitation.

Education and SDGs

Education is a human right and a driver for sustainable development. Every goal in the agenda 2030 needs the education to entitle society to the skills, knowledge, and values to reside in dignity, construct their lives and give to their societies (Nwagwu, 2020). Quality Education (SDG 4) aims at ensuring equitable, inclusive, and quality education for everyone and is directly related to SDG 1 – No Poverty, SDG 2 – No Hunger, SDG 5 – Gender Equality, SDG 6 – Clean water and Sanitation, SDG 9 – Industry, Innovation, and Infrastructure, SDG 10 – Reduced Inequalities, SDG 12 – Responsible production and consumption and SDG 13 – Climate Change. HDI also measures education, using parameters of expected years of schooling and mean years of schooling, which relate directly to targets of SDG 4, namely 4.3 – which aims at ensuring equal access for all men and women to quality technical and affordable tertiary and vocational education, including university by 2030, and 4.6 - ensuring that all youth and a substantial proportion of adults, both women and men, achieve numeracy and literacy by 2030. The World Bank has pointed out that as levels of education for youth in rural areas improve, they can enter a comprehensive array of non-agricultural employment. As women obtain more education, they progressively move out of traditional households or agricultural activities and enter wage-earning work. In regards to the backward linkage impact that sanitation and water have on education, progressions in global energy and water infrastructure can directly affect the opportunities for education to the poorest.

Income and SDGs

Per capita income is an estimate of the amount of money earned on average in a nation or region. Per capita income can be used to measure the average per-person income for an area and to determine the quality and standard of living of the population (Asadullah, 2020). HDI measures the standard of living using per capita income, which is directly related to SDG 8- Decent Work and Economic Growth which aims at promoting sustained, comprehensive, and sustainable economic development, complete and productive employment, and decent work for everyone, more specifically 8.5 - Attaining complete and productive employment and decent work for all men and women, including for persons with disabilities and youth, and equitable pay for work of equitable value, by 2030. Per capita income is also closely interlinked to the other SDGs like SDG 1 - No Poverty, SDG 2 – No Hunger, SDG 3 - Good Health and Wellbeing, SDG 4 - Quality Education, SDG 5 – Gender Equality, SDG 7 – Affordable and Clean Energy, SDG 9 – Industry, Innovation, and Infrastructure, SDG 10 – Reduced Inequalities, SDG 12 – Responsible production and consumption and SDG 14 – Life Below Water. The necessities that the SDGs aim at achieving like access to water and sanitation and access electricity (Alcamo, 2019). Linkages from both sides arise as attaining the SDGs would directly contribute to inequality in income and enable better access to necessities like water, sanitation, education, and clean energy fuels.

Conclusion

This study aims to analyze the interlinkage between two approaches to development namely, human development and sustainable development. Both these approaches to development are interdisciplinary and cannot be dealt with independent of each other, especially, about achieving the SDGs. United Nations General Assembly defined 17 SDGs containing 169 related targets which aim at achieving socio-economic and environmental stability that should be achieved by the year 2030. The agenda 2030 for Sustainable Development identifies that putting an end to poverty and other inadequacies must go alongside strategies that better education and health and drive economic growth. The socioeconomic and environmental concerns regarded in the definition of these 17 goals cannot anymore be dealt with individually and independently. Human development is essentially what is promoted by sustainability and human development is not realistic without sustainability. If human development is related to empowering people to lead educated, healthy, and long and fulfilling lives, then sustainable development stands for assuring that future generations can continue doing the same. According to the literature, we can observe that SDGs and HDI are interlinked as not only HDI as a whole but all of its dimensions like Education, Health, and Income, relate to the SDGs in an interdependent manner.

References

Alcamo, J. (2019). Water quality and its interlinkages with the Sustainable Development Goals. Current opinion in environmental sustainability. 2019 Feb 1, 36, 126–40.

Antoniades, A., Widiarto, I., Antonarakis, A. S. (2020). Financial crises and the attainment of the SDGs: an adjusted multidimensional poverty approach. Sustainability Science. 2020 Nov, 15(6), 1683–98.

Asadullah, M. N., Savoia, A., Sen, K. Will South Asia achieve the sustainable development goals by 2030? Learning from the MDGs experience. Social Indicators Research. 2020 Nov, 152(1), 165–89.

Astudillo, I. L., Cacay, J. C., Blacio, A. M. (2020). Impact of socioeconomic variables in the human development index of the Latin American economies. Universidad y Sociedad. 2020 Apr 20, 12(2), 400–4.

Bissio, R. (2019). SDG Indicators and BS/Index: The power of numbers in the sustainable development debate. Development. 2019 Dec, 62(1), 81—5.

Nwagwu, I. Driving sustainable banking in Nigeria through responsible management education: The case of Lagos Business School. The International Journal of Management Education. 2020 Mar 1, 18(1), 100332.

Opoku, E. E., Dogah, K. E., Aluko, O. A. (2022). The contribution of human development towards environmental sustainability. Energy Economics. 2022 Feb 1, 106, 105782.

Impact of COVID-19 on Domestic Workers with Special Reference to Pune Region

Ranjit Kumar[*,a] *and Dr. Bharati Kumar*[b]

[a]Savitribai Phule Pune University of Pune, India. [b]Sinhgad Business School Pune, India
E-mail: *bharatiattal@gmail.com

Abstract

People across the globe experienced unprecedented levels of disruption in their homes as well as in the community and society at large in the year 2019 due to the outbreak of COVID-19 pandemic. The economy had come to a standstill. Many people lost their jobs, and some were retained in their job but were not able to sustain the lifestyle adopted prior to the pandemic. The impact of this pandemic was more seen in the unorganized working class like the domestic workers. Being weakly organizes and lacking any sort of institutional support, domestic workers are extremely vulnerable to exploitation and violations of human rights. Although domestic workers provide essential services, they rarely have access to rights and protection. The current study was undertaken by the researcher to understand the problems and challenges faced by domestic workers during the pandemic. The study is confined to domestic workers from Pune region of Maharashtra state.

Keywords: Covid-19, domestic workers, livelihood

Introduction

The World Health Organization declared the Covid-19 disease a pandemic in 2020. The Covid-19 pandemic is one of the first and foremost human tragedies that is faced by the human population across the globe. Almost every person had experienced unexpected levels of disruption in their homes, jobs, and society. This virus had even caused many challenges to the public health system globally, when the countries announced partial or complete lockdown, the whole global economy came to a standstill resulting in social and economic distress. Not a single social class was left untouched by this pandemic, in India, the employment protection and social security of those in the informal sector who constitute 86% of the workforce were most affected as they struggle to meet the necessities as well as the threat of infection. Though different groups of workers experienced similar constraints during the pandemic in terms of livelihood and social safety, the women domestic workers faced total or near unemployment and economic insecurities due to social distancing and lockdown restrictions,

Domestic workers: Prior to and during pandemic

Domestic workers are the ones who perform work in or for a private household or household. They basically provide direct and indirect care services. The current number of domestic workers in India ranges from an official estimate of 4.2 million to an unofficial estimate of more than 50 million. Two third of the total domestic workers in India live in urban areas and about 75% of them are women (Ghosh,2013). The increasing number of domestic workers are constantly growing in the informal sector of urban India. The family financial crisis has also compelled women to become domestic workers and protect the interest of the family. The employers extract maximum work from the domestic workers without extending minimum hospitality. Domestic work is looked upon as unskilled because most women have traditionally been considered capable of doing the work and the skills they are taught by other women in the home are perceived to be innate. When paid, therefore, the work remains undervalued and poorly regulated (Roberts, 1997:30). The women workers work as part-time or stay-at-home domestic workers, and their household expenses are predominantly met through the income they earn. Despite the large worker population, domestic work is not recognized as "work" and is always treated as the lowest in the occupation hierarchy. This structural issue of injustice where society doesn't recognize domestic workers as "worker" has left them at the mercy of their employers (Chandramouli,2018). Domestic workers face tough working conditions and are left without any social security protection. They are not protected by labor legislation except for the Unorganized Workers Social Security Act, 2008 (Ghosh, 2013), the benefit of which most do not reach the workers due to the informality of the contracts and ignorance about the social security provisions. National Policy for Domestic Workers has been drafted by the Government of India to ensure the rights and social protection of the workers formally through legislative measures but has yet not materialized. In India, women's participation in the domestic work sector is a common feature that indicates the feminization of domestic work (Augustine and Singh, 2016). Low wages and lack of legal protection (Neetha and Palriwala, 2011), unpaid overtime and occupational health problems (Paul et al., 2018), poor bargaining power, working without leave under coercion, child care issues, and health ailments such as back pain and skin allergy (Moghe, 2013), physical and sexual violence (Hamid, 2006; Paul et al., 2018), exploitative working conditions and human rights violations (Chandramouli, 2018), absence of a formal organizational framework for domestic workers, lack of representations, exclusion of domestic workers from legal rights for minimum wages (Bhattacharya et al., 2010; Chandramauli, 2018) were found to be the issues affecting social justice of women domestic workers in various studies. In the context of India, caste, religion, and gender dimensions also play a major role in determining the features of domestic work as well as the nature of exploitation which also influences the bargaining power of domestic workers (Raghuram, 2001). In Pune city, many domestic workers working in upper-middle-class houses were sent back home on unpaid leave during the lockdown (Parth, 2020). In many cases, domestic workers complained of wage cuts and non-payment for March-April 2020 and later loss of their job. Those who were not listed in the ration registry based on 2011 census and migrant laborers were left out of public provisioning (Goel et al., 2020). Whereas the majority of 180,000 migrant workers in Delhi without documents were unable to access social security and relief measures offered by the government during the pandemic, the native workers in Kerala could avail of free ration and community kitchen services instituted by the state (Self Employed Women's Association, 2020). The absence of formal registration with the social security board denied women domestic workers of any government relief during the pandemic times whereby 51% of women had difficulty buying essential food items and 36% had difficulty with healthcare access (Institute of Social Study Trust, 2020). Telephonic survey with 500 domestic workers in Jaipur showed that 51% of the workers were paid a salary for the work they did in March and 44% of the workers ended up borrowing money from money lenders at exorbitant interest rates (Bharti, 2020). Due to the lack of uniformity in the wage structure, the marginalization and vulnerability of domestic workers in Cuttack doubled during the pandemic (Nanda, 2016). About 80% of the high-volume domestic workers ended up almost jobless and half of them reported poor access to medical care during the period (National Domestic Workers Alliance, 2020). Anand and Deepa based on their study of 2020 in Chennai city concluded that working in high-risk conditions without safety measures and lack of access to health care made domestic workers most vulnerable among the informal sector employees.

Methodology

In the current study, the researcher interviewed around 150 women domestic workers from Pune city peth areas, on issues concerning working conditions, livelihood and household dynamics, health, and government support during the pandemic during March-June 2020. Equal number of respondents were selected from Peth areas like Somwarpeth, Mangalwarpeth, Rastapeth, Shukrawarpeth, and Raviwarpeth. The study was conducted using mixed methods of quantitative and qualitative dimensions. The survey mainly focused on socio-demographic details, working conditions, employer-worker relationships, the impact of the pandemic on livelihood and family conditions, and health care. Snowball sampling technique was used to collect the primary data and secondary data was collected from various books, newspapers, and internet sites related to domestic workers. Analysis of data was done using excel.

Analysis and Findings

On studying the general profile of the women workers, the researcher found that 40% of respondents were engaged in this nature of work for the last 1 to 3 years, 28% were working in this field for the last 3 to 6 years and 32% women were employed as domestic workers for the last 6 to 9 years about 60% of domestic workers that were contacted for this study belonged to the age group of 35 to 45 years, 30% of the respondents belonged to the age group of 25 to 35 years and 10% were below 25 years of age group. When asked about the working hours or timings of their work about 80% were working part-time and 20% were working full-time. Out of 150, 20% worked for more than 5 hours a day, 40% worked for 3–5 hours and 40% worked for less than 3 hours a day. About 80% worked in multiple houses whereas 20% worked only in one house. 90% of respondents were engaged in cleaning tasks, 5% in child-rearing help during office hours, and 5% were involved in cooking, and cleaning together. It was noticed that 3% of them traveled more than 15 minutes to reach the workplace, covering a distance of more than 5–8 kilometers. It was observed on analysis that about 84 (56%) workers did not receive any type of help from their employer however 4% of workers said that their employers assured them of work post lockdown. On the other hand, at the same time, some of the workers reported favorable experiences with their employers during the difficult times of the pandemic. This included providing material support, for example, groceries 8%, assurance to retain for work after lockdown period, at 8%, and payment of advance salary 8%. Assurance for future hikes in salary was received by 16% of workers. Another important observation made by the researcher was that the majority of the domestic workers had an earning of only 2–4 thousand rupees a month. Hardly 10% of them earned a sustainable amount of about 6-8 thousand monthly. It but natural that these women are doing domestic work to support themselves on the financial front, however, they faced a lot of reduction in their monthly income during the lockdown period. 80% of workers experienced reduced salaries up to a certain limit which was almost half the regular income, increased workload, or both together. Unexpected unemployment or reduction in income affected the livelihood prospects of many domestic workers resulting in the loss of capability of their families for fulfilling even the basic needs of the members. Other earning members' loss of jobs also aggravated the situation. 56% of respondents faced termination from job by their employers with no future assurance that they will be hired back or not. 20% of workers left the job themselves owing to the fear of their health and 14% were the victims of the social stigma associated with COVID-19 virus, Only 10% of workers felt no change in their job. All family members staying together in congested living conditions the whole day caused increased working, strife, and privacy issues in the house, this all had a direct effect, especially on the women of the house. Taking care of children who remain at home for the whole day also was an added concern for many women. All the respondents said that their quality of life slipped down. For many going to the workplace was a chance to get away from day-to-day family hassles and have some social interaction. Relaxation experienced while conversation with friends or neighbors too reduced in pandemic. The inability to pay for basic things like television cables or mobile phones took away day-to-day recreational activities increasing domestic violence and stress in relationships. 64% of respondents said they faced domestic violence and 26% agreed that the household work increased 10% of women face a loss of self-respect due to the loss of their job. Domestic workers especially women domestic workers do not have good socioeconomic conditions and face problems both at home and at the workplace. They migrate from rural areas to urban areas in search of employment opportunities. They have a heavy workload with less recognition and remuneration. They work hard for a better future (Hazarika et al., 2002, 11). They spend a large part of their time accessing essential services such as water and toilets. They do not have access to institutional care facilities that provide quality care at affordable rates in their neighborhoods (Jagori, 2004:14). Domestic workers live and work in appalling conditions and are vulnerable to abuse. Their self-esteem suffers considerable damage after prolonged periods of maltreatment, abuse, and humiliation. They feel inadequate, powerless, and worthless. They do not have the opportunity to raise their voice and avenue of redress before the competent organs (Mantouvalou, 2006:20). The women domestic workers face the daunting challenge of combining paid work with their maternal role and long hours of unpaid care work. The women domestic workers are more likely to resort to unfavorable coping strategies, such as leaving children alone at home, enlisting the help of an older sibling or young relative, or taking children to work, if allowed, with adverse consequences on children's health and education as well as worker's productivity (Cassier and Addati, 2007:04). The unorganized sector plays a vital role in terms of providing employment opportunities to a large segment of the workforce in India

Conclusion

Till date, the Central Government has not placed any separate law for the protection of domestic workers. However, as a result of pandemic effect the Ministry of Labour & Employment is considering introducing a National Policy on Domestic Workers which is currently in the draft stage. The salient features of the proposed draft National Policy on Domestic Workers will be to include domestic workers in the existing legislation and to give them the right to register as unorganized workers so as to facilitate their access to rights and benefits. The policy will also give them the right to form their association or union and claim the minimum wages and right to enhance their skills. More emphasis will be given to the protection of Domestic Workers from abuse and exploitation and have access to courts, and tribunals for grievance redressal. The government of India had come up with the Domestic Workers (Registration, Social Security and welfare) Act in 2008 to regulate payment and working conditions and check the exploitation and trafficking of women and young household workers however in practice the implementation is zero. Covid pandemic has shown us that we need to pay attention to the emotional, psychological, social, and well-being of domestic workers by handling their issues like irregular labor, exploitation, and undefined wage and working conditions. The State Governments of Andhra Pradesh, Jharkhand, Karnataka, Kerela, Odisha, Rajasthan, Haryana, Punjab, Tamilnadu, and Tripura have included domestic workers in the schedule of Minimum Wages Act and the Workers are also entitled to file case before the concerned authorities in case of grievance in this regard. The matter of constitution of State Domestic Workers Board is under the jurisdiction of the State Governments.

References

Anandi, S., and Deepa, E. (2020 June) Protecting livelihood, health and decency of work paid domestic workers in the time of COVID-19 (MIDS Occasional Policy paper, COVID-19 Series)

Bharti, M. (2020, June) Helping the helpers. The Hindu. Retrieved 25 December 2020 from https://www.thehindu.com

Bhattacharya, D., Sukumar, M., and Mani, M. (2016). Living on the margins: A study of domestic workers in Chennai, India.

Chandramaouli, K. (2018). Women domestic workers in India: An analysis. International Journal of Innovative Technology and Exploring Engineering, 8, 1–5.

Deshpande, A. (2020). The Covid-19 pandemic and lockdown: First effects on gender gaps in employment and domestic work in India? (Working paper30). Department of Economics, Ashoka University

Entrepreneurial Education and Entrepreneurial Intentions: Mediation of Entrepreneurial Mindset and Moderated Mediation of Creativity

T. Ravikumar,[a] N. Murugan,[2] J. Suhashini,[c] and D. Lavanya[d]

[a]School of Business and Management, CHRIST (Deemed University), Bangalore, Karnataka, India
[b,c,d]Department of Management Studies, PSNA College of Engineering and Technology, Dindigul, Tamil Nadu, India
E-mail: drravibhagavath@gmail.com

Abstract

Entrepreneurship creates employment and contributes to the economy and enhances the competitiveness of the country. Entrepreneurial education is one of the most important issues highlighted today Entrepreneurship education (EE) is a policy initiative to promote entrepreneurial activity. An entrepreneurial mindset reflects a person's eagerness to take up entrepreneurial activities. Entrepreneurial intention (EI) is a person's wish to the entrepreneurial events related to self-employment or new business. This study has been descriptive and examines the effect of EE on Eis and examines the intervention of entrepreneurial mindset and moderated mediation of creativity. The survey method is applied for collecting primary data. A structured questionnaire is used. Respondents of the study are the students who pursue the business program in Bangalore. This study is a cross-sectional one. The study has found that there is a linear and statistically substantial relationship between EE and Eis of business students.

Keywords: Entrepreneurship, Education, Entrepreneurial Mindset, Entrepreneurial Intentions, Creativity

Introduction

Education promotes a country through its contribution to the quality of human resources (Usman and Nia, 2020). Entrepreneurship creates employment and contributes to the economy, and enhances the competitiveness of the country (Seth, 2020). Entrepreneurial education is one of the most important areas highlighted today. It is described as a significant part of sustainable growth. Entrepreneurial education is vital for the efficient and effective performance of productive activities using commitments, initiatives, and innovative efforts (Usman and Widyanti, 2020).

Entrepreneurship is largely about invention and ingenuity in the business, while entrepreneurship education is the foundation for polishing the invention. Entrepreneurship is essential for education. Entrepreneurship is the determinant of the development of the economy (Nowiński et al., 2019). So, interest in entrepreneurial education has been growing (Ndofirepi, 2020). Entrepreneurship looks for new opportunities and exploits the new opportunities identified. Integration of EE motivates the students to become entrepreneurs rather than to be job seekers (Prajapati, 2019).

Entrepreneurial education (EE) has been researched extensively in recent times, especially on the impact of EE on the development of the economy (Boahemaah et al., 2020). EE refers to "the structured formal conveyance of entrepreneurial knowledge" (Al Ghafri and Malik, 2021). Entrepreneurship education is a policy initiative to promote entrepreneurial activity (Nowiński et al., 2019).

Entrepreneurship programs in colleges and universities are the realization of the need for entrepreneurial education and more research is needed particularly in terms of the types, objectives, and outcomes of these courses (Seth, 2020). Entrepreneurial education revolves around the students starting a company or creating a business, instead of making them more innovative, opportunity-oriented, productive, and risk-taking (Usman and Widyanti, 2020). Entrepreneurial education should intend to induce the students' entrepreneurial intentions (Eis). Entrepreneurial education can motivate willingness toward entrepreneurship among students significantly. The entrepreneurial mindset of the students will make a difference in EE and Eis. Furthermore, creativity affects EE and Eis. This study aims at studying EE and Eis relationship. Also, this research work focuses on the mediation role of the entrepreneurial mindset and moderated mediation role of creativity in the above-said relationship.

Review of Literature

Unemployment has been a perennial problem in almost all countries, especially in developing and least-developed countries (Khalifa et al., 2020). One of the determinants of unemployment is the "job-seeking mindset of the students" (Mahendra et al., 2017). Entrepreneurial skills can be inculcated through education (Lv et al., 2021).

Entrepreneurship is believed an effective approach to overcoming this problem (Mahendra et al., 2017). Entrepreneurship is "a process, action, or activity to convert an idea into a value-added product or service" (Hattab, 2014). Entrepreneurial activities contribute growth and development of the national economy (Lv et al., 2021). Entrepreneurial education has been gaining attention and growing swiftly as it resolves several important social problems and entrepreneurial education promotes entrepreneurial intentions (Lv et al., 2021). Entrepreneurial education improves "knowledge, abilities, skills, and perceptions towards entrepreneurship" (Moses et al., 2016). Entrepreneurship teaching is part of EE and EE significantly explains variance in Eis (Lv et al., 2021).

EI is psychological desire that motivates individuals toward new business activities (Hattab, 2014). Entrepreneurial intention is a person's eagerness to set up and operate his own business (Prajapati, 2019). "Entrepreneurial attitudes, behaviors, and perception towards viability and success of innovative ideas have an impact on students' intentions to become entrepreneurs" (Prajapati, 2019). The EE programs and students' entrepreneurial intentions are matters of academic interest (Moses et al., 2016). The mindset of the entrepreneurs determines their success and failure of entrepreneurs (Wardana et al., 2020). The mindset of the entrepreneurs grows over a period (Jiatong et al., 2021).

Creativity is a distinct feature of an individual's cognitive process and it creates new ideas and plans through knowledge and information (Jiatong et al., 2021). Creativity plays an important job in Eis (Cardon et al., 2013; Gubik and Farkas, 2020; Lv et al., 2021; Weeransinghe, 2020).

Methods and Materials

This study has been descriptive and examines the effect of EE on Eis. The survey method is applied for collecting primary data. A structured questionnaire is used. The research instrument has 2 parts. Part A deals with the demographic characteristics of the respondents and Part B has statements in 5 point Likert scale to measure the core variables such as entrepreneurial education, entrepreneurial mindset, entrepreneurial intentions, and creativity. Respondents of the study are the students who pursue business-related programs in Bangalore. This study is conducted at a stretch.

The students pursuing business programs at undergraduate and graduate levels in India form the population of the study. However, on account of thinking about the feasibility, the target population for the study is decided as students pursuing business programs at the undergraduate and graduate level in Bangalore. Bangalore has been a start-up hub and provides ample job opportunities. Bangalore accommodates many prestigious educational institutions and universities. There are more than 5 lakh business students in Bangalore. The sample size has been determined using Krejice and Morgon's (1970) formula. According to Krejice and Morgon's table, when the population of the study is above 5 lakhs, the sample size will be 382 at a 95% level of significance. So, the sample size considered for the study is 382. Based on the nature and requirements of the study, the judgment sampling technique is applied to gather the data from the respondents. Variables considered for the study such as EE, entrepreneurial mindset, creativity, and entrepreneurial intentions of business students are measured using appropriate scales available. Based on a thorough literature survey, measurement scales of the variables are finalized and presented in the following table – 1.

Table 1. Measurement Scales

Variable	Scale Name	Author (s) and Year	No of Items
Entrepreneurial education	Entrepreneurial education scale	Wardana et al. (2020)	6
Entrepreneurial mindset	Entrepreneurial mindset scale	Wardana et al. (2020)	6
Creativity	Creativity scale	Biraglia and Kadile (2017)	6
Entrepreneurial intentions (Eis)	Entrepreneurial intentions scale	Liñán et al. (2011)	6

Results and Discussions

Primary data collected was checked for normality. Both the normality tests reveal that the core variables of this study are not normal. Further, reliability scores of the scales were measured and they have presented in Table 2. α scores are satisfactory as they are above 0.7.

Table 2. Reliability Scores

Scale	α Score	N
Entrepreneurial education	0.718	6
Entrepreneurial mindset	0.709	6
Creativity	0.828	6
Entrepreneurial intention	0.761	6

Male respondents have 54.1% and female respondents have been 45.9%. Male respondents are the dominant group in the sample. 49.7% of the sample units belong to 17 years to 20 years of age and 50.3% of respondents are in 21 years to 24 years of age. Most of the participants study in undergraduate programs (212 participants) while 170 of the participants study in the postgraduate program. 54.9% of the students study in private institutions and 45.1% of students study in government institutions. Most of the students are from urban areas (75.13%) and the rest of the students are from a rural area (24.87%).

To measure and analyze the variances in entrepreneurial education, entrepreneurial intentions, creativity, and entrepreneurial mindset relating to demographic characteristics of the sample units, non-parametric variance analysis statistical tools have been applied. Entrepreneurial education varies based on age and location where the business program is pursued. Entrepreneurial intention changes based on age group and location where the business program is pursued. Creativity differs based on age group and location where the business program is pursued. Entrepreneurial mindset varies based nature of the business school.

Andrew F Hayes PROCESS model 4 has been employed to test the mediation effect of entrepreneurial mindset in EE and Eis relations. Tables 3 and 4 exhibit the direct relation between EE and Eis. Entrepreneurial education significantly impacts Eis of the business students to the extent of 24.16% (Table 3) and this model is a significant one (Table 4).

Table 3. Summary of the model

r	r^2	MSE	F- value	P value
0.4915	0.2416	0.2915	121.1611	0.000

Dependent variable: Entrepreneurial intention

Table 4. Model

Particulars	Coefficients	SE	t	P
Constant	2.2706	0.1623	13.987	0.000
Entrepreneurial education	0.4478	0.0406	11.0300	0.000

Dependent variable: Entrepreneurial intention

Tables – 5 and 6 indicate the indirect relationship between EE and Eis through the mindset of the entrepreneurs. The indirect relationship between EE and Eis is a significant one and EE impacts Eis of business students to the extent of 62.76% (Table-5) when this relation is mediated by an entrepreneurial mindset and this model is a significant one (Table-6).

Table 5. Summary of the model

r	r^2	MSE	F value	P value
0.7922	0.6276	0.1520	321.089	0.000

Dependent variable: Entrepreneurial intention

Table 6. Research Model

Specifics	Coefficients	SE	t	P
Constant	0.2659	0.1441	1.8448	0.000
Entrepreneurial education	0.3552	0.0337	10.5513	0.000
Entrepreneurial mindset	0.5497	0.0369	14.8796	0.000

Dependent variable: Entrepreneurial intention

EE impacts the Eis of the students more strongly in the indirect relationship (62.76%) than in the direct relationship (24.16%). So, it is quite evident that an entrepreneurial mindset significantly mediates the EE and Eis relation. To check whether the entrepreneurial mindset mediation is moderated by creativity or not, Andrew F Hayes PROCESS model 7 has been applied. Table 7 indicates that the direct effect is a significant one.

Table 7. Direct effect

Effect	SE	t	P
0.3552	0.0337	10.5513	0.000

Outcome variable: Entrepreneurial intention

The conditional indirect effect of creativity on the indirect EE and Eis relation is displayed in Table 8. As "zero" is not there between the "Bootstrap Lower-Level Confidence Interval score" and "Bootstrap Upper-Level Confidence Interval score" for -1SD, Mean, and 1SD levels, the conditional indirect effect of creativity on the EE and Eis indirect relation is significant.

Table 8. Conditional Indirect effects

Creativity	Effect	BootSE	BootLLCI	BootULCI
3.3333	0.14	0.031	.0876	.2123
4.0000	0.10	0.021	.0637	.1470
4.6667	0.06	0.026	.0109	.1147

Table 9. Index of moderated mediation

Particulars	Index	BootSE	BootLLCI	BootULCI
Creativity	-.062	0.030	-0.1261	-0.0060

The omnibus test presented in Table 9 indicates that the moderated mediation effect exists in the EE and Eis indirect relation as zero is not there between "Bootstrap Lower-Level Confidence Interval score" and "Bootstrap Upper-Level Confidence Interval score."

The study reveals that EE, Eis, and creativity of the sample students differ based on their age and the location where they pursue their education. On the other hand, the entrepreneurial mindset differs only according to the nature of the business school where they study. Mean scores reveal that "entrepreneurial education, entrepreneurial intention, and creativity" are more among undergraduate students than postgraduate students. Further, "entrepreneurial education, entrepreneurial intention, and creativity" are more among the students studying in urban areas. So, policymakers of the educational institutions may take necessary actions to bring the same level of "entrepreneurial education, entrepreneurial intention, and creativity" among rural students as well. This study confirms the linear and statistically substantial relationship between EE and Eis.

The linear and statistically substantial relationship between EE and Eis had been documented in numerous studies (Abdulrasheed et al., 2019; Mahendra et al., 2017; Seth, 2020; Usman and Widyanti, 2020; Weerasinghe, 2020). The entrepreneurial mindset mediates the EE and Eis relation and in fact, the entrepreneurial mindset enhances the impact magnitude of entrepreneurial education. So, the entrepreneurial mindset promotes the impact of EE on entrepreneurial intentions. Creativity significantly mediates the indirect relationship between EE and Eis through the mindset of the entrepreneurs.

Conclusion

This study aims at measuring and analyze the direct relationship between EE and Eis of business students in India. Further, it studies the mediation of the mindset of the entrepreneurs in the direct relationship between EE and Eis and moderated the mediation role of creativity in the indirect relationship. The study results found that there is a linear and statistically substantial relationship between EE and Eis of business students. Furthermore, the results found that the entrepreneurial mindset of the students significantly and positively mediates the direct relationship between EE and Eis. Creativity significantly plays moderated mediation role in the EE and Eis indirect relations with business students in India.

References

Abdulrasheed, J., Suleiman, Y., Bolaji, H. O., and Tunbosun, L. A. (2019). Assessing the Impact of Entrepreneurship Course on Entrepreneurial Intention: A Case Study of Al-Hikmah University Undergraduate Students. Amity Journal of Entrepreneurship, 4(1), 1–14.

Al Ghafri, F., and Malik, M. (2021). Entrepreneurship Education and Its Effect on Entrepreneurial Intentions of Omani Undergraduate Students. SHS Web of Conferences, 124, 05005. https://doi.org/10.1051/shsconf/202112405005

Biraglia, A., and Kadile, V. (2017). The Role of Entrepreneurial Passion and Creativity in Developing Entrepreneurial Intentions: Insights from American Homebrewers. Journal of Small Business Management, 55(1), 170–188. https://doi.org/https://doi.org/10.1111/jsbm.12242

Boahemaah, L., Xin, L., Dobge, C. S. K., and Pomegbe, W. W. K. (2020). The Impact of Entrepreneurship Education on the Entrepreneurial Intention of Students in Tertiary Institutions. International Journal of Management, Accounting & Economics, 7(4), 123–146. https://search.ebscohost.com/login.aspx?direct=true&AuthType=ip,sso&db=bth&AN=144332680&site=ehost-live&custid=s1020214

Cardon, M. S., Gregoire, D. A., Stevens, C. E., and Patel, P. C. (2013). Measuring entrepreneurial passion: Conceptual foundations and scale validation. Journal of Business Venturing, 28(3), 373–396. https://doi.org/10.1016/j.jbusvent.2012.03.003

Gubik, A. S., and Farkas, S. (2020). Entrepreneurial Intention in the Visegrad Countries. The Danube, 10(4), 347–368. https://doi.org/10.2478/danb-2019-0018

Hattab, H. W. (2014). Impact of Entrepreneurship Education on Entrepreneurial Intentions of University Students in Egypt. Journal of Entrepreneurship, 23(1), 1–18. https://doi.org/10.1177/0971355713513346

Jiatong, W., Murad, M., Bajun, F., Tufail, M. S., Mirza, F., and Rafiq, M. (2021). Impact of Entrepreneurial Education, Mindset, and Creativity on Entrepreneurial Intention: Mediating Role of Entrepreneurial Self-Efficacy. In Frontiers in Psychology (Vol. 12). https://doi.org/10.3389/fpsyg.2021.724440

Khalifa, A. H., Dhiaf, M. M., Abdulrasheed, J., Suleiman, Y., Bolaji, H. O., Tunbosun, L. A., Mahendra, A. M., Djatmika, E. T., Hermawan, A., Fadli, ., Muchtar, Y. C., Qamariah, I., Gubik, A. S., Farkas, S., Hattab, H. W., Li, Z., Islam, A. Y. M. A., Kong, H., Kim, H., ... 宗成庆. (2020). The Effect of Entrepreneurship Education on Entrepreneurial Intention. Frontiers in Psychology, 12(1), 970–985. https://doi.org/10.2478/danb-2019-0018

Liñán, F., Rodríguez-Cohard, J. C., and Rueda-Cantuche, J. M. (2011). Factors affecting entrepreneurial intention levels: a role for education. International Entrepreneurship and Management Journal, 7(2), 195–218. https://doi.org/10.1007/s11365-010-0154-z

Lv, Y., Chen, Y., Sha, Y., Wang, J., An, L., Chen, T., Huang, X., Huang, Y., and Huang, L. (2021).

How Entrepreneurship Education at Universities Influences Entrepreneurial Intention: Mediating Effect Based on Entrepreneurial Competence. Frontiers in Psychology, 12(July), 1–12. https://doi.org/10.3389/fpsyg.2021.655868

Mahendra, A. M., Djatmika, E. T., and Hermawan, A. (2017). The Effect of Entrepreneurship Education on Entrepreneurial Intention Mediated by Motivation and Attitude among Management Students, State University of Malang, Indonesia. International Education Studies, 10(9), 61. https://doi.org/10.5539/ies.v10n9p61

Moses, C. L., Olokundun, M. A., Akinbode, M., Agboola, M., and Inelo, F. (2016). Entrepreneurship Education and Entrepreneurial Intentions: The Moderating Role of Passion. The Social Sciences, 11(5), 645–653.

Ndofirepi, T. M. (2020). Relationship between entrepreneurship education and entrepreneurial goal intentions: psychological traits as mediators. In Journal of Innovation and Entrepreneurship. 9(1). https://doi.org/10.1186/s13731-020-0115-x

Nowiński, W., Haddoud, M. Y., Lančarič, D., Egerová, D., and Czeglédi, C. (2019). The impact of entrepreneurship education, entrepreneurial self-efficacy, and gender on entrepreneurial intentions of university students in the Visegrad countries. Studies in Higher Education, 44(2), 361–379. https://doi.org/10.1080/03075079.2017.1365359

Prajapati, B. (2019). Entrepreneurial Intention Among Business Students: The Effect of Entrepreneurship Education. Westcliff International Journal of Applied Research, 3(1), 54–67. https://doi.org/10.47670/wuwijar201931bp

Seth, K. P. (2020). The impact of Entrepreneurship Education on Entrepreneurial Intention : An empirical study of entrepreneurship education's four key characteristics, 244.

Usman, O., and Nia, S. T. (2020). The Impact Of Entrepreneurship Education, Self Efficacy, Creativity, and Gender On Entrepreneurial Intentions.

Usman, O., and Widyanti, J. (2020). The Impact of Entrepreneurship Education, Entrepreneurial Self-Efficacy, and Gender on Entrepreneurial Intention Osly.

Wardana, L. W., Narmaditya, B. S., Wibowo, A., Mahendra, A. M., Wibowo, N. A., Harwida, G., and Rohman, A. N. (2020). The impact of entrepreneurship education and students' entrepreneurial mindset: the mediating role of attitude and self-efficacy. In Heliyon (Vol. 6, Issue 9). https://doi.org/10.1016/j.heliyon.2020.e04922

Weeransinghe, R. N. (2020). Entrepreneurial education and entrepreneurial intention among ordinary level students in Kelaniya education zone in Sri Lanka. International Journal of Multidisciplinary and Current Educational Research, 2(4), 122–131.

Role of Corporate Social Responsibility in Achieving Sustainable Development Goals

A. Menaga[a] *and S. Vasantha*[*,b]

[a]School of Management Studies, Vels Institute of Science, Technology & Advanced Studies (VISTAS), Chennai, India.
[b]School of Management Studies, Vels Institute of Science, Technology and Advanced Studies (VISTAS), Chennai, India
E-mail: [a]Menagalokesh@gmail.com; [b]vasantha.sms@velsuniv.ac.in

Abstract

The purpose of the Research is to determine how important corporate social responsibility is in accomplishing Sustainable Development Goals. The study identified three major Features of Corporate social responsibility (Job Creation, Labor practice, and Energy usage) which can lead to attaining the sustainable development goal. This quantitative study focuses on the Features of Corporate social responsibility for achieving the SDGs. For the study, both primary and secondary data were consulted. To collect primary data on sustainable development in corporate social responsibility, structured questionnaires were used. The Research concluded that corporate social responsibility helps with energy use, labor practices, and job creation to achieve the objectives of sustainable development. CSR initiatives are a type of management approach that helps organizations accomplish SDGs, which are essential for addressing the needs of both the present and future generations.

Keywords: Corporate social responsibility, Energy usage, Job Creation, Labor practice, SDG

Introduction

As a continuation of the Millennium Development Goals, On September 25, 2015, the Sustainable Development Goals (SDGs) were endorsed by 193 countries. The Sustainable Development Goals (SDGs) are a new agenda for sustainable development that emphasizes ending poverty, protecting the environment, and providing prosperity for all. A total of 17 goals and 169 targets are anticipated to be accomplished by 2030, and their accomplishment will call for the cooperation of the executive branch, private sector, and civil society organizations. The SDGs for sustainable growth are discussed in this research as a way to leverage corporate social responsibility (CSR) in a way that benefits profit, people, and the environment.

The Research focuses on SDGs Goal 8 which formally covers CSR features of social, economic, and environmental goal 8 promoting measures to help businesses expand and create jobs, enhancing resource efficiency in production and consumption, promoting full employment and fair work with equitable pay, and promoting healthy and sustainable tourism. The overall aim of this goal is to the beginning of your sustainable development path on business commitments.

The Role of CSR in achieving sustainable development goals was measured by the research using primary quantitative data, and the findings were consistent with one another. It listed the three main SDG components, including energy use, job creation, and labor standards. The report provides a thorough grasp of how CSR contributes to sustainable development objectives.

Objectives

- To study the role of corporate social responsibility in achieving the SDGs
- To Analyze the Features of Corporate social responsibility in achieving the SDGs

Literature Review

Promoting full and productive employment, decent work for all, and sustained, inclusive, and sustainable economic growth are three of the eight SDGs goals that are underlined by Jeremy Moon (2007). These three goals are comprised of CSR components, which are economic, social, and environmental. Grosser and Moon (2005) state that to Achieve the goal, Implement comprehensive frameworks for employment policy, including assistance for the institutions of the labor market. Implement minimal living wages with full participation from social associates. Espousing a "Just Transition" strategy to change over to a low-carbon economy and develop green jobs. The article highlights the significance of ongoing attempts to boost CSR initiative capacity globally as well as providing value for the business, people, and the environment for the complete realization of SDGs through CSR.

Methodology

The study focused on features of corporate social responsibility, employment, labor practices, and energy usage. To gather data and examine the CSR activities of the firm, a structured questionnaire has been adopted. A five-point Likert scale was used to score a total of 30 questions.

For collecting Primary data Purposive sampling was used to choose 200 employees from various CSR-initiated organizations The minimum sample size was determined to be (30*5=150) respondents using Hair et al. (2007) criteria. This study's research design is descriptive.

The reliability of the questionnaire is evaluated using SPSS, demonstrating its consistency. Economic growth, high-quality education, and innovation all have reliability (Cronbach's alpha) values of 0.971, 0.821, and 0.999, respectively. The questionnaire's overall reliability is 0.842. Equation modeling has been used to determine the relationship between the independent variables (Job Creation, Labor Practices, and Energy Usage) and the dependent variable (Achieving SDG)

Data Analysis and Results

Hypothesis

H_1 Job Creation is positively related to achieving the SDGs.

H_2 Labor Practices are positively related to achieving the SDGs

H_3 Energy Usage is positively related to achieving the SDGs

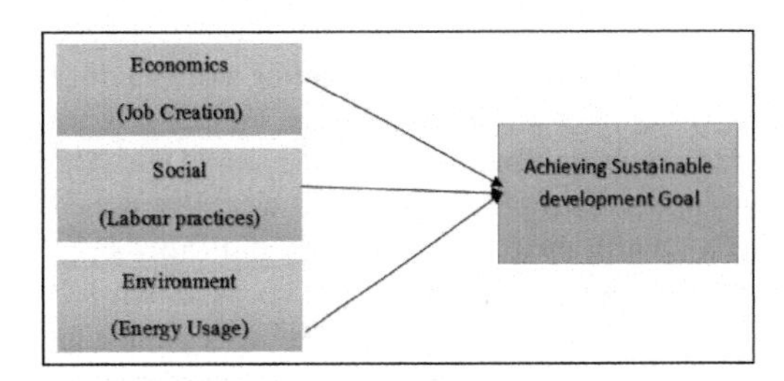

Fig. 1. Proposed Conceptual framework on Features of CSR in achieving the SDGs

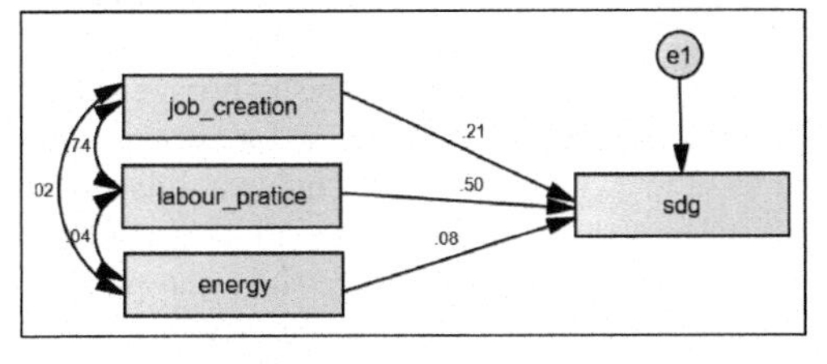

Fig. 2. Structural Equation Model (SEM) for the Aspects of CSR in achieving SDGs

Table 1. Variables used in structural equation model

Dependent Variables		Independent Variables	Unstandardized Coefficient	S.E.	Standardized Coefficient	T Value	P - Value	Result of Hypothesis
Achieving SDG	<---	Job Creation	.212	.076	.206	2.719	.007	H1 is supported
Achieving SDG	<---	Labor practices	.503	.065	.419	6.456	.000	H2 is supported
Achieving SDG	<---	Energy usage	.078	.050	.076	1.507	.032	H3 is supported

The following variables are utilized in the conceptual framework::

I. Observed, endogenous variables: Job creation, Labor practices, and energy usage on Achieving SDG.

II. Unobserved, exogenous variables: e1

The path analysis model is drawn for testing endogenous and exogenous variables using AMOS.

Table 1 explains the causal relationship between Job creation, Labor practices, and energy consumption in Achieving SDGs.

Job creation has a positive effect on Achieving SDGs. Job creation has an unstandardized coefficient value of 0.212, indicating a satisfactory impact. Job creation will upsurge by 0.212 times for a single unit increase in Achieving SDGs. The unstandardized coefficient value was determined to be significant at a 1% level of significance since the p-value > 0.05. This study has proved that Job creation is positively associated with achieving the SDGs. Hence, H1 of this study is accepted.

Labor practices have a positive outcome on Achieving SDGs. Labor practices have an Unstandardized Coefficient value of 0.503, indicating a positive effect. The Labor practices will increase by 0.503 times for every unit increase in Achieving SDGs. The unstandardized coefficient value was determined to be significant at the 1% level of significance since the p-value was less than 0.05. H2 has been accepted.

Energy usage has a direct positive effect on Achieving SDGs. The unstandardized coefficient value of 0.078 energy is used. According to the predicted positive indicators, the positive impact of energy consumption will rise by 0.078 times for every unit that the SDGs are attained. The asymmetrical coefficient value was determined to be significant at the 1% level of significance since the p-value was less than 0.05.

H3 is also accepted and indicated the direct relationship between Energy usage and achieving the SDGs is significant for the study.

The Sustainable Development Goals (SDGs) offer a motivating vision for bettering the state of the planet. The achievement of the SDGs will heavily rely on ethical business. Businesses and corporate executives who do this will not only assist create a world that is affluent and more sustainable, but they will also help usher in the next business era by allowing new global markets to flourish.

To change unsustainable patterns of consumption and production and move toward more sustainable patterns, governments, international organizations, the business sector, and other non-State entities and individuals must contribute. With its creativity, investment, and invention, private business activity plays a critical role in resolving difficulties associated with sustainable development. An important factor behind sustainable development can be a vibrant, well-functioning corporate sector that upholds labor rights, environmental protections, and health standards in compliance with pertinent international standards and agreements (United Nations General Assembly, 2015).

Implications

The managerial strategy will prefer to adopt SDG achievement for Job creation, labour management and energy consumption (Profit, people, planet) as their priority in CSR iniativities because it is regarded as the CSR's next important precursor. For socially conscious corporate enterprises, the importance of the role that job creation plays (Bhandarker, 2003). Thus this research supports a sustainable CSR initiative for the organization for promoting better job creation, proper labour management and safeguarding of the planet.

Conclusion

There are significant economic, social, and environmental concerns facing by our globe. 193 nations agreed and supported a set of goals as part of a new global sustainable development agenda on September 25, 2015. The Sustainable Development Goals (SDGs) present an opportunity to chart a sustainable course for the globe and to specify global objectives for 2030. There are a total of 169 targets spread among the 17 goals, each of which has several targets. To achieve the global goals for Sustainable Development, the public, corporate sector, civil society, and government all have significant roles to play. The goal of creating a highly-skilled, competitive economy and accomplishing global goals across all sectors will support advantages to people around the world. Additionally, it will assist us in realizing our objective of establishing a reputation as a Center of Excellence for moral and environmentally friendly business operations. In-depth information on CSR and SDG accomplishment is provided in this study. The organization expanded its knowledge of sustainability in all associated dimensions. Incorporating CSR and SDGs into business strategy models opens up new options for creating cutting-edge goods and services, greater employment chances, an understanding of labor, and resource conservation that meet the highest standards of sustainability.

References

Grosser, K., and Moon, J. (2005). Gender mainstreaming and corporate social responsibility: Reporting workplace issues. Journal of business ethics, 62(4), 327–340.

Bhandarker, A. (2003). Building corporate transformation new HR agenda. Vision, 7(2), 1–23.

United Nations General Assembly (2015). Transforming our world: the 2030 Agenda for Sustainable Development. Resolution adopted by the General Assembly on 25 September 2015. http://www.un.org/ga/search/view_doc.asp? Symbol=A/RES/70/1&Lang=Downloaded: 2016. 01. 15.

Moon, J. (2007). The contribution of corporate social responsibility to sustainable development. Sustainable development, 15(5), 296–306.

Hair, J. F., Money, A. H., Samouel, P., and Page, M. (2007). Research methods for business. Education+ Training.

Feasibility of DREAMS Afterschool Intervention to Implement SDG – 4, 5, and 11 in Rural India

G. S. Prakasha,[*,a] *Lijo Thomas,*[b] *and Jestin Joseph*[c]

[a, b, c]Christ University of Bangalore, India
E-mail: *prakasha.gs@christuniversity.in

Abstract

DREAMS stands for Desire, Readiness, Empowerment, Action, and Mastery for Success. It is an after-school intervention (ASI) to strengthen psychosocial development of youths. Present study aims to test the feasibility of DREAMS ASI in developing knowledge and awareness of UNs Sustainable Development Goals 4, 5, and 11 among rural youths of India. The intervention had 100 youth participants aged from 11 to 14 years. Mentors led the intervention under the supervision of community leaders of the villages. DREAMS core members trained mentors and community leaders on the intervention activities. Study employed a mixed-method study design to test the feasibility of ASI as an innovative alternative approach to spreading knowledge and awareness on SDGs. Study included a post-test and semi-structured interview with participants, mentors, and community leaders to test the feasibility. Study found ASI as an innovative alternative and effective approach to spreading knowledge and awareness on SDGs through post-test results and inductive analysis of the interview data. Study recommends similar studies to confirm the same.

Keywords: DREAMS, SDG 4, 5, & 11, Afterschool intervention

Introduction

UN summit leaders adopted 17 sustainable development goals (SDGs) as the agenda 2030 in September 2015. Though developed countries took immediate action to answer this call, the developing, middle, and low-income countries could begin their actionable plans at a slower pace. Geographically, India is occupied mainly by rural human settlements. Indian census 2011 reports that 68.84% of the population lives in rural areas. This population suffers various challenges, such as access to food, shelter, transportation, education, and other basic amenities. Education is one of the main means to reach out to them for any newer understanding of policies and rights. Taking UN's 17 SDGs to them is not easy unless they lend themselves an ear to understand it. Although schools in urban areas do introduce SDGs in one or the other forms within their school curriculum, we do not find such events in semi-urban or rural schools of India. Thus, one of the alternatives to introduce SDGs can also be through after-school programs' (ASP).

ASP gained its popularity in India in 2010 under the rural education and development (READ) project or village education improvement project (VEIP). Almost 90% of the ASP focuses on supplementary activities to support the school curriculum and very few focus on other knowledge areas such as psychosocial development. The present study looked at after-school program intervention as an alternative to spread knowledge and awareness on UN's sustainable development goals. After-school intervention is an innovative approach to spreading knowledge and awareness on SDGs, especially to rural population.

Background

Children usually spend their time watching TV or playing with friends after school hours. Participation in ASPs enhances their academic performance and social adaptation (Posner and Vandell, 1994). Deliberate designs and skillful delivery of ASP provide students with an opportunity to shape their future social, educational, and professional aspirations (Miller, 2012). Significant physical, emotional, and cognitive changes occur from early childhood to adolescence and communities have to channel this navigation for successful adulthood. Thus, time spent after school hours needs attention (Miller, 2003). ASPs are ineffective when they solely offer leisure or non-academic activities (Gaffney et al., 2021). Children in the age group of 5 to 18 and teenagers display sedentary behavior in ASP many a time (Arundell et al., 2016). Thus, there is a need for systematically planned ASP activities to engage the children meaningfully. Developing healthy authentic relationships between youth and adult mentors helps in meaningful learning. Being non-directive and providing a collaborative approach to student engagement support building social and emotional development (Chapin et al., 2022). ASPs focusing on developing socio-emotional learning of children must focus on developing positive youth-staff relationships and appointing competent staff (Hurd and Deutsch, 2017).

Longer participation in an after-school program will result in improved academic performance (Budd et al., 2020). Afterschool program activities on reading skills of 3-grade students showed a significantly mild effect on their performance (Baker et al., 2019). A systematic review study reveals that ASP effects have a linkage to better academic performance, stronger school bonding, and prosocial behavior (Lester et al., 2020). Children and young people living in public housing have the potential to perform better academically and behave better at school when exposed to community-based after-school activities (Jenson et al., 2018). Participation in organized after-school activities in middle school helps teenagers develop a stronger work orientation in ninth grade and better academic success in high school (Liu et al., 2021).

Objectives

- To measure the level of knowledge and awareness on SDG goals 4, 5, and 11 developed through DREAMS after-school intervention.
- To analyze the feasibility of DREAMS after-school intervention in developing knowledge and awareness of SDG Goals 4, 5, and 11.

Methods

The present study employed an exploratory research design involving a quanti-quali mixed method approach to address the research objectives raised. Figure 1 presents the pictorial form of the study design.

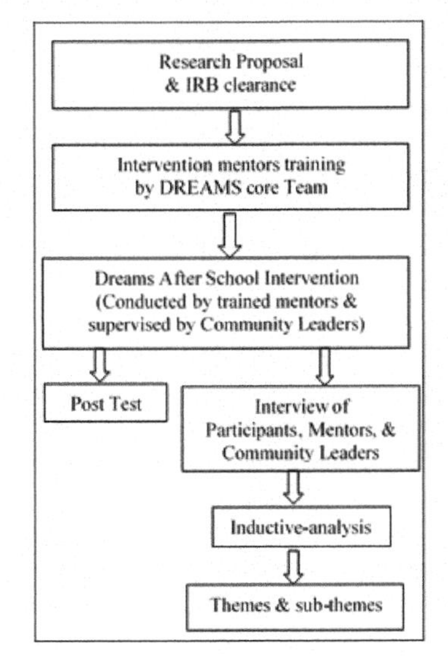

Fig. 1. Showing the study design

DREAMS core members trained the after-school intervention mentors and community leaders on the intervention module. Mentors conducted the intervention under the supervision of community leaders. Researchers constructed a valid post-test questionnaire to measure the achievement level of knowledge and awareness of SDGs 4, 5, and 11. The intervention was held in four different rural locations in India in May 2022. Altogether, 100 youths aged from 11 to 14 years participated. Each intervention lasted for five days. Intervention gave equal weightage to the UNs SDGs quality-education-4, gender-equality-5, and sustainable cities and communities-11. Experiential learning experiences in the intervention allowed students to construct knowledge on their own and internalize SDGs for a greater cause. Researchers administered the post-test questionnaire at the end of intervention.

Study conducted semi-structured interviews with intervention participants, intervention mentors, and community leaders, who directed and monitored the intervention. The interview with intervention participants focused on knowledge and awareness of SDG Goals 4, 5, and 11. Interviews of intervention mentors focused on the planning, execution, and challenges faced during the intervention. Interviews with community leaders focused on the effectiveness of intervention in providing knowledge on SDGs and their viewpoints on replicating and sustaining this intervention program for rest of the SDG goals. The study selected eight student participants, two from each intervention site, four intervention mentors, one from each site, and four community leaders - one from each site. The senior researcher, with a Ph. D. degree, conducted all the interviews in face-to-face mode. Assigned pseudonyms for all interview participants to ensure anonymity. Researchers subjected the interview transcripts to inductive content analysis. Table 2 presents the themes and subthemes (Maykut and Morehouse, 1994) emerged out of inductive analysis.

Results and Discussions

Post-Test Results

Post-test included a questionnaire measuring the participants' knowledge and awareness on SDGs 4, 5 and 11. The average score of all the four intervention sites is 84.5% indicating high level of knowledge and awareness. Figure 2 below presents the results of post-test scores.

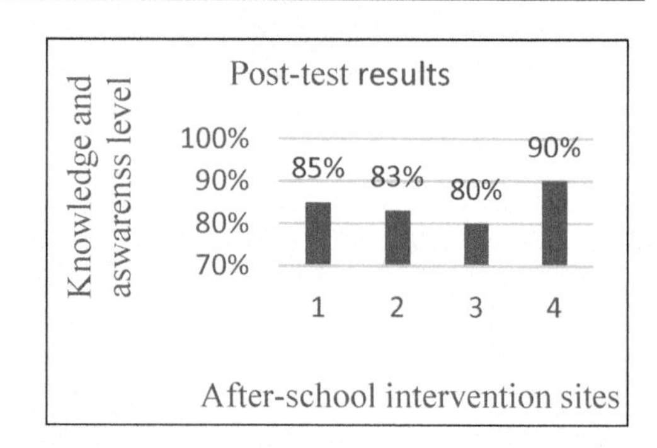

Fig. 2. Showing the results of post-test from intervention sites

Interview Results

Interview with Students: The interview with intervention participants' mainly focused on the intervention provided at the intervention site regarding knowledge and awareness on SDGs 4, 5, and 11. Students were inducted on UN's mission on SDGs in the interventions. Expert talks on quality education, gender equality, and sustainable cities and communities helped children understand the meaning of each goal.

Main theme: SDG awareness. Almost all the students who took part in the interview mentioned that they were not aware of the UN's sustainability goals. Student lacked knowledge about the existence of an intergovernmental organization, or the United Nations. Though the social sciences text talked about the UN, students lacked practical knowledge of the UN's roles and functions.

Subtheme: Values to reach goals. Several activities were conducted in the intervention to teach values like dedication, determination, resilience, willpower to excel, aiming high, and pursuing life's dreams. Students actively participated in most of these activities. An activity on inclusivity taught them caring, empathy and compassion.

Subtheme: Changes in self. The intervention activities and the experts' talk on SDG goals made students realize what was happening in society with respect to the quality of education they were receiving. The students were able to understand the differences in gender equality within their local context. They have a fair understanding of women's empowerment.

Subtheme: Gender equality & empowerment. Students mentioned that they do not see women's representation in every occupation. There is a need for women's representation in all kinds of occupations and work. Theatre activities, such as street plays on the dowry system, women's education, sexual harassment, women's trafficking, and infanticide taught them the value of gender equality and women's empowerment.

Main theme: Purpose in life. Students believe that they have understood the purpose and meaning of their lives. What they must do when they grow up, how to help humankind, how to be self-reliant, and so on are important. A summer intervention was a revelation for many to understand human life.

Subtheme: Lifelong learning. Students realized that education is a lifelong process and we learn from birth through old age, experiencing each day as a lesson for life. Schooling and education are the requirements for livelihood, whereas learning never stops throughout life.

Subtheme: Safe and Happy community. Students shared that they have to help themselves to be safe and happy. Keeping surroundings clean, saving resources for the future, healthy practices and safe routing of life practices are important. They realized that in a democratic country like India, living amid differences matters and they must make adjustments for harmonious living.

Interview with Intervention mentors: Interviews of intervention mentors mainly focused on the planning, execution, and challenges at the intervention site. The inductive analysis gave rise to the main theme - Planned activities. Which included Knowledge of SDGs, Awareness, Attitude, and Emotions of the participants as subthemes. Another main theme was Leadership, which included subthemes of Self-concept and Self-esteem. Mentors shared that sessions on who-am-I, self-realization, and self-actualization helped the children to understand their self-concepts and redefine themselves and their roles as human beings.

Interview of Community leaders: Interviews with community leaders focused on the effectiveness of intervention in providing knowledge on SDGs and their viewpoints on replicating and sustaining this intervention program for the remaining SDG goals. Community leaders expressed that teaching SDGs, as an intervention program is an innovative approach. It is effective, as opposed to embedding SDG knowledge within the curricular framework of the school. Thus, community leaders' response gave rise to two main themes; Effectiveness and Alternative innovative implementation plans to spread SDG awareness.

Conclusion

As intended, the present study was able to measure the feasibility of DREAMS ASI as an innovative approach to spreading knowledge and awareness of UN's SDGs 4, 5, and 11. An average score of 84.5% on the post-test indicated a high level of knowledge and awareness attainment. The inductive analysis of the interview revealed themes and sub-themes about knowledge, awareness, planning, and implementation. Study confirms that ASI is an innovative alternative approach to teaching SDGs to rural schoolchildren than a school hours curriculum. Study concludes that after-school interventions are an effective means to achieve the spread of knowledge and awareness on SDGs to support the UN's call. Future researchers' may explore conducting more robust interventions covering all the 17 SDG goals at ASPs.

References

Arundell, L., Fletcher, E., Salmon, J. Veitch. J., and Hinkely, T. (2016). A systematic review of the prevalence of sedentary behavior during the after-school period among children aged 5–18 years. International Journal of Behavioral Nutrition and Physical Activity, 13(93). https://doi.org/10.1186/s12966-016-0419-1

Baker, S. K., Kamata, A., Wright, A., Farmer, D., and Nippert, R. (2019). Using propensity score matching to estimate treatment effects of afterschool programs on third-grade reading outcomes. Journal of Community Psychology, 47(1), 117–134. https://doi.org/10.1002/jcop.22104

Budd, E. L., Nixon, C. T., Hymel, A. M., and Tanner-Smith, E. E. (2020). The impact of afterschool program attendance on academic outcomes of middle school students. Journal of Community Psychology, 48(8), 2439–2456. https://doi.org/10.1002/jcop.22416

Chapin, L. A., Fowler, M. A., and Deans, C. L. (2022). The role of adult facilitators in arts-based extracurricular settings: Perceived factors for success of adult–youth relationships. Journal of Community Psychology, 50(1), 176–190. https://doi.org/10.1002/jcop.22513

Gaffney, H., Ttofi, M. M., & Farrington, D. P. (2021). What works in anti-bullying programs? Analysis of effective intervention components. Journal of school psychology, 85, 37–56. https://doi.org/10.1016/j.jsp.2020.12.002

Hurd, N., and Deutsch, N. (2017). SEL-focused after-school programs. The Future of Children, 27(1), 95–115. https://doi.org/10.1353/foc.2017.0005https://doi.org/10.1353/foc.2017.0005

Jenson, J. M., Veeh, C., Anyon, Y., St. Mary, J. S., Calhoun, M., Tejada, J., and Lechuga-Peña, S. (2018). Effects of an afterschool program on the academic outcomes of children and youth residing in public housing neighborhoods: A quasi-experimental study. Children and Youth Services Review, 88, 211–217. https://doi.org/10.1016/j.childyouth.2018.03.014

Lester, A. M., Chow, J. C., and Melton, T. N. (2020). Quality is critical for meaningful synthesis of afterschool program effects: A systematic review and meta-analysis. Journal of Youth and Adolescence, 49(2), 369–382. https://doi.org/10.1007/s10964-019-01188-8

Liu, Y., Simpkins, S. D., and Vandell, D. L. (2021). Developmental pathways linking the quality and intensity of organized afterschool activities in middle school to academic performance in high school. Journal of Adolescence, 92, 152–164. https://doi.org/10.1016/j.adolescence.2021.09.002

Maykut, R., and Morehouse, P. (1994). Beginning qualitative inquiry: A practical and philosophic guide. London: Falmer.

Miller, B. M. (2003). Critical hours: Afterschool programs and educational success (ED482794). Educational Resources Information Center.

Miller, P. M. (2012). Community-based education and social capital in an urban after-school program, Education and Urban Society, 44(1), 35–60, https://doi.org/10.1177/0013124510380910

Posner, J. K., and Vandell, D. L. (1994). Low-income children's after-school care: Are there beneficial effects of after-school programs? Child Development, 65(2), Spec No, 440–456. https://doi.org/10.1111/j.1467-8624.1994.tb00762.

Crowdsourcing: A Technique to Sustain the Educational Industry

Dr. A. Dunstan Rajkumar[a] *and Dr. K. Kishore*[b]

[a]Vellore Institute of Technology, Vellore, India
[b]Voorhees College, Vellore 2, India 2
E-mail: *dunstanrajkumar.a@vit.ac.in

Abstract

Education is the foundation of our wisdom, character, ethics, and radical development in our personal development and in promoting our complete society. Today, the pandemic period has changed the delivery methods of teaching and learning. The sustainability of the educational industry is a question mark, starting from small tutoring institutes to larger universities in India. This paper attempts to review crowdsourcing (CS) as a remedy for these institutes to sustain and, as well, to adopt digital platforms to facilitate teaching and learning. The initial steps toward adopting crowdsourcing as a formal learning and teaching method have been taken by many private institutes in India. It has the potential to be a very useful and dependable educational paradigm if it is correctly conceived and implemented. Empirical investigations and theoretical knowledge structures are lacking. This paper systematically reviews the available sustainable methods for educational institutes. Future research can focus on a higher-principal framework as a result.

Keywords: Crowdsourcing, Content development, Practical experience, Education development, Feedback enhancement

Introduction

Traditional educational institutions are being transformed by IT-enabled teaching practices and open learning techniques that change students from passive content consumers to primary content creators and curators (Baggaley, 2013). The term "crowdsourcing for education" refers to the use of digital platforms to engage both learners and educators in the process of knowledge co-creation and sharing. This new type of crowdsourcing manifests itself as an IT-enabled educational strategy in which educators engage a large number of internal and external players in teaching and learning. Massive open online courses (MOOCs) are a recent online learning phenomenon that has made a significant contribution to the evolution of education. Open educational resources (OER) share some of the ideals of crowdsourcing. In addition to being open and free to use, some of them also allow for customization based on individual needs. Due to this enormous development in information and technology, traditional education is challenged. Many organizations have started to use the widespread wisdom and knowledge available worldwide. Crowdsourcing for education has come to the rescue as a way to sustain the field of education in institutes. "Crowdsourcing is *the collection of information, opinions, or work from a group of people,* usually sourced via the Internet. "Crowdsourcing work allows companies to save time and money while tapping into people with different skills or thoughts from all over the world" (Hargrave, 2021).

Concepts of Crowdsourcing

Crowdsourcing (CS) is the act of taking a job that was previously completed by a specified agent who is typically an employee and outsourcing it to an indeterminate, generally large group of people through an open call. Crowdsourcing is an important concept in research since it collaborates significantly by expanding the pool of scientific collaborators. Many crowd-sourced educational initiatives from a variety of fields have drawn thousands of volunteers and reached millions of people. They took a lot of time, energy, and money, but they also brought together a lot of people who were interested in making and sharing knowledge through constant social contact and cooperation (Katerina, 2020).

Several fascinating instances demonstrate how educators have utilized crowdsourcing to improve learning and teaching. The social media network Better Lesson, popularly known as the "Facebook for teachers," helps over 500,000 registered educators create, organize, and share instructional materials and lesson plans with other educators (Better Lesson, 2018). Crowdsourcing is also used to hire ad hoc online labor to complete various jobs, and it has become a popular outsourcing vehicle. However, crowdsourcing for "microtasks" fails to capitalize on the power of people to solve more complicated issues.

Crowdsourcing is a new business model in which work is performed by members of the general public, or the crowd. Crowdsourcing has been employed in a range of fields, including information system development, marketing, and operationalization. It has been demonstrated to be a viable paradigm in reference systems, software design and assessment, database design, and search engine assessment. Despite growing academic and industrial interest in CS, there is still a great deal of variation in how the notion is interpreted and applied.

Table 1. Overview of the hermeneutic approach of analysis and interpretation of Crowdsourcing for Education

Activity	Description	Identifications
Reading	Using analytical reading, the researcher obtained the ability to recognize essential concepts, results, and theories, as well as their interpretations, to deduce assumptions even when they were not expressly mentioned in the literature.	• Crowd voting • Crowdsourcing creative work • Crowd solving • Crowdfunding
Mapping and classifying	It was concerned with the systematic examination and classification of significant ideas, conclusions, and contributions to the knowledge contained within a body of literature.	• Additional Richness • Open Innovation • User Innovation
Critical assessment	The corpus of literature is assessed critically based on what is known, how knowledge is obtained, what forms of knowledge are created, how beneficial different types of information are in comprehending and describing a situation, and where current knowledge's bounds and weaknesses are.	• Contribution • Acquisition • Assessment • Interaction • Content • Interface • Legislation • Platform

Argument development	It drew on mapping and categorization to create a gap or problematization that motivates future investigation. Argumentation is used to generate future research paths and particular research issues.	• The taxonomy of crowd-oriented formal education • MOOC • Open Educational Resources • Sustainability
Research Problem / Questions	The gap in the literature logically leads to a more abstract research issue. Concrete questions developed from the theoretical issue.	• Adoption of crowdsourcing by conventional institutes • Lack of technical knowledge by the traditional faculties • Lack of Advanced Web technologies by the institutes.
Searching	The process of searching results in the identification of other books that may be read further.	• Advancement in Electronic Teaching & Learning Applications • Untapped Potential of CS

Crowdsourcing is a cooperation paradigm enabled by people-centric online technology that uses a dynamically created crowd of individuals who respond to an open call for investment to address individual, corporate, and societal problems. Crowdsourcing is rooted in the innovation process and co-creation research, so it is focused on how a lot of individuals—the "crowd" —can effectively participate in a firm's creative processes. CS originated in co-creation research, implying that it makes sense to broaden the pool of people who participate in the process of value creation. Because of the proliferation of broadband connections and advancements in web technology, teaching and learning have transformed substantially over the last decade. Crowdsourcing in education has been widely embraced and used by the worldwide community because it allows tough and time-consuming tasks to be done incrementally and by a large number of people working asynchronously at their leisure.

Literature Review and Methodology

In the new e-learning environment, there are new ways to collaborate and create knowledge. In a student-centered educational environment, collaborative learning links students, faculty, educational resources, and activities. The importance of students participating in teaching and learning has been elevated to the top of the list. This paper systematically reviews the available sustainability ways for educational institutes through CS, IT platforms, and learner motivators. Conventional Institutions, Open Online Institutions, and Individual Educators will be able to figure out how they will deliver the content to the learners. Sebastian and Dubravka (2014) used a method called a "hermeneutic literature review" for this study.

The method of performing literature reviews relied on a variety of sources to acquire a more in-depth knowledge of crowdsourcing in education, which was the goal of this study.

A keyword-based search was used to begin the literature review. A key phrase included an "AND" between "crowdsourcing" and "education," as well as a collection of "crowdsourcing" variants or alternative names (education, education development, practical experience, knowledge exchange, feedback enhancement). Snowballing and citation analysis was used to identify articles that did not include the search keywords but, in terms of substance, were included in the review. In the backward search, all previously recognized studies" bibliographies were searched, whereas the forward search employed the "cited by" feature of many databases. To begin, broad introductory works were examined from literature studies that mostly drew on Investopedia, Wikipedia, dictionaries, and Google searches to get information. As a result of these works, we were able to gain a general understanding of how different authors see crowdsourcing.

For the second part of the study, researchers looked at articles that dealt with crowdsourcing in the context of content production, practical experience, knowledge sharing and feedback enhancement, and information system development. To complete our search, we looked through the interdisciplinary research databases Scopus, Web of Science, and Google Scholar for any more study articles. These searches were conducted to uncover research on crowdsourcing, with a special emphasis on crowdsourcing for educational purposes. Searches were bolstered even further by the use of citation monitoring software and consultation with peers for new material. It was necessary to employ Digital Object Identifiers to get the essential research articles, hence the website https://sci-hub.mksa.top/ was utilized.

Analysis and Discussion

During this COVID-19 pandemic, higher education is faced with a choice: either raise tuition fees or cut costs. The expense of private higher education, on the other hand, has increased at a rate that surpasses that of healthcare. Students' wallets have been the worst hit by the rising expenses. Tuition fees at private colleges and universities have risen dramatically in recent years. Crowdsourcing might be utilized to improve existing educational procedures in a university setting (Heusler and Spann, 2014). The use of crowdsourcing may be included in classes that need students to work together to find answers to specific challenges. Some jobs need close cooperation among team members, while others call for little or no collaboration at all. Crowdsourcing in education, therefore, encompasses multiple methods such as group work and project-based learning as well as the provision of open educational resources (OERs). The benefits of crowdsourcing are analyzed and discussed below.

Benefits of Crowdsourcing

Content Development

Agarwal et al. 2021: Higher education institutions are increasingly embracing crowdsourcing tactics to help students succeed. The use of crowdsourcing in education helps institutions to better manage their budgets and make better use of students' available learning time, ultimately leading to improved student results. With the help of UGC, AICTE and the COVID-19 pandemic, institutes that would like to sustain themselves in the field of education should focus on implementing IT infrastructure, administrative assistance, and crowdsourcing technology for classroom learning (CTCL). With the help of UGC, AICTE, and MHRD, the development of educational resources can be concentrated (Ramonsito, 2020).

Practical Experiences

The use of computers in education has spawned a slew of innovative approaches. The web and adaptable educational hypermedia systems have had a significant impact on learning. Yang et al. (2020): Studying 295 solvers from a major Chinese crowdsourcing site revealed that points were positively linked to both intrinsic and extrinsic incentives, but quick performance feedback only enhanced intrinsic motivation.

Knowledge Exchange

When students work as interns, and when they leave the institute to find a job or establish a new business, knowledge is transferred. Making universities think about the long-term benefits of information exchange and collaboration in teaching and research should be on the immediate agenda. Policy-makers from different departments might have to work together to execute these programs. Stakeholders such as industry, regional and municipal governments, research institutes, non-profit organizations, should be involved in defining this vision for higher education institutions.

Feedback Enhancement

A wide range of information and viewpoints will be available to writers of new ideas who want input from a wide range. Visual feedback, for example, may be used to motivate and sustain knowledge exchange. According to empirical research, visual feedback given by a system is only helpful if the participants see it as positive (competence enhancing).

A Taxonomy for Crowdsourcing for Education

Using the taxonomy, it is clear that the key crowdsourcing for education (CfE) efforts enhance education in four ways through crowdsourcing. Educational resources (resources), activities (experience), and further knowledge exchanges (support) are all part of this process (evaluation). The creation of instructional content is the primary goal of the first set of developments. Among these are textbooks, assessment materials, and annotated learning resources. In other cases, crowdsourcing approaches are used to establish a platform that accepts instructional information from all around the world. Manual authorship by subject-matter experts is the current state-of-the-art approach for creating instructional content. There is a lot of time and money invested in this process, and the results are undoubtedly skewed. Despite their best efforts, natural language approaches and artificial intelligence are still unable to produce information that is sufficiently innovative and varied. The idea of getting information from online audiences to help people learn is a good one.

It's the second category of development that provides students with hands-on experience. Crowdsourcing assignments that are related to learning are collected and used to provide students with hands-on educational experience. To create high-quality results, these crowdsourcing assignments are frequently innovative, complicated, or require unique knowledge. Typical crowdsourcing platforms like Mturk, are unsuitable, but for industrial-strength training for students, they are great. This kind of involvement in cutting-edge research allows students to work as apprentices while they're still in school.

Complementary knowledge is the focus of the third category of projects. To help students, crowdsourcing can be used. Learners and audiences alike can benefit from each other's expertise. The delivery of large-scale virtual education can also be solved by the online tutor community. As part of this support, there are answers to learning challenges and recommendations to help students learn.

Crowdsourcing is used in the fourth category of initiatives to provide learners with a large amount of input (evaluations). It is difficult for instructors to keep track of grades and comments when there are large numbers of enrolled students. As a result, a more accurate assessment of learning performance may be achieved through the use of crowdsourced assessments by peers or an external crowd, which both lessens the burden on the graders as well as improves the quality of their evaluations. Students who give feedback to their peers are exposed to new ideas and techniques utilized by their peers, and they are encouraged to "strive for a more advanced and deeper understanding of the subject matter and abilities, and to engage students in critical thinking."

Table 2. Crowdsourcing of Education: Taxonomy Model of Rajkumar & Kishore

Content Creation	*Text Books*
	Assessment Materials
	Other Learning Resource
Experiences	*Hands-on Training*
	Apprentice
Knowledge	*Sharing Expertise*
	Recommendations
Evaluations	*Assessments*
	Feedbacks

Findings

This research attempted a thorough assessment of crowdsourcing in education. In addition to serving as a solid platform for future research, this review serves as a useful reference for practitioners. According to our research, which included a study of over 40 relevant publications, we have developed a systematic definition of crowdsourcing for education and suggested a CfE taxonomy.

We first describe crowdsourcing for education as a sort of online activity in which an educator or educational institution suggests to a group of people, via a flexible open call, that they can directly assist in learning or teaching in some way. Before this work, people used CfE in accordance with their understandings, and there was no consensus on how to define this phenomenon. Our definition serves as a starting point for making the debate on CfE more efficient, succinct, and exact, as well as more precise.

As a second step, we create a taxonomy that describes what crowdsourcing may achieve for education, including the creation of educational materials and the provision of practical experience; the exchange of complementary knowledge, and the enhancement of plentiful feedback. Educational organizations currently lack the information, skills, and understanding necessary to develop a crowdsourcing-enhanced learning and teaching strategy that will allow them to take full advantage of the opportunities provided by CfE and other organizations. This taxonomy assists them in identifying the areas in which crowdsourcing can be beneficial for education. In the meantime, it assists scholars in distinguishing between existing and prospective CfE activities, and it contributes significantly to the development of CfE-related theoretical frameworks.

Limitations and Future Research

There are several drawbacks to this study. Education is a right for everyone. CfE, on the other hand, is not a notion that applies to all situations. The use of crowdsourcing will not deliver an effective solution if we crowdsource the wrong jobs. The issue of copyrights is one of the most significant issues associated with crowdsourcing. Future research on crowdsourcing for education should focus on the rationale of which crowdsourcing approach is best suited for a certain educational scenario, as well as the extent to which crowdsourcing can be used to reach the intended beneficiaries of education.

Conclusion

Crowdsourcing can keep higher education institutions on the cutting edge of educational and research innovation while preparing students, faculty, researchers, lecturers, and administrators for the online world. Crowdsourcing may also be utilized to get ideas, evaluations, and comments from students. Crowdsourcing may also be utilized to improve communication between students, instructors, administration, and the general public. Crowdsourcing has been utilized by libraries and researchers for data collection. It appears that both lecturers and students have used crowdsourcing to create content on any topic or domain, including textbooks and other class materials. Crowdfunding is also a great way to sponsor educational initiatives. Crowdsourcing may also be used to get a huge group's opinion on an issue. To conclude, to sustain or provide a competitive outcome, the institutes should be able to adopt the crowdsourcing approach. Various aspects of crowdsourcing have been covered in this paper, which proves that crowdsourcing is cheaper than outsourcing. In this context, this study has listed several positive outcomes of crowdsourcing.

References

Adducul, R. B., and Gumabay, M. V. N. (2020). Crowdsourcing technology for classroom learning. International Journal of Advanced Trends in Computer Science and Engineering. https://doi.org/10.30534/ijatcse/2020/133942020.

Agarwal, V., Panicker, A., Sharma, A., Rammurthy, R., Ganesh, L., and Chaudhary, S. (2021). Crowdsourcing in Higher Education: Theory and Best Practices. In Crowdfunding in the Public Sector, 127–135. Springer, Cham.

Baggaley, J. (2013). MOOC rampant. Distance education, 34(3), 368–378.

BetterLesson. 2018. BetterLesson: The leader in personalized professional development. BetterLesson Retrieved 3 May 2018, from https://betterlesson.com/

Boell, S. K., and Cecez-Kecmanovic, D. (2014). A hermeneutic approach for conducting literature reviews and literature searches. Communications of the Association for information Systems, 34(1), 12.

Hargrave. (2021, May 16). Business Essentials. Retrieved from Investopedia.

Heusler, A., and Spann, M. (2014). Knowledge stock exchanges: A co-opetitive crowdsourcing mechanism for e-learning.

Wang, Y. M., Wang, Y. S., and Wang, Y. Y. (2021). Exploring the determinants of university students' contribution intention on crowdsourcing platforms: a value maximization perspective. Interactive Learning Environments, 1–23.

Yang, C., Ye, H. J., and Feng, Y. (2021). Using gamification elements for competitive crowdsourcing: exploring the underlying mechanism. Behaviour & Information Technology, 40(9), 837–854.

Zdravkova, K. (2020). Ethical issues of crowdsourcing in education. Journal of Responsible Technology, 2, 100004.

Identity of Scheduled Tribes in India - A Systematic Review

Nanthini Balu[*,a] *and Maya Rathnasabapathy*[b]

[a]Vellore Institute of Technology, Chennai, India
[b]Vellore Institute of Technology, Chennai, India
E-mail: *nanthinibalu2103@gmail.com

Abstract

A strong sense of identity helps and strengthens a person's sense of self-worth, social support, belief, and values. The identity of Indian tribes or Scheduled Tribes (ST) is founded on their strong ties to their ancestral homeland. Every national group's mental health is strongly correlated with how people perceive their identities. So, this paper outlines a systematic review that focuses on the identity of scheduled tribes in India. Literature published from January 2000 to June 2022 about ST people was systematically searched in reputed journals to identify the relevant articles. Additionally, appropriate Indian websites, and hand-searched articles, which focus on psychological perspectives of identity also reported according to MeSH terms and PRISMA guidelines. Sixteen studies are included in the final review. There are few resources available to address the psychological perspective of the Indian ST population. Promoting ST identity should be a priority in order to meet their larger need.

Keywords: Scheduled tribe, Indigenous, Identity, Cultural identity

Introduction

According to Waterman (Guardia, 2009), self-realization is a reflection of an individual's highest potential. Self-realizing activities that an individual perceives as self-defining, engaging, stimulating, purposeful, and obliging in achieving their life goals are how identity is represented. Erikson (1968) defined identity as a Continuous development of the fundamental organizing principle throughout our lifespan. It is a formation of a normative developmental process for adolescents and emerging adults (Schwartz et al., 2013). Evidence suggests that Achieved identity is linked to fewer psychological and neurotic symptoms (Chen et al., 2007) as well as fewer symptoms of anxiety (Crocetti et al, 2009) depression, and suicidal tendencies (Ramgoon et al., 2006). In another study, it was found that achieved identity is linked to improved psychological health, emotional growth, and emotional stability (Sandhu et al., 2012; Crocetti, Rubini, Luyckx, and Meeus, 2008; Dumas et al., 2009). Being additional content in love relations (Klimstra et al., 2013) and feeling less anxious or competitive in social situations (Adams et al., 1985) are both positively correlated with reached identity. Identity is both what distinguishes one person from another and what unites them with others (Beijers, 2015).

Scheduled Tribe Identity

Previous theoretical and empirical research (Cotten, 1999; Stryker and Burke, 2000) discuss how people's social networks serve as the contexts for managing their identities and as a source of identity validation by serving as a reminder of people's beliefs, values, and skills. Similar studies (Fornara et al., 2016; Steg, Dreijerink, and Abrahamse, 2005; Kaiser, Hubner, and Bogner, 2005) proved that the relevance of environmental self-identity and values is highlighted in research using social science frameworks like the value-belief-norm (VBN) hypothesis that are intended to explain pro-environmental behaviors. In this perspective, ST's are notified as culturally unique and their belief and identity differ from non-tribal people. Even though located within the same state, ST groups are recognized under the Indian Constitution as having distinct cultural or ethnographic characteristics and frequently reside in designated geographic areas known as ST areas (Brewer and Gardner, 2004; Marcus and Cross, 1990).

Distribution

After South Africa, India is the second-largest country in the world. There are about 90 million STs or Adivasis and they make up 8.6% of the population in India (Ministry of Tribal Affairs, 2014). In comparison to nontribal groups, the scheduled tribes (ST) population is a relegated group that experiences relative social isolation and worse health outcomes (Tewari et al., 2017).

Significance of the Study

However, no systematic review has been conducted to explore the identity of scheduled tribe people in India. Generally, Tribal peoples are frequently referred to be "marginal, minority, and vulnerable." A deficit attitude to Tribal peoples or a disregard for the good features of tribalism, while frequently probably statistically true, can accentuate or even magnify the hardships of Tribal people in a way that is not always beneficial (Craven et al., 2016; Jacobs, 2019). Although each Tribal group has its distinctive cultural traits, many Tribal peoples around the world have shared socio-historical experiences, such as colonial oppression and eviction (United Nations Department of Economic and Social Affairs, 2017, 2019). In Allwood (2018) studies limitations reported that lack of research focused on Scheduled Tribe people's identities and experiences. However, no systematic review has been conducted to explore India's ST population identity. Tribal people are very rich in their culture, belief, and ethnic identity, and they are the ones also facing discrimination compared to non-tribal people. So there is a need to review the identity of ST population in India.

Objective

This systematic review aimed to understand how India's scheduled tribe people perceive their identity and how it is related to their mental health.

Method

As per Preferred Reporting Items for Systematic Reviews and Meta-Analyses (PRISMA) the systematic review framed (Moher et al., 2009). It attempts to find, relevant and concise information in order to answer a particular question, and to produce a scientific summary of the evidence.

Information Sources

For collecting relevant studies by using Pubmed, Web of Science, PsychInfo, Embase, Scopus, Pro-quest, and reputed social science index studies. Based on high regard in educational institutions and the direction of Dalhousie University librarians who are experts in the methods of systematic reviews, Scopus and Web of Science databases were chosen (Kivimaa et al., 2016; Gjaltema et al., 2020).

Search Strategy

Literature published from January 2000 to June 2022 on scheduled tribes published in English was systematically searched in reputed journals to identify the relevant studies. Additionally, relevant Indian websites and hand-searched articles focused on psychological perspectives of identity were also reported according to MeSH terms and PRISMA guidelines. To filter relevant studies which meet the purpose of the systematic review, a Boolean search string was used. Search string employed databases were "Trib*, OR Aboriginal*, OR Tribal*, OR Adivasi*, OR First nations*, OR Social indicators" with the second keyword "Identity." Eligible studies references were also reviewed for selecting a suitable article for this systematic review.

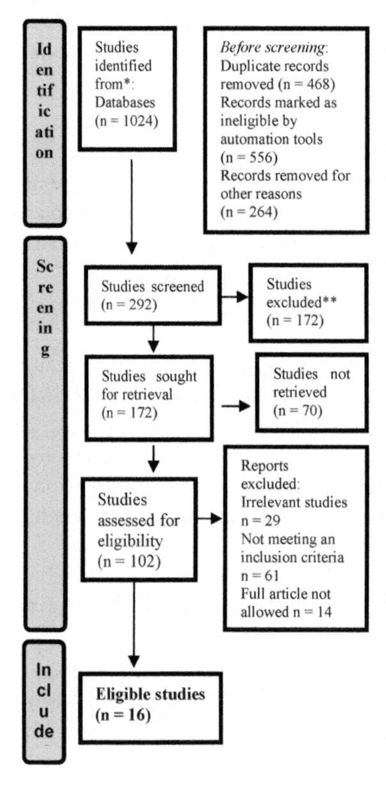

Fig. 1. Selection of eligible studies

Study Selection

Studies published from January 2000 to July 2022 (the time of search) and the highest number of articles published in 2022. Identity studies which are focused on Scheduled tribal people and .both quantitative and qualitative studies published in English were included. Articles related to Indian Scheduled Tribe identity are only included. Identity studies that are not focused on Scheduled tribes in India were excluded from the analysis.

Quality Assessment

There are 1024 hand-searched articles found. After removing 468 duplicate studies 556 studies were collected. During the abstract screening step, 264 articles were excluded for the reason of not focusing on the research objectives. 172 studies were removed because of tribal people in a sentence inside the paper it is not specifically mentioned or makes in-depth information about their life. Some of the studies are irrelevant and do not mention their sustainable life. So, 102 studies were excluded. Once these studies were filtered out, finally 16 studies are included in this systematic review.

Results and Discussion

Literature Selection

Outcome of the screening and searching process is detailed in Table 1. Of the 16 eligible studies, 7 were descriptive study, 3 were structured and semi-structured interview method, 2 were longitudinal research, 1 was case study, 1 is empirical research and 3 were observational qualitative study.

Scheduled Tribe Identity

An early study indicates that positive correlation between identification and psychological outcomes. This analysis covers the ethnic and cultural identities of India's scheduled tribes. The ethnic identity of India's Scheduled tribes has only been the subject of one empirical study. Insurgent groups fighting for their community, the protection of one's identity, customs, and traditions, government policies, the guard of regional limitations, the perception of fetching subgroups, and other factors are the causes of the contradiction of political autonomy, the rise of ethnic and religious tensions in response to the intensifying competition for economic and radical chances, control over land, the establishment of the ethnic native land, the establishment of the ethnic homeland, and these factors and others (Gohain, 2020). Assam's Assertions for the separate identity of the tribes of the plains currently dominate Assam's political landscape. The plains tribes' identity movement began with a non-political concern before morphing into a political one. They did so because they understood that communal growth on all fronts would be impossible without political power. Additionally, it should be noted that in a multi-ethnic community like Assam, every varied group has the right to keep and protect its linguistic and cultural characteristics. Their claim of separate individuality is therefore significant in the truest meaning of the word (Chetia, 2018).

However, because every group strives to maintain its identity, it is impossible to completely eradicate ethnic statements from modern culture. Therefore, it should be necessary to preserve each ethnic group's identity rather than eradicate it to preserve nationalist sentiment. The politics of northeast India have gradually been shaped by identity politics through religion, language, culture, area, caste, etc. during the past few decades. Therefore, both the federal government and state governments must take action to address these identification challenges (Gupta, 2018).

Therefore, it is a pressing need that both the federal government and state governments to take action to address these identification challenges (Gupta, 2018). Over time, ethnic identity centrality significantly improved life happiness, wellness, and self-esteem. Additionally, over time, growths in in-group warmth and national identity significance are predicted by both personal-wellbeing and self-esteem, demonstrating the mutually reinforcing nature of in-group warmth, personal well-being, and self-esteem. Social and cultural factors may be to blame for this tendency. A sense of belonging and connection to the wider ethnic identity group is promoted (or devalued) through interactions with family and friends, as was previously said. Ethnic identity is shaped through relationships and is experienced through these interactions. Ethnic identity reflects social conditions in this way.

Conclusion

Majority of studies in this review reported that ST identity and culture should be preserved adequately. Because their way of life and identity depends on accessing and owning their lands, and natural resources in a traditional way. Basic component of the country's knowledge skill system is ST people's knowledge. It includes the skills, experiences, and insights of people, applied to progress their livelihood.

Strength and Limitations

This review is based on identity which makes positive psychological outcomes for our entire lifespan. Scheduled Tribes are facing identity discrimination compared to non-tribal people. Indeed, this review addresses the importance of preserving their identity in a systematic approach. However, tribal lifestyle and their approach to society were not examined.

Future Research

As mentioned above, there are very little psychological empirical research was conducted and it is not focused on their identity which is play a major role of enhance their self-esteem and psychological well-being. So, future research needs to conduct more research in the field of indigenous psychology to know more about Scheduled tribes and to preserve their culture and identity.

References

Bria, M. (2003). The Tribes Within: The Search for Identity in the Modern State, Studies of Tribes and Tribals, 1, 1, 29–34, doi: 10.1080/0972639X.2003.11886481

Erikson, E. (1968). Identity: Youth and crisis. New York: W. W. Norton & Company.

Gohain, C. (2020). The Ethnic Identity of Karbi: Challenge of Ethnic Identity. Jetir, 7(10).

https://www.jetir.org/papers/JETIR2010390.pdf

Kujur J., Irudaya Rajan S., and Mishra U. S. (2020). Land vulnerability among Adivasis in India. Land Use Policy, 99, Article 105082. https://doi.org/10.1016/j.landusepol.2020.105082

Lucky, C. (2018). Politics of Identity Assertion of the Plains Tribes in Assam. International Journal of Science and Research. 2319–7064. https://www.ijsr.net/archive/v8i9/ART20201570.pdf

Neha, G. (2018). Ethnicity in north-east India: A challenge to identity. International Journal of Advanced Educational Research, 3, 292–294. http://www.educationjournal.org/archives/2018/vol3/issue2/3-2-118#:~:text=North%20%2DEast%20India%20has%20been,the%20Himalayan%20State%20of%20Sikkim.

Patrik, O., and Siddharth, S. (2020). Adivasiness as Caste Expression and Land Rights Claim-Making in Central-Eastern India, Journal of Contemporary Asia, 50, 5, 831–847, doi:10.1080/00472336.2019.1656277

Shrikanth, N. (2020). A Study of Social Transformation, Identity of Indian Tribes in Recent Time: An Anthropological Prospective. International Journal of Creative Research Thoughts, 8, 10, 2320–2882.

Tyser, J., Scott, W. D., Readdy, T., et al. (2014). The Role of Goal Representations, Cultural Identity, and Dispositional Optimism in the Depressive Experiences of American Indian Youth from a Northern Plains Tribe. J Youth Adolescence, 43, 329–342. https://doi.org/10.1007/s10964-013-0042-2

Wilson, A. R., Johnson, R. L., Albino, J., Jiang, L., Schmiege, S. J., and Brega, A. G. (2021). Parental Ethnic Identity and Its Influence on Children's Oral Health in American Indian Families. International journal of environmental research and public health, 18(8), 4130. https://doi.org/10.3390/ijerph18084130

Strategies Employed to Acquire and Reflect Political Knowledge

A. Fredrick Ruban[*,a] *and Dr. P. V. Arya*[b]

[a,b]Department of English and Cultural Studies, Christ (Deemed to University), Bengaluru, India.
[a]Asst. Prof. of English, Department of Humanities, Krupanidhi Degree College, Bengaluru
E-mail: *fredrick.ruban@res.christuniversity.in

Abstract

Acquiring political knowledge is essential in today's educational set-up as there is an influence of politics on education. There are various sources available to acquire knowledge therefore students need to pay enough care to choose the right source to acquire authentic knowledge. This underpins them to contemplate the acquired knowledge and then reflect and express their views. The present study intends to study how the target group acquires political knowledge and shares their political views. The present research has been conducted among a specific group of undergraduate students of Tamil Nadu who are engaged in sharing political views and acquiring political knowledge. The study consists of 192 undergraduate students including both boys and girls as the target group.

Keywords: political knowledge, political views, students, Tamil Nadu

Introduction

Knowledge has a significant position in academia and political knowledge is in no way lesser than it. Political knowledge has been defined in multiple ways by various political scientists, theorists, and researchers. According to Nick Clark, political knowledge is the "facts about a political system that an individual can recall from their memory to interpret and understand happenings and developments within that system" (Clark, 2013). Clark takes a comprehensive approach to perceive political knowledge as an ability to recollect, interpret, understand and apply within a system. Recalling from memory can happen under two circumstances: a) when it is being learned from sources, b) when is experienced in one's own life. Thus, political knowledge can either be acquired from eternal sources or it can be experienced in one's own life. It is needed for an individual to realize and evaluate the experience. This adds up to the accumulation of political knowledge and thereby paves a way for the expression of one's views and criticism. It can be expressed in multiple ways. The present study focuses on this aspect of acquiring political knowledge and reflecting on the acquired political knowledge.

Objectives

The objectives of the present study are:

- to find out the sources used to acquire political knowledge
- to detect the medium of expression used to communicate political views
- to figure out the factors that influence the students to express their political views

Method and Methodology

The present research is a quantitative study using the survey method. The research has been conducted with a target group of 192 students who are interested in politics and involved in sharing political views. Among these 192 students, 63 were male students and 129 were female students residing in Tamil Nadu. A questionnaire consisting of seven closed-ended questions covering the aspect of acquiring political knowledge and reflecting political views has been placed in front of the target group.

Interpretation and Discussion

The prepared questionnaire emphasizes four aspects: area of interest of the respondents, acquiring political knowledge, expressing political knowledge, and influencing others. The first question placed in front of the respondents is "Are you from a political background?" and the choices are "Yes" and "No." This question was framed to examine whether the students are influenced by their family background to deal with political knowledge. Even hereditary can contribute to political knowledge therefore the present research attempted to explore whether the respondents are from a political background. It has been found that 13% of the respondents have agreed to the fact that they are from a political background and 87% of the respondents have registered that they are not from a political background. It is very evident that the majority group of respondents is not from a political background, yet they acquire and share political knowledge. this fact is revealed through the question "Which is the source of your political knowledge?" and the choices are "printed material," "E-material," and "Both." The chart:3 given below clearly describes the response received for the aforementioned question.

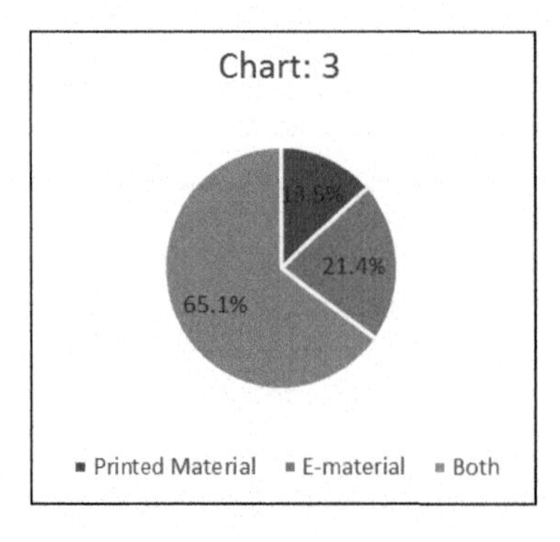

Printed material has been chosen as a source of political knowledge acquisition by a group of 13.5% and E-material has been used as a source of political knowledge acquisition by a bigger group of 21.4% and there is a huge group with 65.1.% of respondents who have registered that they use both "printed" and "E-material" to enhance their political knowledge. There could be several reasons why the respondents do not stick to a single source to acquire political knowledge. first and foremostly, the respondents are undergraduate students who keep themselves engaged with their mobile phones to read and listen to political knowledge in addition to getting into the library to read newspaper and magazines. In connection to this, the present research attempted to figure out the area of interest of the respondents for which the question "Which politics interests you?" has been placed in front of them. The choices are "Politics of Tamil Nadu (PTN)," "Politics of South India (PSI)," "Politics of North India (PNT)," and "International Politics (IP)." A group with a maximum strength of 35.4% of respondents has chosen to say that the politics of Tamil Nadu interest them. The chart:2 given below describes it vividly.

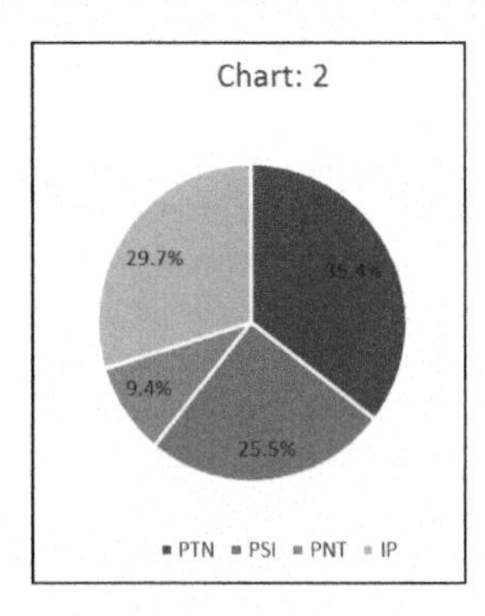

The politics of South India has been selected by 25.5% and the politics of North India have been chosen 9.4% of respondents. A group of 29.7% of respondents has registered that international politics interests them. The students of Tamil Nadu have come forward to say that Tamil Nadu politics interests them, and the second maximum group has confessed that international politics interests them. As the respondents got an interest to acquire political knowledge, they have also been prompted to express their political views. The research minted to detect the medium of expression of their political knowledge. the framed question is "Which medium of expression do you prefer to express your political views?" and the choices are "memes," "trolls," "spoken," and "written." The chart given below sheds light on the medium of expression of the respondents.

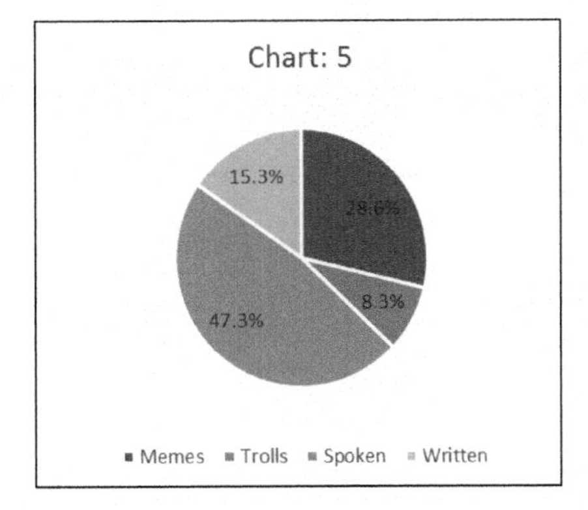

From the above chart:5, it can be interpreted that a majority group of respondents has used "spoken" as a medium of expression to share their political views. It is found that 47.3% of respondents have used the traditional medium of expression to express their political views. The reasons could be that firstly, the target group is comfortable with the conventional medium of expression, and secondly, the target group is not interested in using technology to express their political views. It is due to the lack of technical skill to produce memes and trolls. It is also evident that this majority group is not interested in writing their political view. Though "written" and "spoken" are the conventional medium of sharing political views, the majority strength has chosen "spoken" as a medium of expression. This "spoken" as a medium of expression can be perceived in two ways: a) sharing one's political thoughts and views in person or phone, b) using technology to either record in video form or audio form. The second largest group consisting of 28.6% uses memes to express their political views. The respondents of this group are considered to be technically good at creating and sharing memes. Finally, the group with the least strength is the one that prefers to express their political views using trolls. It has been detected that 8.3% of respondents prefer this medium of expression.

In connection to the medium of expression of political views, there is a need to determine how often the respondents express their political views. The question: "How often do you share political views?" was given to respondents with the choices "frequently," "infrequently," and "habitually." This question was framed to figure out how well the respondents are engaged in acquiring political knowledge and expressing their concerns and views in return. A person without political knowledge would not be able to express his comments on any issue because would sound strange to him. Therefore, it is necessary to update the memory with political knowledge so that the ability to criticize and comment gets developed over a while. Expression of political views happens under two circumstances: i) when a person is politically educated, ii) when a person is interested and prompted. For the given question, 29.2% of respondents replied they frequently express their political views and 51.6% of respondents have documented that they infrequently share political views. It is quite astonishing to find that 19.3% of respondents share political views habitually. There is a need risen to find out who has prompted the respondents to share political views when the 29.2% registered that they share their political views frequently. In this regard, the question: "What prompted you to express political views?" the options are "family," "friends," "society," "political climate," "political culture," "self-interest" and the final option is "None of the above." The chart given below gives a transparent picture of the reply of the respondents.

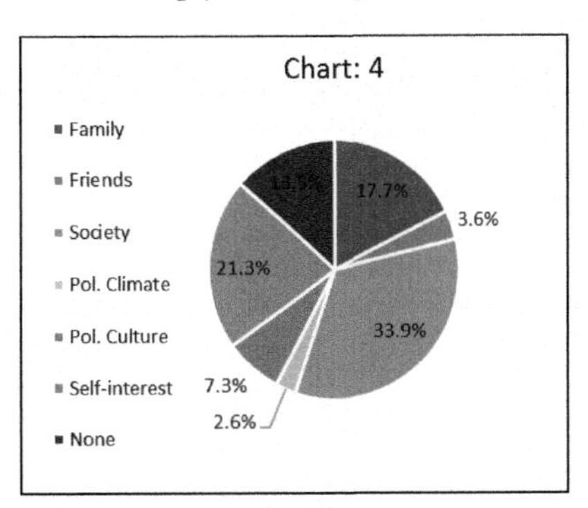

The chart:4 depicts that 17.7% of the respondents are being prompted by their family to express their political views. This finding aligns with the finding of the first question. The first question in the questionnaire found that 13% of the respondents are from a political background. Although only 13% of the respondents are from a political background, 17.7% of respondents have agreed to say that they express political views. From this, it is evident that even though the respondents who do not have a political background have been engaged in expressing their political views. It can be studied from the above chart that 3.6% of the respondents are prompted by friends to express their political views and 33.9% of the respondents are encouraged by society to give political views. It is clear from the chart that 2.6% of the respondents are encouraged by the political climate and 7.3% of the respondents are prompted by political culture. It is found that 21.4% of respondents have agreed that it is out of their self-interest they are involved in sharing political views. Overall, 13.5% of respondents have accepted that none of the above options has underpinned prompt political views. A vast majority (33.9%) of the respondents have agreed to say that "society" has prompted them to express political views. In connection with this, the present research aimed at exploring if the political views of the target group have influenced the other group. It has been found that 44.8% of respondents have come forward to register that their political views have influenced others and 55.2% of respondents have registered that their political views have not influenced others.

Conclusion

The present research has found that the majority group (65.1%) of respondents rely on both printed material and E-material to acquire political knowledge. It is found that a very minimal percentage of respondents use printed material for this purpose. From the above discussion, it is evident that 47.3% of respondents have used the "spoken" medium to express political views. The maximum percentage of respondents have chosen the "spoken" medium. The research has also found that there are various factors like family, friends, society, political climate, political culture and self-interest play a vital role in influencing a person to express his political views. Among all these, society occupies a prominent place in the matter of influencing the students to express their political views.

References

Barabas, J., Jerit J., Pollock W., and Rainey C. (2014). The Question(s) of Political Knowledge. American Political Science Review. 108(4), 840–55.

Chilton A. Paul and Schaffner, C. (2002). Politics as Text and talk. The Netherlands: John Benhamins Publishing Co.

Clark, N. (2013). Contextual Dynamics and Political Knowledge: The Role of Institutional Quality in an Informed Citizenry. Susquehanna University.

Creswell, John W. (2009). Research Design: Qualitative, Quantitative, and Mixed Methods Approaches. America: Sage Publications, Inc.

Galston, A. William. (2001). Political Knowledge, Political Engagement, and Civic Education. Annu. Rev. Polit. Sci. 4, 217–34.

Owen, D. and Soule, S. (2015). Political Knowledge and Dimension of Political Engagement. American Political Science Association.

Pastarmadzhieva, D. (2015). Political Knowledge: Theoretical Formulations and Practical Implementation. Trakia Journal of Science. 13(1), 16–21.

University Libraries. (2018). Research Methods Guide: Research Design & Methods.

Impact of Select Vocabulary Learning Strategies (VLS) on Vocabulary Acquisition of Tertiary Level Learners

S. Kavitha,[*,a] P. Jayakumar,[b] P. Paulsy Diana,[c] and I. Jane Austen[d]*

[a,b,c,d]Sathyabama Institute of Science and Technology, India
E-mail: *kavithasankaralingam5@gmail.com

Abstract

In learning a second language, vocabulary is extremely important. The current study is to enhance tertiary-level students' vocabulary knowledge through explicit vocabulary strategy instruction. The study investigates the efficacy of the chosen VLS under discovery technique, such as analyzing an unidentified word utilizing word part analysis, and contextual cues. The research seeks to show that the chosen VLS will aid students in deciphering the meaning of unknown words, which will foster learner autonomy and word retention. The information collected from diagnostic tests and questionnaires allowed the researcher to understand the learners' prior knowledge and to tailor the practice sessions or treatments according. It briefs on how the researcher used backward design to create six modules to teach vocabulary learning strategies. The purpose of the study is to find out whether or not students can independently analyze unknown words without the help of a teacher or technology.

Keywords: VLS, Retention, Explicit teaching, Learner's autonomy

Introduction

A robust vocabulary builds the overall communication skills of a language with a natural ability to listen, speak, read and write. Vocabulary, once a neglected area, in recent times, has taken a steep learning curve in Second language acquisition. Gifford (2013) puts forth that "a person having more breadth and depth of vocabulary has wider competence to communicate and to understand a communication, and is to be considered intelligent" A good knowledge of vocabulary helps a person to have better articulation, comprehension and it adds up to the power of persuasiveness.

Second language learners by gaining a lot of vocabulary words tend to have Word consciousness. This curiosity makes them learn not just the meaning but about the word's history, its origin, and other aspects of a word. Vocabulary is not a developmental skill and cannot be fully mastered. Vocabulary of any language is a lifetime and ongoing process with endless elaboration and expansion. (Heibert and Kamil, 2005) A learner's vocabulary is meant to be the range of words known either by incidental or intentional learning. The impact of Covid-19 and the curfew have fueled the scope of being an autonomous learner and it demands to know independent language learning strategies. Therefore, teaching vocabulary learning strategies will give a huge influence on the learners to widen their vocabulary knowledge.

Literature Review

Oxford (1990) stated that "Learning strategies are specific action taken by the learner to make learning easier, faster and more enjoyable, more self-directed, more effective and more transferable to a new situation" O'Malley and Chamot (1990) defines LLS as "the special thoughts or behaviors that individuals use to help them comprehend, learn, or retain new information." Gu (2003) puts forth that learning VLS is an add-on to teaching in the classroom.

Mastering vocabulary is one of the challenging tasks that the learner undergoes while acquiring a language and is also a challenge for the teachers to choose the best way of instructing vocabulary. Despite the availability of numerous vocabulary teaching methods, techniques, and activities, it is always a puzzle to choose the suitable method and presentation of words, allocate specific time for vocabulary, choose the suitable words, etc. Despite teaching vocabulary words by teachers during class hours, the availability of words in a language is abundant and so it calls for the need for other ways of teaching vocabulary (Sedita, 2005).

A good language learner is judged by the several types of strategies he uses regularly. To determine the meaning of unknown words and to consolidate the words learned, Discovery strategies and Memory strategies play an important role.

Schmitt's taxonomy of VLS is widely known, as he conducted an extensive study on the vocabulary learning strategies employed by Japanese language learners, using a detailed questionnaire of VLS. He classified the data based on the frequency of its usage and benefits. Schmitt's taxonomy of VLS is broadly classified into two major categories such as Discovery strategy and Consolidation strategy. Discovery strategy is employed when the learner comes across a new word in the learning process. Strategies such as identifying the parts of speech, analyzing the unknown words using word parts, checking for cognates, and guessing from the context are found.

There is a difference between vocabulary teaching and Vocabulary strategy instruction. Vocabulary teaching deals with implicit or explicit instruction of words with several methods, techniques, and activities. While Vocabulary strategy instruction explicitly introduces the different Vocabulary learning strategies available so far.

Significance of the Study

The study emphasizes that vocabulary learning strategy instruction will motivate the learners to hold responsibility for improving their communication skills. No matter how many words the students learn in a classroom, they need to upskill their vocabulary knowledge by implementing vocabulary learning strategies. Learning Discovery strategies will help the learners to decode the meaning of unknown words without referring to the dictionary or any online resources. One of the main barriers to effective reading is the lack of linguistic competence. Many students fail to decrypt the lexical units and it hampers the smooth reading when they verify the meaning all the time. Thus the study is significant as learning Vocabulary learning strategies helps students in two different ways. One, they decode the meaning of unfamiliar words on their own, and two, they constantly improve their vocabulary knowledge by VLS.

Strategies Chosen for the Study

Analyzing the Parts of Speech

When words are unknown, one of the preliminary strategies employed is to find the parts of speech. Despite knowing the meaning of a word, finding the parts of speech can help the learner to communicate and remember it better. In this way, analyzing the parts of speech can be both a Determination strategy and a Memory strategy according to Schmitt. (2000) Rodger (1969) conducted a study on the effects of parts of speech in helping learners to guess the meaning of unknown words and he concluded that verbs and nouns are easier to guess than adjectives and adverbs. It gives clarity to both the writing and speaking skills of language learners.

Analyzing the Word Parts/Morphology

Morphemes are meaningful word parts that the learners can identify and analyze to determine the meaning of unknown words. (Carlisle, 2014)

Researchers are interested in teaching prefixes, suffixes, and roots to Second language learners to improve their vocabulary in two possible ways. Learning the meaning of several prefixes, suffixes and roots will help them to decipher the meaning of one. It also becomes a good strategy to start learning words consciously.

Guessing from the Contextual Clues

Guessing the meaning from the context is one of the widely promoted strategies under the Determination strategy. Schmitt says "guessing from the context most commonly refers to inferring a word's meaning from the surrounding words in a written text." A good reading habit with the knowledge of context clues strategy will increase vocabulary proficiency to a greater extent. A study by Reardon (2011) on English learners of both 5th grade and adults who were instructed about the different types of contextual clues such as synonyms, antonyms, definitions, examples, phrases, punctuations, and appositive words showed a positive output after the intervention. They find the contextual clues strategy to be helpful as they were able to derive the meaning of the unknown words from the given text or sentences.

Design of the Modules

The researcher has designed six modules to accommodate all three strategies. The modules are designed based on the backward design which is generally used as a pattern to set curriculum. Each module is designed in such a way that it has a reference sheet, worksheet, and test sheet. The researcher introduced the strategies first with the aid of a reference sheet and they were asked to work out as pairs with the worksheets. This is followed by a short test which helps to assess whether they get hold of the strategies.

Technology and digital learning have been great sources to practice self-learning, especially during the pandemic (Jayakumar and Ajit, 2016). A root word for the day was sent through WhatsApp and the students came up with numerous words.

Statement of the Problem

Teaching vocabulary becomes the most discussed part of language teaching, yet it seems to be a problem, why? A Nationwide survey commissioned by the National Book Trust (NBT) says that youth have poor reading habits. This might be one of the reasons for poor vocabulary knowledge. Many learners find it difficult to retain the meaning of words. Vocabulary learning is an endless process and one cannot learn the maximum only through classroom learning. Despite having offline and online classes on vocabulary, students lack sufficient knowledge of vocabulary. The purpose of Teacher-centred learning alone in the field of vocabulary is questionable. Lack of sufficient time to teach vocabulary in the classroom is one of the main concerns of the research. Adding on curfew had given ample time for the students to do self-learning. However, students making use of it in a desired way influence the researcher to come up with the Vocabulary strategy instruction.

Needs of the Study

Vocabulary learning is one of the fundamental steps in the language learning process. English in a second language and foreign language setup, vocabulary plays a vital role and becomes the basis for the other language skills (Schmitt, 2000).

Despite the availability of innumerable techniques to teach vocabulary, it is always a puzzle for teachers to choose the appropriate technique to be implemented. Even if the most suitable method is chosen, teachers could spend limited time teaching vocabulary while they focus much on improving the other skills. Vocabulary teaching just becomes a part of the lesson plan at the tertiary level. Teaching vocabulary alone for the entire period may lead to boredom.

Learners are taught certain strategies that they can implement on a day-to-day basis to improve their vocabulary. The students also become independent learners and learn vocabulary using a student-centered approach than teacher-centered approach. Vocabulary of the language is vast and it is a lifelong learning process that the student cannot depend on the teachers always, thus if a vocabulary learning strategy is introduced, they can practice even without the support of the teachers.

Research Objectives

The objective of the study is as follows:

- To improve the vocabulary knowledge of tertiary level students by teaching them specific vocabulary learning strategies.
- To analyze whether specific VLS instruction helps the learners of heterogeneous classrooms to have language fluency to a considerable extent.
- To promote learners' autonomy over the learning process. To shift the learning from Teacher oriented to learner-oriented.
- To make learners apply the VLS they had learned to infer the meaning of unfamiliar words.

Research Questions

Based on the objectives of the study, the research looks to answer the following questions.

- Does Specific VLS called Discovery strategies help tertiary-level learners to build their vocabulary?
- Does VLS instruction and application enhance the student's retention of words and make them autonomous learners?

Research Methodology

To find the effectiveness of the selected strategy instruction in a heterogeneous classroom, this study uses a Quasi-experimental method to find the relationship between the two variables. The researcher conducted a Quasi-experimental study with a group of 146 students. The students were asked to fill in the Questionnaire which helps the researcher to be aware of their previous knowledge and also aids the researcher to design the practice session or treatment accordingly. A pretest before the treatment and a post-test after the treatment has been employed to derive a conclusion to test the effectiveness of specific vocabulary strategy instruction to the target group. The intervention of select VLS was encompassed in six modules focusing on three strategies such as Identifying the parts of speech, analyzing the word parts, and guessing from the context.

Analysis and Interpretation

Analysis and Interpretation of Data From Student's Questionnaire

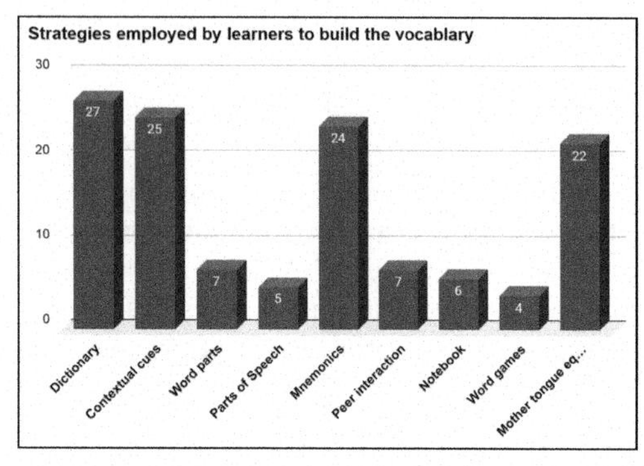

Fig. 1. The strategies employed by learners to build the vocabulary

The researcher intended mainly on improving their vocabulary through vocabulary learning strategies. Everybody who is learning a language consciously or unconsciously uses certain vocabulary learning strategies and so basic vocabulary learning strategies are listed out to know their familiarity with it. This helps the researcher to select unique and effective vocabulary learning strategies which they're not aware of or never used regularly. Dictionary, contextual clues, mnemonics, and mother tongue equivalents were strategies used by them predominantly while other strategies such as dividing unknown words into word parts, looking for parts of speech, peer interaction, maintaining separate notebooks for vocabulary, and playing vocabulary games were least picked ones.

The target group was tested based on three different strategies. The first strategy is analyzing the meaning of words using the parts of speech. In the pretest, the average mark was 5.1 for 10 while in the post-test the average increased to 8 for 10 with a 29% increase. The second strategy is to analyze the word using word parts and the performance of the students in the pretest is 8.28

for 15 while the average score in the post-test is 12.6 with 28.7% increase. The third strategy is to guess the meaning using contextual clues and the average score in the pretest was 2 for 5 marks while in the post-test, the average score increased to 3.9 with 38% increase. There is a significant improvement seen in all three strategies. More than 80% of the students scored in the range of 21 to 28 out of 30 marks in the test. The performance of the students in the post-test, when compared to the pretest, is much more visible after the intervention of the VLS by the researcher.

Graphic Analysis of the Average Performance of the Students

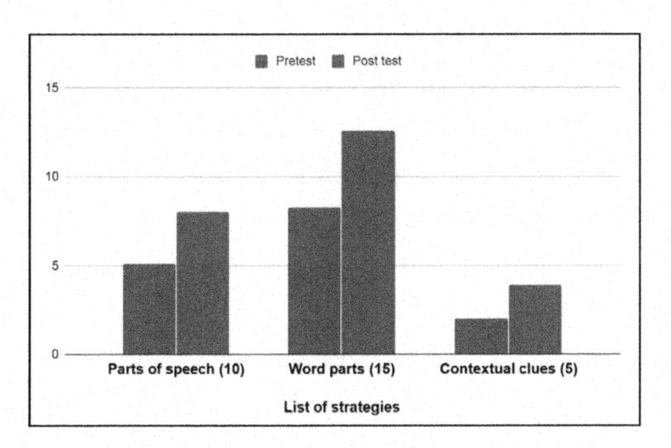

Fig. 2. Average performance of the students on the selected vocabulary learning strategies

Among the three strategies, a substantial increase is found in the use of the contextual clues strategy. The student's performance in the pretest of contextual clues was very less but in the post-test, 38% increase is visible. This ensures that the students were capable of performing well and it is triggered after the intervention of the vocabulary learning strategies. The students gained confidence when they confronted new words.

Graphic Analysis of the Improvement in Scores in Percentage of the Students

The difference in the performance of the students in the pre-test and the post test ensures that there is a significant improvement found in the vocabulary knowledge of the target group after the intervention of various VLS modules. It is shown in the following bar chart.

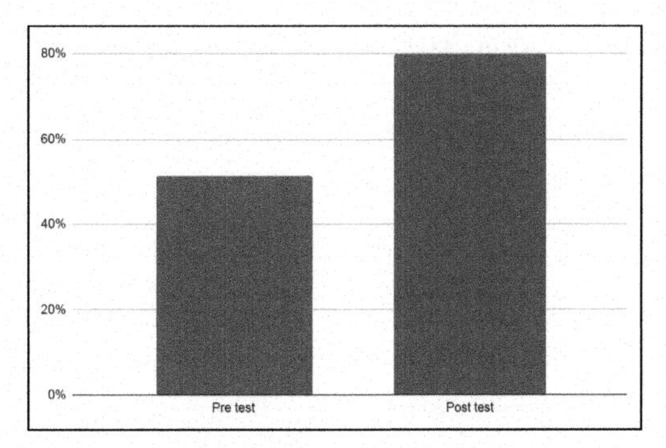

Fig. 3. Comparison of pre-test and post-test scores of the students

The difference in the performance of the students in the pre-test and the post-test ensures that there is a significant improvement found in the vocabulary knowledge of the target group after the intervention of various vocabulary learning strategy modules.

Conclusion

Vocabulary becomes important in every stage of language learning and also becomes a foundation upon which other skills are built. Students have to make a conscious effort to improve their vocabulary knowledge apart from the input from the classroom as vocabulary learning is a lifelong process. Once the teacher makes the students familiar with the suitable VLS, good learners make an effort to improve their vocabulary, which helps them to improve their other skills as well. Thus, the selected vocabulary learning strategies have a positive impact on the students.

Findings of the Study

- A study group of 146 students was introduced to the vocabulary learning strategies and application of such VLS and training had a significant improvement in their vocabulary knowledge. The analysis of the pre-test and post-test performance of the students shows the effectiveness of the selected VLS.
- Awareness of Vocabulary Learning Strategies increased and the students are familiar with selective VLS after the intervention. Interest in learning such strategies helped in learners' participation.

- Students have become autonomous learners and they can retain the words effectively after the application of the VLS.

References

Carlisle, Joanne F., and Fleming, Jane. (2003). Lexical Processing of Morphologically Complex Words in the Elementary Years. Scientific Studies of Reading 7.3, 239–253. Scientific Studies of Reading. Web.

Gifford, F. (2013). How to Enrich Your Vocabulary?, Cyber Tech Publications, New Delhi.

Gu, P. Y. (2003). Vocabulary learning in a second language: Person, task, context and strategies, TESL-EJ, 7(2), pp. 1–25.

Jayakumar, P., and Ajit, I. (2017). The pedagogical implications on the root and route of English basic verbs: An extensive study through android application. The Social Sciences, 12(12), 2244–2248.

Kamil, M. L., and Hiebert, E. H. (2005). Teaching and learning vocabulary: Perspectives and persistent issues, In E. H. Hiebert and M. L. Kamil (eds.), Teaching and learning vocabulary: Bringing research to practice, Mahwah, NJ: Erlbaum.

Oxford, R. (1990). Language learning strategies: What every teacher should know. Boston: Newbury House.

Reardon, K. T. (2011). To what degree will learning to use context clues impact students' reading comprehension scores. University of Wisconsin, River Falls, Retrieved from https://minds.wisconsin.edu/bitstream/handle/1793/./Reardon.pdf./

Schmitt, N. (2000). Vocabulary in language teaching. New York: Cambridge University.

Sedita, J. (2005). Effective vocabulary instruction - keys to literacy. Insights on Learning Disabilities, 2(1), 33–45. Retrieved (November 12, 2016) from: https://keystoliteracy.com/wp-content/uploads/2012/08/effective-vocabularyinstruction.pd

Socialization of Culture: Sociopolitical and Sociocultural Contexts Ensuing Cultural Transition and Hybridity

Fredrick Ruban Alphonse,[a] Dr. Arya Parakkate Vijayaraghavan,[b] Joy Christy L.,[c] and Dr. M. Angeline[d]*

[a]Research Scholar, Department of English and Cultural Studies, Christ (Deemed to University), Bengaluru, India; Assistant Professor, Department of Humanities, Krupanidhi Degree College, Bengaluru

[b]Assistant Professor, Department of English and Cultural Studies, Christ (Deemed to University), Bengaluru, India

[c]Research Scholar (FT), Department of English, PSGR Krishnammal College for Women, Coimbatore

[d]Assistant Professor, Department of English, PSGR Krishnammal College for Women, Coimbatore

E-mail: *fredrick.ruban@res.christuniversity.in

Abstract

In cultural socialization young minds get to learn about a culture and try to develop a sense of belonging thereby associating themselves with a cultural group. It underpins the young minds to notice and mark the cultural transition occurring in the culture. Acquiring the knowledge of one's culture has significance in determining one's identity. Therefore, emphasis on acquiring cultural knowledge needs to be stressed among young minds. The present study attempts to determine whether young minds are culturally socialized to notice the cultural transition. It also attempts to explore the factors and contexts that have underpinned the hybridization of Tamil culture. The survey has been conducted among 215 college students in Tamil Nadu. The study has used a random sampling method to collect the data and the interpretation is based on Homi Bhabha's concept of hybridity.

Keywords: culture, hybridization, socialization, transition, sociocultural

Introduction

Cultural socialization primarily deals with the parents teaching their race and ethnicity to their children (Aldoney et al., 2018). It is a salient aspect of child-rearing among ethnic minority families. This process fosters a strong sense of identity and belonging to a child's cultural group (Aldoney et al., 2018). Cultural socialization teaches the positive aspects of the child's race, ethnic heritage, traditions, and cultural customs. Firstly, in the process of cultural socialization, the child develops the ability to cope with discrimination. Secondly, the child is taught to mistrust and distrust other people who are not part of the child's ethnic group. Thirdly, the child is educated to value others over their racial identity, in this process, the child is taught not to mention racial identity during his/ her interaction with the other children (Aldoney et al., 2018). In this course of learning the child can learn about his/ her cultural heritage and others' perception of his/ her culture. In addition to practicing his/ her culture, the child is taught to acquire knowledge about the mainstream culture so that the child accumulates more knowledge about his/ her society thereby getting prepared to challenge and succeed in the mainstream society (Wang et al., 2015). It is vivid from this claim that children ought to be educated about their culture and the culture practiced in mainstream society so that the children will be able to figure out the prominence of their culture and its transition. It further underpins to detection of the factors involved in hybridizing the existing culture and defiling it with the influence of high culture. The present research paper intends to explore the factors that add up to cultural transition and hybridization. The objectives of the present study are to detect whether the target students are culturally socialized and to explore the factors involved in promoting cultural transition and hybridity. In addition, the study aims to find the perception of the students about Tamil culture and the hybridization of Tamil culture

Method

The present research uses the quantitative method. The survey has been conducted among the undergraduate college students of Tamil Nadu. The study consists of 215 students including male and female students from Tamil lineage and non-Tamil lineages. These students were selected using a random sampling method and were invited to respond to a questionnaire consisting of 12 closed-ended questions.

Methodology

The price of hybridization includes the loss of regional traditions and local roots (Burke, 2014). Homi K. Bhabha perceives hybridization as the creation of new transcultural forms within the contact zone produced by colonization (Mambrol). Bhabha perceives the term "hybridity" from the perspective of colonization, wherein there is a colonizer and a colonized society. It gives birth to the idea of mimicry. The colonized society mimics the culture of the colonizer who values it greater than the existing culture. This act of hybridity defiles the purity of the colonized culture. In such a case, it gives something and takes back something that it wants. An exchange of customs happens here as a result of hybridity. Bhabha explicates that the idea of hybridity results in the mutilation of cultures within colonial and post-colonial contexts. The word hybridity can have at least three meanings – in terms of culture, ethnicity, and biology. Bhabha argues that there is a space in-between the designations of identity (Easthope, 1998). He advocates hybridity as the position between existing positions.

Perspectives on Cultural Socialization

In this section, the study figures out the perspective of students on hybridization. The first question placed in front of the target audience is "Have you noticed cultural hybridization within the context of Tamil Nadu?" and the responses are "Yes" and "No." It is found that 89.3% of the students agreed that they have noticed the hybridization of Tamil culture within the context of Tamil Nadu. A meager group of students that is 10.7% of students have selected "No" to convey that they have not noticed the hybridization of Tamil culture. A majority group of students has noticed Tamil culture hybridizing with other cultures. The next question aims at detecting if the students agree with the hybridization of Tamil culture. The startling result is that 80% of the students have agreed that Tamil culture needs a change, and it has to mingle with other cultures to move to the state of hybridization. The present study attempted to detect if hybridization can contribute to the growth of an individual and a community, and the choices are "Yes" and "No." It has been found that 81.4% of students agree that the hybridization of Tamil culture spurs the growth of an individual and community and 18.6% of students have rejected this view. In connection to this response, the present research has attempted to detect the knowledge of respondents about Tamil culture thereby figuring out their perspectives about Tamil culture hybridizing with other cultures. The question placed in front of the respondents in this regard is "Mark your knowledge of Tamil culture on a scale of 0–5." The survey has disclosed that

1.9% of the students have little knowledge and 2.8% of students have chosen point 1. On a scale of 5, 9.8% of students have selected 2 to signify the quantity of knowledge that they have about Tamil culture. The study also found that on a scale of 5, 30.7% of students have chosen point 3 and 28.4% of students have chosen point 4 and 26.5% of students have chosen point 5 to indicate the degree of their knowledge about Tamil culture. It is evident from this data interpretation that the majority group of students has selected point 3 to imply their knowledge of Tamil culture. This data makes the fact vivid that a majority group of students does not have proper knowledge about Tamil culture though they live in Tamil Nadu.

The study intended to determine how the knowledge of Tamil culture is being disseminated across Tamil Nadu. Hence, the question: "Which of these propagate the knowledge of Tamil culture to the maximum degree?" the options are "literature," "movies," "political discourse," "social practices," and "sports." Figure 1 pasted below illustrates the responses of the target group.

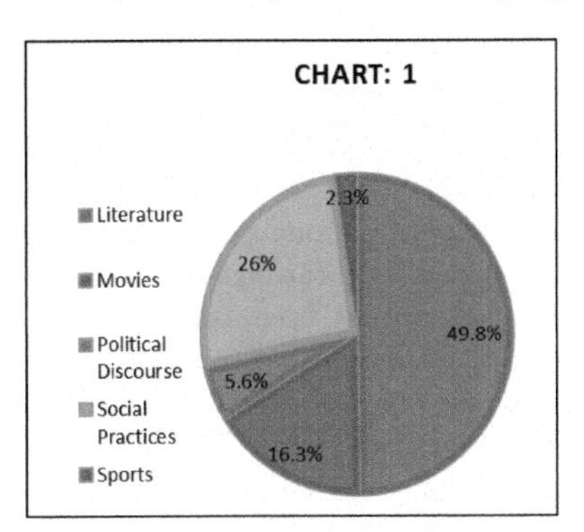

Fig. 1. Responses of Target Group: Cultural Socialization

It is apparent from the chart that a majority group (49.8%) of students have registered that literature disseminates the knowledge of Tamil culture and a minority group (2.3%) of students have chosen to say that sports underpin sharing the knowledge of Tamil culture. Social practices are considered the tool for spreading the knowledge of Tamil culture by 26% of students and 16.3% of students believe that Tamil movies play a significant role in disseminating knowledge. It has been found that political discourse also plays a significance in spreading the knowledge of Tamil culture which has been agreed by a group of 5.6% of students.

It is necessary to know who contributes to the socialization of Tamil culture. The question framed in this regard is "Who contributes more to cultural socialization?" and the choices are "People" and "E-people." The majority group (75.8%) of students preferred to say that people contribute to the socialization of Tamil culture and the minority group (24.2%) of students have chosen to say that E-people contribute more to the socialization of Tamil culture. It can be read from this response that the students believe face-to-face interaction underpins the dissemination of the knowledge of Tamil culture rather than meeting the E-people on online platforms.

Culture sustains based on purity and current relevance. A culture can remain pure if only it does not accept the influence of other cultures. Tamils living in New Zealand and America retain the purity of Tamil culture by spreading its essence. The popularity of Tamil culture is seen in America, which has been recorded by *PuthiyathalaimuraiTV* in the video "Increasing popularity of Tamil Culture in America." The recording says that Tamil culture has been spread in America by the immigrant Tamils through an art form. Though the Tamils have adopted American culture in their temporary space, they have not forgotten the legacy of their culture therefore they reflect it through plays and other forms of art. The present study has focused on determining the factors that preserve the purity of Tamil culture. The framed question is "Which of these retain the purity of Tamil culture?" and the options are education, politics, religion, social practices, and none of the above. Education supports retaining the purity of Tamil culture is the response from 21.9% of students and 3.7% of students chose to say that politics helps to disseminate cultural knowledge. It is found that 31.2% of students have chosen to say that religion retains the purity of culture and 31.2% of students have chosen to say that the aforementioned factors can retain the purity of Tamil culture. A group of students (24.2%) have chosen "None of the above" to convey that education, politics, religion, and social practices do not retain the purity of Tamil culture. There could be two reasons for choosing "None of the above" by 24.2% of students. It can be hypothetically stated that the students are either not convinced by the given options or have believed that purity in culture is unfeasible.

Factors Ensuing Cultural Hybridity

Cultural hybridization can happen due to numerous factors. The present study has attempted to explore this aspect by placing five questions about the factors that can influence other cultures. The question "Which of these factors promote the hybridization of Tamil culture?" and the choices are language, food, dressing, social practices, and all of the above. The target group has registered that language, food, dress, and social practices are the factors that ensure cultural hybridization in Tamil Nadu. It has been found that 79.1% of the target group agreed to the same notion by choosing "All of the above." The next question is seen as an extension of the dealt question. The framed question is "Which one of these contexts promotes the hybridization of Tamil cultures?" and the options are social, political, religious, and economic. Figure 2 given below illustrates the response of the target group.

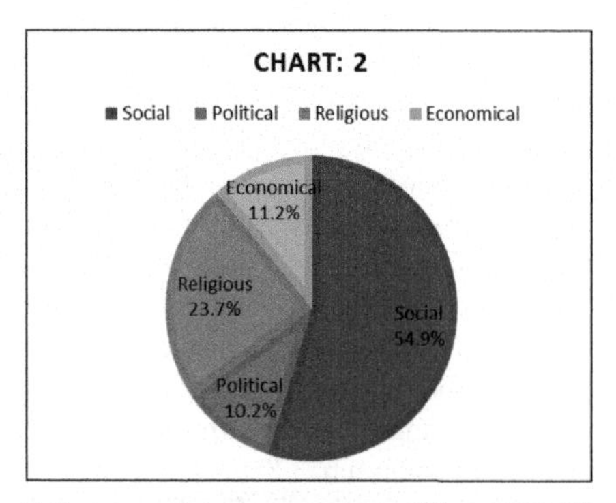

Fig. 2. Response of the target group to Cultural Hybridity

It is apparent from the above chart that a majority group of students has chosen to say that social context promotes the hybridization of Tamil culture. This majority group consists of 54.9% of the respondents. Religious and economical contexts are other setups wherein cultural hybridization is being promoted. The above chart makes it explicit that 23.7% of the students have agreed to say that religious context can underpin cultural hybridization and 11.2% of the students have chosen to say that economical context can lead to cultural hybridization. The study has proved that even cultural hybridization can be promoted in a political context as well. In addition to contexts and factors, even other cultures can propel the action of hybridity. In this regard, the question "Which of these cultures have influenced Tamil culture?" the options are Western culture, other foreign cultures, North Indian culture, and South Indian culture. The study has found that South Indian culture influences Tamil culture. South Indian culture refers to the culture that is practiced in Kerala, Karnataka, Telangana, and Andhra Pradesh. The South of Tamil Nadu is influenced by the culture of Kerala and the North of Tamil Nadu is influenced by Andhra and Karnataka. Western culture has influenced Tamil culture to some extent, and it has been agreed by 35.8% of respondents. It is found that 12.1% of students have agreed to say that Other foreign cultures have influenced Tamil culture and 10.7% of students have chosen to register that North Indian culture has influenced Tamil culture.

Bertrand Russell has stated that education spreads a sense of nationalism far and wide. This view can be perceived through the lens of culture. The target group has been asked to respond to the question "Urbanization and civilization are the causes of cultural hybridization." and the choices are "Yes" and "No." The survey makes it transparent that 80.9% agree urbanization and civilization have ensued hybridization of Tamil culture and 19.1% of students have documented that these factors do not contribute to cultural hybridization.

Conclusion

From the above discussion, it can be indisputably argued that the Tamil culture has undergone cultural hybridization. Hence, the present study has framed the question: "Purification of Tamil culture is urgently needed." and the options are "Yes" and "No." A majority percentage (70.2%)

of the target group has stated that there is an urgent need for the purification of Tamil culture and 29.8% of students have registered that they have not felt the urgent need for the purification of Tamil culture. Young minds have noticed the transition of culture. In addition, the study has surveyed and discovered that the students are culturally educated therefore they are open to the hybridization of Tamil culture. They have understood the fact that in the modern era, the world is heading towards global culture in the matter of language, food, dress, and social practices as a result hybridity in culture is inevitable therefore cultural hybridization has to be accepted. The present study has transcended to explore how Tamil culture is influenced by Western culture and how the knowledge of Tamil culture is spread in America to show the probability of Tamil culture influencing American culture.

References

Mambrol, Nasrullah. (2016). Homi Bhabha's Concept of Hybridity. Literary theory and criticism.

Wan, Y., Benner, A. D., and Kim, Su Yeong. (2015). The Cultural Socialization scale: Assessing Family and Peer socialization toward Heritage and Mainstream Cultures. Psychot Assess. 27(4), 1457–1462.

Aldoney, Daniela, Kuhns, C., and Cahrera N. (2018). Cultural Socialization. The SAGE Encyclopedia of Lifespan Human Development. SAGE Publication, Inc.

Lee, M. R., Grotevant, Harold D., Hellessted, Wendy. L., Gunnar M. R., and The Minnesota International Adoption Project Team. (2006). Cultural Socialization in Families with Internationally Adopted Children. J Fam Psychol., 20(4), 57–580.

Burke, P. (2009). Cultural Hybridity. United States: Polity Press.

Marotta, V. (2009). Cultural Hybridity. The Blackwell Encyclopedia of Sociology. Researchgate. 1–4.

Kraidy, M. M. (2002). Hybridity in Cultural Globalization. Communication Theory. 12(3), 316–339.

Bhabha, Homi K.(1994). The Location of Culture. London: Routledge.

Easthope, A. (1998). Homi Bhabha, Hybridity and Identity, or Derrida Versus Lacan. Hungarian Journal of English and American Studies. 4(1/2), 145–151.

Parables of the Lost and Found: A Semiotic Dissection of Religious Discourse

Fredrick Ruban Alphonse,[a] Job Sam Benjamin,[b] and Dr. S. Theporal[c]*

[a]Assistant Professor of English, Department of Humanities, Krupanidhi Degree College, Bengaluru
[a]Research Scholar, Department of English and Cultural Studies, Christ (Deemed to University), Bengaluru
[b]Department of Commerce, Krupanidhi Degree College, Bengaluru
[c]Assistant Professor of English, Department of Arts and Humanities, Christ Academy Institute for Advanced Studies, Bengaluru
E-mail: *fredrick.ruban@res.christuniversity.in

Abstract

A parable is an earthly story with a heavenly meaning. The term "parable" is closely associated with Jesus Christ. It is the chief teaching method that Jesus used to convey the truth about the kingdom of God. Jesus employed a simple style of teaching to deliver lofty messages. The parables of Jesus contain socio-historic relevance. The parables selected for the present study are the lost sheep, the lost coin, and the lost son. The trilogy advocates for the reconstruction of the Jewish perception of a sinner. It can be analyzed and interpreted from multiple perspectives. The present article takes a semiotic approach to the parables documented in Luke chapter 15. Semiotic are the systematic study of signs and symbols. This article studies the signs and symbols present in the parables. In addition, the identified signs and symbols are interpreted in the light of socio-historical aspects.

Keywords: parables, trilogy, symbols, Jewish, lost, sinner

Introduction

Jesus Christ adopted a parable as a technique to preach the gospel. The parables "are pictorial, easily grasped, quickly remembered and attention holders" (Doerksen, 1983). The trilogy of parables in Luke chapter 15 is an answer to the problem aroused when Jesus was associating himself with socially secluded people. Each parable of Jesus in the Gospels has a specific message to the audience. These parables can be categorized by each writer of the gospel by stating a specific subject. A similar trilogy is also stated in the gospel of Mathew. From "the preceding narrative and indeed from with the trilogy itself, the reader is given ample reason to conclude that the judgment announced by Jesus will be directed solely against the Jewish leaders" (Olmstead, 2003). The purpose of parables is to vividly teach the lofty message of spirituality. Luke 20:19 explicitly conveys that the parables are understood by the chief priests and the scribes. Vernon D. Doerksen states that "No doubt, the full implication of the parables and certainly the prophetic utterance, they did not understand, but it was sufficiently clear for them to desire to kill him (1983). Thus, it is understood that the purpose of narrating the parables was grasped by the audience. It was necessary to express the message in the parabolic form rather than as a dogmatic exposition because of the darkened and rebellious nature of the audience.

In the trilogy, Jesus teaches the subject of "lost." Even though the subject is "lost and found, there are various subunits. A corrupt pseudo-theology was prevailing during the time of Jesus in Israel. Pharisees and the scribes were regarded as those who observe the Law of Moses and took pride in their deeds. They considered themselves righteous because of their religious deeds, as the law prescribes, "a man may live if he does them" (Leviticus 18:5). This setting made people of other sectors inferior to those in the higher authority. Falsely assured of their righteousness derived from their works, they took pride in themselves. At the same time, they were strictly admonishing the people. These religious people were oppressing the gentiles by putting heavy burdens, which they could not carry. The tax collectors are also considered very deprived among the Jewish community as they "had identified themselves with the Roman conqueror" (Youtie, 1967). The prayer of the Pharisee and the sinner exposes the despicable nature of the tax collectors. The prayer shows how the Jews consider the tax collectors: "God I thank you that I am not even as rest of men – thieves, cheats, adulterers nor even this tax-farmer" (Luke 18:11). In the prayer the tax collectors pose that he is better than the other men in the society. Thus, all the evident sinners in the community were isolated and were considered with contempt, whereas the pious people were regarded as high. Hence, the Jewish political and religious system was corrupt with hypocrisy, lacking the true knowledge of God. Even the Pharisees and other Jewish leaders were not hesitant to bring the adulterers to Jesus to stone her as per the law. Thus, the Jewish perspective on sinners was extremely far from the reality of God's love and knowledge.

Jesus expressed his love for society through his religious discourse. The parables are considered the religious discourse of Jesus Christ. Religious discourse is a form of social interaction wherein "the connection to religion is foregrounded" (Pihlaja, 2021). In religious discourse, the speaker focuses on the issues of religious belief and practice (Pihlaja 2021). The claim of Pihlaja is evident in the religious discourse of Jesus. His religious discourse deals with religious belief and practice. It underpinned to the transaction of the knowledge of God and other ideas associated with Christianity. Chiefly, His discourse expressed the love of God for society and attempted to redefine the picture of God that society had already sketched.

Theoretical Framework

Semiotics studies signs and symbols. It helps in interpreting the symbols within social and cultural contexts. The semiotic model encompasses three concepts: signs, context, and meaning. According to Saussure "the combination of a concept and a sound image is a sign" (1959). A sign can be exhibited through smell, body language, sounds, etc. Saussure explains sign as the composition of the form of the physical reality and the conception or interpretation of the signifier by its viewers. The idea that is being conceived or interpreted is called signified. In the semiotics model, the second concept is context. According to Bowcher, context indicates those aspects in a conversation that renders specific and relevant meaning to the particular exchange that is occurring. It enables the recipient in the exchange to make proper sense of the interaction and acquire the intended meaning from it. The final concept in the semiotics model is "meaning." In cognitive semiotics, Zlatev has stated that meaning is the connection between the recipient of a sign and their personal experience. He articulates that meaning is produced when the recipient makes sense of the sign by linking and interacting with the surrounding reality. Semiotics helps in understanding why certain signs are interpreted incongruously in different cultures and geographies. It assists to think deeply about the meaning that is associated with sounds, images, colors, and events. It also sheds light on how perceptions might have been predetermined by the recipient.

Parable of the Lost Sheep

The three parables in Luke chapter 15 contain various signs and symbols. These are employed from different perspectives following the context.

They are used on the common grounds of "lost and found." Also, there are various subgroups in the passage which add to the central theme of the parables.

The parable of the lost sheep is detailed in Luke 15: 3-7. In this parable, Jesus gives the picture of a sheep that is lost and separated from the rest of the ninety-nine sheep. In certain instances, the people of Israel were compared to scattered sheep. Here, the lost sheep signifies the sinners and the tax collectors who were disregarded by the Jewish community. A sinner is analogized in the scriptures in many instances. Here, a sinner is referred as a person who is lost. In some passages, a sinner is illustrated as a person who is in dark. Whereas the Pauline epistles emphasize more the aspect of a sinner being dead in the spirit. Even "in Hebrew as in several ancient languages, including Greek and Sanskrit, the word for 'lost' is the same as for destroyed," even though we cannot do the same (Derrett, 1979). Thus, the lost sheep symbolizes the sinner. The sheep is described by Jesus to give a glimpse of the happiness when it is lost and found. The imagery of the shepherd laying the sheep on his shoulders, rejoicing, along with the communal celebrations expresses immense joy. It rejoices over the one sinner who repents. This is contrary to the knowledge which is prevailing among the Jews.

Parable of the Lost Coin

The Bible describes this parable in Luke 15: 8–10. This parable also shares a similar structure, but the symbols and signs are employed from a different social perspective. One can see the different priorities of people from different social backgrounds. In the present case, the woman is the main character. Usually, women are presumed to be engaged in domestic chores. Even though it is the husbands who work and earn, women play a role in saving money J. Duncan M. Derrett gives a quirky remark on this:

> The sum (ten coins) would do very nicely for a trip of no very great distance for a family to spend a week. It was natural for the wife to report whether she had saved enough for their purpose, while the substantial savings must be made, if at all, by the husband putting aside money for his daughter's dowries: a task that would not be entrusted to the wife's care. (1979)

Woman plays a significant role within the household. They are very keen to fulfill their little dreams. As a person who is confined to work within the four walls, the woman's dreams and aspirations are fulfilled by the coins that she saves. The ten coins symbolize the fulfillment of her dreams. When one of the coins is lost, it shatters the entire hope of the woman. The possibility of woman planning for any extra purchases during the Sabbath or any festivals cannot be ruled out (Derrett, 1979). The loss of one coin signifies the loss of her entire dreams, without which she cannot fulfill her dreams. The loss of one coin is equated with the loss of all the coins. Later, the woman lights the lamp and sweeps the house. The lamp is lightened to reach the darkest corners of the space. The coin lost in darkness and dust metaphorically signifies the lost in the darkness of the heart and entangled in the dust of sin. Here, the light of the lamp "signifies the knowledge about God to lead a disciplined life." (Pallathadka et al., 2021). The darkness represents the lies that the devil propagates to the people. Thus, light is absent in every lost sinner in the world. To overcome the darkness and to be found, the light should prevail. Without communicating the truths through words, a sinner cannot be saved. The phrase "sweeping the house" indicates the need to clean to find the coin. The similarities in the color of the coin and the dust will make it harder to find the coin. The imagery lost coin among the dirt symbolizes the sinner entangled in the dirt and filth of the world. The filth takes us far away from God. The nature of a sinner going astray from the presence of God is mentioned here.

Later, the woman finds the coin, she invites her friends and neighbors to rejoice with her. She does not call her husband to rejoice with her. Here, the communal celebration is very genuine. This also indicates the same joy that heaven has when a sinner repents. This parable also shows how much God is wanting a sinner to go to him. This is contrary to the present knowledge of the

Jews who are trying to seek condemnation for the sinner's deeds instantly without mercy. Without much change in the rhetorical structure, the parable of the lost sheep and the lost coin conveys the same ideas from different perspectives.

Parable of the Lost son

The parable of the lost son is mentioned in Luke 15: 11–32. It gives a magnified view of the progressive stages of a person being led to sin, and how he repents and comes back. The parable of the prodigal son predicates the value of a soul in the eyes of the heavenly father. Even though the younger son went rebelling the father, the father was very compassionate to go and receive his son back. The passage can be divided into two wherein "verses 11–24 deal with the father and the younger son and whereas the verses 25–32 focus on the father and the elder son (Forbes, 1999). After the first part, the story focuses on how the father deals with the arrogant elder son.

The prodigal son symbolizes the sinners. The words "squandered his estate with loose living" indicates the momentary pleasure one gets when he is engaged in the act of sin. These pleasures are "like the snowfall in the river, a moment white, then melts forever" (Brown). It will gradually take away the inherent heavenly riches like joy and peace and will finally put every man in the pit of guilt and desperation. Verses 14, 15, and 16 exhibit the transition of a sinner. The lost son was desperate to take a job that was detestable for any Jewish person as "to feed pigs, unclean animals, was degrading work for a Jew (Forbes, 1999). Verses 17 to 20 directly signify the act of repentance in a sinner. The guilt and conviction drive him to his father's house. The confession of the lost son is recorded twice within the passage; calling the attention of the target listeners, who are the Jewish hypocrites to true repentance.

In the parable, the love and care of God towards humanity irrespective of their former deeds is symbolized by the words "felt compassion for him and ran and embraced him and kissed him." Here, the father of the prodigal son symbolizes God and his attitude towards his previously rebellious son shows his kindness and acceptance for any person who is willing to seek him. Without seeking any revenge for squandering his wealth, he joyfully accepts his lost son. He is not only willing to accept him, but also accepts the prodigal son and gives the "best robe," "a ring" and "sandals." All the adornments are provided by the father without any hesitation. The parable of the wedding feast conveys the importance of grooming oneself, before entering the kingdom of God (Mathew 22).

The last part of the parable deals with the elder son, who symbolizes the religious Jews, who were infuriated by the kindness of Jesus toward the sinners. The character of the elder brother resembles the character of the Jews. But without completely ignoring the interest of the elder son, the father consoles the elder son and sheds light on the truth of the resurrection. The final discourse between the father and elder son signifies the care and consideration which God exhibits to everyone.

Conclusion

The subjects "lost sheep," "lost coin" and the "prodigal son" symbolize the sinners. "Shepherd", "woman" and the "father" symbolize the characters and expressions of God. All these characters are in contrast with each other. The concept of sin is signified throughout the parables. The parable of the lost sheep analogizes wandering and finally being lost. The parable of the lost coin also proposes the subject being lost in the dust to convey the act of sinning. In the parable of the lost son, the act of sin is directly analogized with the rebellious son. Finally, it is vivid that the communal celebration is present in every parable to portray the love of the heavenly father.

References

Sebeoks, Thomas A. (2001). Signs: An Introduction to Semiotics. Toronto Press Incorporated.

Saussure, Ferdinand De. (1959). Course in General Linguistics. The Philosophical Library INC.

Sebeoks, Thomas A. (1986). Advances in Semiotics. London: Indiana University Press.

Olmstead, Wesley G. (2003). Matthew's Trilogy of Parables. London: Cambridge University Press.

Doerksen, Vernon D. (1983). The Interpretation of Parables. Grace Journal. 3–19.

Youtie, Herbert C. (1967). Publicans and Sinners. Zeitschrift Für Papyrologie Und Epigraphik. 1–20.

Derrett, J. Duncan M. (1979). Fresh Light on the Lost Sheep and the Lost Coin. New Testament Studies, 36–60.

Harikumar, P., Tiganlung R P., and Shoraisam K S. (2021). A Semiotic Approach to Unfold the Metaphorical Interpretation of the Select Parables of the Son of God from the Book of Matthew. European Journal of Molecular & Clinical Medicine, 7(10), 4131–4140.

Forbes, Greg W. (1999). Repentance and Conflict in the Parable of the Lost Son (Luke 15:11-32). Journal of the Evangelical Theological Society, 42(2), 211–14.

Pihlaja, Stephen. (2021). "Analysing Religious Discourse: Introduction." Cognitive Linguistics and Religious Language: An Introduction. UK: Routledge.

Incorporating Research-based Pedagogical Implications in Grammar Through the Android Application: An Experimental Study

P. Jayakumar,[a,*] *S. Kavitha,*[b] *P. Paulsy Diana,*[c] *and I. Jane Austen*[d]

Sathyabama Institute of Science and Technology, India

E-mail: *jaikmabed@gmail.com, kavithasankaralingam5@gmail.com, dianadas2@gmail.com, janemaan03@gmail.com

Abstract

The current work is an effort to evaluate the instrument used to teach English grammar, particularly the production of negative sentences. Being an experimental study, it includes a pilot study, a pre-test, a teaching intervention, and a post-test. Additionally, both control and experimental group settings are used. When gathering data, a questionnaire is utilized as a research tool and distributed. This study looks at the mistakes that students made on the pre-test. As the preferred sampling technique, judgmental sampling is used. The investigator then uses the Cognitive Code Method and Mobile Assisted Language Learning to combine pedagogical intervention after classifying and tabulating the errors. Ultimately, a post-test that acted as an achievement exam is used to evaluate the impact of the instructional intervention and the performance gap between the experimental group and the control group. The customized Android app has shown to be a successful educational tool for enhancing the grammar abilities of teacher candidates in the Chennai area, according to the study's conclusions.

Keywords: Android app, CC Method, MALL

Introduction

English language is spoken almost all over the world. It is considered to be the universal language. Torres professes that English is to be upheld as the second largest language which people use in the world and it is officially accepted in 70 countries. In India, English language is declared as the second language and it is used for all academic purposes. No language is possible without grammar. Grammar of the language enhances productive skills. Undoubtedly grammar helps the learner to speak and write well. Mobile Assisted Language Learning (MALL) has gained massive attention in the field of English Language Teaching. In the present decade, MALL and teaching English through android applications have established themselves as interesting, remarkable, popular, important, and innovative areas for research.

Brief Review of Literature

Jie Chi Yang and Kanji Akahori (1998) focused on error analysis in Japanese. The remedial implementation developed was based on NLP through CALL programming. It was a survey-based study with a questionnaire to collect data from the learners. Totally 29 foreign students participated in two Japanese classes. The study tested the errors in using passive voice and led to the development of a tool for the World Wide Web.

Suja (2009) investigated the application of ICT for rural development in Kerala. The researcher selected three out of fourteen districts in Kerala to collect data. A stratified random sampling method was used to gather the required information from four panchayats of each district. Questionnaire, interview, and literature searches were the methods used in the study. The result of the study showed that ICT made a positive environment in the rural community in Kerala.

Bhuvaneswari, Swami, and Jayakumar (2020) conducted a study on the perspective of digital learning. Jayakumar and Ajit (2016) and (2017) made a study on grammar learning using digital tools in their study. From these studies, it is understood that there is a necessity for a digital learning environment in the contemporary teaching-learning process.

Chun, Kern, and Smith (2016) researched language use about teaching and learning. The researchers and English faculty members could benefit from this study in their efforts to integrate technology into education. The feasibility, applicability, and impact of integrating technology were also assessed in the study.

Bhuvaneswari, Borah, and Hussain (2022) examined the effects of both in-person and online language instruction. The study placed a strong emphasis on the direction of ESL education. The study's primary goal was to concentrate on the student's motivation and willingness.

Statement of the Problem

Even though there are numerous ways and approaches to teaching English, ESL students still encounter some issues. The children cannot form sentences because their first language places the components of speech. Students must be able to construct sentences employing the new words they have learned as soon as they have mastered the language's lexicon. The areas that must be learned include the placement of nouns, verbs, adjectives, pronouns, adverbs, prepositions, conjunctions, and interjections. The goal of this research is to impart knowledge. This study aims to impart the ability to construct declarative, interrogative, and imperative phrases utilizing positive and negative sentences.

Methodology Adopted for the Study

The study conducted a pilot study to know the trials and tribulations of B. Ed. students administered a questionnaire method. It was processed with 60 students in four colleges adopting a judgmental sampling method. The study was based on the experimental design so it had a pre-test and a post-test to test the performance of the learners. It was then divided into the control group and the experimental group to measure the level of achievement gained by teaching intervention. A prepared pedagogy based on the error analysis in their pre-test was analyzed by the investigator.

Aims and Objectives

The primary aim of teachers when teaching grammar is to methodically explain the language's structure and help their students develop a strong command of the language so they can accurately generate the grammatical structures they have learned.

The objective of the study is as follows:

- To acquire a wide range of sentence structure skills,
- To get an understanding of linguistic conventions about LSRW skills,
- To recognize and pinpoint sentence components, and
- To identify and apply the rules of sentence structure skills.

Selection of the Testing Items

Negative sentence formation is selected to identify the errors of the students when they are changing positive sentences to negative sentences. Based on this, they can have a better understanding of the use of auxiliary verbs.

Results and Discussion

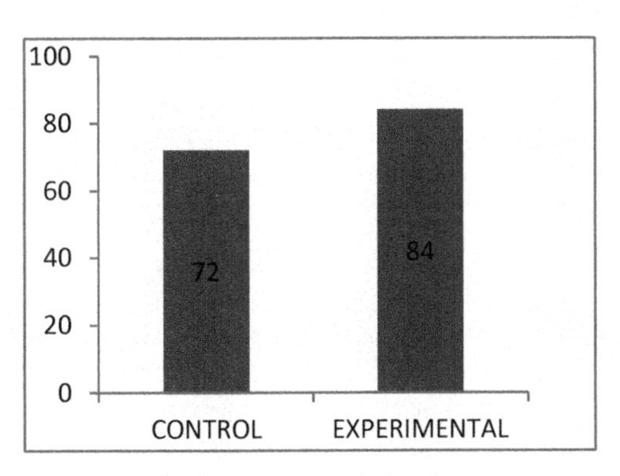

Fig. 1. Comparison of the Control group and the Experimental Group in the Pre-test

The contrast between the experimental group and the control group during the pre-test is shown in the graph. As seen in chart 1, the experimental group received 84 correct answers out of 180, compared to the control group's 72 correct answers out of 180 for the grammatical unit NEG.

H1- There is no significant difference in the performance between the experimental group and the control group in the pre-test regarding negative sentence formation.

Table 1. Pre Test

Testing Items	Group	Number	Mean	t - Value	Significance
NEG.	EG	30	2.8000	1.051	.298
	CG	30	2.4000		

The table displays the levels of performance for the experimental and control groups during the pre-test, as well as the mean difference and its level of significance between the two groups (before intervention). The table demonstrates that the experimental group outperformed the control group, with a mean value of NEG for the experimental group of 2.8000 and 2.4000 for the control group. It is clear from the mean value of the experimental and control groups and their significance in the pre-test that their grammatical knowledge was nearly identical before the intervention.

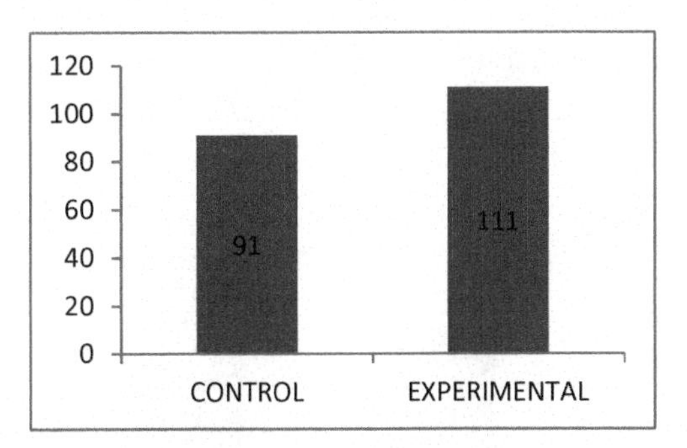

Fig. 2. Comparison of the Control Group and the Experimental Group in the Post Test

In the post-test, the chart compares the experimental group and control group. In the grammatical unit NEG, the control group received 91 out of 180 correct answers, compared to 111 out of 180 in the experimental group. After the Android app intervention, the experimental group outperformed the control group in terms of the number of correct responses. The experimental group and the control group differ significantly from one another. Therefore, it is plain to see that the Android App intervention has given the group a solid foundation for verb learning.

H2- There is no significant difference in the performance between the experimental group and the control group in the post-test with reference to negative sentence formation.

Table 2. Post Test

Testing Items	Group	Number	Mean	t - Value	Significance
NEG.	EG	30	3.8667	3.077	.003
	CG	30	2.9667		

The table includes the post-test performance levels for both the experimental and control groups as well as the mean difference and its level of significance between the two groups (after intervention). The table reveals that NEG. has a mean value for the experimental group of 3.8667 and a mean value for the control group of 2.9667, indicating that the experimental group outperformed the control group. The experimental group's mean value was greater at the 0.01 level than the control group's, indicating a significant mean difference between the two groups. This suggests that the Android app intervention has helped them perform better.

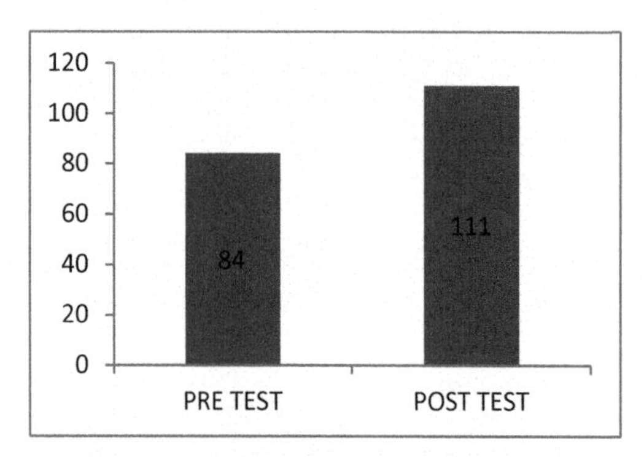

Fig. 3. Pre-test Experimental Group vs. Post-test Experimental Group

The graph displays how the experimental group fared on the pre-and post-tests. Chart 3 shows that in NEG, the experimental group had 84 correct answers out of 180 on the pre-test while having 111 correct answers out of 180 on the post-test. The graphic makes it clear that there are now significantly more correct answers than there were on the pre- to post-test. The Android app educational intervention has spread through the development of negative sentences.

H3- There is no significant difference in the performance between pre-test and post-test in the experimental group test with reference to negative sentence formation.

Table 3. The Experimental Group in the Pre-test and the Post-test

Testing Items	Experimental Group	Number	Mean	t - Value	Significance
NEG.	Pre-test	30	2.8000	-2.827	.008
	Post-test	30	3.8667		

The mean difference and degree of significance between the experimental groups for the pre-test and post-test are statistically analyzed in the table. The table illustrates that in NEG, the experimental group has a mean value of 2.8000 and 3.8667 for the same group in the post-test, in which the group did better. The table also displays the experimental group's mean differences between the pre-test and the post-test. The difference is 1.0667 in NEG.

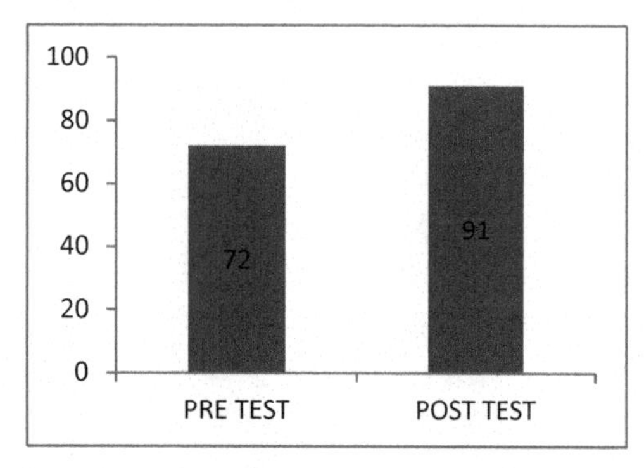

Fig. 4. Pre-test Control Group vs. Post-test Control Group

The chart compares the results from the control group's pre-and post-tests. Chart 4 shows that in NEG, 72 out of 180 correct answers were given in the pre-test while 91 out of 180 correct answers were given in the post-test.

H4- There is no significant difference in the performance between pre-test and post-test in the control group with reference to negative sentence formation.

Table 4. Control Group in the Pre-test and the Post-test

Testing Items	Control Group	Number	Mean	t - Value	Significance
NEG.	Pre-test	30	2.4000	-2.482	.019
	Post-test	30	2.9667		

The performance of the control group in the pre-test and post-test is statistically analyzed, along with the level of significance of the mean difference. The table reveals that NEG. performed better in the post-test, with a mean value of 2.9667 for the same group compared to the control group's mean value of 2.4000.

Errors Committed by the Target Learners

Using a questionnaire method for data collection, the researchers examined the students' sentence construction abilities regarding assertive and negative phrase kinds. Since the test was developed, it is clear that students' mistakes were caused by their lack of familiarity with the target language.

- Students were not able to differentiate the be form verbs in simple present, present continuous, and simple present in passive voice construction.

  Example 1: I *am* a clever boy. I am not a clever boy.

  Example 2: I *am doing* my work sincerely. I am not doing my work sincerely.

  Example 3: I *am loved* by the students. I am not loved by the students.

- Students found it difficult to change the sentence from assertive to negative when the sentence has modal auxiliary verbs.

  Example 1: He needs water. He does not need water.

  Example 2: He needs to go to the hospital. He needs not to go to the hospital.

- Placing the "not" in an interrogative sentence was not so familiar in some cases.

  Example 1: Was it clear? Was it not clear?

  Example 2: Are you going to school today? Aren't you going to school today?

Pedagogical Implications

The students can learn the following outcomes. They are as follows:

- In grammar, there are two sentences to express positive ideas and negative ideas in the verb phrase.

  Example: 1. I have a tool. (Positive) 2. I don't have a tool. (Negative)

- Adding "not' in the negative sentence may differ from sentence to sentence in terms of tense.

  Example: 1. We become friends. (Positive) 2. We don't become friends. (Negative)

  Example: 1. We are friends. (Positive) 2. We are not friends. (Negative)

Discussion and Findings

The findings of the study offer several pedagogical implications for teaching English grammar regarding negative sentence formation. To summarize, grammar learning through technology, especially mobile-assisted language learning, is effective. It is evident in this experimental study that the use of the Android application in teaching would be more productive than the conventional method of teaching. In the present study, the researcher has attempted to evolve a process to teach grammar by combining the existing methods. The researcher earnestly wishes that the present study may be of use to the teachers of English language in the ESL contexts in implementing appropriate methods conducive to their language teaching environment.

Conclusion

At the outset of the findings, the Android application influenced the target learners in their achievement level. Based on the analysis and interpretation, it is found that the experimental group acquired positive learning and secured better marks in comparison with the control group. On account of the Android pedagogical intervention, it made a significant impact on the target learners. It is observed from the findings that learning grammar through the Android application has resulted in more commendable results than the conventional teaching method. It is evident from the analysis that the learners have shown considerable improvement from pre-test to post-test. For the present study, the learners exhibited their performance in the post-test after the pedagogical intervention assisted by the Android application. It served as a tool to bring out their proficiency towards learning grammar.

References

Bhuvaneswari, G., Borah, R. R., Hussain, M. M. (2022). Willingness to Communicate in Face-To-Face and Online Language Classroom and the Future of Learning. In: Hamdan A., Hassanien A. E., Mescon T., Alareeni B. (eds). Technologies, Artificial Intelligence and the Future of Learning Post-COVID-19. Studies in Computational Intelligence, vol. 1019. Springer, Cham. https://doi.org/10.1007/978-3-030-93921-2_14

Bhuvaneswari, G., Swami, M., and Jayakumar, P. (2020). Online classroom pedagogy: Perspectives of undergraduate students towards digital learning. International Journal of Advanced Science and Technology, 29(04), 6680–6687.

Chun, D., Kern, R., and Smith, B. (2016). Technology in language use, language teaching, and language learning. The Modern Language Journal, 100(S1), 64-80.

Jayakumar, P., and Ajit, I. (2016). Android app: An instrument in clearing Lacuna of English grammar through teaching 500 sentence structures with reference to the verb eat. Man in India, 96(4), 1187-1195.

Jayakumar, P., and Ajit, I. (2017). The pedagogical implications on the root and route of English basic verbs: and extensive study through android application. The Social Sciences, 12(12), 2244-2248.

Perumal, M., and Ajit, I. (2022). An Exploratory Study on the Difficulties Faced by First-Generation Learners in Writing Skills. Journal of Higher Education Theory and Practice, 22(1).

Suja, K. (2009), "Application of information and communication technology for rural development in kerala."

Yang, J. C., and Akahori, K. (1998). Error analysis in Japanese writing and its implementation in a computer assisted language learning system on the World Wide Web. Calico Journal, 47-66.

Legacy and Evolution of Panchayati Raj Institutions and Tribal Self-Governance in India

Shamsher Alam[*,a] *and Priyanka Thakuri* [*,a]

[a]Central University of Jharkhand, India
E-mail: [*]shamsher@live.com; thakuripriyanka.1919@gmail.com

Abstract

The Indian Constitution exhibits elements of centralization as well as federalism. The Directive Principles of State Policy mandates that the State shall take steps to organize village panchayats and endow them with such powers and authority as may be necessary to enable them to function as units of self-government. The inclusion of such an article in the Constitution corresponds with the vision of the makers of the Constitution and their aspiration to strengthen democracy at the grass root level. Such a grass root mechanism was strongly advocated by Mahatma Gandhi, and his ideals were enthusiastically supported by other leaders as well, which eventually resulted in legislating various laws related to Panchayati Raj. Reflecting on the history of India, it can be asserted that Panchayati Raj Institutions have existed for a long time, exhibiting great similarities to the present system of local governance. The institution has manifested itself after channeling through various timelines and has finally adapted to its current form. The paper attempts to understand the nature and eminence of the Panchayati Raj System in India, and delves into the pages of history to sail through its legacy and evolution.

Keywords: Indian Constitution, Panchayati Raj, Self-Governance, Local Administration

Introduction

Indian civilization is one of the oldest civilizations in world history and so is the system of local governance in the country. It was prevalent ever since indigenous communities had their kind of administrative system (Shah, 2006), and continued later with Aryans, Mughals, and finally the Britishers. The contemporary setup of the local governance system today is grounded on the principle of decentralization, which relates to sharing of powers to craft healthy governance (Sondhi, 2000). To comprehend the need and importance of Panchayati Raj in modern Indian polity, it is essential to probe into the past and divulge through the nature and function of traditional governance systems and their democratization. In earlier times, *chaupal* was an integral part of rural life, whereby villagers, generally elderly, sat together and discussed local affairs pertaining to community and village affairs (Chandhoke, 1990).

All such matters were decided in such assemblies, which later on became known as Panchayats. Etymologically, the term Panchayat is composed of *ayat* (assembly) and *panch* (five community elders); and has its roots derived from the Sanskrit words *Pancham* and *Ayatanam*, wherein the former refers to gathering of five prominent local personalities and the latter denoted an abode for gathering, thereby meaning an office of *Panch* (Alam and Raj, 2018).

Legacy of Panchayati Raj

Ancient India

In Vedic Era (1750 BC–500 BC), the rural body constituted of primary territorial unit of administration in which the local villagers participated in communal decision-making. *Sabhas* (meetings) were a popular platform, in which the common populace had the authority to express their views and had a say in local affairs. This is how the rural community was always a structural, functional as well as a self-dependent unit of authority during the Vedic Era. It generated its assets and had its types of machinery and working sphere, which rarely had any conflicts with the larger state functionary. Rather they were complementary to each other. The pieces of literature carved during the Vedic Period like Rig Veda, Manusmriti, Dharmashastras, Upanishads, and Jatakas mention the distant evolutionary past of the present-day archetypal of the Panchayati administrative system of the country. The Shanti Parva of Mahabharata also conveys insinuations to the sustenance of Grama Sanghas or village councils, as referred to in the other Vedic Age literature. Adding to the list is Kautilya's Arthashastra and Valmiki's famous epic Ramayana (Sharma, 1994) which also provides an all-inclusive description of the system of village administration prevailing during that time.

During this period, the village administration was carried under the supervision and control of Adhyaksha or headman (Desai, 1990). There were officials such as Samkhyaka (accountant), Anikitsaka (veterinary doctor), Jamgh Karmika (village couriers), and Chikitsaka (physician); who assisted the headman. The Adhyaksha, was also accountable for the collection of state taxes, controlling the activities of offenders, and delivering justice in consultation with peers. There were groups of families in the village and the head of the village, popularly known as Gramini, acted as the administrative head. The group of villages worked under a key officer Gopa. Generally, the King himself held the position of Gopa. Hence, it is evident that during the Vedic period, monarchical system was in existence, but the essence of communal participation in local self-governance co-existed as well.

The Buddhist period (around 5900 BC–600 AD) is remarkable in Indian history, well-known for its non-turbulent tenure. Local governance was also present during this period (Majumdar, 2009). Administration during this period was quite democratic and even in nature. The commoners were involved to carry out tasks of administration. For efficient administration, empires were divided into various provinces known as *Deshas* or *Bhuktis*. Governors who belonged to the royal family directed these provinces. The villages were the smallest unit for administration and the headman was called Gramic. He was responsible for the maintenance of law and order in the village and was directly answerable to the king.

Medieval India

The early phase of the medieval period (750 AD - 1200 AD) was dominated by the Chola rulers. Here also there was the existence of autonomous local self-governance. The Cholas strongly believed in the mode of self-rule, in which people were involved in the process of governing the empire. Village administration was carried out by villagers. In every village, there was a village assembly known as Kudavolai, which selected the Sabhas and its members. In addition, there were committees, named Variyams to look after the governance. Although the king had a council of ministers and the supreme power vested with him, the sentiments and the involvement of the people in governance were never underestimated.

However, during the reign of the Mughals during the late medieval period (1200 AD–1700 AD), local governance suffered a setback due to over-centralization of administration. Due to centralism, villages lost much of their autonomy. Law and order of villages were controlled through *Mugaddams*. The judicial powers of Panchayats were also curtailed. A new kind of Zamindari system was put in place in which taxes were collected by administrators who were directly responsible to the king. This robbed the autonomy of Panchayats and curtailed their erstwhile significance (Singh, 2009).

Modern India

Later on, during the British period also the village administrative system continued receiving setbacks. Since powers and responsibilities were largely concentrated in the hands of the Governor General of India, pre-existing village-level administration and revenue collection were adversely impacted. Rules focusing on an oppressive collection of revenue and taxes from land came into existence (Reed, 1955). In this course of action, village land record officers were replaced by *Patwaris*. Similarly, the existent village police system was also dismantled and the trend of Magistrate - as the regional controlling officer, came into being. These measures drastically shattered the local administrative setup in rural India.

Nevertheless, in 1870, a landmark plan initiated by Britishers, known as Mayo's Resolution, was initiated for the decentralization of local governance in the country for ensuring organizational competence. Subsequently, Lord Ripon set up another benchmark in 1882 by splitting administrative units into further smaller units. To ensure public participation, election at local levels was introduced (Arora and Goyal, 1995). However, this was only limited only to urban areas. Rural areas remained neglected in this reform.

In the year 1907, a Royal Commission on Decentralization was ordained under the chairmanship of C.E.H. Hobhouse, which recognized the importance of panchayats at the village level. Almost a decade later, in 1919, the Montagu Chelmsford reforms were introduced. Under this reform, Dyrachy System was introduced. In this, there were two broad subjects of administration, namely, the reserved subjects and the transferred subjects. Governors were responsible for control over reserved subjects while transferred subjects were under ministers responsible to the governing body. This reform, however, could not yield many results, as it had several administrative and financial drawbacks. Finally, the Government of India Act 1935, in limited ways, mooted for the progression of Panchayats in India (Gupta, 2004). According to this Act, popularly elected governments in different provinces were empowered to enact legislation for a democratization of institutions of local self-governance. However, in majority of cases, the system of responsible government at grass root levels was least interested in delivering good governance, and the situation remained more or less unchanged until liberation of India from the Britishers.

Post - Independence India

After independence, the accountability and responsibility to provide the nation with a set of governing rules was the next immediate challenge for the then leaders. For this, the constituent assembly was constituted for drafting the country's Constitution. Although Mahatma Gandhi advocated for decentralization and wanted to strengthen local governance system, his idea was opposed by B.R. Ambedkar, the chairman of the drafting committee of the Constitution. He opined that villages of the nation, represented a divided India and hence it shall act as a source of oppression for them in the coming times (Chitkara, 2002).

Nevertheless, local self-governance was placed under the Directive Principles of State Policy (DPSP) of the Constitution. Because of such a disposition, only a few states like Rajasthan and Karnataka initially adopted the mission of decentralization of power at village-level, whereas in majority of the states, Panchayati-Raj was not looked upon with enthusiastic eyes.

The situation remained unchanged for the coming four decades. However, various advisory committees were constituted to explore different aspects of decentralization, both in rural as well as urban settings. The Balwant Rai Mehta Committee of 1957 reported that the mission of community development could be fulfilling and enduring only if the community was involved in planning, decision-making, and implementation. It further recommended for establishment of a three-tier system of local governance, wherein elected bodies should enjoy the autonomy of essential resources, power, and authority.

Later in 1963, K. Santhanam Committee examined the financial issues of existing Panchayati Raj Institutions in the country. It observed that the financial capability of PRIs was very limited as a major part of revenues were pre-empted at higher levels of governance. Therefore, the committee recommended that Panchayats should have special financial powers bearing in mind that people should not be burdened with too many taxes.

Later in 1977, another committee under the chairmanship of Ashok Mehta was constituted to bring forth the weakness of Panchayat system and suggest measures to strengthen them. The committee laid stress on the developmental role of Panchayati Raj Institutions in areas of agricultural activities, forestry, small cottage industries, and other welfare activities, and suggested a 2-tier Panchayati system; Zilla Parishad at the district level and Mandal Panchayat below it. It also added that seats for members of Scheduled Castes and Scheduled Tribes be reserved based on their population at both levels.

Furthermore, two more committees were constituted in this regard in 1985 and 1986. The first was G. V. K. Rao Committee which opined that the main foray of PRIs should be rural development. For that, they must be activated on a pan-India basis and provided with all requisite support to become effective organizations of local governance. The second committee was L. M. Singhvi Committee which studied the existing models of local governance and recommended that institutions of local self-government should be constitutionally recognized, protected, and preserved through the inclusion of a new chapter in the Indian constitution, with a separate Finance Commission. All these recommendations assisted in nurturing local self-governance within the democratic setup of India.

The 73rd Constitutional Amendment Act, 1992

It was in the late 1980s, when Rajiv Gandhi, the then Prime Minister of India initiated a bill in the Parliament on 15th May 1989 with the vision to make Panchayati Raj Institutions, a truly effective and representative system (Sharma, 2000). Unfortunately, the bill could not become a law, as it failed to pass through the upper house of the Parliament. Later, in the year 1992, the 73rd Constitutional Amendment Bill was mooted once again, during P.V. Narasimha Rao's government. This time, the ruling side was successful in getting clearance, from both houses of the parliament. The Act, after passing, granted Constitutional status to the Panchayati Raj bodies on April 24, 1993. It was notified by the government in the Gazette of India, published on 20th April 1993, after ratification by state legislatures and assent of the President of India. This marked a new era in the federal setup of the country and provided constitutional status to Panchayati Raj Institutions in India.

Self-Governance in Scheduled Areas

After having the landmark amendment to uplift and expand the Panchayat system in the country, it was discovered that it assisted a lot in ameliorating the condition of rural areas, but as per Article 243 M of the 73rd Amendment Act, Panchayati Raj setup was not applicable in Scheduled Areas; which were pre-dominantly tribal belts in terms of demography. To secure the interest of tribals dwelling in such areas, the Bhuria Committee was instituted to make recommendations for extending the provisions of the Act to Scheduled Areas. It led to the formulation of *The Provisions of the Panchayats (Extension to Scheduled Areas) Act, 1996* (in short, PESA Act).

It extended Panchayati Raj to 5th Schedule Areas located in the states of Andhra Pradesh, Chhattisgarh, Madhya Pradesh, Jharkhand, Himachal Pradesh, Gujarat, Maharashtra, Odisha, Rajasthan, and Telangana. The Act gave sweeping authoritative powers to tribal communities and recognized their traditional community rights over local natural resources. The Gram Sabhas and Panchayats of these regions were granted a wide range of power, functions, and responsibilities, especially in domains pertaining to approval of projects for social and economic development; poverty alleviation programs; utilization of funds; acquisition of land; management of minor water bodies; grant of prospecting license or mining lease for minor minerals; regulating sale and consumption of intoxicants; manage village market; and control over money lending activities.

Conclusion

Despite several hiccups and hindrances, Panchayati Raj Institutions, which was once an inevitable part of Indian civilization, has adapted itself to its current form by the passing of 73rd Amendment Act 1993. It has deeply impacted local self-governance and autonomy and has brought about bliss in the lives of rural people who can now legitimately govern and take decisions for themselves. Later on, the enactment of PESA Act 1996 added to the wave and paved path for the participation of tribals in local self-governance as well. Viewing these developments, it can be conceded that although slow, governance models are steadily inching towards decentralization, and further changes are expected in near future as well. However, the only catch is that these legislations must fulfill their objectives in letter and spirit and that people truly get to live a life of their own choice, rather than based on somebody else's dictates.

References

Arora, R. K., and Goyal, R. (1995). Indian Public Administration: Institutions and Issues. New Age International (P) Ltd.

Chandhoke, S. K. (1990). Nature and Structure of Rural Habitations. Concept Publishing Company.

Chitkara, M. G. (2002). Dr. Ambedkar and Social Justice. APH Publishing.

Desai, V. (1990). Panchayati Raj: Power to the People. Himalaya Publishing House.

Gupta, D. N. (2004). Decentralisation-Need for Reforms. Concept Publishing Company.

Majumdar, R. C. (2009). Corporate Life in Ancient India. BiblioLife.

Reed, S. (1955). The Times of India Directory and Year Book Including Who's who. Bennett Coleman & Company.

Shah, A. (2006). Local Governance in Developing Countries.

Sharma S. (1994). Grass Root Politics and Panchayati Raj. Deep & Deep Publications.

Alam, S., and Raj, A. (2018). The Academic Journey of Witchcraft Studies in India. Man in India, 97(21), 123–138.

Sharma, R. D. (2000). Administrative Culture in India. Anamika Publishers & Distributors (P) Ltd.

Singh, U. B. (2009). Decentralized Democratic Governance in New Millennium. Concept Publication Company.

Sondhi, M. L. (2000). Democratic Peace: The Foreign Policy Implications. Har-Anand Publication Pvt. Ltd.

Prospects of PESA Act and Inhibitions in its Implementation in Scheduled Areas of Jharkhand, India

Shamsher Alam[*,a] *and Priyanka Thakuri*[*,a]

[a]Central University of Jharkhand, India
E-mail: [*]shamsher@live.com; thakuripriyanka.1919@gmail.com

Abstract

Forty-fifth year of Indian independence witnessed an event so momentous that it would be remembered by generations to come as the commencement of the 73rd Amendment Act, 1992. It was helmed as a milestone for Panchayati Raj in post-independent India. The dream, once envisioned by Mahatma Gandhi of a decentralized government, catering to the hopes and aspirations of the rural population, was finally materialized. However, given the dichotomy of diversity vs. inclusiveness of the Indian population, an exclusive act was enacted for scheduled areas having tribal inhabitants, called the Panchayats (Extension to Scheduled Areas) Act, 1996. The PESA Act aimed to safeguard and preserve the traditions and customs of tribal communities and empower them with a legal instrument for their development. The present paper attempts to demonstrate the prospects of this Act, as enshrined through its provisions. A case study of Chouli village in Ratu Block of the capital city (Ranchi) of Jharkhand was convened to bring forth the striking contrast between desired results and realities on the ground. It endeavors to examine barriers hindering the accomplishment of specified objectives along with exploring outcomes of such lacunas followed by recommendations to improve the current situation.

Keywords: Decentralised Tribal Governance, Development, Jharkhand, Panchayati Raj, PESA Act

Introduction

Since the days of Indian freedom struggle, Mahatma Gandhi has been an ardent believer of Gram Swaraj (village autonomy) and has championed for this cause even after India became an independent state. His dream saw the light of day as 73rd and 74th Constitutional Amendment Acts, 1993 were passed by the Parliament. This step was helmed as a milestone for Panchayati Raj in India, because it not only granted constitutional status to local self-governing bodies, it also gave them powers to take their own decisions, rather than obey the mandates of officers sitting miles away. The amendment led to the addition of a separate "Part - IX," entitled "Panchayats," comprising Article 243 to 243-O and a new "Schedule-11" having reference to Article 243G, enumerating 29 different powers and functions of PRIs in the Constitution (Laxmikanth, 2016). However, since article 243 M of the 73rd Amendment Act restricted the setup of Panchayati Raj in 5th and 6th Schedule Areas (tribal demographic region), Ministry of Rural Development, Government of India, constituted the Bhuria Committee to recommend ways to extend provisions of Part IX of the Constitution to the Scheduled Areas. It led to the passage of Panchayats Extension to Scheduled Areas Act (in short, PESA Act) in 1996. It covered the first category of Tribal Areas envisaged in Article 244 of the Indian Constitution, i.e., Fifth Schedule Areas, under which, a total of ten Indian states were included- Andhra Pradesh, Chhattisgarh, Gujarat, Himachal Pradesh, Jharkhand, Maharashtra,

Madhya Pradesh, Orissa, Rajasthan, and Telangana. The Act gave sweeping authoritative powers to tribal communities and recognized their traditional community rights over local natural resources. It not only accepted the validity of customary laws, social and religious practices, and traditional management practices of community resources; but also restricted the concerned state governments to make any law inconsistent with these, making it a powerful legal instrument for tribal self-management and development.

Review of Literature

Panchayati Raj system in India is not a novel phenomenon, but its contemporary disposition i.e., as a means of decentralization can be attributed to Gandhian philosophy (Jathar, 1964), which is aptly reflected in the Indian Constitution (Pal, 2004). However, its journey so far has not been smooth sailing. Panchayat elections have been irregular throughout the country. Either they are not carried out in accordance with the Constitutional provisions and or they are being delayed because of one reason or the other. Despite getting directions from the High Courts and Supreme Court, elections have not been conducted. Sometimes the State Election Commissions have to fight lone battles against the State Governments, to hold timely elections as per the provisions of law (Mathew, 2001; Sharma, 2001). Very often, the elected representatives exhibit technical and administrative incompetence (Kumar, 2006; Kumar, 2010). Sharma (2001) reasons that the State Government has done little to facilitate for achievement the broader democratic goals of decentralization of power and authority. The political parties are equally lethargic towards promoting the objective of local self-governance. The author has also exemplified that, the people of the upper castes, still dominate the Panchayat elections and the lower castes are ousted, humiliated, beaten up, and not allowed to contest. Lack of financial devolution might be the cause behind it (Alam and Raj, 2018). He observes that the economic provisions as incorporated in the Constitution and in the State Panchayati Raj Acts, have not been followed sincerely. Moreover, the local bodies have now become mere spending agencies and are always looking to the state government to finance them. The kinds of literature on the functioning of Panchayati bodies point out that although political decentralization has taken place true decentralization is yet to be achieved (Bandhopadhyay, 2002). Women's representation remains inadequate. The consequences of such deterrents are borne by the common people and their issues like elementary education, tribal development, rural employment, etc. remain unresolved. Despite such roadblocks, the truth cannot be denied that the creation of Panchayati Raj has created great hope among the people. But the real motive of PRIs shall come out only when there is a measure of accountability involved with power as well. Therefore, with strong political will and resilience; the goal of true decentralization can be pursued (Malik, 2005).

Research Methodology

The paper, having empirical backing from a decadal-long study (2010–2020) conducted in the Chouli village of Bajpur Panchayat of Ratu Block of Ranchi District (Jharkhand), attempts to bring out prospects and problems of implementation of PESA Act in the region. Snowball sampling method was used for identifying the key sample, after establishing a good rapport with the villagers. Primary data was gathered through various means such as observation, interview schedule, photography, and case study techniques. Genealogical table helped in the quick gathering of information about the family members of villagers. Secondary data was collected from reports, plan drafts of village panchayats, annual reports of the district panchayats, committee reports, online journals, publications of the Ministry of Panchayati Raj and Rural Development, websites of concerned ministries and books (both online and hard print). During the study plentiful data was collected, some relevant and some irrelevant. Hence, the collected data were sorted according to the aims and objectives of the study and were organized, categorized, and analyzed to bring out relevant results.

Panchayati Raj in Jharkhand

Jharkhand, one of the Scheduled Areas, was carved from the southern part of Bihar and came into existence on 15th November 2000. With an outstanding tribal population, it was expected that the state would play a prominent role in the upliftment and development of the tribal population. To achieve this dream, the Jharkhand Panchayati Raj Act 2001 (JPR Act) was successfully passed by the first cabinet of the state. This act was in accordance with the PESA Act 1996 which authorized Scheduled Areas to create their laws. It consisted of 17 chapters, and 164 schedules, including the constitution of Gram Sabha and Panchayats, their power and functions, funds and properties, taxation and budget, rules and bye-laws, and miscellaneous other items.

Unfortunately, owing to the unstable political situation of the state, things did not materialize smoothly and people were left empty-handed. A decade later, the Jharkhand Panchayati Raj (Amendment) Bill, 2010, was mooted and it entered into force on 15th April 2010. The major amendments comprised of elevation in women reservation percentage, i.e. from 33% to 50% in all categories; subtle changes in reservation rules for the posts of Mukhiyas and Up-Mukhiyas, Pramukhs and Up-Pramukhs, Adhyakshas and Upadhyakshas of Zila Parishad; and permitting the State Election Commission to act in case complaint of irregularities in Panchayat elections and also appoint a General Expenditure Observer, post consultation with the State Government to supervise and submit a report on the entire electoral process. Thereafter, in the month of November-December 2010, the first-ever panchayati elections were conducted in the state.

Development Through Panchayati Raj Institutions

There is no universally accepted definition of development. Its meaning changes with time, place, culture, and context. In simple terms, improvement in the present quality of living can be termed as development. The Panchayati Raj Institutions were conceived with this mission and vision for people living in rural and tribal/ scheduled areas. It was designed to function as a powerful tool to ensure that it reaches those at the grass-root level. However, field visits represent a story that does not conform to the aforesaid mission. For instance, the observations pointed out that Panchayats failed to deliver their duties regarding land improvement, potable water management, irrigation facility, and management of primary activities such as animal husbandry. The only positive outcome was that tribals, as well as non-tribals, thoroughly enjoyed their rights on Minor Forest Produce. The economy of the village is more or less based on agriculture. Most of the villagers are agriculturists. Owing to their meager financial circumstances, they cultivate on their own and do not hire laborers on wage. Often, villagers go to the neighboring town of Ranchi or other urban areas in search of work as laborers in building and road construction sites. In off-seasons, most villagers work under the MGNREGA job scheme.

Looking towards educational institutions, all we can observe is the dire state of school buildings, insufficient number of teachers, poor enrolment ratio, and irregularities in the mid-day meal scheme; resulting in diversion from the prime objective of delivering quality education to village children. Owing to such a pathetic situation, some well-off villagers prefer to send their children to nearby private schools. There exists a primary school in Hisri, a secondary school in Bajpur, and for higher secondary education, the most accessible colleges are Kartik Oraon Inter College, Ratu, and Sanjay Gandhi Memorial Collage, Kamre. Most of the mid-age villagers are educated only till the primary level. It is the younger generation of the village which is more interested in having quality and proper education. Unfortunately, present education conditions and infrastructure facilities act as a hindrance to their dream. Transport and communication are also at a nascent stage. For communication, people mostly rely on radio and television (some better-off families) or simple mobile handsets. Gudu is the nearest post office, located about 4 kilometers from the village, and provides the facility of speed post, money order, and delivery of parcels. It also facilitates people with banking services as well.

People use bicycles for traveling and some dwell at the city use motorbikes. Other than individual transport, public transport services are availed by villagers.

Similar is the case with health facilities. The Primary Health Centre is located at Gudu, which is about 4–5 km away, with no active transport facility to reach there. The infrastructure of Health Centre is contemptible. The condition of the newly constructed health care Centre is even worse than older one. There are three doctors in the center, among which, two of them are general physicians and third is an AYUSH doctor hired on contract. Doctors, too, agree with the fact that health facilities are not up to mark. Not to mention emergencies, sometimes, even common medicines are unavailable. Child delivery and treatment of small non-serious ailments are the only facilities available in the Centre. Provision of having an ambulance for pregnant women is also missing. In the name of staff, there are two ANMs. (Auxiliary Nurse and Midwife), with a dresser and a peon. Health awareness programs to be conducted by Health Care Centre is also not carried out regularly. Yet, amidst such deprivations, the village Aanganwadi Kendra, mandated for the development of women and children, exhibits a ray of hope. It runs a few development schemes efficiently. Every month a medical check-up camp is organized. Polio and other vaccinations are administered by Kendra. Packet food and iron tablets are provided for pregnant women. It also provides health supplements to newborn children.

Inhibitions of the Act

Institutional Issues

Lack of awareness can be attributed as the major reason behind the failure of this act. It is expected that elected representatives must know what to do and how to do it. They must have an arduous understanding of their functions and power. In Jharkhand, the SIRD (State Institute of Rural Development) and CTI (Central Training Institute) has been entrusted with the responsibility to conduct regular training programs for all those who are directly or indirectly involved in the process- officials, non-officials, and elected representatives. An empirical study showed that most members lacked awareness of their powers, functions, roles, and responsibilities, resulting in inactivity. The elected representatives complained that during their tenure, they had only attended only a couple of One-Day training programs. The level of unawareness can be judged by the fact that the majority of representatives had not even heard about PESA and its provisions. They vehemently wished to contribute to the development of their village and community, but because of such a hindrance, they are unable to perform their duty aptly.

Another shortcoming observed during the field study was the lack of coordination between the three tiers of the Panchayati Raj. The state of hierarchical domination and control was found predominant among the Panchayat officials as well. Additionally, many people were left unsatisfied and unhappy with their elected representatives. A serious lack of miscommunication is visible among different levels of Panchayats. Minutes of the meetings at Zila Parishad or some Block level Committee is never percolated down. Likewise, official notices or orders relating to the devolution of Panchayats or bye-laws are unknown by elected representatives at ground level, though they exist on the government website. This creates a situation of gap and mis-coordination in the functioning of the Panchayats. It is another major institutional loophole in the system. Coupled with the fact that bureaucratic red tape in India is a meddlesome hindrance that hampers public welfare, such issues dampen the spirit of democratic decentralization. This institutional lethargy is reflected through misappropriation of funds, non-utilization of funds, increasing corruption, delay in the approval of schemes and programs and their overdue implementation, thereby leading to poor quality work.

The PESA Act empowers Gram Sabhas to play a key role in approving developmental plans and controlling all social sectors. Therefore, it becomes the duty of Gram Pradhan to convene and preside over the meetings of the Gram Sabha. But according to the Gram Pradhan, people are mostly disinterested mainly because of dismal performance of elected representatives and such meetings hinder their daily chores. Even if they do attend, women's attendance is next to zero, and no formal procedures such as maintaining a register to record meeting proceedings are followed. Similar is the situation with annual Gram Sabha meetings. They are not held regularly which is why all those schemes meant for development remain on the paper itself, without actual work. Neither do there exist any committees (Agriculture Committee, Development Committee, Health Committee, and others) as defined by the Act. Due to the nonexistence of Village Development Committee, plans and strategies for the developmental projects have never been discussed in the Gram Sabha. In the absence of health committee and infrastructure committee, Gram Sabha alone is unable to evaluate the functioning of Health Sahiyas, and medical facilities and take care of the infrastructural loopholes. Due to the lack of an education committee and vigilance group, Gram Sabha fails to inspect the irregularities in the school and check corruption.

To identify the selected beneficiaries for poverty alleviation programs and other schemes is the role of Gram Sabha. Barring the case of MGNREGA, it has not played any significant role in achieving its mandate. Even worse was the situation wherein, the Below Poverty Line or BPL cards (Laal Card) were given to some of the affluent families, while those entitled remained cardless. Villagers had reported the matter to their elected ward members, but the problem remained unresolved. Irregularities were also found in the payments of vridha and widow pension schemes as well; majority being the case of non-payment or wrongful payment to fictitious individuals. To inspect the functioning of Panchayats, post of Inspection Officer has been mandated. But the villagers had neither heard nor seen such personnel. There was no audit of the financial records of Gram Panchayat or Gram Sabha. All of these lead to the misfunctioning of PRIs. At present, the country is gearing up for a Digital India, but people of this area lag far behind their counterparts. Despite being launched in 2004, the e- panchayat portal is yet to be introduced in the area of field study. The elected representatives are unaware of anything as such, which indicates how the village is yet to embrace digitization.

Legal Inconsistencies and Shortcomings

Various legal shortcomings also contribute as a hindrance to the implementation of law. Various incongruities raise questions about the intention of State's legislation. For instance, the JPR Act fails to dichotomize the working zone and jurisdiction of the involved stakeholders. All four bodies (namely Gram Sabha, Gram Panchayat, Panchayat Samiti, and Zila Parishad), have been given the power to operate in common areas, along with some developmental committees whose functions overlap with the above mentioned bodies, resulting in a conflict of interest. And in such a situation, whose decision shall prevail, remains unanswered. The PESA Act demarcates the powers between three tiers of Panchayats, but unfortunately, the JPR Act contravenes it by conferring the same authority to higher and lower levels of Panchayats. For example, it allows all three tiers of PRIs to administer the village markets, which can be handled at village level itself. All three tiers of PRIs are given the authority of initiating local plans including tribal sub-plans, without any clear demarcation. Furthermore, both Panchayat Samiti and Zilla Parishad have been given the right to own minor water bodies. Such unthoughtful actions reflect that the relationships between different tiers of Panchayats haven't been addressed thoroughly. A similar contradiction between state and center law is visible in the case of land. The PESA Act makes it mandatory for the functionaries to consult Gram Sabha before the acquisition of land in the Scheduled Areas for development projects or rehabilitating persons affected by such projects. However, the JPR Act is completely mouth-shut on this crucial issue, leading to many cases of poor land acquisition.

Non-Institutional Concerns

In villages, there are a few people who are politically active but are not elected representatives. Usually attached to some political party; they often act in a superseding manner in meetings and even indulge in unscrupulous manipulations to secure financial gains by acquiring tenders and work orders. Such tout leaders, showing their political contacts, impair the functioning of the PRIs. The PESA Act mandated reservation for women in elections to increase the participation of women in the decision-making process as well as pave a way for their empowerment. However, a glaring truth was revealed through empirical pieces of evidence. At the time of the election, once the seats were declared to be reserved for women, the politically active male members stood no chance to intervene directly in the local governance. So, they made the female members of their household to contest the elections and they were once again reduced to puppet rubber-stamp signatory, who listens to either their husband, brother, brother-in-law, father or father-in-law. The intervention of the relatives of the elected female representative is prominently visible in the meetings, signing of documents, and planning of programs. Her male family member takes all the decisions, and she just signs the relevant documents. Therefore, once again, males are dominating the scene, under the guise of a female. As a result, terminologies like Mukhiya Pati / Mukhiya Pita, etc. are cropping up in villages.

Another unbelievable yet true concern is lack of unity among people. At a glance, it is hard to accept that lack of unity may also be a reason for the clambering of Panchayati raj institution, but it is true. There are various issues where people are frightened to stand with each other to have their rights. It was observed that due to a lack of consensus among villagers, some of the public welfare works did not materialize. For instance, there were a couple of cases in the village in which there was disagreement in the planning of some development project. Due to non-cooperation and lack of unity, the project never began and the money returned unutilized. Had there been a common minimum consensus, the work would have begun smoothly and people would have benefitted. But things did not materialize for good. Adding to this misery is the fact that on several occasions, suggestions and recommendations given by villages were not forwarded for requisite action. This is by far the most glaring concern leading to the failure of panchayat system on the nongovernmental side.

Conclusion and Recommendations

It can be rightfully concluded that apart from a few notable and worthwhile contributions, in totality, the Panchayati Raj Institutions have failed to achieve their designated objectives. Various issues, as mentioned above, have hindered people's growth and development, thereby rendering the process of establishment of PRIs as futile. For major part, the failure can be attributed to the government, as it did not attempt to understand the concerns of people at ground level. Without training or adequate knowledge, people were suddenly chosen as elected representatives, desirous of performing functions of bureaucratic capacity. Therefore, as the state gears up for an upcoming Panchayati election, it is expected that the challenges and inhibitions of the first two tenures of Panchayati Raj will be addressed and the aspirations of people will be fulfilled. To achieve the same, the following are a few suggestions for policymakers and the Government for enhancing the functioning of PRIs:

1. Elected representatives as well as the common people need to be made aware of the powers, functions, and duties of the Panchayati Raj. For this, phase-wise and area-wise training programs, ensuring mandatory attendance, must be conducted.
2. Mainstreaming the gender perspective in the developmental process must be ensured. All the policies, programs, and the system should be established with a motto to not only empower women but also to mainstream them, to act as a catalysts in the developmental process.
3. To remove the legal inconsistencies in the state legislation, there is a further need to incorporate appropriate amendments in the JPR Act, to ensure that there is a clear-cut dichotomization of powers and functions for all the tiers of the system.
4. The concept of free prior informed consent concerning land acquisition must be mandatorily incorporated in the JPR Act (within the powers of Gram Sabha) for fortifying the rights of tribal people on their land.

5. The state government must take measures to reduce unwanted checks and balances in the system. The entire paperwork and mode of operation must be simplified, keeping in mind the average literacy level of the elected representatives.
6. Proper communication mechanism, for ensuring 2-way flow of information (both up to down and down to up) must be made such that orders/notices/rules/regulations/resolutions etc. from the top are transmitted to the lowest level and vice-versa, without any delay. Use of Group SMS, Group E-mail, WhatsApp, Facebook, Toll-Free helpline numbers, etc., may be handy in this regard.
7. A punishment mechanism for the non-utilization of the sanctioned grants must be developed. It must be equally applicable to both the elected as well as the associated government personnel.
8. Digitization and "e-governance" has already been initiated by the Government. However, its pace needs a boost. Therefore, it must be augmented as early as possible.

References

Jathar, R. V. (1964). Evolution of Panchayati Raj in India. Institute of Economic Researcher.

Kumar, G. (2006). Local Democracy in India: Interpreting Decentralization. SAGE Publications.

Laxmikanth, M. (2016). Indian Polity for Civil Services and Other State Examinations. McGraw Hill.

Malik, A.S. (2005). Local Self Government at Village Level-An Assessment. The Indian Journal of Political Science, 66(4).

Mathew, G. (2001). Panchayat Elections: Dismal Record. Economic and Political Weekly, 36(3).

Pal, M. (2004). Panchayati Raj and Rural Governance: Experiences of a Decade. Economic and Political Weekly, 39(2).

Alam, S., and A. Raj. (2018). The Academic Journey of Witchcraft Studies in India. Man in India, 97(21), 123–138.

Ram, D. S. (2007). Panchayat Finances in India: A Macro Study. Panchayati Raj and Empowering People, New Agenda for Rural India, Essays in Honour of Shri Mani Shankar Aiyar. Kanishka Publishers & Distributors.

Sharma, M. (2001). Making of a Panchayat Election. Economic and Political Weekly, 36(19).

Exploring Health Information Seeking Behaviour Among Young Oraon Women in Jharkhand

Kumari Vibhuti Nayak,[*,a] *and Shamsher Alam*[*,b]

[a]Centre for Studies in Social Sciences Calcutta, Kolkata, India
[b]Central University of Jharkhand, India
E-mail: [*]vibhuti.nayak@cssscal.org; shamsher@live.com

Abstract

This paper aims to analyze various sources of health information and their impact on the health-seeking behavior of women from an indigenous (also known as Adivasis or tribal) community, namely Oraon, living in Jharkhand, India. Young Oraon women revealed that due to different sources (i.e., interpersonal, institutional, and commercial) of health information their choices of health treatment have increased. Contrary to previous generations, young Oraon women were more aware and informed of the modern health care system due to institutional (like educational institutes and hospitals) and commercial (such as media and internet) sources. Nevertheless, young Oraon women did follow the traditional cultural belief system and represented a balance between traditional cultural beliefs and modern knowledge of health in a way. Findings point to the fact that young Oraon women make treatment choices at the intersection of traditionalism and modernity due to various health information sources.

Keywords: Sources of Health Information, Treatment Choice, Young Oraon Women, Jharkhand

Introduction

Globally, indigenous populations experience poorer health status than their counterparts from non-indigenous communities (Delbridge et al., 2018). Ample of empirical literature highlighted the health inequality between the indigenous and non-indigenous groups. These studies identified several socio-economic, geographical, cultural, and political parameters affecting the health status of the indigenous population. For instance, empirical research studies have attributed lower health literacy, lack of competent health care workers, poor access to health services, and language barriers along with cultural and traditional beliefs as the key variables leading to relatively poor health of indigenous people (Alam and Raj, 2018).

In India, several researchers focused on the efficacy of government interventions program and also noted the varying degree of success. While such studies were informative, it falls short in terms of developing the connections between informational sources and health-seeking behavior – crucial variables in determining the success or failure of government health interventions program among indigenous communities. Still, there exists a research gap on how the indigenous people process different sources of health information and decide terms of choosing a health treatment.

The current article seeks to analyze the relationship between sources of health information and the health-seeking behavior of the Oraons (one of the indigenous communities in India) in

Jharkhand. Hence, the study analyses the causes in a routine life that affect information-seeking behavior and its outcome on young Oraon people, especially among women.

Information-Seeking Behaviour Among Indigenous Communities

Studies on indigenous communities predominantly focused on documenting the health-seeking behavior, role of culture, and traditional beliefs in shaping health-related practices. Concerning health-seeking behavior, indigenous people were intertwined with their traditional beliefs and practices, nature of interaction with the physical environment, and changing social, cultural, and economic environment. Consequently, the dependency of indigenous people residing in rural or remote areas on the traditional healers and members of the community was frequent for information-seeking behavior regarding their health-related issues, as compared to the health care team.

However, modern medical practitioners were the first point of contact for indigenous people living in urban places (Chandra, and Patwardhan, 2018). These conflicting findings suggest that the availability and accessibility of health-related services in specific locations play a major role in determining how and from where indigenous people quest for health information. This should be taken into account while designing and imparting health information among indigenous communities.

In recent years, various new platforms were added to health communication such as text and voice messages on mobile phones, video calls on smartphones, the Internet, online social network tools, and various other devices. These tools provided an opportunity to reach the people living in remote areas. However, many people don't have the knowledge to access these tools so various health information seems to be undelivered among the older generation as compared to the younger generation. At present, there is limited research on how and to what extent the indigenous young generation, particularly women utilize various sources of information (especially in remote areas) and how these sources influence their treatment care. Therefore, this article attempted to analyze how young Oraon women access various sources of health information and thereby decide on a particular choice of treatment method.

Sources and Channels of Health Information

Most of the young Oraon women actively encountered health information by having conversations with the immediate people around them. Family members especially mothers, grandmothers, spouses, relatives, friends, and traditional healers were easily accessible and approachable people, therefore, these were active sources. Also, from an early age, they used to get health information on general and preventive health care in their routine life. As young Oraon women grew up, they start to learn about healthy habits, health matters and issues from other family members, friends, educational institutes, doctors and Non-governmental Organisations (NGOs) were other sources of health information, but due to distance and time constraints, less accessibility, language and cultural differences they had limited conversation occur between them on health matters and issues, and hence these sources were passively considered.

Again, currently, young Oraon women who have the accessibility to smartphones and the internet actively used social media like Facebook, Youtube, WhatsApp, and other online platforms for receiving information on health regarding healthy lifestyle, symptoms, health conditions, and various diseases especially Covid-19. Further, radio, textbooks, newspapers, and posters were passive sources of health information, of which radio was common among participants.

Topics and Agencies Concerning Health Information

Various topics concerning health information were identified by the young Oraon women, among which curative and preventive measures with regard to minor health issues (like cold, cough, fever, headache, toothache, skin infections, conjunctivitis, diarrhea, and body ache) was the most recurrent topic that was actively and passively received during information seeking behavior. The next emerging topics were information on experiencing health issues with symptoms, self-care, a side-effect of drugs, alternative medicine, nutrition, exercise, reproductive health care, and disease complications. The young Oraon women preferred to have all this information in their local language.

Further, the dependency of the young Oraon women was noted to be high on family members especially mothers and grandmothers because they shared lived experiences and were aware of cultural beliefs along with home remedies. Both mothers and grandmothers were viewed as the cultural gatekeepers and influencing agents of cultural beliefs associated with health. For instance, preventive measures in early life such as keeping themselves and their surroundings clean, and wearing clean clothes for keeping diseases away. Eventually, they learned the importance of having seasonal and leafy vegetables, fruits, and a nutritious diet, which were good for their health.

The traditional healer, also known as Bhagats, was an important source of health-related information because he possesses extensive knowledge of treating infirmities and diseases. Although practicing without any educational qualification, they preferred medical doctors because it was easy to approach them, and had to pay less. It was widely believed by the young Oraon women that Bhagats smear magico-religious practices in their treatment processes which is indicative of the cultural notions.

Regarding information provided through health care team, the young Oraon women opined those doctors and other allied personnel, often stressed on carrying out free medical check-ups available in the hospital stating that numerous diseases could be avoided if diagnosed at an early age. They also opined that prevention of disease and illness at an early stage could be central to its treatment. However, this information was not provided until enquired specifically. Apart from these, measures for prevention of diseases like HIV-AIDS, tuberculosis, polio, filariasis, malaria, dengue, and mother-child care depicted on the walls of hospital premises, served as means for providing preventive healthcare information to the female adolescents, consequent upon their reaching the hospital. The young Oraon women noted in government hospitals, although being less maintained had posters, banners, and wall paintings for creating awareness and spreading health-related information. Contrarily, the private ones were relatively better maintained in terms of cleanliness and hygiene but did not provide any such information for people at large.

Apart from the formal role of teaching and learning, schools and colleges also play important in providing health-related information to their students. The young Oraon women mentioned the information provided to them by their teachers. It included information, knowledge, and skills that were needed to make informed decisions, practice healthy behavior, and create conditions that were conducive to maintaining healthy living. Teachers made respondents understand the importance of healthy life practices and persuaded them to adopt those at school, as well as, at home.

Front Line Health Workers (FHWs) comprising ASHA, Anganwadi workers, and ANM, play a very pertinent role in the dissemination of health-related information among young Oraon women. These FHWs were found to be very active in the study area wherein they remained very approachable to the community members. They assisted in securing supplementary nutrition for pregnant and lactating mothers, provided preschool education to children below five years of age, conducted immunization and health check-ups for mothers and newborn children, and imparted health education and referral services for young women.

Observations were made that the young Oraon women passively heard and watched health information about health consciousness through radio and television. For them, health information oriented, held stronger health beliefs, and were more likely to engage in healthy activities. Further, Internet users recognized the importance of maintaining their health and considered it a reliable source for disease prevention and health promotion. Fitness, prevention, and healthy lifestyle information were the most common categories of wellness and prevention information.

As a result of their Internet access, the young Oraon women also learned about symptoms, what their medical conditions are, what treatments are likely to be effective, and how to prevent further illness so that they can speak more effectively with their healthcare providers. The Internet provided them with information about their treatment options, as well as whether they should seek medical care as a first step.

However, barriers about information-seeking behavior the institutional and commercial sources were observed among the young Oraon women. The barriers include trust issues due to cultural differences, the bulk of information, language issues, and various distractions. It was obvious that they viewed commercial sources of health information with caution but trusted interpersonal sources.

Influence of Sources on Health-Seeking Behaviour

The young Oraon women had both the modern and traditional sources of information available to them regarding health and ill health. They tried to choose between the available health options and sometimes amalgamated both the traditional and modern treatments simultaneously. However, the Bhagat always remained the first line of reference for them due to the deep belief of their parents, husbands, or in-laws in warding off the supernatural effect causing ill health. At the same time, they relied on modern medical treatment, especially when the traditional treatment methods were ineffective. The young Oraon women remained indifferent towards the mode of treatment, as long as it worked for them. It was found that early adolescents were exposed primarily to the traditional mode of health-related perceptions and age-old cultural values associated with the same; whereas in their late adolescence became aware of modern healthcare practices through schools and other social institutions. Also, with the intervention of various government programs, such as Kishori Swasth Yojna (an adolescent health program), medical health practitioners started imparting specialized knowledge about health issues and treatment among young Oraon women. They in a way understood and appreciated differences in the cultural and modern approaches toward health matters and rather than honing one and outrightly rejecting the other, they tried to walk on the middle path by maintaining a balance between the two.

In close-knit communities of the tribal population, social networks and peer groups play a vital role in determining health seeking behavior of the individual members. The social network in health studies is considered to be interactions and exchanges between socially bounded groups, where the mutual relationship of community members could be identified and defined. From the context of this research, the interpersonal sources of young Oraon women greatly influenced their health-seeking behavior.

The Oraon tribes, where the young women were positioned, had a significantly high degree of social cohesion with an abundance of cultural capital in the form of information, gifts, goods, and ideas bred through mutual interactions. It was also very common among the Oraon community to seek advice from knowledgeable cultural entities, such as Bhagats on topics related to health and health treatment. This provided fertile ground for the dissipation of collective knowledge related to diseases and in a way largely affected the young Oraon women's views on perceiving it and its treatment. Researchers, in the past, have also highlighted the significance of social networks in determining health seeking behavior of individuals (Kim et al., 2015). Through social networks an individual shares certain health-related activities, exchanges knowledge or information about health, and decides the place for receiving health treatment.

From the study, it could be noted that both the health-seeking behavior of young Oraon women had different constructs and was often multi-dimensional. While some young Oraon women connected health with the ability to work or functional capacity, others defined health more in terms of nutrition and hygiene. Further, differentiation in conceptualizing health was determined by cultural agents such as Bhagats, community members, and family members. There were significant overlaps in different forms of health-seeking behavior, where the young women were required to undergo two types of treatment or interventions simultaneously. Besides, a difference in the pattern of health-seeking behavior was traced between respondents who went to school and those who remained at home. However, Bhagat, by distinguishing between minor and major health issues had a profound impact on the health-seeking behavior of respondents.

Therefore, he not only acted as a bridge between the young Oraon women and the supernatural but also became a bridge between them and the modern health care practitioner. Due to this, they were found to be standing at a crossroads of traditionality and modernity regarding healthcare and striking a balance between traditionality and modernity through their health-seeking behavior. It helped them in dealing with the ailments which hampered their day-to-day life.

Conclusion

By analyzing various sources of health information, this study provides a detailed overview of various daily life situations in which the young Oraon women were occupied with health information. By combining the interpersonal, institutional, and commercial sources, this study shows how especially food and nutrition are important aspects in the daily lives of young Oraon women and that family and friends play a large role in health-seeking behavior. The study recommends that young Oraon women belonging to the homogenous group had the same cultural notions and beliefs which strengthen the social relations among them. Social relationships had a significant impact on health, especially through behavioral pathways throughout the life course of an individual.

References

Delbridge, R., Wilson, A., and Palermo, C. (2018). Measuring the impact of a community of practice in Aboriginal health. Studies in Continuing Education, 40(1), 62–75.

Chandra, S., and Patwardhan, K. (2018). Allopathic, AYUSH and informal medical practitioners in rural India–a prescription for change. Journal of Ayurveda and Integrative Medicine, 9(2), 143–150.

Alam, S., and Raj, A. (2018). The Academic Journey of Witchcraft Studies in India. Man in India, 97(21), 123–138.

Kim, W., Kreps, G. L., and Shin, C. (2015). The role of social support and social networks in health information-seeking behavior among Korean Americans: A qualitative study. International Journal for Equity in Health, 14(40), 1–10.

Evolution and Implementation of Land Acquisition Legislations in India

Shamsher Alam[*,a], *Priyanka Thakuri* [*,a]

[a]Central University of Jharkhand, India
E-mail: [*]shamsher@live.com; thakuripriyanka.1919@gmail.com

Abstract

Although land is a natural resource, by rule of law which governs the people residing in any region, all land belongs to the state and it can reclaim it whenever it is required. This highly draconian power is vested in the notion of eminent domain and allows any sovereign government to claim land for all public purposes, remunerating the land giver with a fair value of his land, and some additional bonuses. Given the enormity of the task, it is inevitable that the process of land acquisition will be a non-aggressive one. Large-scale protests and violence enshroud it. This applies to tribal regions as well. Owing to special provisions laid down for these regions, the complexities of the process escalate further. Dispossession of land is not an easy feat, which is why the land acquisition is still heralded with stumbling blocks. Therefore, it becomes all the more necessary to have a better understanding of the process and clogs present within.

Keywords: Dispossession, Eminent Domain, Land Acquisition, Tribal Regions

Introduction

Land acquisition is the process by which government acquires land from private property owners for specific purposes. In every single case of land acquisition, the lone reason quoted is "development" (Punj, 2017). Residents of the concerned land have to helplessly part away with their land, which has been part of their lives even before they took their first breath. Each time this happens, a recurring set of issues arise which are debated over and over (Ghatak, Mitra, Mookherjee and Nath, 2013). These include - the identification of those whose lands were acquired and disposed of (Anthony and Chakraborty, 2017); whether they were poor cultivators or wealthy landowners and what was the amount of land acquired, whether the compensation abided by the legislation relating to land acquisition, what measures were adopted for rehabilitation of the displaced population and lastly, the impact of the acquisition on those involved.

In India, the Right to Fair Compensation and Transparency in Land Acquisition, Rehabilitation and Resettlement Act, 2013, serves as the basic guide to the land acquisition process. Earlier, the Land Acquisition Act 1894 (Hoda, 2018), a colonial-era law, allowed the government to acquire land. Since it was a contested act, which failed to give a clear picture of a variety of issues, a need was felt to bring in new legislation that could dissolve the discrepancies, which resulted in the passage of 2013 Act. The act though explicitly mentioned that no land acquisition shall take place in the tribal regions, but in case, such an action is required, it also specified conditions that would have to be met, and allowed the concerned state to create their laws, provided that they do not stand in conflict with the central legislation.

As a result, many tribal states which had their state legislations related to the process of land acquisition, continued with the pre-existing legislations, while some of them proceeded toward the creation of new ones. A few notable existing legislations consist of the Andhra Pradesh Scheduled Areas Land Transfer Regulation, 1959, in Andhra Pradesh, the CNT Act, 1908 and the SPT Act,1949 in the state of Jharkhand, the Orissa Scheduled Areas Transfer of Immovable Property (by Scheduled Tribes) Regulation, 1956, in the state of Orissa, the Tripura Land Revenue and Land Reforms Act, 1960, in the Tripura state (Ashokvardhan, 2006).

Eminent Domain and Tribal Peoples

The concept of Eminent Domain has an appeal internationally, for it is a powerful sovereign tool for all countries alike. It enables the government to occupy any private property for public purposes (Ramesh and Khan, 2015). Such an absolute and potent power is enjoyed thoroughly by governments everywhere. The popularity of eminent domain laws can be gauged by the fact that different countries have coined beautiful terms to address them. In United States and the Philippines, it is called eminent domain, while in United Kingdom, New Zealand, and Ireland, it is referred to as compulsory purchase. It is named as a resumption in Hong Kong, resumption/ acquisition in Australia, and expropriation in South Africa. Its first usage can be traced back to 1625, when Hugo Grotius used it in his work De Jure Et Pacis (Chowdhury and Chowdhury, 2016). Since then, the notion of eminent domain spread like wildfire in not only in European states but also trickled down into colonies established by them, like India and China. The Indian government can very well credit the British colonial rule for introducing English land laws and related ideas into our soil. Their legacy has been continuing to date and is shrouded by myriad problems, especially for the inhabitants of tribal regions, as land is not just an entity, but a heritage to them. Since the colonial period, much of tribal land has been acquired by the government, and it continues even after independence, with much stronger legal backing. However, the pains associated with parting of such lands are immeasurable, which is why strong opposition to land acquisition exists.

History of Land Acquisition in India

The first ever law related to land acquisition in India was enacted by the colonial government in 1824. It was known as the Bengal Resolution I of 1824, and applied to the whole of Bengal province subject to the presidency of Fort William. This law gave the right to obtain, at a fair valuation, land for public usages such as the construction of roads or canals. In 1850, the regulation was extended to Calcutta, through the Act I of 1850. Furthermore, similar land acquisition acts were legislated in other provinces in India, such as the one in Bombay in 1839, the Building Act XXVII, and Act XX of 1852 in Madras. After the Sepoy Mutiny in 1857, all of these acts were replaced by the Act VI of 1857. Under this Act, the Collector could fix the amount of compensation by agreement or by referring it for arbitration. This was the first law that applied to the whole of British India. Later on, a few other insignificant acts were passed such as Act II, 1861, and Act X, 1870. In eventuality, all of these acts were subsumed into the Land Acquisition Act, 1894 (Chowdhury and Chowdhury, 2016). This law, however, was applicable only for British Indian provinces, as the native states enacted their land acquisition laws (Tokas, 2021). The subsequent legislations such as the Government of India Act 1919 and the Government of India Act 1935, gave the power to provinces to legislate with respect to the compulsory acquisition of land, and no new legislations relating to the land acquisition were further enacted.

Post-independence in 1947, the independent government did not initiate any new law for land acquisition, rather it adopted the 1894 Act, just like it did with many other colonial period legislations. However, the Constitution makers declared the Right to Property as a fundamental right in the Constitution of India and included it in Chapter IV of the Indian Constitution (Laxmikanth, 2021). The Supreme Court, however, in 1972, gave the basic structure doctrine of the Constitution, following which, the property right no longer remained as a fundamental right. Another article – Article 300A was added to the Constitution as a replacement regarding property rights, which dictated that no one could be deprived of his property, except by the authority of law (Pandey, 2020). Such an amendment gave sweeping authority to the government to gain control of any private property, by virtue of the land acquisition process. It eventually led to frequent clashes between the land givers and the government, in the form of violent protests, threats, and killings. The situation remained the same in tribal as well as non-tribal regions. That is why, as the century came to an end, the legislators, too, felt the need to design a comprehensive law catering to the demands of the present situation, repealing the anarchic colonial land acquisition law.

LARR Act 2013

Some of the most significant protests against disputed land acquisition were related to "The Narmada Bachao Andolan (a movement seeking redress for families displaced as a result of the Narmada Valley project), the Tarapur agitations (born in the wake of land acquisition for the Tarapur Atomic Project), violence in Nandigram (as a result of forcible acquisition of land from farmers)," which jolted the government to take prompt action and the desire to repeal old existing laws (Ramesh and Khan, 2015). The Government of India believed that "a combined law was necessary, one that legally required rehabilitation and resettlement necessarily and simultaneously followed government acquisition of land for public purposes as since 1947, land acquisition in India has been only done through the British-era act. Therefore, in 1998 that Rural Development Ministry initiated the actual process of amending the act. The result was the enactment of the Right to Fair Compensation and Transparency in Land Acquisition, Rehabilitation and Resettlement Act, 2013."

According to the 2013 Act, "land can be acquired for land needed by the government for its use, including public sector undertakings and any public purpose including strategic purposes, infrastructure projects, or urbanization or housing projects. Land acquisition under the Act can also be made for public-private partnership (PPP) projects and for private companies for public purposes, but prior consent of landholders is required. Important provisions in the Act relate to the steep enhancement of the scale of compensation to land-owners and other project-affected persons, the requirement of consent of land-owners for acquisition on behalf of private companies and public-private-partnerships, the need to undertake a social impact assessment of the project for which the acquisition is being undertaken and limits on acquiring multi-cropped and other agricultural lands."

In addition to the stipulation for the consent of the majority of landholders, "there is a requirement for a social impact assessment of the project, somewhat similar to the environmental impact assessment under environmental laws. The social impact assessment begins with a study of all aspects of the project and its impact on the livelihood of affected families and the facilities and amenities enjoyed by them. A public hearing and involvement of the local body are also mandated. Among other things, the government has to consider whether the potential benefits and public purpose outweigh the social costs and adverse social effects as determined by the Social Impact Assessment. The Act limits the invocation of the urgency clause, whereby possession can be taken even before the award of compensation is made, only to the minimum area required for the defense of India or national security or any emergencies arising out of natural calamities or any other emergency with the approval of Parliament."

The act strictly stated that "as far as possible, no acquisition of land shall be made in the Scheduled areas. If it is unavoidable, then it shall be done as a demonstrable last resort. In case of acquisition or alienation of any land in the scheduled Areas, prior consent of the concerned Gram Sabha or the Panchayats or the autonomous District Council shall be obtained, in all cases of land acquisition in such areas, including acquisition in case of urgency, before the issue of notification under this Act, or any other Central Act or a State Act for the time being in force provided that the consent of the Panchayats or the Autonomous Districts Councils shall be obtained in cases where the Gram Sabha does not exist or has not been constituted. In case of a project involving land acquisition on behalf of a Requiring body that involves involuntary displacement of the Scheduled Castes or the Scheduled Tribes families, a Development Plan shall be prepared. The Development Plan shall also contain a program of development of alternate fuel, fodder, and non-timber forest produce resources on non-forest lands within 5 years, sufficient to meet the requirements of tribal communities as well as Scheduled Castes."

In case of land being acquired from members of all the Scheduled Castes or the Scheduled Tribes, "at least one-third of the compensation amount due shall be paid to the affected families initially as first installment and the rest shall be paid after taking over of the possession of the land. The affected families of the Scheduled Tribes shall be resettled preferably in the same Scheduled Area in a compact block so that they can retain their ethnic, linguistic, and cultural identity. If the affected families belonging to the Scheduled Castes and Scheduled Tribes are relocated outside of the district, then, they shall be paid an additional 25%, of rehabilitation and resettlement benefits to which they are entitled in monetary terms along with a one-time entitlement of Rs. 50,000." With such special provisions, the act, nevertheless, brought in a way to penetrate the land of tribals as well.

Though the features of the act look all beautiful and fulfilling, there have been rampant reports about the non-payment of compensation, and unfulfillment of the rehabilitation policies, and the struggles continue. Furthermore, many private companies, on whose behalf the government had acquired lands, through the construction of industries, have not only created pollution but also passed many steps, thus, rendering the process of corruption among officials. In many such cases, whereby initially, the land givers were entitled to employment in the industries, were denied any opportunity, post the process of land acquisition was completed. Therefore, in such situations, the whole process of land acquisition not only disposed of and deprived the people of their precious assets, but their economy and livelihood were also destroyed, and additionally, they were exposed to pollution and other health maladies. If this continues, there can be no permanent cure for such a precarious situation.

Conclusion

The path of land acquisition is not a smooth one. It is rather curvy and has a lot of bumps on it (Alam and Raj, 2018). There is no denying that India needs to grow and develop and shed away its identity as a third-world country, for which a multi-sectoral approach towards the growth of the country is required. The stakes are high but resources are lacking. Inevitably, the government has no choice but to use the much-feared eminent domain law to acquire more land for further infrastructural building. The idea that one has to sacrifice for the greater good is usually promulgated at large for those who have to part away with their land. The intent is a good one, and that is why people, though unwilling, give away their lands, with the hope that their land would be put to good use and they would receive fair compensation for their deeds. But what follows are long-awaited delays, coupled with corruption and apathy on part of implementing agencies, which is nothing but a form of harassment for already distressed people. As described before, for tribal people, land is a God-like entity to them. They are forced to give up their sacred lands, which is already a painful experience, and many a time they do not receive their due compensation or other benefits listed for them. That is why, till date, no peaceful land acquisition has occurred. Unless the crux of the problem is solved, the situation is going to remain unaltered in the coming times as well, with unknown complexities of the future.

References

Chowdhury R. I. and Chowdhury R. P. (2016, January–December). Holdout and Eminent Domain in Land Acquisition. Indian Economic Review, 51(1/2), 1–19.

Laxmikanth, M. (2021). Indian Polity for Civil Services and Other State Examinations (6th ed.). McGraw-Hill.

Ramesh J., and Khan A. M. (2015). Legislating for Justice: The Making of the 2013 Land Acquisition Law. Oxford University Press.

Tokas J. (2021, October 27). Analysis of Land Acquisition in India. Law Insider. https://www.lawinsider.in/columns/analysis-of-land-acquisition-in-india

Pandey, A. (2020, August 28). Doctrine of Basic Structure and Judicial Review. Law Times Journal. https://lawtimesjournal.in/doctrine-of-basic-structure-and-judicial-review/

Punj A. (2017, April-June). Partners In Development Under the New Land Acquisition Law. Journal of the Indian Law Institute. 153-177

Alam, S., and A. Raj. (2018). The Academic Journey of Witchcraft Studies in India. Man in India, 97(21), 123-138.

Hoda A. (2018). Land use and Land Acquisition laws in India. ICRIER.

Ghatak M., Mitra S., Mookherjee D., and Nath A. (2013, May). Land Acquisition and Compensation: What Really Happened in Singur? Economic and Political Weekly, 48(21), 32-44.

Ashokvardhan, C. (2006). Tribal Land Rights in India. Centre for Rural Studies.

Anthony P. D. C., and Chakraborty A. (2017). The Land Question in India-State, Dispossession, and Capitalist Transition. Oxford University Press.

Change in Gender Relations: Re-Visiting Gender-Based Violence in Tribal Communities of India

Kumari Vibhuti Nayak[*,a] *and Shamsher Alam*[*,b]

[a]Centre for Studies in Social Sciences Calcutta, Kolkata, India
[b]Central University of Jharkhand, India
E-mail: [*]vibhuti.nayak@cssscal.org; shamsher@live.com

Abstract

Gender-based violence (GBV) within tribal communities has caught less attention among social researchers. The National Family Health Survey 2019 showed that violence is much higher against women from the scheduled tribes than women outside these categories. Over the last few decades, the passage of the Joint Forest Management Act (1996), interventions of developmental projects, and other socio-cultural influences have led to significant socio-economic changes and urban migration that changed the egalitarian concept of gender relations. Such changes fabricated gender relations which resulted in GBV. Regardless of conventional understanding, gender equality in tribal communities was poorly understood, especially in the context of changes. It created disempowerment and economic instability among tribal women and men leading to GBV. Hence, this article provides an overview of how gender relations changed over time with the experiences of tribal women who faced multiple and intersecting kinds of violence due to gender and distinct socio-economic characteristics.

Keywords: Gender relation, Gender-based violence, Tribals

Introduction

Gender-based violence (GBV), especially against women, has been defined by the United Nation as "any act of violence that results in, or is likely to result in, physical, sexual or mental harm or suffering to women, including threats of such acts, coercion or arbitrary deprivation of liberty, whether occurring in public or in private life." Globally, the Fourth World Conference on Women in Beijing (1995) proposed to provide an action plan to enrich women's socio-economic and political empowerment as well as to deal with obstacles like health, education, equal provisions, wages, and violence against them.

In India, evidence of GBV against women was noted across communities in various forms. Domestic violence is considered to be the most predominant form of GBV which comprises verbal, physical, emotional, sexual, and economic abuse of women by their partners or family members. According to NFHS-4 (2015–16) report, 30% of women belonging to the age group of 15–49 have experienced physical violence since the age of 15 and 6% of women in the same age group have experienced sexual violence at least once in a lifetime. The prevalence of GBV varies across different states, such as 55 % in Manipur followed by 45–46 % in Telangana, Bihar, Andhra Pradesh, and Tamil Nadu, 30-35% in Meghalaya, West Bengal, Jharkhand, 7% in Himachal Pradesh and 4 % in Sikkim (NFHS-4). Pieces of evidence on GBV were uneven in various geographical locations due to differences in economic (poverty), social (caste and class), and cultural factors (gender inequalities).

Pieces of evidence concerning GBV were limited/underreported with respect to certain communities such as women with disabilities, Adivasis, and Dalits. Among these communities, Adivasis women experienced multiple and intersecting kinds of violence and discrimination due to gender, culture, and distinct socio-economic characteristics. Hence, this article focuses on women belonging to Adivasis (also known as tribal) communities. Tribal women make up 4.3% of India's population.

Despite of low literacy rate (i.e., 50%), tribal communities are known to have a more balanced sex ratio (990), than the majority population (943), and have been historically viewed as more egalitarian in comparison to their non-tribal counterparts (Xaxa, 2004). However, 31% of tribal women experience GBV and the reasons might be different from the general population.

A review of the literature initially addresses gender relations among tribals. Traditionally, gender equality was based on economic equality and predominantly skill-based. Even under these conditions, women were culturally discriminated such as the practice of witchcraft. Other than witch-craft, how the implementation of Joint Forest Management (JFM), displacement, and migration paved the way for gender disparities and positioned tribal women in a disadvantageous situation, requires the spotlight. Yet, before discussing JFM, displacement, and migration, the pervasiveness of gender relations and gender equality among tribal communities needs to be understood.

Gender Relations and Gender Equalities Within Tribal Communities

Researchers have noted that tribal communities had fairly gendered egalitarianism in terms of social, cultural, economic, and political milieus. Based on physical capabilities and age, both genders participated equally in each task of the community. Young ones, irrespective of their gender, were trained by assigning various tasks and were regarded as the community's economic asset. Gender roles and positions were contested based on knowledge, skill, maturity, resource engagement, and production among tribals, in which, tribal women excelled in many aspects and thus occupied influential positions and respect within the community.

While emphasizing the tribal community as gender egalitarian, the economic aspect played a major role in it. In earlier days, tribals were dependent on the forest for livelihood and each one had to work. Dependency on forest produce and resources attributed to equal gender roles which not only promoted but also maintained gender equality within tribal communities. Besides depending on the forest, in the economic context, tribals were also noted to be engaged in a wide array of livelihood activities ranging from agricultural farming, livestock rearing, plantation, and fishing, where both genders equally participated.

At this juncture, a strand of literature noted the gender-based division of labor within tribal communities. For instance, tribal men from agricultural groups, such as Munda, Oraon, Ho, Santhal, Bhil, and Bharwas engaged in preparing and plowing fields while women helped in domesticating animals, water sourcing, harvesting, sowing, and weeding. It depicted the equal participation of both genders in their economic activity.

In the context of social life, tribal women were free to choose their life partner, work, move in the market and other public places, participate in socio-religious events, had the power of decision-making within the family, and were economically independent. They equally shared family responsibilities, authority to manage household resources, and possessed a higher degree of responsibility (Xaxa, 2004). Other than this, the absence of widow burning and occupational segregation was noted. It depicted that within the family and social structure tribal women were in a better position.

Since the social life of tribals was intricately linked with political, therefore, women's political status moved according to social status. Within the family, tribal women were seen as decision-makers but in public spaces, they were hardly noticed by the administrative powers. On the contrary, Xaxa (2004) highlighted among Nagas, Sema Nagas, and Baiga women hold an esteemed position in the tribal council for decision-making, managing community resources, and having a strong say on community affairs.

Tribals also possess a rich cultural life as they have a wide range of folk tales, songs, and dances. Both women and men equally participated in religious activities, festivals, social events, and occasions. Despite widespread equality, the tribal communities do suffer from instances of GBV and gender inequality. In different tribal communities, women and men were discriminated against on basis of their social position, family influence, property, and wealth. Documentation of witch-hunting in various tribal societies in Jharkhand, Odisha, Andhra Pradesh, Madhya Pradesh, Chhattisgarh, Haryana, West Bengal, and Bihar, shows clear signs of coordinated violence against tribal women (Alam and Raj, 2018).

Witch-hunting was purposely conducted against a widow for grabbing property and abusing her sexually. Such activities depict how superstition, greed, and lust resulted in exploitation and dignity loss for tribal women. Moreover, the gender-relation narrative essentially has misled scholars as they tend not to explore the prevalent gender-related issues in tribal communities. Therefore, the next section uncovers diverse research that reveals changes in gender relations that lead to the prevalence of gender inequality and GBV within tribal communities.

Changes in Gender Norms and Relations

A growing body of literature illustrates economy as a noteworthy aspect of change in the gender norms and relations, that varied across time and context. Economic factors affected gender relations as they changed the roles and behavior of tribal men and women which resulted in a change in expectations for each other. Hence, this section illustrates changes that occurred in the tribal community and how they influenced and affected tribals about gender relations.

Joint Forest Management Act: The Economy Changer

In 1990, the Government of India (GOI) implemented Joint Forest Management (JFM) (currently known as, the Forest Right Act, 2006), which initiated changes in the economic life of tribals. The provisions of this act not only led to changes in gender-based relations of tribal communities but also severely affected their livelihood opportunities. JFM was introduced for developing relationships between the forest department and communities residing in the forest for sharing the benefits, rights, authority, and costs related to forest production and management. While introducing such a partnership, JFM started creating gender differences and gender-based division of roles and responsibilities among tribals by ignoring the meaningful incorporation of women. While issuing rights to harvest forest produce, women were not a part of joint decision-making and therefore have to bear the impractical decisions taken on their behalf.

Tribal women were prohibited to collect firewood from the designated region of the forest, forcing them to travel long distances, which often leads to high opportunity costs for them. Another major concern was whether women in the household should be given equal rights over the distribution of profits (either through community-instituted funds or direct transfer of cash to households), where men of the household were identified as beneficiaries.

The forest management policies did not address the pre-existing gender relations realities of tribal communities and arbitrary moves. Only tribal men were authorized to have policing rights in the forest, which further undermines women's rights and promotes social unacceptability within their community. Such a scenario reinforced the dominance of propagating gender-based roles in managing and harvesting forest resources. In a way, the JFM Act changed the traditional economic system, activities, and gender relations in tribal communities and re-defined gender relations in daily life that undermined women's rights and representation.

Development Projects: An Influencing Factor

In the name of development projects (such as educational institutes, dam construction, and hospitals), tribals were either displaced or were made foreigners in their land. Due to displacement, tribals experienced various hardships that were caused by governmental officials, forest officers, guards, landlords, money lenders, contractors, and dealers. For example, government and forest officials snatched traditional rights of forest from tribals through JFM and made them unemployed; moneylenders or landlords exploited and oppressed tribals as bonded laborers, when they were unable to return the money; contractors and dealers were not even paying minimum wages, particularly to tribal women.

Because of displacement, tribal women (belonging to the states of West Bengal, Orissa, and Jharkhand) not only lost their livelihood but also land ownership, influence within the household, and independent status. Consequently, displacement brought multiple changes that affected tribals' life in ways that made women subordinate and dependent on men (Xaxa, 2004). Such shifts also affected the prevailing egalitarian gender norms and eroded the authority or influence of tribal women both within the household and the larger community.

In Orissa, Andhra Pradesh, Jharkhand, Goa, Kerala, and West Bengal, according to government rules, the displaced family were offered semi-skilled or un-skilled jobs in government projects. Government officials seldom consider tribal women as a candidate for the job and therefore only men were considered for jobs on concerned grounds. Hence, tribal women's status changed from household earner to housewife category and entitled men to become sole bread earners. Displacements have changed the social fabric of tribal society where women suffered dis-proportionality higher than men. The such adjustment also eroded the previously established cultural norms of work and gender-based equality.

Migration: An Employment Opportunity

Tribal migration was observed, during the 1990s, due to the labor market regarding industry construction for economic development. The construction industry was the second largest industry in India to generate huge employment opportunities for poor, unskilled/manual workers who could get work in drought times. The construction industry was found to be unique as 98% of women were employed as casual laborers in comparison to other industries.

In pursuit of survival and better economic conditions, tribals tend to migrate to neighboring urban settings and join there as physical or manual laborers. Later on, the prevalence of tribal migrant communities and their social network facilitated migration in general and particularly with families including females. Unskilled tribal laborers still prefer manual work like making bricks, material loading and unloading, road construction related, and soil digging whereas few skilled laborers prefer some other work as per their skills.

Hence, the eclectic mix of change in forest usage, displacement, and migration has created a previously unknown tension between tribal men and women. While the tribal men started drinking and purchasing expensive materialistic things to cope with tension, the women remained mostly dependent, non-earning members or housewives. The drinking habit among tribal men was found to be common in displaced areas like Orissa, Jharkhand, and Andhra Pradesh, which instigated household violence against women like wife-beating.

Almost all Pradhi and Gond women experienced trauma or emotional torture due to quarrels or verbal spats at home and a few women even reported physical violence by their husbands. Dhal (2018) too observed gender violence against tribal women in Orissa and attributed it to the prevalence of gender inequality in the community. Tribal women were found to be vulnerable as they witness discrimination and deprivation at both household and larger community levels like unsafe work environments including economic and sexual exploitation of tribal women in the workplace.

Present Scenario

Due to colonization, militarism, racism, migration, displacement, urbanization, and social exclusion, tribals of Jharkhand, experienced numerous forms of GBV varying from verbal to non-verbal. GBV among tribals was often considered a minor issue compared to the violence that tribal women face in public spaces (for example, violence by the dominant groups of people). Such a scenario depicts, often, GBV against tribal cases repercussions by the dominant feudal groups against the tribal's proclamation of self-respect, identity, and engagement in the development process. Thus, tribals especially women were subjected to extreme social discrimination and disentitlement, leading to GBV in various ways.

Historically, tribals came into tenuous socio-economically and politically, not only due to cultural and language differences but due to displacement and migration. In recent times, the situation of tribals worsen due to state policies that gave preference and job opportunities to tribal men, while actively discriminating against tribal women. Hence, development projects did not benefit everyone in tribal society, they created disempowerment and economic instability among tribal men and women leading to GBV.

Conclusion

Research on GBV within tribal communities is limited. However, past research on tribal communities noted pain and trauma experienced by tribals during displacement, adaption to the new economy, acculturation, and influence of modern life. During this process gender inequality between tribal men and women was amplified, however, the documentation on how it results in GBV against women is still under-researched. It is, however, evident that poor economic conditions, lack of social support, alcohol abuse, and high-stress level within the household led to increased instances of conflict within the household, putting tribal women on the receiving end.

References

Alam, S., and A. Raj. (2018). The Academic Journey of Witchcraft Studies in India. Man in India, 97(21), 123–138.

Dhal, S. (2018). Situating tribal women in gender discourse: A study of socio-economic roots to gender-violence in Odisha. Sage, 64(1), 87–102.

National family health survey (NFHS-4) 2015-16 (2017). Mumbai: International Institute for Population Sciences (IIPS) and Macro International. Retrieved from: http://rchiips.org/NFHS/NFHS-4Reports/India.pdf Accessed on 26 June 2022.

Xaxa, V. (2004). Women and Gender in the Study of Tribes in India. Indian Journal of Gender Studies, 11(3), 345–367.

"Sarna Adivasi" Religion Code: Contextualizing Religious Identity of Tribals in India

Shamsher Alam,[*,a] *and Kumari Vibhuti Nayak* [*,b]

[a]Central University of Jharkhand, India
[b]Centre for Studies in Social Sciences Calcutta, Kolkata, India
E-mail: [*]shamsher@live.com; vibhuti.nayak@cssscal.org

Abstract

This article explores the tug-of-war between the right of recognition of tribals and the so-called preservation of cultural unity of India's largest religious denomination. Tribals adore and worship nature. Due to their distinctive way of living, tribals desire to be recognized as their religion which they name as Sarna religion, particularly in the Jharkhand state. The tribal populace constitutes 8.8 % of India's population; however, they have no separate religious code in the census. Consequently, they have been forced to identify themselves by subscribing to either Hinduism, Christianity, or even Islam to a few extent. It is important to note that tribals of Jharkhand have constantly asserted that they belong to the Sarna religion. Hence, the current research article investigates the struggles, scenarios, and apprehensions of adopting Sarna as a religious code in the census, which tribals lost post-independence somehow.

Keywords: Recognition, Identity Politics, Sarna Religion, Tribes, Jharkhand

Introduction

In India, religion is a sensitive issue, and now, religious conversion and interruption have become touchy subjects. Concerning tribal communities in India, currently, the religious intrusion has a significant impact not only on tribal life but also on their surroundings and ecosystem with profound consequences. Tribals have been always nature worshippers. In colonial times, tribals were religiously recognized as animists by various British anthropologists and scholars documented in the census. However, post-independence, somehow tribal's religious identity was not documented in the census. In other words, tribals lost their religious identity. They were forced to be clubbed under either "others" column or had to identify themselves as belonging to one of the six specified religions – Hindu, Muslim, Christian, Buddhist, Jain, and Sikhism. Such a scenario depicts tribals having no separate religious identity. After being aware of this information, present-day educated and conscious tribals demand their lost religious identity by including Sarna as a religious code in the census. Therefore, this article explores the struggles, scenarios, and apprehensions of adopting Sarna as a religious code in the census.

Research Context

The devotees of the Sarna religion, i.e., tribals, adore nature. Hence, tribal's "Jal, Jungle, Zameen," i.e., water, forests, and land which includes trees, mountains, ponds, hills, and nature, have been worshipped, and their protection remained at the central point for tribals across states. Here, "Sarna" refers to sacred groves of the Sal tree (etymologically identified). Spot of Sarna, i.e., presence of Sal tree, has been regarded as sacred by tribals of Jharkhand state. Within the village premise, Sarna Asthal is generally marked with several trees ranging from Sal to Banyan. Concerning the Sal tree, there have been different cultural beliefs that act as protectors for tribal communities, especially among tribes of Jharkhand. For instance, among the Munda tribes, it has been believed that once a man was chased by a lion and he got away by covering himself in a hedge of Sal tree. After that, he committed to offering sakhua flowers and leaves just as a living creature. The procedure during the celebration involves the priest bringing three water pots to the Sarna asthal. If it is found that the water level in the pot diminishes, then they expect that a storm would fizzle, yet if it remains similar, it would mean that a storm is about to come. Hence, the followers of Sarna respect nature as a sacred complex of their life and pledge to pour deference to the woodland where they live.

The disciples of Sarna, i.e., the tribal population, constitute around 8.3 percent of India's total population. It is surprising to see how such a big chunk of the population still struggles to have their religious code. It is important to note that many communities are less in number in comparison to tribals that have managed to secure a separate code under the census. It has always been a matter of concern for the tribal that, due to the absence of their particular code, have to either describe themselves as belonging to Christians or Hinduism. Tribals, however, claim they don't belong to either group. Hence, the current scenario portrays that religious identity has become an essential concern for tribals.

Motive for Religious Identity

Each tribe in India has its own unique cultural beliefs, history, language, and way of living and practices its way of worshipping nature. All these became significant aspects in the formation of identity for tribals and created a sense of ethnicity that differed from one another. However, with the changing situation and times, tribals had undergone various changes from the pre-colonial to post-colonial period concerning the protection of their land, environment, rights, and formation of identity. Still, their struggle for identity has not ended, and now they are fighting for documentation of religious code in the census to have their own religious identity.

Earlier struggles of tribals were advocated by "others" to restore their ancient heritage, desired to integrate them as citizens of nation-sate, and to assimilate into the Hindu fold (Xaxa, 2005). In recent times, however, with the advent of education, the demand for development projects, and the feeling of being discriminated against by others, tribals initiated to take charge of their protection and development of rights, identity, language, culture, and customs. So, this section uncovers how the prior struggle of tribals for history, culture, and identity has experienced various transformations in changing situations and times.

Concept of Identity

The identity of an individual and community differs based on the area of place of residence, language, culture, religion, and other socio-cultural and economic practices (Nayak, 2021). Regarding tribals, there seems to be no end to the identity debate. It has been evident that each person carries their identity and shares a community identity with others. So, it describes essence as a fluid concept frequently constructed in socio-cultural and political relations to serve the interest of the ruling and dominant class, caste, and other social and cultural groups.

Tribal Identity in Colonial Times

During the colonial period, the British rulers were shaken by the heat of the revolution of the tribal heroes. Again, taking account of Jharkhand tribes, evidence of the 1807 rebellion, Munda's Tamad revolt in 1835, an uprising of Budhu Bhagat, and in 1830 Manki Munda revolt in Kolhan is still remembered. Further, in 1855–56, under the leadership of Sidhu Kanu, Chand Bhairav, Phoolo, Jhano initiated Santhal Hul, a tribal movement for land acquisition. In 1919, Birsa Munda's Ulgulan created a history of such a high mark that even today, after independence, it is remembered by all as their quest to protect water, forest, land, the identity of tribals, and to preserve the existence of tribal society struggle. Similar kinds of revolts had taken place in other states.

The abovementioned revolts brought various forms of tribal identity formation, which is unchallengeable. It created a different problem for tribes because, during colonial times, they were recognized as brutal, savage, illiterate, and unsophisticated people. By claiming that territorial, tribal identity is a colonial fabrication, historians like Damodaran unintentionally defend colonialism's role in providing tribal people with a sense of self and identity. In the writings of postcolonial historians like Damodaran (2006), as well as in the postcolonial development planning for tribal people in India, Dalton's colonial legacy continued, ensuring that they would continue to experience marginalization, exploitation, and inequality through the politics of co-option and domination. Tribal identities had been put in danger by this historical construction.

Tribal Identity After the Colonial Period

Tribal identity ignores the subjective and objective interactions involved in the formation and generational diffusion of tribal identity. Various international and national organizations, governments, social scientists, and scholars, particularly anthropologists, created the entire discourse of tribal identity based on indigeneity (Conklin and Graham 1995), which aimed to develop an identity of "self" in contrast to "the other." This process began under colonial control, was maintained during the anti-colonial national uprising, and was institutionalized under postcolonial succeeding regimes.

In postcolonial India, the idea of tribal identity institutionalizes, normalizes, and consolidates the colonial paradigm of identity construction based on area and location. India's domestic politics, which give a platform to the regional political formations essential to national politics, are responsible for establishing territorial identity throughout the postcolonial era. This process has strengthened and rationalized the notion of identity creation based on region and territory, contrary to the idea of a unitary national polity. It has been promoted further by a neoliberal development policy model.

Building tribal identities based on land is not only reductionist in philosophy but also invalidates the concept of "identification" altogether. By fostering distinct territorial identities among tribe members, the territorial, tribal identity creation aids elites in maintaining the politics of marginalization. Territorial, tribal identity is a controversial concept that, in turn, territory identity brought crises in tribal society. Without considering the involvement of outside forces like the government, the market, and popular culture, this dilemma is presented as being innate to tribal society. In postcolonial India, there is still a trend toward weakened neoliberal resistance among tribals.

Territorial, and tribal identities, therefore, are necessary for the process of emancipation and preservation of tribal identity and the growth of their society, culture, and economy. In addition to assisting tribe members in confronting and challenging the exploitation and inequality inherited, sustained, and perpetuated by internal and external elites, such a process can preserve tribal unity in diversity. The social, economic, cultural, and political environment of the tribal people in India has altered due to mining-led industrialization, NGO and civil society-led rural and tribal development, and the Hinduization of the tribes by Hindu right-wing groups. Due to territorialization and de-territorialization initiatives driven by neoliberal Hindutva forces, their identities are shifting in their daily lives.

Why Religious Code is Essential for Tribals?

An essential fact was noted during the British regime that tribals were allotted religious codes. The codes were: 1871 – Aborigines, 1881 – Aboriginal, 1891 – Aboriginal, 1901 – Animist, 1911 – Animist, 1921 – Animist, 1931 – Tribal Religion, 1941 – Tribes, and 1951 – Scheduled Tribe. These censuses witnessed their presence because they provided a specific column for tribal people. The scheduled religion was given a short code; however, tribal people's identity was omitted secretly in the 1971 census.

The result was published regarding only the scheduled religion and not those of the people belonging to the tribal community. In 1981, the first letter of the religion was indicated, and from 2001 to 2011, the schedule religion was given a code from 1 to 6 in roman numerals. Tribals were categorized under the category of "others" but were not given any specific code as provided to other religions. Such a scenario depicts that after 1951 somehow, these codes got omitted, and tribals were forced to identify themselves as either subscribing to the religion of Christianity or Hinduism. This had a prolonged effect on abstracting certain and real data about the populaces of tribes across the country. Although the tribes' specific faiths undoubtedly distinguished them in colonial writing, they were also understood in the context of other factors, particularly their seclusion from the rest of civilization.

Politics Around the Code

The present government promotes Hindustan (territory), Hindi (language), and Hinduism (religion) as a central narrative to undermine the natural and multicultural variety of modern India. This process of re-embodying identities after disembodying coincides with the alignment of global capital. The way social relations have been mediated is changing as a result of this reconfiguration of uniqueness, which is accelerating India's transition to capitalism.

Further, the growth of the Christian population in several sections of the state for a long time was a significant concern for the present government. According to them, an increase in population reflects an organized and targeted change action by some groups with personal stakes. For instance, it was claimed that missionaries' schools were established for the social and economic uplift of tribals by educating, providing healthcare services, giving employment, and providing other facilities. While providing the facilities to the tribals, Christian lecturers were misleading tribals about their religious identity and changing their minds. All these came at a cost for tribals by abandoning their religion. In Jharkhand, Christians make up about 4.5% of the total population.

The main grievance of tribals and the present government is that most Christians in the state of Jharkhand are tribal people who were lured into Christianity by Christian Missionaries through allurement and otherworldly things. They demand that those who were converted to Christianity should not be granted the privileges granted to Scheduled Tribes under the Indian Constitution. The state's anti-conversion statute, the Jharkhand Freedom of Religion Act, 2017, exacerbated the religious conflict. It is incredible to see how these sentiments are stoked even by the present government.

According to the current government, the words Santana and Sarna share some similarities, explaining why the Sarna is closer to Hinduism than any other faith. A section of the Hindu holidays, such as Durga Puja, is praised by the locals with the same enthusiasm as any Hindu. It should be noted. According to the present government, "Tribals are not Christians; rather, they are Hindus or Santana." Although the Indian Constitution (Articles 25–28) provides freedom of religion, the Census denies many Adivasi (original people) communities the right to self-identify with their religious traditions.

Religious identity classification in the Census is inconsistent with the UN Declaration on the Rights of Indigenous Peoples and the spirit of the Constitution. The right of indigenous people "not to be subjected to forced assimilation or destruction of their culture" is expressly recognized in the declaration. Therefore, the Sarna code resolution passed by the Jharkhand Assembly is a partial corrective action. Partial because many Adivasis follow other traditional traditions with different names, such as Sari Dharam, Mari Dharam, and Adi Dharam, and not all Adivasis adhere to the Sarna code.

Apprehensions of Adopting Sarna as a Religious Code

The United Nations wanted tribe members to spread and promote their culture to other people. Hence, 9th of August has been celebrated as an indigenous or (tribal) day globally (Kraft, et al., 2020). The protection of tribal people's rights and their use of forest, water, and land resources was emphasized. Their culture, educational, and self-respect-related facets must be popularised and spread across society.

The primary motivation behind the creation of the current State of Jharkhand in November 2001 was to save, conserve, and advance tribal cultures within India and beyond the world. People must fully understand that protecting tribal interests was the primary motivation behind the establishment of the state of Jharkhand.

After two decades of Jharkhand state formation, tribals noticed although they received territorial rights, however, somehow, they were losing their religious identity or were represented incorrectly in census data (Alam and Raj, 2018). In addition, the religious population of the tribal community was decreasing, according to data. Hence, to protect and regain their religious identity, which was already documented during colonial times, tribals of Jharkhand, with four other states, demanded Sarna as a religious identity.

Conclusion

Tribals are requesting to add a section in the enumeration categorizing the tribal religion as a code. There is not an iota of doubt that government will not allow this as the statistics depend on self-recognizable proof. However, the fact remains that merely getting up to code on the census would not amount to the addition of religion to the list of religions recognized by law. Still, it could begin to be a legitimate claim over time. Maintaining tribal character and characteristics this way may be a viable solution. The depletion of the tribal way of life is one of the driving forces behind such a claim.

References

Conklin, B. A., and Graham, L. R. (1995). The shifting middle ground: Amazonian Indians and eco-politics. American Anthropologist, 97(4), 695–710.

Damodaran, V. (2006). The politics of marginality and the construction of indigeneity in Chotanagpur. Postcolonial Studies, 9(2), 179–196.

Alam, S., and Raj, A. (2018). The Academic Journey of Witchcraft Studies in India. *Man in India*, 97(21), 123-138.

Kraft, S. E., Tafjord, B. O., Longkumer, A., Alles, G. D., and Johnson, G. (2020). Indigenous Religion (s). Routledge.

Nayak, B. S. (2021). Colonial world of postcolonial historians: reification, theoreticism, and the neoliberal reinvention of tribal identity in India. Journal of Asian and African Studies, 56(3), 511–532.

Xaxa, V. (2005). Politics of language, religion and identity: Tribes in India. Economic and political weekly, 1363–1370.

Sustainable Fashion: "Form Leisure"- Deconstructing Men's Formal Shirts Into a Women's Wear Collection

Kanaka Pandit[a] and Anahita Suri[,b]*

[a]Student, Unitedworld Institute of Design, Karnavati University, Gandhinagar, India
[b]Assistant Professor, Unitedworld Institute of Design, Karnavati University, Gandhinagar, India
E-Mail: [*]anahita@karnavatiuniversity.edu.in

Abstract

The pandemic introduced us to the "Work-From-Home" concept which was practiced in almost every corner of the world. This led to many people taking workcations with their family and working at their convenience and leisure. It somewhere made them not want to slip into those formal shirts and just work in lounging clothes. While men might not have missed dressing up during the lockdowns, most of the women did. This also resulted in women shopping more than usual, looking for stylish and one-of-a-kind pieces. Another noticeable factor of the pandemic for most people was the fluctuations in their weight. This has resulted in many garments in the wardrobe not fitting well anymore. This research studies the combination of these factors of the pandemic to come up with a solution to produce something useful and purposeful.

Keywords: Casual Wear, Deconstruction, Formals, Sustainability, Upcycling, Workcation

Introduction

Fashion industry suffered a huge loss during the Covid 19 pandemic. From people losing jobs to businesses being shut down and collections being canceled, it faced it all. The restrictions that this pandemic imposed on the entire population made everyone want to travel, go on vacations, and getaways at the first chance. Though the tourism industry had a hard hit, it bounced back with travel packages following all the Covid 19 safety guidelines for people to travel stress-free again. Everyone missed dressing up, wearing good clothes, and taking great photographs. While men might not have missed dressing up, most of the women did. This has also resulted in women shopping more than usual and taking more interest in presenting themselves with an extra effort which they missed during the lockdown.

Another noticeable factor of the pandemic for most people was the fluctuations in their weights, which resulted in many garments in their wardrobes not fitting them well anymore. This has created a wide gap between the two ends where on one side there is excessive shopping, while on the other side, there is a wardrobe full of ill-fitting garments ready to be thrown away. The sustainability factor is deeply affected in this scenario as the circular life-cycle of a garment is missing. This has resulted in the need for a sustainable solution.

Background

The Covid 19 pandemic introduced us to the "Work-From-Home" concept which was practiced in almost every corner of the world. In some parts of India, people are still working from home. They looked at this as an advantage of having the liberty to work from anywhere. This also led to many people taking vacations with their families and working at their convenience and leisure. It made them habitual to their lounging clothes, with a huge drop in the need for formal wear.

People who missed occasions of dressing up started shopping as soon as stores opened up, as a way of filling the major gap in their routine shopping that they faced during a lockdown. With increasing consumerism, this is a perfect spot for sustainable fashion to step in and make itself accessible to hungry consumers. Now women take more interest in being fashionable than they did before and make sure their wardrobe is filled with unique, interesting, stylish, and one-of-a-kind pieces.

Due to lifestyle changes, some people put on some weight because of lack of physical exercise while on the other hand, some people turned into fitness enthusiasts and lost a few kilos.

Connecting it back to the beginning, formal shirts are one of those garments that need to fit oneself perfectly. Formals exude discipline and perfection. Hence, their fits matter much more than any other garment. A little tight or a little loose can break the wearer's impression for the viewer. This project focuses on 2 aspects, i.e. 1. the reduced demand for formal wear (specifically shirts), and 2. the already existing ill-fitting shirts waiting to be discarded.

Aim and Objective

This project suggests a sustainable solution for tackling the issue of the drop in demand for formal shirts as well as the excess ill-fitting shirts in the wardrobe. The solution is sustainable as the ill-fitting formal shirts aren't discarded or disposed of directly. Instead, those are put to use for fashionable women who collect interesting pieces for their wardrobe by deconstructing and making them into womenswear garments- stylish & one of a kind.

Methodology

Initial research was done through secondary sources like books and news articles.

Primary research was then conducted through tool of questionnaire & survey with a sample size of 50 respondents. The questionnaires contained closed-ended questions with three different types of scale; the scale of agreement, multiple choice, and five-point scale. The data was collected, analyzed, and represented through tables and/or graphs.

This sample was further divided into,

i. 20 men in the age group of 25–55 years (working class). A questionnaire was developed to understand the perspective and situation of these men pertaining to aspects like their employment status, their work environment, dress code, preferences of work dress codes, their frequency of shopping, and their use of formal shirts.
ii. 30 women in the age group of 18–40 years to continue with the second part of this research. The purpose of this questionnaire was to understand women's opinions on sustainable fashion, their acceptance, and familiarity regarding upcycling, and their willingness to embrace a deconstructed garment.

Survey Results

Identifying the Problem

During the pandemic, about 80% of people were working remotely full-time. This led to the realization that the workforce could be equally productive in work from home format. Almost anyone can work from nearly anywhere. The workforce realized that a 9–5 office routine is not everyone's cup of tea. Studies reveal that flexible ways of working improve productivity, happiness, loyalty, and job satisfaction. Creative professionals rely on a change of scenery, new experiences, and inspiration to produce their best work (Metcalf, 2020). The increase in number of professionals working remotely is likely to be one of the lasting changes after the Covid 19 pandemic. Research says that employers plan to make this a permanent pandemic response plan. A survey by the research & advisory firm Gartner suggests that approximately 80% of organizations plan to allow employees to work remotely at least part of the time upon reopening from the Covid-19 pandemic (Golden, 2020).

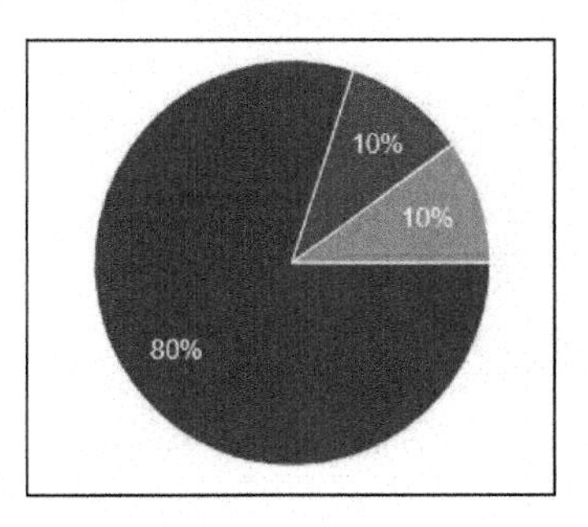

Fig. 1. 80% Men said "Work from Anywhere" when asked their mode of work during the pandemic

In the United Kingdom, sales of formal clothing dropped by approximately 25% in 2020 - the largest annual drop in 23 years. The picture was similar in the USA, where companies saw a 90% decline in profits in 2020. The business-fashion sector was hard hit as more workers opted for remote working, leading to a change in the requirement of clothing styles (Lufkin, 2021).

As some companies started calling employees back to the offices, there were expected to be some dress code changes. Employers' attitudes had started to become more functional. More importance was given to how much is accomplished rather than how many hours one sits in an office and what one wears. Moreover, almost all employees preferred wearing casuals to work (Clarey K, 2020).

Formality's Rise and Fall

During the initial days of the pandemic, organizations encouraged people to dress up for work video meetings to bolster mental health and increase a sense of purpose and productivity. Comfort was the key to dressing for most people, while also ensuring productivity. The idea that "to do your best work and cultivate the best impression, you need to look the part' dates back to the Victorian era, when educated professional men dressed in velvet and fur, signaling status and influence. The 1960s saw the popularity of "casual Fridays" featuring blue jeans and Hawaiian shirts, which grew with the rise of Silicon Valley. Mark Zuckerberg with his hoodie popularized the idea that success didn't need to come buttoned up in a suit and tie. Suits, a symbol of professionalism, now became associated with professionals from IT industry and law firms redefining the look of a CEO (Lufkin, 2021).

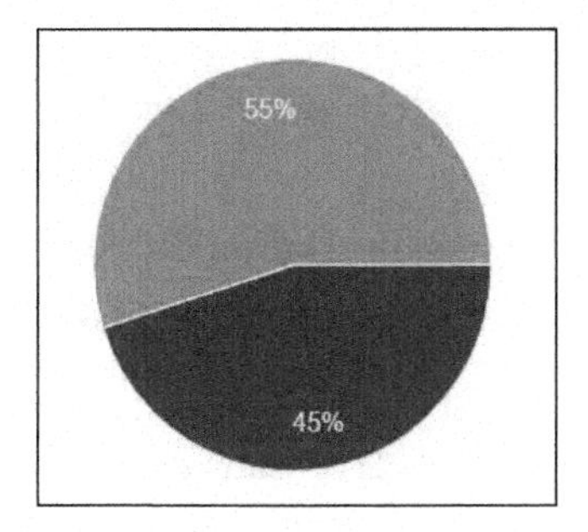

Fig. 2. 55% Men responded they would prefer wearing a mix of casuals and formals to work whereas 45% responded they would choose only casuals. No one chose to wear only formals to work

The Hybrid Factor

The sales in the work-fashion sector have seen a decline post pandemics many companies plan to embrace a hybrid future. Implications of this transition are a decline in the requirement for office wear & more importantly, a significant shift in employer thinking. Dress codes have toned down and become more relaxed even in the most formal of sectors like finance. A survey conducted in 2019 suggested that almost 50% of companies in USA allow casual dress. With only 5% of people stating they enjoy wearing formal to work, there is a great number of people who prefer casuals as a work dress code, and don't think that formalities signify anything other than a company protocol (Lufkin, 2021).

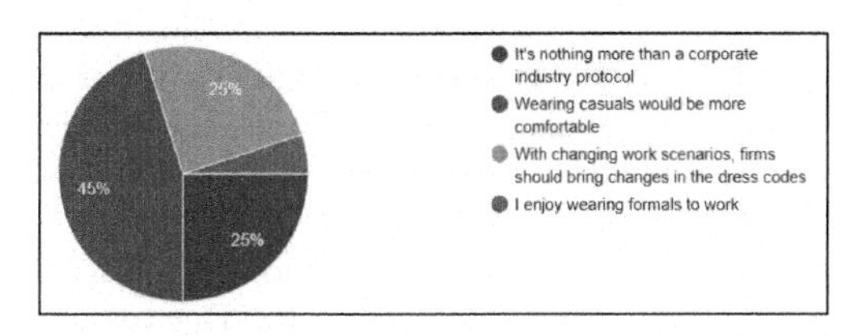

Fig. 3. Responses of men when asked about their views regarding having to wear formals to work

Organizations today are embracing a more functional approach leading to a significant shift in employer mindset. More importance is given to what is accomplished rather than how many hours one sits in office and what they wear. A rise in start-ups that already have no formal dress code has been observed (e.g. Advertising agencies, Design studios, etc.). Many youngsters prefer freelancing & content creation as their career choice, which requires no formal dress code.

Weight Fluctuations During the Pandemic

Since this drop in the formal dress code has been recently observed, about 55% of Indian men were required to wear formals to work on a daily basis earlier and 25% only during important days. Because of this, the frequency of buying and owning formal shirts was higher before the pandemic hit. On an average, men own 20–40 formal shirts out of which they only use 30%–50% of them regularly.

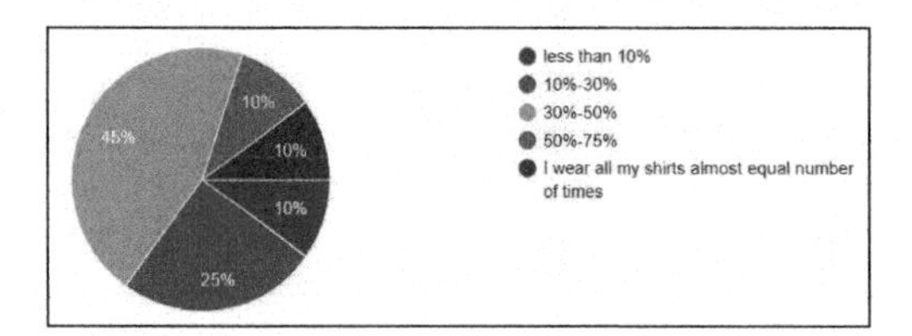

Fig. 4. Responses of men when asked about the regularity of wearing their formal shirts

Apart from a change in the professional space, during the pandemic, many people were confined to their homes, leading to unhealthy food habits, as a means to distract themselves or learn something new. With reduced or no physical exercise and mental & emotional stress, there was a noticeable situation of unwanted weight changes. As a result, a sudden boost was also noticed in the fitness industry. People invested their time at home into fitness training, home workouts, and online gym sessions. This also resulted in great changes in body weight (Begdache, 2021). 70% of men surveyed reported that they had experienced weight changes during the pandemic. Due to this, about 65% men stated that their formal shirts didn't fit them well anymore. Considering the shift in the dress code of the corporate sector and weight fluctuations, these aspects indicate that there are already enough shirts in the wardrobe that might go unused shortly.

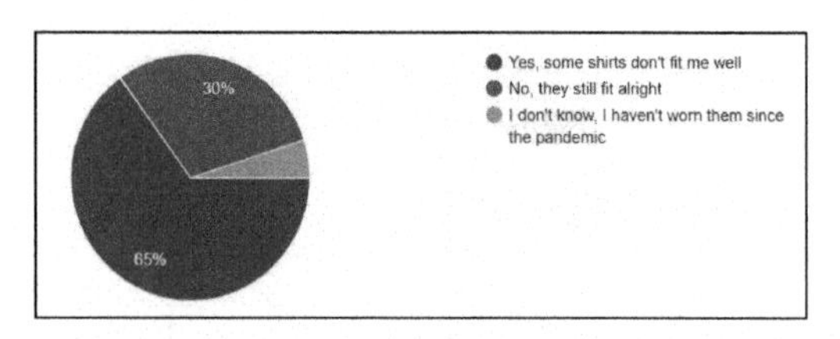

Fig. 5. 65% Men stating that their formal shirts don't fit them well anymore

What About the Existing Shirts in the Wardrobe?

Generally, women show more engagement and interest than men in shopping. There comes a wider variety and options to choose from for women when it comes to silhouettes (Chevalier, 2021). With the travel restrictions that the pandemic imposed, many people missed traveling. This has made them want to grab every opportunity that they get to travel, and take a vacation. This also gave rise to the concept of "work-cation" which was working while taking a vacation. Hence, the market for men's formal shirts has already begun to reduce. Changes in men's weights have reduced the probability of them wearing some of their old shirts again, and they don't know what to do anymore.

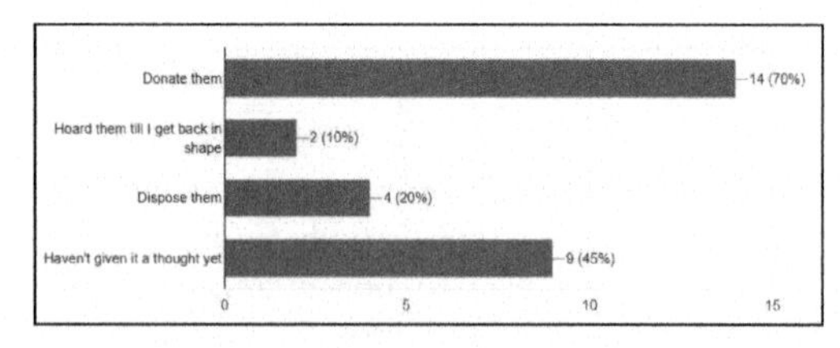

Fig. 6. Responses of men when asked about what they would do with the shirts that don't fit them well anymore

Solution for the Problem

Upcycling men's formal shirts for creating a summer wear collection for women.

There is a growing demand for casual women's wear in the market. Research shows that cotton is the most purchased fabric by women, which even formal shirts are made of. Women are interested in sustainable fashion if it is made more accessible, unique, and fashionable. There is a huge scope in the arena for sustainable women's fashion if presented interestingly. 90% of women are positive regarding the concept of purchasing garments upcycled from men's used formal shirts. 30 women above the age of 17 were the respondents for this research, with the maximum number of women falling in the age group of 18-25 years.

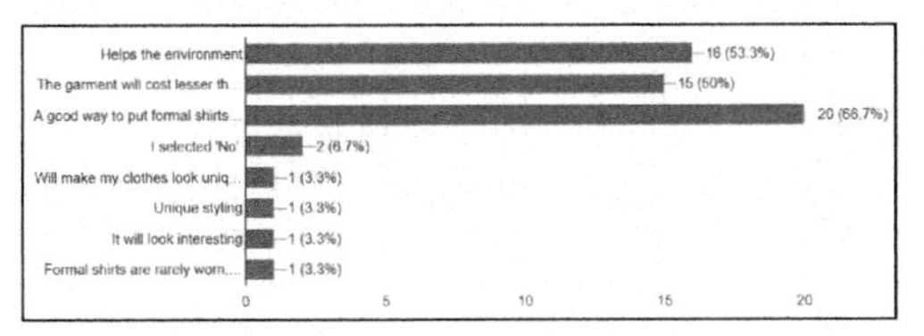

Fig. 7. Women stating their reasons for being interested in buying clothes upcycled from men's formal shirts

Project

Men's Formal Shirts + Upcycling = Women's Casual Vacation Wear is the solution ideated for this issue.

With "Work From Home" being on the rise and tourism starting to gear up again, this concept has the potential to be a sustainable option for the recent future. As of now, if we focus on only formal shirts for this collection to not make it complex by including other formal garments in it, we would also be able to see what kind of clothes can a single type of garment bring after being upcycled. Therefore, this collection is titled "Formaleisure" to form an amalgamation of the words "formal" and "leisure" as these clothes are formal shirts transformed into vacation wear.

A range of women's casual garments upcycled from discarded men's formal shirts was designed. This collection shows the versatility of garments and emphasis the fact that there could be a sustainable and pocket-friendly solution to this problem of post-consumer garment waste.

Fig. 8. Range of Deconstructed, Upcycled women's casual collection- fronts

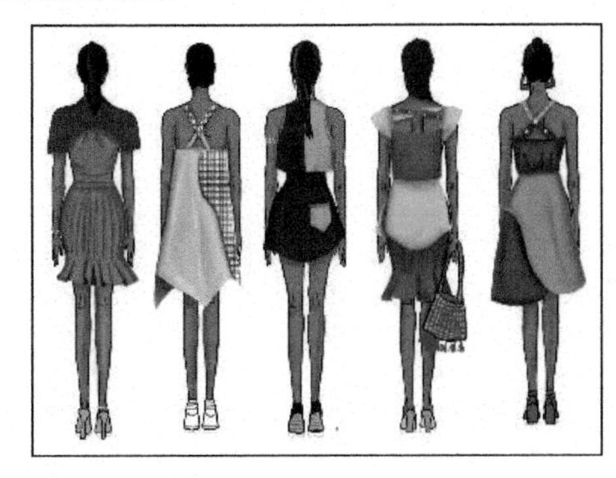

Fig. 9. Range of deconstructed, upcycled women's casual collection- fronts

One ensemble was executed and realized as an exploration.

Fig. 10. Two menswear shirts that have been deconstructed

Fig. 11. New Womenswear garment created out of the two men's shirts

Conclusion

This project highlights the problem of post-consumer garment waste created especially during and after the Covid-19 pandemic and lockdown due to lifestyle changes, workplace changes, relaxation in dress codes, and weight fluctuations. It also looks at finding a sustainable solution for the same. Upcycling and deconstruction provide an effective solution in reducing waste, reusing it in a new avatar that is stylish, pocket friendly, and one of a kind, catering to the buying behavior of the youth today. This solution could also be applied as a possible solution to the problem of excessive buying and waste generation of garments.

References

Begdache, L. (2021). Unwanted weight gain or weight loss during the pandemic? Blame your stress hormones. The Conversation. https://theconversation.com/unwanted-weight-gain-or-weight-loss-during-the-pandemic-blame-your-stress-hormones-157852

Chevalier, S. (2020). Online clothing purchasing by demographic Great Britain 2020. Statista. https://www.statista.com/statistics/286096/clothes-and-sports-goods-online-purchasing-in-great-britain-by-demographic/

Clarey, K. (2020). Half of HR pros say they'd opt to continue working from home.

HR Dive. https://www.hrdive.com/news/poll-hr-professionals-remote-work-coronavirus-pandemic/578812/

Golden, R. (2020). Gartner: Over 80% of company leaders plan to permit remote work after pandemic. HR Dive. https://www.hrdive.com/news/gartner-over-80-of-company-leaders-plan-to-permit-remote-work-after-pande/581744/

Lufkin, B. (2021). Is the formal "suited and booted" office dress code extinct? BBC page. https://www.bbc.com/worklife/article/20210713-is-the-formal-suited-and-booted-office-dress-code-extinct

Malek, MohammedShakil S., and Bhatt, Viral (2022). Examine the comparison of CSFs for public and private sector's stakeholders: a SEM approach towards PPP in Indian road sector, International Journal of Construction Management, DOI: 10.1080/15623599.2022.2049490

Metcalf, M. (2020). What is a "Workation?" Timetastic - A blog about taking and managing time off work. https://timetastic.co.uk/blog/what-is-a-workation/

Stuart, O. (2019). How do men & women shop for clothes differently? Freedonia Focus Reports. https://www.freedoniafocusreports.com/Content/Blog/2019/10/01/How-Do-Men--Women-Shop-for-Clothes-Differently

Kondapalli Toys: White Woodcraft of Andhra Pradesh

Vibha Kalaiya[*,a] *and V. Prashanthi Priyakumari*[b]

[a]Assistant Professor, Unitedworld Institute of Design, Karnavati University, Gandhinagar, India
[b]Student Unitedworld Institute of Design, Karnavati University, Gandhinagar, India
E-mail: *vibha@karnavatiuniversity.edu.in

Abstract

The toy heritage of Andhra Pradesh - Kondapalli Toys was famous for its lightweight, vibrant colors and age-old production techniques. Themed around mythology, rural life, and animals, these toys exhibit joyous and realistic expressions. The present research paper focuses on the rich heritage of Kondapalli toys. The researcher intends to document in detail the legacy of Kondapalli toys with their special significance. The methodology adopted was the personal interviews conducted along with the observation method through field surveys. A multi-visit approach was implemented for authentic data collection. The findings found that a 400-year-old custom has been passed down from generation to generation, with every member participating in the toy-making activity in Kondapalli's Toy Colony. The village had also gained popularity for the manufacturing of dancing dolls, also known as Thanjavur Dancing Dolls – a form of art that had been adopted from the state of Tamil Nadu. The making of these wooden Kondapalli toys involved great craftsmanship at each stage of the production. They were made out of white wood which was soft in nature, and used as the main raw material. The toys displayed various themes from daily household work to the theme of dashavatar. The art has a significant Islamic influence, and the pointed noses of the human figures are reminiscent of the 17th-century Rajasthan style. It has been observed that the Kondapalli toy has evolved from a child's toy to an appealing showpiece throughout the years. Furthermore, the move eroded the repurchase market because, unlike children's toys, collections require little replacement. Before the traditional toys got redundant even by the younger generation the researcher conceptualized the study given its relevance to the current times.

Keywords: Andhra Pradesh, Dashavatar, Kondapalli, Thanjavur Dancing Dolls, Traditional Craft, Wooden toys

Introduction

This village is known for more than 100 years to be occupied by these artisans since the Mughal Emperor. This 400-year-old custom had been passed down from generation to generation, with every member participating in the toy-making activity in Kondapalli's "Toy Colony." Here everyone was an expert in making wooden toys. As the researcher walked through the small roads of this village, the researcher saw toys displayed in many shapes on each corner including Deities, which had been in this village for the last 10 years. The village had also gained popularity for the manufacturing of dancing dolls, also known as Thanjavur Dolls – a form of art that had been adopted from the state of Tamil Nadu.

Methodology

The data pertaining to the craft was collected through the multi-visit approach to the Kondapalli village. The primary data was collected through personal interviews with the artisans using the structured interview schedule comprising open and close-ended questions.

Results & Discussion

Artisans

The artists who work on making these wooden toys are known as "Arya Kshatriyas," as mentioned in the "Brahmanda Purana." It was reported that they relocated from Rajasthan to Kondapalli in the 16th century and credited their dominion to Muktharishi, a sage bestowed with arts and crafts abilities by Lord Shiva. Mostly men engage in chiseling and making the statues and once it's done finally the toys were sent to different homes for painting which was done by the women.

Fig. 1. (a) Kondapalli village (b) Artisan Hanumanth Rao shows his Creations of numerous sculptures like the Garuda, Nandi

Simha and the Vahanas in the many temples in Andhra Pradesh. Kondapalli toys had evolved from playthings to collectibles throughout the years. The shift harmed the buyback market since, unlike children's toys, collections require little replacement. The "Bommala Koluvu" or "Kollu," when toys are collected and ceremoniously presented, is an important feature of the Dusshera and Sankranthi celebrations. Most children and women would compete to have the most impressive and elaborate collection. Every home in this village had small shops with no specific names to it. They also take orders from other stores and engage only in making the toys and painting them. The Government, Government departments, certain institutions, and organizations were also lending a helping hand to develop this industry.

Raw Material

The artisans required Tella poniki tree, and softwood as raw material for their produce. The highly treasured, brightly colored Kondapalli toys were under threat from deforestation, and in particular the rampant exploitation of the Tella Poniki Tree. These trees were becoming extinct as the wild pigs were eating all the seeds in the forest.

Fig. 2. (a) Tella poniki tree (b) Tamarind seed powder (c) Knife and a wooden basket as tools

Process

Fig. 3. (a and b) Artisan chiseling the wood block according to the design model (c) smootheningthe surface

The trees were chopped into logs and dried. When cut, it weighed greater and after drying it weighed less. The basic block was chiseled according to the design model. The chiseled blocks were heated by placing them on a metal mesh basket to drain the remaining moisture in them. Now the different parts of the toys were made according to the theme they were designing for and the model. A paste made of tamarind seed and gum was used to stick the joints together. After making the toy, these were sent to the women at home for painting.

Heating Process

The toys were placed on the basket for heating. The basin with dried leaves, wood chunks, and sticks was burnt for heating the chiseled blocks. The metal mesh basket protects the blocks from catching fire. Each Slow heating was used to extract all of the moisture from the hardwood piece. Each limb was carved separately and then attached to the body with tamarind seed adhesive paste and lime glue. After that, it was painted with either water or oil paints.

Fig. 4. (a) Chiseled block model (b) Heating process of the toys for finishing

Painting Process

Fig. 5. (a) Paper glued with an adhesive paste of tamarind seeds over the toy for painting (b) Women artisans painting the toys with enamel paints

The paints used were Natural and Enamel paints for painting these toys. Natural paints give a rough look and go away when used or wiped with water. Enamel paint gives the glossy look and can be washed with water.

Women engage in painting the toys, they give the first coating, base color, and dry it for one day or a few hours depending on the weather. Once the base color was dried, the other colors were painted. Likewise, they color a bulk of toys every day with different colors.

Themes

Fig. 6. Various themes of Kondapalli toys (a) Daily household work (b) Bullock cart

Various themes inspired by the surroundings such as daily household work, village scenarios, spinning fiber, cutting vegetables, pulling water from a well, Dashavathr, Sankranti Haridhas, Elephant Chariot, Wedding Ceremony, Krishna with Gopi's, etc. were most popular for the Kondapalli toys.

More than wooden toys, the settlement became well-known for the production of dancing dolls, popularly known as Thanjavur Dolls – a form of art which had been adopted from the state of Tamil Nadu. The Thanjavur dancing dolls had become an integral part of the renowned Kondapalli craft.

Fig. 7. Thanjavur dancing dolls

In Kondapalli, an artisan named Chavala Uma Maheswara Rao began producing these dancing dolls in 2002. Dancing dolls became nearly associated with Kondapalli art throughout time, and today the demand for these dancing dolls across Andhra Pradesh is so high that they are often unable to fulfill the aim.

Conclusion

The Kondapalli toys were crafted from a locally manufactured wood known as Tella Poniki, which could be found in the nearby Kondapalli Hills. Tella Poniki was mostly preferred for making these toys because of its lightweight and can be carved easily. The process was quite comprehensive and exhaustive, which required observation of minute details else anything could go wrong.

To the comfort of Kondapalli craftsmen, the State Government decided to establish a "Wood Bank" under the supervision of the forest department. Tella Poniki's cutting would be assisted by the Capital Region Development Authority (CRDA). According to the most recent information available, 211 artisans from 80 families were participating in the toy-making process. Four Vana Samrakshna Samitis (VSS) were operating in the Kondapalli, G Konduru, and A Konduru forest reserve areas. They took orders from the private sector and government sectors. Depending on the season and demand in the market they got orders. They sold their stock according to the manpower, work, and time taken.

However, due to a lack of profit, this old art style that received patronage from monarchs is diminishing and it's time taking process. There were barely 200 nimble-fingered artisans now who were retaining this glorious art of toy-making in Kondapalli. The younger generation was reluctant to pursue this art further due to the economic crisis. Despite the efforts of numerous organizations to preserve this unique craft, only time will tell whether our future generations will visit to see this type of art or not.

References

Hanumanth, R. (2019). Interviewed by Prashanthi Priyakumari V. Kondapalli, Andhra Pradesh.

N. A. (2017). Wooden Painted Toys, viewed on June 2019, <https://lepakshihandicrafts.gov.in/kondapalli-toys.html>

N. A. (2018). India's centuries old toy- making trade whittled by deforestation, AFP, viewed on June 2019,<https://www.dailysabah.com/environment/2018/05/16/indias-centuries-old-toy-making-trade-whittled-by-deforestation>

Rajamahendravaram. (2018). Kondapalli toys: Administration comes to rescue of artisans, to set up "wood bank" in Andhra Pradesh, The New Indian Express, viewed on June 2019, <https://www.newindianexpress.com/states/andhra-pradesh/2018/apr/04/kondapalli-toys-administration-comes-to-rescue-of-artisans- to-set-up-wood-bank-in-Andhra-pradesh- 1796771.html>

Somya, J. N. (2019). Traditional Toys of Kondapalli, D'Source, viewed 10 June 2019, https://www.dsource.in/resource/traditional-toys-kondapalli/kondapalli-toys>

Malek, M. S., and Gundaliya, P. J. (2020). Negative factors in implementing public–private partnership in Indian road projects. International Journal of Construction Management, DOI:10.1080/15623599.2020.1857672.

Borikar, H., Bhatt, V., and Vora, H. (2022). Investigating The Mediating Role Of Perceived Culture, Role Ambiguity, And Workload On Workplace Stress With Moderating Role Of Education In A Financial Services Organization. Journal of Positive School of Psychology, ISSN, 9233–9246.

Malek, MohammedShakil S., and Bhatt, Viral (2022). Examine the comparison of CSFs for public and private sector's stakeholders: a SEM approach towards PPP in Indian road sector, International Journal of Construction Management, DOI: 10.1080/15623599.2022.2049490

Hiral Borikar, M., and Bhatt, V. (2020). A Classification of Senior Personnel with Respect to Psychographic and Demographic Aspect of Workplace Stress in Financial Services.

Borikar, M. H., and Bhatt, V. (2020). Measuring impact of factors influencing workplace stress with respect to financial services. Alochana Chakra Journal, ISSN, 2231–3990.

Role of Consumer Perception on Genderless Fashion in Deconstructing Gender Stereotypes in Indian Society

Nitanshi Tripathi,[a] Taruna Vasu,[,b] and Dr. Vibha Kalaiya[c]*

[a]Student, Karnavati University, Gandhinagar, India
[b]Head of the Department, Karnavati University, Gandhinagar, India
[c]Assistant Professor, Karnavati University, Gandhinagar, India
E-mail: [*]taruna@karnavatiuniversity.edu.in

Abstract

This study will look at how Indian consumers view Gender Neutral Fashion. It seeks to discover the significance of Genderless Fashion as perceived by people in Indian society, as well as how they relate it to gender stereotypes. India has similarities in clothing for both genders since its mythology, history and among tribes but this has gradually got decreased. It's true that women have started wearing bifurcated garments for a long time and has now become common but the author wants to know through this research that even today straight men have not inclined towards Feminine wardrobes. There is a need to understand the psychology behind this perception of genderless clothing. This research uses random surveys, literature reviews, and text analysis in the context of Indian society. This study aims to identify public confusion about genderless fashion and to serve as a resource for designers interested in the limitations and acceptance rate of genderless fashion in India.

Keywords: Consumers, Equality, Genderless fashion, Perception, Stereotypes

Introduction

Fashion in India has been neutral over millennia, we can find facts that our goddesses were bare-chested. Our clothing for men and women was not much divided like kurta, pajama, and dhoti. India has lost that gender-neutral clothing over the ages and it's coming back again with the urge of representing one's personality. Gender-expressive fashion is a style of attire that allows people to show the gender role they are playing that day. Because they do not identify with anyone's sex, someone who identifies as gender-fluid may alter their role every day. But, the way of thinking and Indian Society has lots of stereotypes about clothing and gender.

The role of fashion is to make apparent particular scenarios about this issue and to dismantle norms about forms and colors that society has adapted to people since childhood. It seeks to dismantle gender stereotypes in the context of fashion. Because of the premise that garments have no gender, fashion is attempting to soften the masculine/feminine distinction (Akdemir, 2018).

The primary use of genderless clothes, aside from being a form of expression, is the desire to experiment with style while not being constrained by gender constraints. Gender perception opens up a fresh option for a designer to create a new collection that will improve their performance.

Changes in society and customer behavior perception of gender could lead to fashion designer performance excellence (Tetovo, 2016).

Genderless fashion can be defined as the process of designing apparel without regard for gender. The distinction is in one's physical appearance; androgyny fashion is for people who have both male and feminine characteristics (Pambudi et al., 2019).

Genderless dashi creates a genderless look by blending traditionally feminine elements of beauty, such as make-up, with the male body. Individually, these characteristics can still be interpreted as gendered, and it is up to the observer to determine which elements most strongly suggest conformance to gender identification. Today's youngsters in Japan value fashion as a way to express themselves authentically and challenge traditional norms (Balkon, 2018).

There is a possibility to find more about the views of consumers regarding genderless garments and extend the research to numerical figures through respective tools to make it more personal and stronger. Also, research with people from non-creative fields who don't have much exposure to fashion and trends could discuss more how instead of prioritizing individual characteristics designers still specify gender at first in their client boards.

The research uses random surveys, literature reviews, and text analysis in the context of Indian society. This study aims to identify public confusion about genderless fashion, as fashion is not viewed as an article in Indian society, but as a source of gender identity. People are not ready to accept clothing as per their personality but they wear it as per their gender. Although, women have got fewer eyes judging them while wearing pants than a man wearing a puff-sleeved shirt.

The research will help to understand this clearly as per the research tools used. It will depend on the validation of the response by the target group of people and a small sample size due to a lack of time and resources.

Methodology

A survey was conducted to examine the issues in the consumer perception of Genderless Fashion in India. It consisted of both open and closed-ended questions including multiple-choice questions and Semantic Differential Scale. The sample size was kept between 80-100, and the resultant size was 96 in the age group 18-30 including all genders of the Middle class (upper and lower). The target sample was consumers in north Indian states - Uttar Pradesh, Rajasthan, Haryana, and Delhi. The sample was narrowed down to consumers from non-creative fields as creative fields get more exposure with their creative field institutions and workspaces. They are already exposed to a lot of platforms where people have their individuality and they think out of the box (without depending on what society has to say).

The tools used in the study are all created under the guidance of academic mentors and every help that is taken from online websites is cited. The questionnaire is developed with the objective of the study with the help of Google Forms. All the participants who are taken as a sample agreed to participate and are informed about the purpose of the study.

Analysis and Results

The researchers received 96 responses: 55 male and 41 female. They are as: Uttar Pradesh-75; Delhi-13; Rajasthan-5; Haryana-3. They belong to non-creative fields (Student, Engineer, Doctor, Bachelor of Ayurvedic Medicine and Surgery, student, Government job, Chartered Accountants, Marketing, Executive, Developer, Teaching, Management, Analyst, Advocate, Navy, Supervisor in food tech, Businessman, Coder, Consultant, Cricketer, Senior Research Fellow). 56.3% of them are aware of Genderless Fashion and 22.9% are not aware and 20.8% are not sure if they know what is genderless fashion.

The researcher asked if according to them, there is any connection between gender with some given articles and the responses are shown in Fig. 1. Around half of them i.e., 52.1% think that Fashion is related to gender and 38.7% think that lifestyle is that article while 35.4% think that none of the articles has any relation with gender. When asked the reason for their choice of the article that relates to gender, they gave a variety of responses that were very interesting to know. Some say Fashion and lifestyle accessories are made according to the body types of different genders, while some say it's according to their taste. People also think that due to the habit of seeing this for ages, they cannot think of fashion as genderless. Some of the respondents think nothing is related to gender as these articles are a matter of choice for a person regardless of gender.

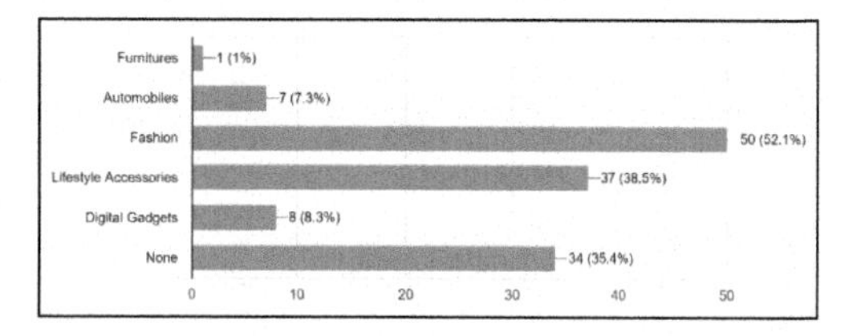

Fig. 1.

They were asked what they think genderless fashion is and they came up with thoughtful responses from them. According to most, genderless clothing is clothing by choice regardless of gender, color, design, or silhouette of the garments. They also gave examples of garments like jeans, trousers, hoodies, and sweaters. Some say they are designed in a way keeping in mind to be able to be worn by both genders before thinking to what sex it belongs to. They say that they are formless, shapeless, and oversized. Some of them take this to gender equality and state that it is the clothing to makes people realize that both genders are equal. They also stated some Indian laws that are made for gender equality and their right to choose. When they were asked if they own any trousers or skirts, the response was quite predictable. 99% of respondents own trousers irrespective of their gender but only 37.5% own skirts. As per the data received, it can be said that 37.5% who own skirts is female portion of the respondents and males do not own any skirt. And it can be seen all of them own trousers for sure be it male or female.

The respondents were asked to rate on a scale of 1 to 5 (where 1 is weird, 3 is neutral and 5 is pleasant), how would they feel if they spot a woman wearing a tuxedo and a man wearing a skirt with the *images A* and *B* shown here as references given to them. As in Fig. 2, 71.9% rated the girl wearing a tuxedo as 5 (pleasant) and only 1% rated it as 1 (weird). Figure 3 shows, 32.3% rated the man wearing a skirt as 3 (neutral), 25% rated it 1 (weird) and 10.4% rated it as 5 (pleasant).

A B

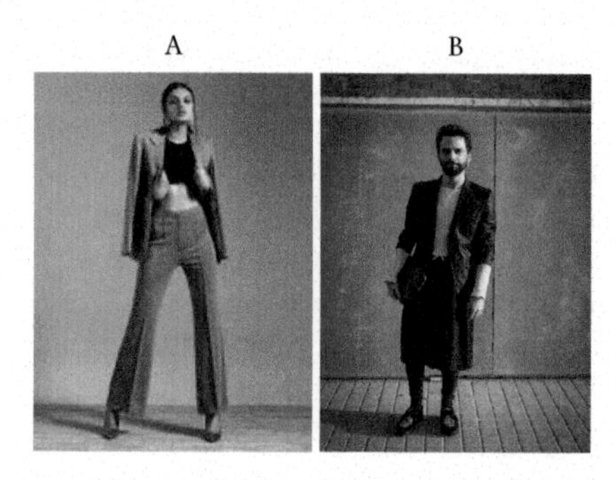

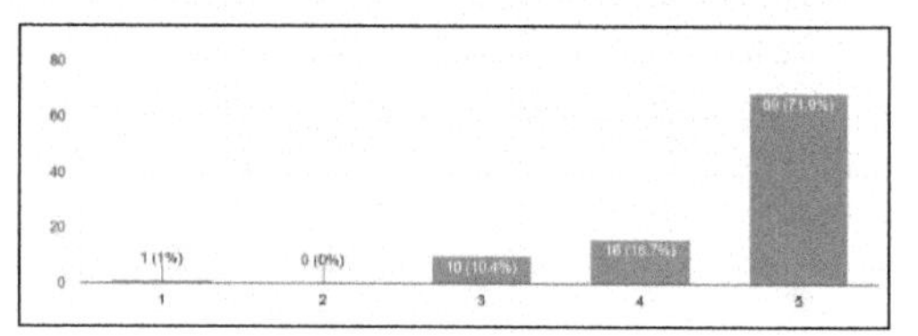

Fig. 2.

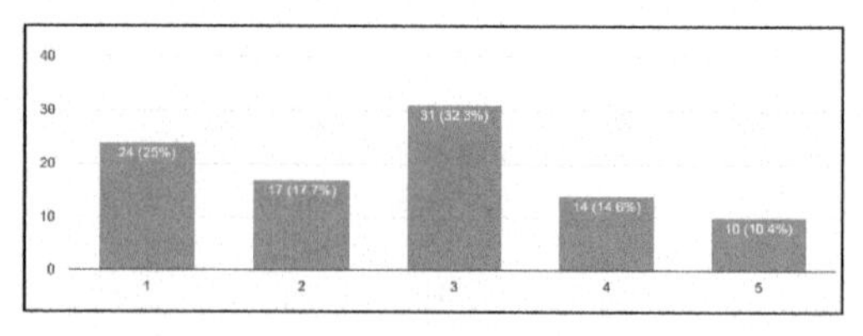

Fig. 3.

When asked about what one's style of clothing should identify, in Fig. 4, 83.3% of respondents chose personality, 46.9% chose the state of mind and 12.5% chose gender. Then the question was asked if they think gender identification is crucial for clothing, most of the people expressed that it's the choice of one's self to what to wear. They think that clothing is something that fulfills people's needs and they should not identify gender through it. Clothing should be according to one's convenience and aesthetics and not what the gender tags describe. Some of them thought it is important as per the mentality of society. The society standards we grew up with, where men and women had to wear certain types of clothing concerning their gender. But this should not be the case now.

When asked if they feel much difference in our ancestors' clothing between different genders, this is questionable as 40.6% of them replied with yes and 30.2% with no, and the remaining 29.2% are not sure and replied with maybe.

When asked about their perception of gender to the sentence "our ancestors wore dhoti," their response was interesting as 74% of people find it more related to male while only 21.9 % think it's related to female. According to most of the respondents, dhoti is something that males used to wear. But when asked how would they like if someone ask them to drape around a dhoti for their party, most of the respondents showed not much of their willingness and 45.8% of them rated the scale to neutral while only 22.9% rated to happy and 7.3% felt sad with this.

Figure 5 shows their response when asked if they think the idea of having separate sections for shopping is perfect.

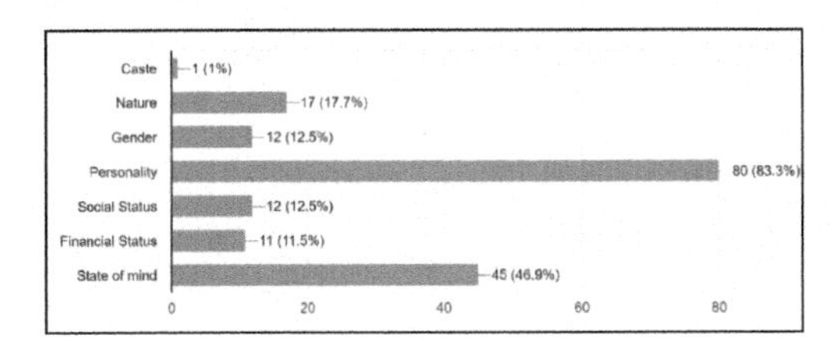

Fig. 4.

When asked, how would you like the idea of having a common clothing section for all genders rather than separate sections of menswear and womenswear in shopping malls and retail stores, Fig. 6 below shows their responses. 22.1% of them think the idea is perfect, other 22.1% also think the idea is good and 37.9% reacted neutrally to this idea and only 5.3% of them think that the idea is stupid.

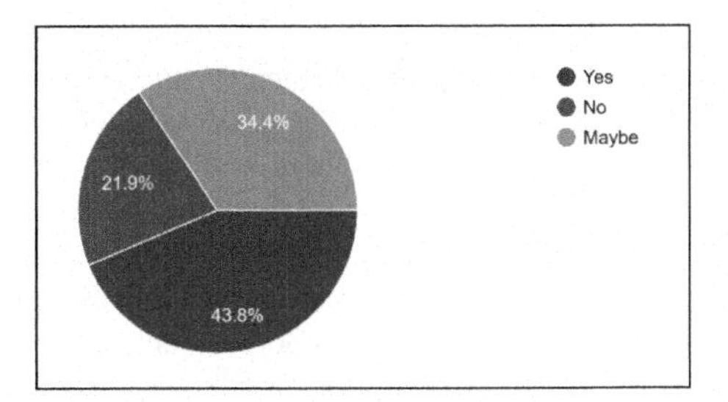

Fig. 5.

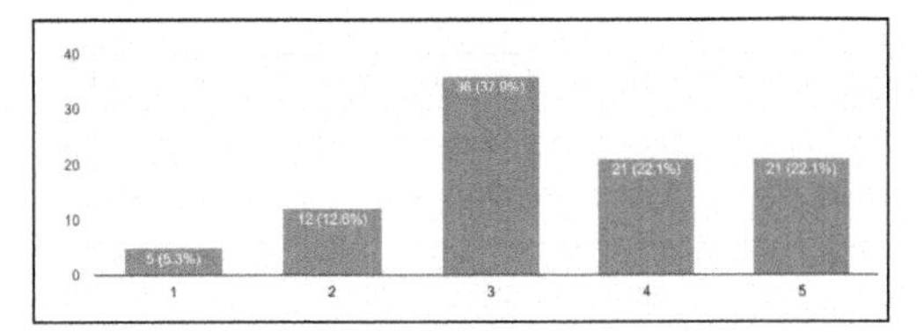

Fig. 6.

They were asked the reason for their reactions to the previous question, some replied with neutral that they don't feel any problem or comfort with the idea. Some of the respondents feel that it's perfect for one can choose whatever they want to wear without looking at the gender labels and according to one's personality. It would be a place to find clothes according to one's comfort. They say they can be separated based on the type of clothes like jackets, jeans, shirts, traditional, western, etc. rather than separated based on gender for having it organized and less chaotic. Usually, girls say that there can be other means of segregation and they don't have to switch from one to another section to find their comfy, relaxed, and oversized hoodies or tees that they usually have to buy from the men's section. Some people think that separation is necessary to avoid chaos and get their type of clothes as they would not like to wear other gender clothes. As easily understandable, men had this kind of response.

Conclusion

The most fashion-active age group thinks their personality and choice are more crucial in clothing than their gender. They have the realization that their way of clothing depends on how society would accept and so they try to adjust to the already existing ways of clothing rather than creating their own. This case is not similar for both males and females. Females are freer to wear their style without facing much weirdness of society while males are still in their silhouettes and if they try something new it's hard for them not to be accepted by society. The basis of everyone's clothing is comfort and personal style. For any garment, people do have this notion that they will wear it only if it's comfortable for them and belongs to their style. But the interference of society in their style and gender identity is also in existence in Indian Society and they are aware of it. Clothes don't come with gender tags, it's the way different people style them, the way they want to represent themselves. A person who wants to wear a piece of garment can wear it easily but if it's something that may be too much for society to digest, they back off from it and choose a little similar way to represent themselves. This helps them not to get conscious or lose their confidence in just thinking about what would people think. It is the way Indian society knows genderless fashion and it will take time to get over that gendered division of clothes. The researcher thought of a notion of common displays in showrooms for both genders, but it still needs to be thought upon the challenges that come with it as per the respondents.

References

Anyanwu, O. (2020). Decoding Genderless Fashion, The Future of the Industry, WWD.

Balkon, N. R. (2018). Genderless danshi: An emerging force against Japan's hegemonic masculinity.

D'Souza, O. (2018). The rise of genderless fashion in India., DNA.

Luna, C. P., and Barros, D. F. (2019), Genderless Fashion: A (Still) Binary Market, Latin American Business Review, 20.

Pambudi, N. S. H., Haldani, A., and Adhitama, G. P. (2019). Jakarta Society Preference Study on Genderless Fashion, Journal Rupa, 4(1).

Raniwala, P (2020). India's long history with genderless clothing, Mint Lounge.

Reis, B., Pereira, M., Azevedo, S., Jerónimo, N., and Miguel, R. (2018). Genderless clothing issues in fashion, Textiles, Identity and Innovation: Design the Future, 1.

Malek, M. S., and Gundaliya, P. J. (2020). Negative factors in implementing public–private partnership in Indian road projects. International Journal of Construction Management, DOI:10.1080/15623599.2020.1857672.

Borikar, H., Bhatt, V., and Vora, H. (2022). Investigating the Mediating Role of Perceived Culture, Role Ambiguity, And Workload on Workplace Stress With Moderating Role of Education in a Financial Services Organization. Journal of Positive School of Psychology, ISSN, 9233-9246.

Malek, MohammedShakil S., and Bhatt, Viral (2022). Examine the comparison of CSFs for public and private sector's stakeholders: a SEM approach towards PPP in Indian road sector, International Journal of Construction Management, DOI: 10.1080/15623599.2022.2049490

Denial of Human Right to Water During Pandemic: Experience of Indian Slum

C. S. Arya,[a] and Nagendra Ambedhkar[b]*

[a]Research Scholar, Department of Public Policy, Law and Governance, Central University of Rajasthan, India

[b]Sole Professor, Department of Public Policy, Law and Governance, Central University of Rajasthan, India

E-mail: *aryacs1000@curaj.ac.in

Abstract

The spread of the pandemic and related control measures such as Lockdowns affected the lives of vulnerable people tremendously. The augmented number of Covid cases in the slum areas of India is caused mainly by a lack of good water and sanitation and social distancing. Water right has a direct relationship with the health and hygiene of slum people. The Covid scenario shows that everyone has no equal right to access water and sanitation. The gender, class, and caste inequality on behalf of water services makes them more vulnerable to viruses. This paper explains how the violation of water rights affects the pandemic more severely in the extremely weaker sections of society. This paper explores the importance of the human right to water, particularly during Covid-19. This paper is based on an analytical survey of literature mainly focusing on secondary sources such as books, journal articles, reports, and online databases.

Keywords: Inequality, Human right of water, Slum, Right to life

Introduction

The Covid 19 pandemic is a new dilemma for global society. The pandemic can delay the accomplishment of Sustainable Development Goals (SDG) by the lack of increasing access to clean drinking water and sanitation (SDG goal-17). Even before Covid 19, the diseases connected to the quality and availability of water and hygienic environment threaded the world's mortality. Data shows that 1.6 million people die due to diarrhea in a year (three people per minute), and 3 million die yearly due to respiratory diseases (6 people per minute). The risk of viral infections such as MERS (2012), SARS (2013), and Coronavirus increases because of a lack of personal hygiene practices and social distancing (Staddon et al., 2020). The universal accessibility of safe and sufficient water with improved sanitation is necessary for health, hygiene, basic needs, the human right to life, and a preventive block for fatal diseases like Covid 19 (Braun). World Health Organization strongly recommends that personal hygiene practices (regular hand washing and bathing) can reduce the spread of Coronavirus (WHO and UNICEF, 2020). The hand washing practices should follow 8-10 times daily for at least 20 seconds. The estimates show that the total amount of running water required for hand washing per person daily is eight to ten liters (Staddon et al., 2020). The lack of adequate availability, quality, affordable, and accessibility of water can increase the terrific effect of the pandemic. The United Nations

"Decade of Water for Life 2005–2015" Report shows that 1.8 billion people will live in water-stressed regions, and two-thirds of people will suffer water scarcity by 2025 (Maestu, 2016). One in three people does not have access to clean drinking water, and fifty percent of people lack sanitation facilities. Promoting "water, sanitation, and hygiene" (WASH) practices among people can mitigate the tremendous effect of the diseases (Donde et al., 2021). The pandemic control measures like self–isolation, avoidance of public places, and self-hygiene are not equally applicable in low-income countries' densely populated urban-rural areas. The joint pipe point, share latrines, and washrooms challenge WASH hygiene practices (Donde et al., 2021).

The right to sanitation guarantees safe, secure, hygienic, affordable, socially and culturally acceptable, and provides privacy and dignity without discrimination. Both rights to water and sanitation affirm the right to an adequate standard of living (UN, 2016). The global concern of denial of the "human right to water, worsens immensely during pandemic situations (Tortajada and Biswas, 2020). The lack of human rights to water is more visible in experiences of slum areas. The informal setting occupied sixty percent of Asia. 17 percent of urban household in India is constituted by slum. The slum has inadequate access to water, overcrowding, poor structure of houses, insecure residence, and poor sanitation facilities. Slum people features like inequality, rapid immigration, poor governance, no job in the formal sector, and exclusionary actions (UN-HABITAT, 2003). Data indicates that the slum people (Example: Dharavi) have severe vulnerability towards Covid 19 Virus (Patranabis et al., 2020).

Vulnerability is a dynamic concept; it can differ from person or group depending on diverse backgrounds. The epidemiological vulnerability (old age, low immunity) and the vulnerability due to socio-economic and political status makes slum various financial, mental, and physical crises during the pandemic lockdown. The most vulnerable community, like the slum, has long-term effects of pandemics compared to other communities (Acharya and Porwal, 2020). Restrictions such as lockdown self-isolation from the government put slum people under tremendous psychological stress. The loss of their social life, daily wages, water, and food increases the stress and anxiety of slum people (Sanganal, 2020).

This paper discusses the experience of slum people during pandemics and Lockdown, focusing on the lack of adequate safe water and sanitation. The paper qualitatively explains how the lack of water availability intensified more struggles in their life during the pandemic.

Methods

Study Setting

This study was conducted in slum areas of Ajmer district in Rajasthan state. Rajasthan is a water-deficient and recurring drought state in India. Rajasthan has semi-arid geography with minimal surface water resources, inconsistent rainfall, drought-prone climate, and metallurgical elements in water, which is why water scarcity (Chakraborty, 2018). There are 41000 villages in the state facing severe drought problems. It affects the life of people and manifests the situation as unavailability of basic needs, safe water, and nutrition. Many poor and landless people face the risk and vulnerability of drought (Drew et al., 2016). Ajmer city is situated 130 km away from the state capital Jaipur. The city has a low density of people with a highly populated inner core. Ajmer city has sixty-seven notified slums, most of which are in the city center. Forty-three are non-notified informal settlements. The Rajasthan State Slum Policy defines a slum in the definition of Dr. Pronab Sen's Committee Report on Slum Statistics, "a slum is a compact settlement of at least 20 households with a collection of poorly built tenements, mostly of temporary nature, crowded together usually with inadequate sanitary and drinking water facilities in unhygienic conditions." The Ajmer slum faces serious issues regarding water, health, and education.

Research Design

The research followed a qualitative study approach. Qualitative data describes and interferes with the characteristics of people, groups, and phenomena in a narrative and descriptive manner.

The qualitative study goes through different stages, such as organizing, explaining, presenting, and analyzing data procedurally. The study could reveal the experience of urban slums during Covid 19 regarding the lack of essential services like water, food, and sanitation. The study qualitatively analyzes the water problems in slum areas and how their hygiene is associated with a water-related pandemic.

The sample is collected conveniently in the slum areas of Ajmer city. A total sample of fifty people was selected from the slum area. Most people work in the informal sector, such as rickshaw drivers, artisans, construction workers, and rag pickers. Ninety percent of people are illiterate, and the remaining 10 percent have primary education. Their income is approximately less than 2000 per month. The socio-demographic profile of people is shown in Table 1.

Table 1. Socio-demographic profile

Characteristics	Number of the Respondent (N= 50)	Percentage
Age	15–30	22%
	31–45	60%
	46–60	18%
Gender	Male	40%
	Female	60%
Education	No Education	90%
	Primary	10%
	Secondary	0%
Occupation	No job	8%
	Informal sector	92%
	Formal sector	0%
Income	Below 2000	88%
	2000–4000	12%
	4000–6000	0%
Socio- economic status	APL	4%
	BPL	96%

Data Analysis

In-depth semi-structured interviews with respondents collect the data. Informed consent was taken from participants with a detailed description of the aim and objective of the study. The photos and audio were recorded with the respondent's permission. The face-to-face in-depth interview per person takes a minimum of 25 minutes to a maximum of 45 minutes, according to the convince of respondents. All interviews were transcribed verbatim in Hindi and translated into English. The systematic analysis of data gave significant themes and sub-themes.

Findings

After analyzing data related to slum life in the absence of water and hygiene practices during Coronavirus, the four major themes are identified. The major themes are water scarcity, awareness, human rights, and challenges during the pandemic. Each central theme is divided into sub-themes. The water scarcity theme includes water availability, water accessibility, water quality, and affordability. The awareness theme consists of sub-themes of awareness about water-related diseases, Covid- 19, about the human right to water. The human rights theme is divided into the right to safe water, the right to sanitation, and the right to basic needs. The challenges in the lockdown theme include social support and socio-economic vulnerability. The themes were analyzed using primary data from the field.

Water Scarcity

The slum area is affected by severe water scarcity issues. The area has a limited number of water supply pipes, and four or five families use one supply pipe. Families far away from pipe points must walk with buckets and canes to collect water. Mostly women come coming to the pipe points to take water. The water comes from pipes two or three times a week. The timing of water flow is one hour or one and a half hours at a time. In this limited period, those who can collect maximum water can collect. The water availability and accessibility vary from the person who is more capable of collecting water from a joint pipe. The people's lifestyle is also connected to water availability; they wake up at 3 am to collect water, and that water is not enough for all family members.

We need to wake up early, 3 am, or 4 am, to collect water from pipes. We will get information one day before from the local representatives. That night, we can't sleep. We will think the whole night about tomorrow morning and water collection. If we do not wake up at 3 am, we will not get water.

(Statement by a 32-year-old woman)

The water scarcity intensity is more visible in the words of young women with period pain,

For washing the period's cloth, we need more water than normal. Some days, the blood smell may not be completely removed from the clothes. If I use more water for cleaning period's cloth, my mother will scold me for wastage of water.

During the pandemic, water usage increased for personal hygiene such as hand washing, bathing, and washing clothes. Nevertheless, the availability and accessibility of water are not increased in slum areas. Precautionary practices such as frequent hand wash and personal hygiene are impossible in the slum. Daily bathing is a myth for them. Most people bathe two or three times a week.

If we get water after four or five days, How will we bathe twice a day? Which day, we are getting the water that day I will bath, remaining days we will not take a bath if Corona or any disease we do not bother.

(Statement by 58-year-old man)

The statement of another person is that,

The water is coming in pipe one hour within five days, and you tell us, how we will bath two times during covid. If the government tells us to hand washing regularly, they have to give water to us. Without giving water, how they can tell us about Corona and hygiene.

(Statement by 35-year man)

Table 2. The primary observation of slum people's water availability

Water and Sanitation	Observations
Number of pipes	No household pipe connection Standard pipe connection for 3 or 4 families One-hand pipe and water is limited
Water availability	Water is available 2 or 3 times a week for one hour
Water storage	No individual tank Canes and Utinels are used for the storage of water
Number of times bathing	A limited number of people bath daily, the majority bath 2 or 3 times a week
Amount of water for bathing	Approximately 10 liters for three children and 10 liters for two person
Availability of soap and sanitizer	Most people use soap for body wash Limited people used sanitizer during the first wave of Covid Some people use oil instead of soap still in the time of covid
Toilet facility	No household toilet Majority doing open defecation
Waste management	No waste management facility is available No particular channel for wastewater

Awareness

The essential awareness of water-borne diseases such as Diarrhea and jaundice is unfamiliar to most of them. Disease prevention methods are unaware of the urban slum because of a lack of education. The limited awareness about the importance of hand washing and social distancing during Corona was understood by Anganwadi workers and NGOs. However, they could follow partially because of a lack of resources and education.

We used mask during Lockdown. Nevertheless, I did not use sanitizer and frequent hand wash. Health workers came and gave us some instructions. But they do not know we are following or not.

Using sanitizer and soap is not a custom of slum people during covid. The frequent use of sanitizer and hand washing is not possible in the limited supply of water. Most people are not familiar with sanitizer and its usage.

> *We did not use sanitizer during Corona. We sometimes used soap for hand washing. But usually we are using Oil for bathing.*
>
> (Statement by a young 15-year-old girl)

The people do not know about the Right to Water as a human right. They do not know the meaning of the word "Human right." After explaining about Right to water concept to them, they are asking,

> *If it is right, then why are we not getting enough water.*

Human Right

Water is the basis of life. The human right to water is related to people's right to life. The slum people lack the fundamental right to water, food, health, and life. The city regions are getting regular water supply from Ajmer municipal authority. However, this disparity is visible in water availability in slum areas.

> *Water is not an easy thing for us. We need to walk this much distance to take water. It is tough for us going and collecting water.*
>
> (Statement by 34 aged women)

An older adult showed her concern about water.

> *We are not getting enough water; then how can our children get things in their life?*

The toilet facility is not available in the slum area. Most people use open defecation. Some people are using standard toilets. However, the water availability and drainage system are not adequately managed.

> *We do not have toilets in our home. We will go outskirts with water. Taking water in buckets and going to the outskirts is also difficult.*

There is no facility for waste management in their homes. The wastewater flows through near houses. The people are bathing and washing near the wastewater channel, which makes their environment unhealthy.

> *We do not have a washroom in our homes. Women bath inside the home and boys bath near to wastewater channel*
>
> (statement by a 37-year-old woman)

Social Support

Social support from the government and other NGOs is significant for their survival during a lockdown. They were getting food items and grains three or four times during the lockdown. However, the people did not know about the authority giving benefits, and they pointed out that food grains were not enough for them. Food items are limited in quantity. Those who are reaching their first will get benefits, and others are not. The social support during the lockdown was not equally distributed.

> *We do not know who coming and giving food is. Whenever we were reaching, It will be finished.*
>
> (Statement by 47 old men)

The lockdown and related regulations gave more stress to slum people. The stress on income, job, food, and water made more dilemmas in their life. Most people didn't recover the impact of Lockdown in their life. The absence of a job and the non-availability of food and water made their life more stressful. An old age woman stated this:

> *We didn't have fear on Corona, we only have fear on Lockdown. We can even get food and water during Lockdown; We can't go for job also. If corona comes again, we have no problem. But if Lockdown comes again we will suffer.*

Table 3. The theme-based findings

Major Themes	Sub-themes	Major Findings
Water scarcity	Water availability Water accessibility Water quality Affordability	Water availability is limited in slum areas. They are getting water 2 or 3 times a week. The lack of water availability is the major challenge for hand washing and personal hygiene during Covid 19.
Awareness	About water-related diseases About Covid 19 About human right of water	Most people are illiterate. Without proper education and awareness about water-related diseases and Corona makes their life difficult in water scare situations. The people don't know the importance of the human rights aspect of water. The lack of awareness about human rights is also a tragedy in their life.
Human right	Right to safe water Right to sanitation Right to basic needs	The right to safe water, basic needs (food, education, health), and sanitation is not guaranteed to slum people. This makes situations such as inequality, poverty, and social exclusion. This makes them more vulnerable and stressed during Lockdown.
Impact of Covid and Lockdown	Social support Socio-economic vulnerability	The government restriction gave more stress to slum people. Most people fear the lockdown situation. Lack of income, food, and water makes the situation more painful for them.

Discussion

The slum people are socio-economically vulnerable to Covid 19. The constructed water scarcity makes them more severe life during and after the pandemic outbreak and related Lockdown. The slum people affect water scarcity entirely in their life. When a particular section of society faces water insecurity for a significant period, it becomes water scarcity. The constructed scarcity has a complex connection with the socio-political and economic context of people. It doesn't affect the entire society, but specifically weaker sections (slums). These areas are affected by the unequal and inappropriate distribution of water and it is depending on social influences (Sasidevan and Santamore, 2018).

The constructed water scarcity is directly connected with the human right to water. Sufficient quality and quantity of water are essential for human health, the standard of living, preventing dehydration, reducing water-related diseases, and hygiene practices. The international convention and agreement on water explicitly show the connection human right water to and the right to food, health, and human development. The lack of basic services and the lack of access to water and sanitation is the major of slum. It can affect life in slum in different ways.

The pandemic and followed lockdown period made a tremendous dilemma in their life. The slum people depend on common water taps or pipes within the primes. The water is coming in two-three times a week, and the group of people needs to wait near to tap. This scenario is a space for the spreading of Covid-19 among slums. The slum people are socially excluded and illiterate. The unawareness about water-related diseases and difficulty of maintaining prevention methods of Corona (isolation, mask, sanitizing, hand washing, and personal hygiene) is against the government control measures of the Pandemic. Open defecation, usage of common toilets, and no proper waste management are also against the control measures of Covid 19. These findings are consistent with the study of Das et al. (2021).

The irregular informal setting, unhealthy living conditions, social exclusion, poverty, inadequate structures, high density is the basis of the vulnerability of slum (UN-HABITAT, 2003). People might not be vulnerable in the initial phases of the Corona pandemic, but could afterward become vulnerable depending on government response (Acharya and Porwal, 2020). The period of Lockdown made dark shadows on the daily life of slum people. All of them are working in informal sectors such as rickshaw drivers, artisans, construction workers, and rag pickers. Lockdown affected their daily income and they didn't get access to basic needs like food, water, and medicines. Poverty was the main challenge during Lockdown. The children (girl children faced more) in slum areas faced challenges such as lack of health, menstrual hygiene, nutrition, lack of education (online is not practical) during Lockdown. This result is similar to the study that happened in slum areas of Delhi, Maharashtra, Bihar, and Telangana (The New Indian Express, 2022).

The old age people, handicapped, and women faced the challenge of safe water and sanitation. The denial of basic human rights features (water, food, health, and sanitation) is not guaranteed in slums. The integration of pandemic situation with inequality, poverty, and social exclusion can make urban slum life a tragedy during Covid 19.

Conclusion

The spread of the pandemic and related control measures such as Lockdowns affected the lives of vulnerable people tremendously. The augmented number of Covid cases in the slum areas of India is caused mainly by a lack of good water and sanitation and social distancing. Water right has a direct relationship with the health and hygiene of slum people

References

Acharya, R., and Porwal, A. (2020). A vulnerability index for the management of and response to the COVID-19 epidemic in India: an ecological study. *Lancet Glob Health*, *8*.

Braun, Y. A. The Human Right to Water and Sanitation in the Age of Covid-19.

Chakraborty, V. (2018, March 21). Water Crisis in Maharashtra and Rajasthan. Retrieved August 1, 2022, from Maps of India: https://www.mapsofindia.com/my-india/india/water-crisis-in-maharashtra-and-rajasthan

Das, M., Das, A., Giri, B., Sarkar, R., and Saha, S. (2021). Habitat vulnerability in slum areas of India – What we learnt from COVID-19? International Journal of Disaster Risk Reduction, 65, 102553.

Donde, O. O., Atoni, E., Muia, A. W., and Yillia, P. T. (2021). COVID-19 pandemic: Water, sanitation and hygiene (WASH) as a critical control measure remains a major challenge in low-income countries. Water Research, 191, 116793.

Drew, G., Jalees, K., and Pandey, P. (2016, April 18). Water Crisis in Rajasthan. Retrieved Aug 1, 2022, from India Water Portal: https://hindi.indiawaterportal.org/content/water-crisis-rajasthan/content-type-page/53102

Maestu, J. (2016). A 10 Year Story: The Water For Life Decade 2005–2015 and Beyond. New York: UN-Water Decade Programme.

Patranabis, S., Gandhi, S., and Tandel, V. (2020, April 16). Are slums more vulnerable to the COVID-19 pandemic: Evidence from Mumbai. Retrieved August 2, 2022, from Brookings: https://www.brookings.edu/blog/up-front/2020/04/16/are-slums-more-vulnerable-to-the-covid-19-pandemic-evidence-from-mumbai/

Sanganal, A. (2020, May 26). Slums are congested, vulnerable spaces. Retrieved August 1, 2022, from Decan Herald: https://www.deccanherald.com/opinion/panorama/slums-are-congested-vulnerable-spaces-841799.html

Sasidevan, D. E., and Santamore, S. D. (2018). The social construction of water scarcity: An exploratory study along the "Bharathapuzha" in Kerala. Glocalism: Journal of Culture, Politics and Innovation.

Dynamics of the Demographic Transition on Economic Development: Evidence From SRS Data in India

Ankita. Srivastava,[*,a] *A. K. Tiwari,*[b] *and Vastoshpati Shashtri*[c]

[a]UWSB, Karnavati University, Gandhinagar, India
[b]Department of Statistics, Institute of Science, Banaras Hindu University
[c]Department of Statistics, Government Science P.G. College, Ratlam (M.P.)
E-mail: [*]ankita.srivastava2007@gmail.com

Abstract

Countries across the globe are passing through the phase of demographic transition due to continuous changes in the global environment and lifestyle. This transition significantly affects the progress of any economy. The association between demographic variation and economic development still is a matter of debate amongst economists and demographers. This study has attempted to explore the long-run association ship in demographics and economic development for India for the period 1980 to 2020, using the regression analysis technique. The analysis found that the total fertility rate and natural growth rate are significantly associated with Gross Domestic Product (GDP) per capita but the infant mortality rate is not. Moreover, infant mortality and natural growth rate in the population have negative coefficients suggesting an inverse relationship between the said variables and GDP per capita growth rate.

Keywords: Fertility, Infant mortality rate, GDP

Introduction

Rapid economic growth has been recorded by India in the last few years despite being the world's second most populous country. This has been majorly achieved due to the high availability of labor in India. Labor intensiveness has made India globally competitive in several low-cost, labor-intensive productions. Indian population has undergone a significant transition in the last few decades. As per the American Demographer Warren Thompson (1929), the demographic transition was confined to a decline in birth and death rate but this word has evolved over time and now conveys more than one meaning. Demographic transitions in a country provide a beneficial condition for its development (Bloom et al., 2001).

India at the national level is in the phase of fertility transition marking a fertility level of 2.1 children per female in 2020 (office of RGI, 2020). The nation right now has the advantage of having more of a youth population since 2005–06 and expects to have it till 2055. In the census of 2020, the average age of Indians was 29 whereas in countries like the USA, Europe, and Japan ranges from 40-47 years.[1] Researches viz. Aiyar and Mody, 2011 projected India as a hub of young people which is believed to have a positive impact on the economy.

There are five stages in demographic transition i. significant drop-in death rate and no substantial decline in crude birth rate (CBR) ii. Natural

1 National Policy for Skill Development and Entrepreneurship Report, 2015

growth rate increases iii. CBR begins to decline significantly iv. The birth rate begins to decline faster than the death rate and the natural growth rate begins to decline. v. Finally, in this step natural growth rate becomes 0.

A large number of researches have been done to analyze the relationship between demographic and economic variables. The neoclassical growth model by Solow (1956), suggests that an increase in the population of a country decreases economic development as there is capital dilution. The study by Yamada (1985) on developed and underdeveloped nations found that a rise in per capita real income decreases infant mortality.

Literature Review and Conceptual Background

Chesnais (1992) considered around three to five stages of the transition of the human population. Similarly, Nolan (2014) describes four stages of this transition as follows:

1. The first stage is before an industrial society which is characterized by more birth and death rates. It is believed that this equality in birth and death rate balances each other which results in a low and stable population.
2. The second stage is marked by a sharp decrease in death rates. This can be attributed to increased sanitation, lessened diseases (better healthcare facilities), and a steady food supply due to improvements in agricultural practices. Throughout this stage, the death rate reduces keeping the birth rate high which creates a significant increase in population.
3. The third stage focuses on the reduction in the birth rate due to an increase in education, especially for women, contraception is accepted, and people are making more money. The value system changes in this stage as the next generations receive education now rather than doing work. Many families migrate to urban areas from rural areas where they go for fewer children in contrast to rural living environments.
4. The fourth stage has low birth and death rates. At times, lower birth than death rates is found because families having multiple children may give way to couples that may choose not to have any children. This may cause a shrinkage in population that causes a shift in the economic load of the country on a lesser number of people and potentially threaten an economy that is based mainly on growth for stability.

Notestein (1945) suggested three phases and Becker (1949) two phases. It is believed that most of the advanced countries lie in the fourth stage whereas developing ones are either in stage two or three. Black, (2002) reported that all countries across the globe have more or less passed the stages of this transition in demography. Hondroyiannis and Papapetrou (2003) studied the association between choice of fertility, death rates of infants, choice of getting married, and real wages and output per capita. The study found a decline in death rate of infants in a long run considering the economy and labor will reduce the fertility rate. Climent and Meneu (2003) examined the relation between demography and variables measuring the economy of Spain and found that total fertility may react against the blows of GDP Further the study concluded that higher wage rates may lead to a negative impact on fertility as well as GDP and no causal association exists between infants death rate and fertility.

Choudhry and Elhorst (2010) extended Solow–Swan model by including demographic variables and analyzed the influence of shifts in demography on the economic condition of the country. The study using the data for 1961-2003 from seventy countries found GDP per-capita growing rate associates positively with differential growth rate amid working age population & overall population while has a negative association with child and aged dependence ratios.

The researchers developed a theoretical framework and analyzed it to measure changes in demography. The analysis found that an increase in the demand for human resources for development contributed to the decrease in fertility and transition to current growth.

The developed economies are already facing a serious challenge in the terms of paying money as retirement income and the medicare cost for the large elderly population. The innovative medical technologies which are extremely costly and save a relatively less number of patients have made their transition more complex. The countries are bound to give thought to what amount of community capital should be used to protect the lives of many senior people.

The studies in past employed the ARDL method (for the long run) and ECM (for the short run) for the period 1974-2011 to understand the influence of changes in demography on economic growth. The growth of working-age, life expectations, and total literateness are contributing significantly toward economic development. The study suggests that demographic transition affects the economy in long term but in the short run, a negative association was found among all variables of the study except literacy rate.

As per the analysis of Jakovljevic (2015), the history of developed countries found that the rate of transition is much faster in developing economies as compared to developed nations. The statistics suggested that it took almost 115 years for France to double the fraction of the elder population i.e. from 7% to 14% but Brazil took only 21 years. It is expected that these developing nations may face larger bills for retirement benefits for the aged population.

Saidi and Zaidi (2019), measured the effect of growth in population on the economy and environment using panel data from ten countries between 1982 and 2013. The results suggest that progression in population has a positive association with economic growth. Jain and Goli (2021), evaluated the benefits of changes in demography on the economy through panel data of thirty years.

A little work in this area is found in India. In this paper an attempt has been made to explore the association ship amid the transition in demography and economic progress in the context of India for the period 1980 to 2020, using the time series technique. This study is imperative as the generations and facilities have changed significantly.

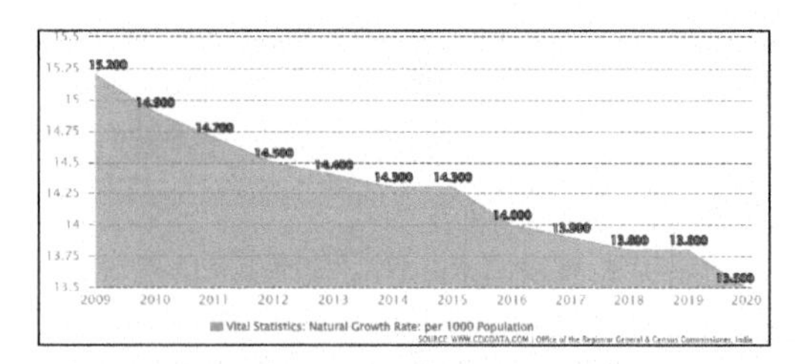

Fig. 1. Demographics and Economic Development of India, Total Fertility Rate (TFR)

India with a population of 1.39 billion people[2] consists of around 17.7% of the world's population. TFR explains the total number of children a woman can have. In 1950 India's TFR was around 6. Since the first five years plan India has tried to control its population growth. In 1983, the goal of the country was to have a replacement level overall fertility rate of 2.1 by 2000. According to National Family Health Survey (NFHS)-5 TFR of India presently is 2.0. Researchers in the past have found mixed results for this relation like Dominiak, Lechman, and Okonowicz (2015) found u-shaped relation between total fertility & GDP per capita.

Infant Mortality Rate (IMR)

It explains the likelihood of death before the first birthday. As per the latest Sample Registration Survey data, IMR is 35 in 2019 (NFHS-5). It has seen a significant improvement but a lot still has to be done. Past studies have found a negative association between IMR and GDP per capita.

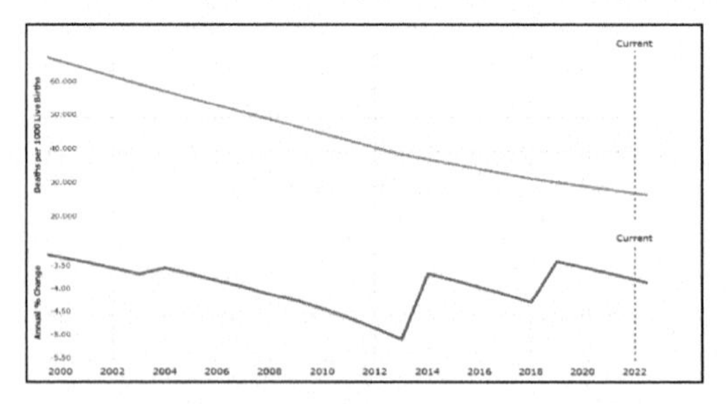

Source: https://www.macrotrends.net/countries/IND/india/infant-mortality-rate

2 Census (2020).

Natural Growth in Population (NGR)

In demography, natural population growth is the difference between the basic birth rate and death rate. If migration is ignored, the positive value of this difference suggests an increase in the population, and a negative one means that the population is declining. As per the SRS data, NGR of the population was 14.9 in 2010. A rapid fall in the death rate from 25.1 in 1951 to 7.2 in 2010 suggests technological advancement and improvement in health- care facilities. The birth rate has also declined from 40.8 in 1951 to 22.1 in 2010. From 1960 to 2021 the population of India increased by **209.3 percent.** The highest and lowest increase in India was recorded in 1974 (2.36%) and 2021 (0.97%) respectively.[3] As per the study of Thuku, Paul, and Almadi (2013) there is a positive association between population growth and economic development whereas Chang, Chu, Deale, and Gupta (2014) found causality between the two depends on the time period under consideration.

Gross Domestic Product Per Capita

In this study GDP per capita growth rate has been used to measure the economy. GDP per capita is the total gross value by all producers in the economy along with product taxes (minus subsidies) not included in the valuation of output and divided by midyear population.

Data and Methodology

Accurate data on vital indices is necessary for monitoring the ongoing demographic transition. The Office of Registrar General established a sample registration system (SRS) in 1964-65 on a pilot basis. In India, the SRS was developed in order to provide annual and state-specific data on vital rates. Data from SRS and Census are used for assessing ongoing demographic transition in the country, but census provides data once in ten years.

The problem of stationarity has been tested by employing the Augmented-Dickey-Fuller (ADF) test which hypothesizes that, a series is non-stationary and if the null hypothesis is rejected it implies stationary characteristics of the data series.

Variables of Study

Demographic: Total Fertility Rate, Infant Mortality Rate, and Natural Growth in Population

Economic Variables: Gross Domestic Product Per capita growth rate

Study Period: The period of study is 1980-2020; about forty years.

Methodology: The study first applied a unit root test through ADF statistics to check the stationary of the data as it is a time series data. As a thumb rule if data is stationary at level the ordinary least square (OLS) method of regression can be applied directly whereas if all the variables are stationary at first difference form we should proceed with co-integration analysis which further may lead to VAR or VECM. If the variables are a mix of stationary at level or first difference ARDL is applied.

Data Analysis and Results:

The dependent variable i.e. GDP per capita growth rate and the independent variables total fertility rate, infant mortality rate, and natural population growth have been checked stationary. The data of all variables were found to be stationary at level form, the model becomes:

$$Y_{(t)} = b1\ X1_{(t)} + b2\ X2_{(t)} + b3\ X3_{(t)}$$

$$GDP_{(t)} = a + b1\ \text{natural growth}_{(t)} + b2IMR_{(t)} + b3\ TFR_{(t)}$$

The results of the stationary check have been presented (table 1). As mentioned earlier, the series is stationary at level, so, we can proceed with the ordinary least square method (OLS) of regression to measure the coefficients.

3 https://www.worlddata.info/asia/india/populationgrowth.php

Table 1. Stationary Test for Variables

Variable	ADF Value at level	Prob.
GDPPCGR	-6.056174	0.0000
NGR	-5.646576	0.0000
IMR	−9.503405	0.0000
TFR	−6.708032	0.0000

Source: Author's own Calculation

Table 2. Ordinary Least Square

DV: GDP				
Variable	Coeff.	SE.	t-Statistic	Prob.
TFR	17.57927	6.647039	2.644677	0.0119
IMR	-0.052406	0.148676	-0.352486	0.7265
NGR	-3.956113	1.238184	-3.195094	0.0029
C	6.211384	0.587605	10.57068	0.0000
R^2	0.333957	F-statistic		5.636585
Adj. R^2	0.279954	Prob(F-statistic)		6.183996
D-W stat	1.967419	Prob(F-statistic)		0.001630

Source: Author's own Calculation

In Table 2, the total fertility rate and natural growth rate with the coefficient values of 17.57927 and −3.195094 are significant at a one percent level of significance. The estimation of coefficients through the ordinary least square method found that the natural growth rate and total fertility rate are explaining the gross domestic product per capita of a country but infant mortality is not contributing much to explain the income of the country as its coefficient turned out to be insignificant. However, the coefficient of the natural growth rate is negative which suggests that with an increase in population, there is a decline in the per capita income of the country. The value of Durbin Watson (DW) statistics is near 2 which suggests the absence of autocorrelation and the F statistics of 6.183996 is significant suggesting the model to be a good one. The R square value is 33.3957 percent which is less but Gujarati and Porter (2009. 5th ed) suggest that if the model is free from autocorrelation and the F statistic is significant, we can accept a low R square value.

The final equation becomes:

GDP = 6.211384 - 3.956113 NGR - 0.052406 IMR + 17.57927 TFR

Conclusions

As the analysis has been done for a longer duration i.e. forty years we can say the natural growth rate and total fertility rate have an impact on the income per capita of a country. These results are in contrast with that of Thornton (2001), that observed no long-run and causal relationship between population growth and per capita income. Per capita GDP association with population growth is significant and negative meaning that there occurs an increase in the per capita GDP actually with lower growth in the population growth of a country. Similarly, though the infant mortality rate turned out to be insignificant with a coefficient of -0.052, this negative sign again indicates an inverse relationship between the infant mortality rate and the per capita income of the country. It means with an increase in infant mortality rate there is a decline in GDP per capita. The total fertility rate turned out to be significant and positive suggesting that an increase in the fertility rate of a country is favored as it increases the GDP per capita. However, now a shift can be noticed in this trend as the education of females nowadays has created awareness among them which has led to producing less number of children per female. Female literateness also mediates the effect of income on child mortality.

References

Bloom, David E. (2001). Canning, David and Sevilla, Jaypee (2001). Economic Growth and the Demographic Transition NBER Working Paper No. 8685 December.

Aiyar, S., and Modi, A. (2011). The Demographic Dividend: Evidence from Indian States. IMF Working Paper 11/38. Retrieved from https://www.imf.org/external/pubs/ft/wp/2011/wp1138.pdf

Solow, R. M. (1956). Source: The Quarterly Journal of Economics, 70(1) (February 1956), pp. 65–94.

Yamada, T. (1985). Causal Relationships between Infant Mortality and Fertility in Developed and Less Developed Countries. Southern Economic Journal 52(2), 364–370.

Chesnais, C. J. (1992). The Demographic Transition: Stages, Patterns and Economic Implications. (Clarendon Press, Oxford)

Notestein, F. W. (1945). Population — The Long View. In: Schultz, Theodore W., Ed., Food for the World. Chicago: University of Chicago Press.

Hondroniyiannis, G., and Papapetrou, E. (2004). Demographic Changes and Economic Activity in Greece. Review of Economics of the Household, 2(1), 49–71

Climent, F., and Meneu, R. G., (2004). Demography and Economic Growth in Spain: A Time Series Analysis. SSRN-Electronic-Journal. http://dx.doi.org/10.2139/ssrn.482222

Choudhry, Misbah T., and Elhorst, J. Paul (2010). Demographic transition and economic growth in China, India and Pakistan Economic System, 34(3), 218–236.

Jakovljevic M. B. BRICS (2015) Growing Share of Global Health Spending and Their Diverging Pathways. Front Public Health 3, 135. doi: 10.3389/fpubh.2015.00135

From Function to Fashion, Face Masks as a Flourishing New Product

Mahima Chandnani[a] *and Sambit Kumar Pradhan*[b]

[a]Unitedworld Institute of Design, Karnavati University, Gandhinagar, India
[b]Unitedworld Institute of Design, Karnavati University, Gandhinagar, India
E-mail: sambitkumar@karnavatiuniversity.edu.in

Abstract

Face masks are considered one of the prime precautionary measures to fight the COVID-19 pandemic as stated by WHO. Masks were not being used for the first time for protection during a pandemic, they did exist for decades, but what made them convert into a present-day attention-seeking product is what these paper talks about— from preventing the spread of COVID-19 to turning into a fashion statement and further. This paper includes various concepts and studies/statements made by influential people in society including designers, business consultants, and medical councils. The main aim of this paper is to relate the visual aspects of a product with its usability. Brand identity plays a vital role in communication and also the expression that visuals depict. Masks as a tool for branding are used to understand how visuals play and change according to needs and situations, and so does a product's evolution.

Keywords: Face Masks, COVID-19, Function to Fashion

Introduction

First identified in the city of Wuhan, China, in December 2019; COVID-19 (coronavirus disease) is now an ongoing pandemic. The spread of COVID led the World Health Organization (WHO) to set precautionary measures to reduce the chances of being infected or spreading the virus any further. One of these primary measures was to cover the mouth and nose since the virus predominantly spreads via respiratory droplets sprayed into the air as an infected person coughs, sneezes, or talks.

To go ahead with the above measure, surgical masks were initially used, as these were traditionally designed to prevent infections. However, as the outbreak progressed, it caused a shortage of surgical masks., and even though new surgical masks (like N95 and FFP) were made to trap smaller-sized virus particles like coronavirus, the health workers that were on the front line were supplied with such masks instead of the rest.

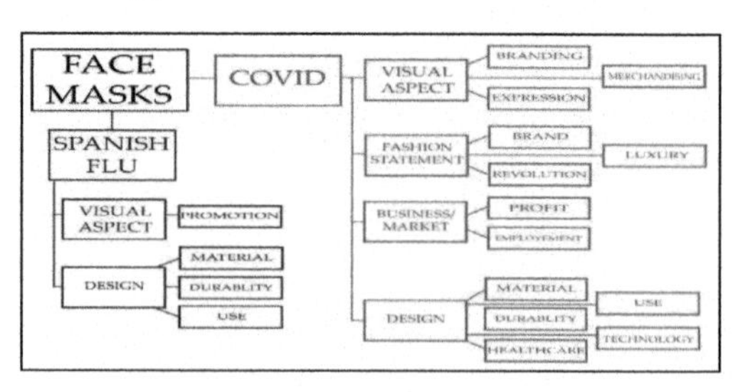

Fig. 1. Mind map of the studies covered up in the paper

While it may have started with healthcare workers, the general populace were forced to wear masks in public, based on the fact that infections spread outdoors could be prevented by everyone wearing masks. Eventually, the CDC approved the use of cloth-based coverings to help slow the spread of the virus. This is where the main twist comes in.

Literature Review

We all are a part of the world where HAI (Human-Animal Interaction) is obvious and it keeps on evolving. Humans tend to adapt to their surroundings to grow, and so do animals. When we look back to the times when technology and medicine didn't exist, humans still managed to survive. Pandemics have been a part of human history since the very beginning. According to Miller (2020), the world has faced the deadliest pandemics for decades, namely, The Great Plague of London, Spanish Flu, Asian Flu, 1968 Flu Pandemic, 2009 Flu Pandemic. All these pandemics have not only caused the deaths of millions of people but also other damage like loss of economy, poverty, and unemployment, as well as social and political effects and at the same time brought in a few aspects which have continued to stay with us. A major aspect that comes with a pandemic is *Acceptability;* how readily we accept the current scenario and the precautionary measures related to the same.

To understand this better, considering history is a good idea. Spanish Flu was a deadly influenza pandemic with 500 million infected victims. Spanish Flu symptoms were the same as that of COVID; chills, fever, and fatigue. The flu was first observed in Europe and spread to other parts of the world like the US and Asia. Citizens were ordered to stay indoors, wear masks, schools, cinemas, and offices were shut. Masks were a great concern then as well. These masks saw a different phase during the Spanish flu. Masks were asked to be worn so that we could prevent ourselves from getting infected. More than that, masks were to be worn so that the infected don't spread the virus. But acceptance of wearing these masks was a huge task then. These masks were a discomfort and still are today, and the awareness amongst people to wear them was spread through visual aids such as posters and advertisements, unlike the present-day masks that are being used as a tool for branding and business. (History.com Editors 2020; Little 2020; FF Bureau 2020)

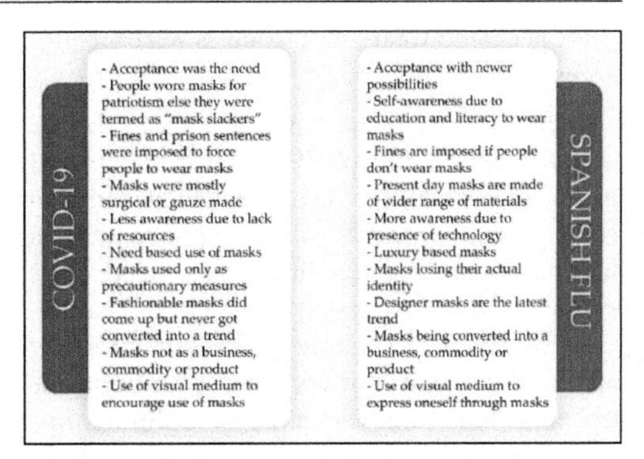

Fig. 2. Comparison Between COVID-19 and Spanish Flu

Face Masks and Its Past

As mentioned in the literature review about the Spanish Flu, history has repeated itself. Spanish Flu and COVID are parallel to one another, with similar spread, symptoms and precautions that are being taken. The major difference is the *acceptance of wearing masks.*

Face masks during Spanish Flu were made of heavy-duty-six-layers cotton gauze, the layers were numerous to avoid transmission to a longer distance. Little (2020) mentioned that people, mainly women and kids, wore these masks readily, while on the other hand, men refused to wear them as they complained about masks being uncomfortable. Moreover, men thought that wearing masks was feminine.

Several ways were adapted to encourage and make people wear face masks. One of them being *advertising campaigns and jingles* which were formulated specifically for this. According to shreds of pieces of evidence taken from The Guardian (Adams 2020), these jingles were –

> *Obey the laws and wear the gauze. Protect your jaws from septic paws.*
>
> *Wear a mask and save your life! A mask is 99% proof against influenza.*

Catchy taglines are always handy with posters, ads, and products so that the consumer recalls them. Most of the present-day brands have a jingle or a tagline. Hence, the next time when someone says the tagline we instantly click to the product. Similarly, instead of using direct words to ask people to wear masks, jingles were a better means of communication.

Another source of communication was the use of visual aids. Little (2020) already mentioned that men were very reluctant when it came to wearing masks, also many of them poked holes in their masks to smoke, reducing the effectiveness of wearing masks and using it as a safety measure. Canales (2020) in her article, has put up a few images of several posters and advertisements depicting men and young boys, uplifting one another to wear masks or use handkerchiefs. Newspapers, which are a great source of communication, helped to spread the message out, and posters and advertisements of such kind show the simple yet effective power visuals have in communication.

Fig. 3. A newspaper cartoon from 1918 (left)

Fig. 4. A newspaper clip from 1919 (right)

Spanish Flu masks were also converted into New Fashion Veils as per Little (2020), where women in Seattle wore Fine Mesh with Chiffon Border masks. But the trend didn't last due to the type of material they were made of. Also, medical and scientific communities believed that multiple-ply gauze was much more effective than one-layered masks made out of non-porous material. Hence from history, we definitely cannot conclude whether wearing masks was effective or not.

Face Masks as a Fashion Statement

With COVID-19, masks have evolved beyond just protection, with the face mask now turning out to be a new "*must have fashion accessory.*" Ever since the CDC approved using cloth-based masks – with some instructions about how to make them – fashion designers across the world got engaged in the same. Not only fashion designers, but big namely brands also started launching their masks with celebrities being their ambassadors.

As per Nandwani (2020) in her article, fashion brands such as Fendi, Louis Vuitton, Dior have now engaged themselves in manufacturing masks and protective gear. We see a wide range of masks these days, from print-based masks; manufactured by Off-white to masks with monogram prints by Fendi, signature prints by specific designers, masks with tied bows, masks with beads, masks made up of gold, silver, and diamond, occasion-based masks, masks for kids and so on. (Kapur 2020; Aubrey 2020)

Fig. 5. Fendi Mask

Fig. 6. Mask made of beads

Fig. 7. Gold Mask

Every day new designer or brand is coming up with a new design. India being a hub of fashion designers has also seen this trend. In the same article by Kapur (2020) it shows that fashion designers like Masaba Gupta, Anita Dongre, Ritu Kumar, have shifted to production of three-layered designer masks. But are these masks only a fashion statement or trend? Or do they convey or act more than that?

In a recent survey with Indian Express (Chaudhary 2020), the famous fashion designer Ritu Kumar said that "*I think now that we have to learn to live with the virus, the problem is that if we go out of our homes, we'd need to take protection, therefore, the mask becomes a necessity. Having said that, it doesn't have to be a very simple looking accessory, therefore, we're creating masks that complement our outfits, with printed fine fabrics which can be mixed and matched with garments in anyone's closet.*" Ronita Mukherjee, executive director for client services at brand management company Landor Associates says that (Kapur 2020) "*These masks are almost a form of self-expression, an extension of fashion.*" This is amply visible in the current scenario, because when we see people around us buying designer masks, why should we wear surgical masks?

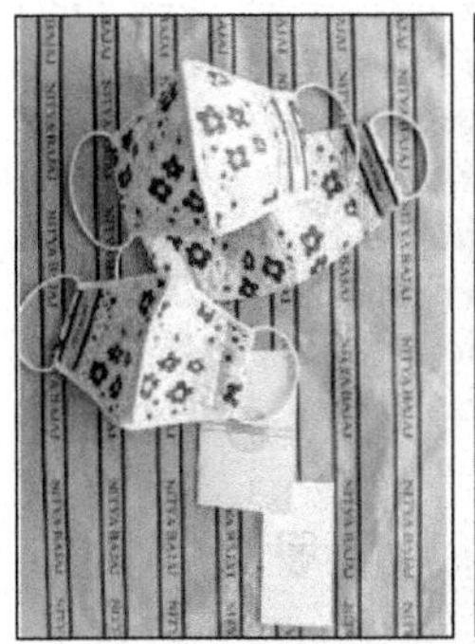

Fig. 8. Masks by Nitya Bajaj
Fig. 9. Mask-making @ Ritu Kumar

Except for wearing face masks as an accessory, fashion designers also state that making these masks are a source of income to the tailors that work for them. Suchita, an Indian resident in her interview with Indian Express (Kapur 2020) stated "*I am skilled in stitching so I just stitched some masks while sitting at home during the lockdown. I was also aware that there were small-scale tailors and migrants who were out of work. So, I decided to rope them in as I started.*" "*We now have more than 20 people working with us. Our team is making 400-500 masks daily. The masks are priced between Rs 30-150.*" The above statement proves that not only designers, but many other non-working people figured out a way to get employed.

On the other hand, the famous Indian Fashion Designer Sabyasachi Mukherjee in his video interview with India Today E-conclave (2020) when questioned that "*Will masks turn into fashion statements*" to which he replied "*I hope not. Because we are talking about a health crisis and I hope we don't put privilege on things that are essential for health.*" He also added, "*I think it is really obnoxious and it is really offensive to do designer face masks. If I have to use my factory to be able to manufacture face masks, which is just a simple mask and where I am only a vendor, I am happy to do that. I will never put my label on a face mask. Ever.*" On agreeing to what Mr. Mukherjee has to say, masks are to be used as a need and not a necessity, but definitely, the designers are converting it into one. It doesn't end here though; many designers are making profits out of selling masks by putting their brand labels on them. To such brands per Amarnath (2020), Sujata Assomull has to say that, "*brands like Off White are trying to make profit out of pandemic.*" She adds "*Now that wearing a mask is a reality, I do see that there is cause for fashion masks. I have friends who run small ateliers and they are making masks as a way of ensuring that their tailors make some income.*" Brand consultant and business strategist Harish Bijoor quotes in the article by Amarnath (2020) "*that the masks are an essential part of COVID couture. You will find masks that straddle every price segment.*" Luxury is for those who can afford it, and Bijoor believes that designers can make profits through selling masks, as long these profits devote to a cause in the current scenario.

As stated earlier, masks as a new way for self-expressionism: it is about reflecting outside what we feel inside. We usually express ourselves through our face mainly eyes and mouth i.e. the way we speak and our tone. Putting a mask on makes it difficult to express and for the listener to understand what you are trying to say. In such cases, we can reflect our thoughts through masks wherein we don't have to say anything but our masks speak for us. Kaushal (2020) evidences the same in the words of Suneet Verma "*Masks are now becoming more about self-protection and self-expression. Everyone knows that they are at risk, so people are speaking their minds. When you wear a mask with a message or either the gay flag or the political satire, you are voicing your morals and sentiments. I think it is a very good time to say without speaking.*"

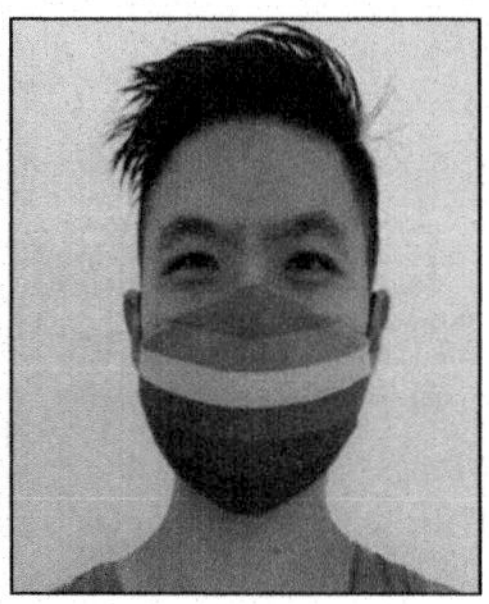

Figs. 10. and 11. Expressing through masks

Masks are also trying to bring in revolution. How? A London based designer, Saran Kohli, manufactures masks with Vitiligo prints. Saran has been suffering from Vitiligo since childhood. Saran Kohli as per Kaushal (2020) states that "*I always wanted to push this topic for a discussion and I took the help of masks, because they are not just a form of protection for us now. They are beyond that and the recent example of the Black Lives Matter masks, which had I Can't Breathe written on it, states the very fact that fashion masks during these times have become a language or a form of expression for many. My hope is to begin open conversations and educate people about vitiligo via these visual representations.*"

Fig. 12. New source of branding & merchandising

Not only fashion-based brands but brands like Disney, Puma, Adidas, and also a few local brands have stamped masks with their logos. Paresh Mishra, senior VP of sales and marketing, Essel World Mumbai quotes "*We have masks for Essel World, Water Kingdom and Essel World Bird Park, which have been designed by our internal creative teams.*" He adds that branded masks are an addition to their other merchandise like t-shirts and caps.

Face Masks and Healthcare

People with hearing loss struggle to communicate in healthcare settings when they have a mask on. According to the researchers' commentary (Iswarya 2020), "*Communication between patient and clinician is at the heart of medical care. Even before masks became ubiquitous, people with hearing loss struggled to communicate in healthcare settings, and poor communication was the likely cause of their documented worse health outcomes. Now masks are blocking lip movements and facial expressions – and these are so important when hearing is marginal. Masks also muffle the high-frequency portions of sound that are essential to speech.*" Hence, to improve communication in such a case, transparent masks are developed for better communication. Such masks should be open for the general public and also for improved communication.

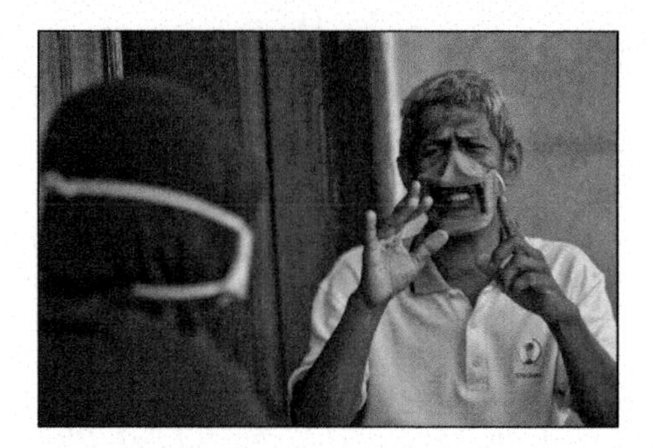

Fig. 13. Transparent masks for healthcare

New technology leading to innovations and designs is helping convert a simple necessity into a long-running product. An NDTV article highlights how Donut Robotics, a Japanese start-up, has developed a face mask, that can record, call, and translate in up to 8 languages and amplify your voice while speaking. This device fits over the normal fabric-based masks and uses Bluetooth to connect to an app that performs these functions.

Another technology-driven mask as revealed by News spot (2020), was developed by LG electronics. This mask has miniature fans that suck in contemporary air and scale back the stuffiness of the one who's wearing it. The filters in the mask are replaceable and the inside of the mask is a silicon layer. Made for people who have to wear masks for longer durations such as health workers, hence it is light-weighted. The mask has a life of 6 hrs and can be reused after charging it for 2 hours.

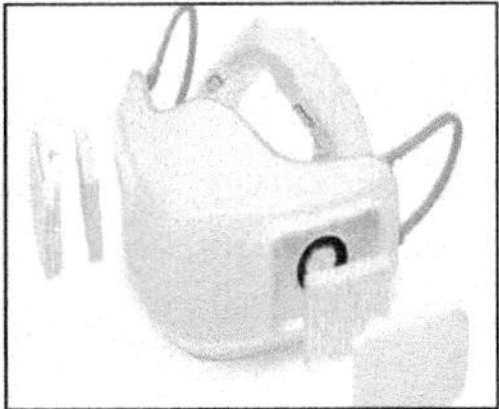

Fig. 14. Smart Face Mask

Conclusion

During the Spanish flu, aesthetics of the mask was not important – the people wearing them were. Hence visual communication played a vital role in spreading the importance of doing so. But the current day pandemic scene is not the same. Today, aesthetics matter, and people are ready to pay huge prices for the same. Masks these days are also used for promoting a brand's identity, this is how visual communication has leaped over the years. Also, with developing technology, the visual aspect of the developing product and the way it is being marketed is important. Hence, regardless of whether it's a fashion or a medical product, we all recognize them, relate to them, and buy them because they visually speak to us.

References

Malek, M. S., and Gundaliya, P. J. (2020). Negative factors in implementing public–private partnership in Indian road projects. International Journal of Construction Management, DOI:10.1080/15623599.2020.1857672.

Borikar, H., Bhatt, V., and Vora, H. (2022). Investigating The Mediating Role Of Perceived Culture, Role Ambiguity, And Workload On Workplace Stress With Moderating Role Of Education In A Financial Services Organization. Journal of Positive School of Psychology, ISSN, 9233-9246.

Malek, MohammedShakil S., and Bhatt, Viral (2022). Examine the comparison of CSFs for public and private sector's stakeholders: a SEM approach towards PPP in Indian road sector, International Journal of Construction Management, DOI: 10.1080/15623599.2022.2049490

Hiral Borikar, M., and Bhatt, V. (2020). A Classification of Senior Personnel with Respect to Psychographic and Demographic Aspect of Workplace Stress in Financial Services.

Borikar, M. H., and Bhatt, V. (2020). Measuring impact of factors influencing workplace stress with respect to financial services. Alochana Chakra Journal, ISSN, 2231-3990.

Bhatt, V. (2021). An empirical study to evaluate factors affecting customer satisfaction on the adoption of Mobile Banking Track: Financial Management. Turkish Journal of Computer and Mathematics Education (TURCOMAT), 12(10), 5354-5373.

Bhatt, V. (2021). An empirical study to evaluate factors affecting customer satisfaction on the adoption of Mobile Banking Track: Financial Management. Turkish Journal of Computer and Mathematics Education (TURCOMAT), 12(10), 5354-5373.

Bhatt, V. (2021). An Empirical Study On Analyzing A User's Intention Towards Using Mobile Wallets; Measuring The Mediating Effect Of Perceived Attitude And Perceived Trust. Turkish Journal of Computer and Mathematics Education (TURCOMAT), 12(10), 5332-5353.

Malek, M. S., and Zala, L. (2022). The attractiveness of public-private partnership for road projects in India. Journal of Project Management, 7(2), 111–120, DOI: 10.5267/j.jpm.2021.10.001

Malek, M., and Gundaliya, P. (2021). Value for money factors in Indian public-private partnership road projects: An exploratory approach. Journal of Project Management, 6(1), 23–32, Doi: 10.5267/j.jpm.2020.10.002

Cartoons and their Visual Aspects Affecting Children

Kirti Ojha[a] *and Sambit Kumar Pradhan*[b]

[a]Unitedworld Institute of Design, Karnavati University, Gandhinagar, India
[b]Unitedworld Institute of Design, Karnavati University, Gandhinagar, India
E-mail: sambitkumar@karnavatiuniversity.edu.in

Abstract

When people think of cartoons their memories take them to their childhood. Most of the people in their 20s were interested in discussing cartoons they used to watch during their childhood, while parents of kids aged 4–11 years were more concerned about the type of cartoons that are being shown on TV. This paper focuses on understanding the role of cartoons and the way their visual content affects children. Cartoons have made a major impact on mental and physical development of children. With the advancement in technology and digitalization children are observed to have low physical participation in indoor games as well as outdoor games. This can have a lasting adverse impact on their upcoming future. The paper aligns cognition factors with the responses received from children, backed by the character analysis of a cartoon followed by the concerns that parents have and the lasting impact of it.

Keywords: Cartoons, Development, Cognition, Content, Interaction

Introduction

Most of the memories of our childhood remind us of not only the games we played but also the cartoons that were being watched by us. With the advent of animation industry in India a new audience of kids started taking interest in the content that was shown during that time.

Lexico Dictionary defines a cartoon in relation to animation refers to a motion picture that uses animation techniques so as to picture drawings that are arranged in a sequence rather than taking shots of real people or objects.

With the years passing by it has been observed that the overall quality of cartoons is degrading. In an article by (Charvi Bhatia (2018) she quoted "After 2007, the content in television saw a downfall. For the most part, television, which at least made sense earlier, now lost its USP. That fake laugh, those forced punches, the fake accent took away the inherent sense of authenticity of cartoons." Therefore, this paper is more focused on understanding the visual aspects of a cartoon and the way these characters are perceived by children, and their impact on them.

The areas that will be covered In this paper Ides the emergence of the animation industry in India, technological advancement and evolution, understanding the making of a good cartoon, cognitive development, and visual aspects of a cartoon.

Literature Review

Children's are going through major psychological health problems. Earlier with a low reach for television kids were more into outdoor games but now with the arrival of number of platforms for cartoons, children are getting hooked to watching cartoons continuously for hours.

This has led to a rise in violent behavior, aggression, usage of foul language, disobedience, etc. Since most of the visual information is understood and noticed more attentively amongst children from the age of 4–11 years, it's the quality of content that has become a major concern for parents with their children consuming these visual languages heavily. All of it is not only impacting their mental development but also lowering their motor development.

Indian cartoons are more trending amongst kids in the current scenario but the problem of quality of visuals, storyline, and voiceovers is something that the Indian animation industry is failing to cater to because of the underlying struggles that are faced by the animation industry itself.

How India Started by Making Their Cartoons?

The discussion with Ram Mohan (2014) in an interview with D'Source brings attention to the traditional technique of animation that was taught by Clair Weeks. The Banyan Deer was created under the guidance and project led by Clair Weeks which was further managed and led by Ram Mohan when Clair went back to his place. "EK Anek Ekta "was majorly the first project that was focused on children with the project assigned by the Government of India.

The major change I got introduced in styling the characters was when one of the Japanese producers

Yugo Sako got flattered by the story of Ramayana and wanted to make an animated movie based on it. Unfortunately, government didn't fund the project in that period thus denying the idea of a joint venture on it. Finally, with a team of Japanese and Indian animators, the film got released in 1992. After the project was over UNICEF wanted Ram Mohan to work on cartoon series called "Meena" which starred in South Asian Children's shows in Bangladesh because of its huge success from then onward India started getting more outsourced projects in animation.

How Should Cartoons Be?

In an article from an e-paper discusses the differences between the cartoons that were aired in 90s and 2000s with the cartoons that are on TV now bringing attention to certain aspects like such a clear premise, an easy-to-follow storyline and yet making a viewer think about the further consequence are the factors that make a cartoon an ideal one and all of it is missing in the cartoons that are currently trending. (Palash Volvoikar 2017).

Cognitive Development

- **Color and Shape Recognition**

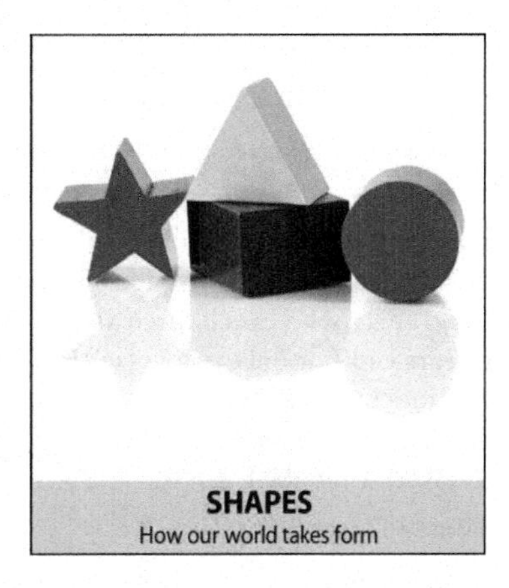

Fig. 1. Colours and shapes (mensaforkids.org 2020)

Colors are perceived differently as per the different stages of development in children., and is one of the first recognition external factors that is stimulated even before emotion recognition. In evident that colors are also used to decode different messages. (Katrin Ann Orbetta, 2020)

Children falling in the spectrum of 3–5 years of age have a steady pace of progress in overall cognitive development. They can count, name colors, and can tell about their name and age. This is more of the initial stage where children begin to enjoy humor in stories and imagine characters and scenarios. Whereas the cognitive skill of children falling in the spectrum of 6-11 years have a more organized and logical way of dealing with concrete information but it's the attention span that is limited and might become stable after the age of 11 years. (Dennett, 1987; Callaghan et al., 2005).

- **Emotional Recognition**

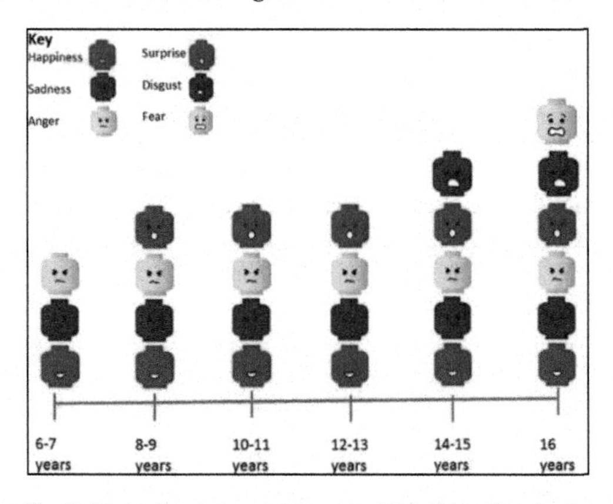

Fig. 2. Chart showing emotion recognition (kids. frontiersin. org2020)

Children below the age of six years can recognize simple expressions like happiness, sadness, and anger whereas children above the age of 6–11 years can not only recognize those three spectrums of emotions but also can recognize the expression of surprise (David Henry, Kate Lawrence, Ruth Campbell, 2016).

- **Age-Based Perception**

A child's Relatability to Cartoons varies as per the responses from children aged 4-10 yrs. Children of different ages, genders, and natures have different ways of relating to cartoons.

In a personal telephonic communication with the respondents, the response stated below gives a basis for the differentiation and relatability they can create along with their reasons of liking or disliking a cartoon character.

Akanksh Sharma (4yrs): The respondent likes the character Oggy from the cartoon, Oggy, and the cockroaches because his favorite color is blue whereas he dislikes Jack because of the green color of his body.

Dhani Saini (6 yrs.): The reason she doesn't like watching Shin Chan is that she feels that Shin Chan is not of her age and is younger than her.

Mrutyunjay Kaushal (6 yrs): He likes watching the cartoon Honey Bunny but comparing the characters of both the brothers who are cats, he dislikes the character of Honey more because he feels that the character is dumb whereas likes the other character and considers him as brainy.

Shraddha Mishra (7 yrs.): She dislikes the character bunny as discussed above because she feels that Bunny is arrogant whereas she likes the Character of Honey because of the innocence and calming nature that he has.

Abhinav Bhardwaj (10 yrs.): He watches cartoons based on how entertaining the jokes are and the style or way in which the dialogue delivery takes place. Currently watching the cartoon Fukrey Boys because of the relatability and similarity in characters from the movie Fukrey.

Therefore, as per the responses it is understood that children of different age groups have differences in the cognition of characters in a cartoon.

Character Analysis

Fig. 3. Motu Aur Patlu (Telanganatoday.com 2019)

Considering the cartoon characters whose likeability is higher in all the age groups mentioned above is "*Motu Aur Patlu*."

- **Color**

Both Motu and Patlu have a fair complexion but what takes the eyes off it is the vibrant color that is used in their attire. Motu wears a Red tunic and Yellow pajamas a combination of warm colors supported by a cool color with a navy-blue colored overcoat.

The character Patlu on the other side wears a yellow tunic and an orange-colored pajama.

The reason for doing so might be to create the pairing by keeping their attire in warm colors. Another reason for using these colors was for drawing the viewer's attention toward the major character.

- **Shape and Form**

The round shape of Motu depicts the nature of a character, round shape is also associated with friendliness. The visual of the character justifies certain personality traits in context with the overall form. The structure of the face is similar to that of his upper body. The round belly signifies and helps in creating the relatability of love for food as per the character.

Patlu on the other side has an Oval face to depict that he is also as friendly as Motu. The characters' overall appearance is slender so as to maintain a visual balance yet complement each other through shape and form aspects and for the sake of semiotics.

- **Facial Features**

A short-heighted Motu has a mustache and hair at the back to make him look elder than Patlu. Motu also has big eyes in comparison to Patlu to balance the heavy face structure with big eyes and to depict a characteristic trait of expressiveness which this character possesses and the other character has but the chances might be low. The character Patlu on the other side has small eyes in comparison to Motu but is accessorized with a round set of spectacles to highlight the nerd or the genius side of his character.

- **The Storyline**

Suhas Kadav (2012) "The story revolves around Motu and Patlu, two friends living in Furfuri Nagar, and their daily activities. The show focuses on the two landings themselves in trouble and comical situations, later being rescued only by sheer luck. Samosas are Motu's favorite food, and he frequently tries to steal them from the local Chaiwala who makes the best samosas in the city. Motu is mainly the black sheep of the duo, unintentionally creating problems due to his incompetency while Patlu is the smart one that always tries to stop him."

- **Observational Concern**

The name itself depicts that physical appearance is all that matters. Many parents claim that the language that has been used in this cartoon is inappropriate.

Parents also highlighted that the character Motu is shown defeating the bad by eating a plate of samosas. This creates a dangerous tendency among children and they may misunderstand that these types of junk food provide strength and sharpness to solve problems. (Wow Parenting 2020).

The Impact of Cartoons on Children and the Concerns of Parents

Fig. 4. A kid watching cartoons on television (hsph.harvard.edu 2020)

Cartoons have more of a behavioral impact on children. (Tatev Derzyan 2019, n.p.) states in the research that "children are more attracted to cartoons than learning from traditional and academic sources, the main reason lies in cartoon scenarios, audiovisual effects, and color solutions. Scientists have proved that children not only learn from cartoons but also memorize and imitate the behavior of cartoon characters due to their cognitive abilities."

There are both positive and negative impacts of cartoons on children

- **Negative Impact**

The books capturing the study of Media and Children discussed the following statements.

1. Children are exposed to media for a major period. Considering their daily activities children spend most of their time watching T.V. rather than playing outside.

2. As per the book and the response related to concern from parents it has been observed that children find it difficult to distinguish fantasy from reality and might do something from the influence of cartoons in real situations that might hurt them or get them into trouble.

   Singer and Singer (2001) and Potter (2003)
3. The research done by American Pediatric found that children who watch cruel cartoons are more likely to become aggressive, disobedient, and angry. Even though cartoon helps children in developing their imagination but their negative impact of it make children stay in their imaginary world for a major period.
4. One of the Professor from Michigan University Hyman noted that the violent scenes in cartoons make it difficult for the children to understand the effect and consequences of violent action and explained the statement further by quoting it with an example:

"When a heavy object falls on the hero's head, the scene makes laugh, the hero is not harmed. Seeing this, the child gradually ruptures the action-related relationship (for example, Tom and Jerry, Teenage Mutant Ninja Turtles)."

- **Positive Impact**

The Scientists distinguish auxiliary role of cartoons in education for children, additional information about the world, cultures, and interesting entertainment. Cartoons also help in enhancing vocabulary (the child can master the language while watching a cartoon in a foreign language), helping to discover the world (Dora. traveler, Dexter's Laboratory) (Tatey Derzyan, 2019).

Conclusion

While going through the information related to research, I came across understanding two different sides of the subject. The dependency of children on a cartoon is indeed booming in the case of most children and it is becoming difficult for parents to understand the reason behind children getting hooked on them. The major reason for this action is that children start relating themselves to life and the character itself.

The overall research focused on providing a glimpse of the way children perceive and understand the visuals of a character. But considering the current scenario with most of the children liking the content of Indian cartoons has affected them more because low content quality, bad jokes, and usage of foul language is overpowering the content and the storyline of the overall cartoon.

Looking into the aspect of the storyline most of the cartoons have the same genre where the villain character is dealt with physical violence by the main character. This might make a child feel that it is normal to thrash or beat people in case they do something that doesn't fit into their reasoning because the essence of emotion like pain is not portrayed, thus normalizing these actions. A perfect example of the same could be "Chota Bheem" where the story of an episode always ends up with Bheem hitting the characters like thieves as a solution to the problem that occurred in the overall episode.

References

(Lexico Dictionary, Available: https://www.lexico.com/en/definition/cartoon [14 Aug 2020]

(Charvi Bhatia (2018) *Indian Folk Network*, [Online], Available: https://www.indianfolk.com/cartoons-vs-now-edited/ [15 Aug 2020]

Ram Mohan, 2014, The story of Indian Animation, online video, 30 Dec, *YouTube*, viewed 14 Aug 2020, https://www.youtube.com/watch?v=T4qirW0KFKE&t=969s.

Palash Volvoikar 2017, "The Apparent Downfall of Animated TV in India," *The Neutral View*, blogpost, June 9, viewed on 15 Aug 2020, http://theneutralview.com/the-apparent-downfall-of-animated-tv-in-india/.

Katrin Ann Orbetta, "Colours and Cartoons," [Online], Available: https://mashandco.tv/en/cartoons-their-importance-in-kids-development/. [20 Aug 2020]

Stages of Development. Authored by: OpenStax College [Online], Available: http://cnx.org/contents/Sr8Ev5Og@5.52:b7opmCF3@6/Stages-of-Development#Figure_09_02_Stages. [20 Aug 2020]

Henry, David, Lawrence, Kate, Campbell, Ruth (2016), "Can Children see Emotion in Faces," *Frontiers for young mind,* article, Aug 25, viewed on 20 Aug 2020, https://kids.frontiersin.org/article/10.3389/frym.2016.00015.

AnimationXpress Team (16 October 2012). "Motu Patlu" Premiers Today on Nick–lodeon - In Conversation with director of "Motu Patlu" Suh's Kadav. AnimationXpress.

Borikar, H., Bhatt, V., and Vora, H. (2022). Investigating The Mediating Role Of Perceived Culture, Role Ambiguity, And Workload On Workplace Stress With Moderating Role Of Education In A Financial Services Organization. Journal of Positive School of Psychology, ISSN, (9233-9246)

Malek, MohammedShakil S., and Bhatt, Viral (2022). Examine the comparison of CSFs for public and private sector's stakeholders: a SEM approach towards PPP in Indian road sector, International Journal of Construction Management, DOI: 10.1080/15623599.2022.2049490

Ergonomic issues faced by transporters of LPG gas cylinders

Rahul Hareesh,[a] Anup Hengade,[b] Jonathan Shajan,[c] and Paramesh Krishnan,[d] and Arunachalam Muthiah[*,e]*

Karnavati University, Gandhinagar, India
E-mail: *arunachalam@karnavatiuniversity.edu.in

Abstract

The study is concerned with ergonomical evaluation of carrying and transporting liquefied petroleum gas (LPG) cylinders by deliverymen and by home users. By using surveys and interviews, data was collected to determine work-related musculoskeletal disorders (WMSDs) experienced by these users, and the working posture of the people transporting gas cylinders was evaluated using Ovako Working posture Assessment System (OWAS) postural analysis tool. It was found that only two out of 10 postures did not require correction (category 1), while one required immediate correction (category 4) according to OWAS. Studies were also conducted on the present use of mechanical aids like trolleys to prevent WMSDs in the target group. Use of trolleys was more prevalent in home users than deliverymen, although the variety of MSDs reported was high in home users.

Keywords: Ergonomics, MSD, OWAS, Cylinders, posture

Introduction

Gas cylinders have been one of the building blocks of a kitchen to cook delicious food. Looking at a highly populated country like India, many people from rural areas or towns still rely on gas cylinders even though the use of gas pipelines has increased. A considerable amount of the population lives in multi-floored apartments, but with an absent or out-of-service elevator given the electricity conditions. Here, the major task involves transporting the gas cylinders right from the truck to the house. It is mostly done with hands or with some cheap solutions that result in ergonomic issues (Abidin, 2017).

A delivery man delivers the filled gas cylinder, which is weighing around 30 kg, and takes away an empty cylinder weighing around 15 kg for refilling and reusing. The deliverymen will have to handle the delivery process physically, doing actions like lifting the full cylinder from the transport vehicle and then distributing it to the customer; then bringing back the blank cylinder to the transport vehicle. The deliverymen have to rely on stairs to manually transfer the LPG cylinder from the ground to the required upper level in the case of apartments with no elevators. The delivery men are exposed to various ergonomics risk factors due to the above-mentioned LPG cylinder delivery process that is associated with excessive force that may lead to MSDs, such as back and shoulder injuries.

Some men suggested they go with manual methods because they get better degree of freedom in the movement, without an assisting device for transport (Roaslni, 2019). In case of customers, they would generally move it inside to the required spot using cheap stands.

The existing cylinder LPG trolley is not safe to use due to the durability and stability problems, but no ergonomic failure was confirmed when RULA analysis was done on the product (Tajwi, 2015). A case study of four workers rolling gas cylinders in an industrial area, where their working postures were analyzed using Rapid Upper Limb Assessment (RULA). The result of this research was that the postures assessed by RULA all exceeded Action Level 3. 1- time/second was the repetitive activity frequency for each wrist. The radial deviation of the wrists and the dorsiflexion on the right wrist approximately reached the highest range of motion. Hence as suggested by RULA, the tasks could cause injury, particularly to the wrists (the upper limbs), and require study and changes to the working postures (Chen and Chiang, 2014).

In a study, it was found that users are not aware of or are neither bothered about ergonomics. In some cases, many workers suffer from MSDs due to the reason that the industrial trolleys are not ergonomically designed, possibly due to the reason that even local workshops manufactured or produced most of the trolleys, who have lack of experience and knowledge about ergonomic facts and designs, hence they are not ergonomically designed and hence they lead to various problems (Talapatra et al., 2019). In another study, it was found that there is a mismatch between industrial trolley dimensions and anthropometric data workers. This has resulted in workers experiencing pain in various body parts. 92% of all the delivery men in the study experienced discomfort in the lower back, 84% felt discomfort in the neck, 80% for upper back, 60% had discomfort in the right shoulder, 8% in the left shoulder, while 28% of them reported uneasiness in both shoulders. The cause of discomfort to the workers was found to be short wheel radius and handle height. Manual transport aids (trucks and trolleys) are in widespread use throughout most industries, but their use does not always result in the anticipated reduction of workload or musculoskeletal stress because of problems of overloading, overexertion, difficulty in steering, use of cheap or low-quality wheels and insecure loads (Al Amin et al., 2015).

Most critical accidents were caused by slips on steps, walkways or floors and posture, bodily reaction, and overexertion due to heavy cylinders as well as falling from floors, walkways, or steps during loading and unloading from vehicles (Kim et al., 2017).

Handle ergonomics should also be considered important in the case of ergonomics. It was found that during the starting phase, significant wrist deviations and high stresses at the elbow occur. The design of the trolley handle significantly affected elbow stresses, as well as the tilt of the trolley while moving forward with the weight. 35° and 1.0 m were proved to be the best configuration for the handle angle and handle length respectively. The arm postures adopted in the task were influenced by the orientation of the handle Okunribido and Haslegrave (1999, 2003). For steady movement, 50°C handle proved most desirable. Changing the handle material from nylon to polypropylene to increase the durability of the product can also be done (Tajwi, 2015).

Most of the trolleys currently used in industries were not ergonomically correct and hence they resulted in several accidents. Weight of the trolley has to be reduced. Amount of load carried at a time should be reduced to prevent overloading. Caster wheels have to be chosen such that they don't swivel too much and cause difficulty in moving the trolley in a straight line. Appropriate securing mechanism is to be provided to fix the gas cylinders to the trolley and prevent it from falling off.

Problem statement

Carrying LPG cylinders results in MSDs and accidents due to incorrect postures or incorrectly designed trolleys.

Aim

To study the ergonomic evaluation of transporting gas cylinders.

Objectives

To evaluate the working posture of the people transporting gas cylinders using OWAS postural analysis tool, and to determine the WMSD problems of LPG cylinder deliverymen and LPG cylinder users using questionnaire.

Methodology

People including deliverymen and home users of LPG cylinders from a sample size of about 60 people from different parts of India were interviewed or given surveys to get to know about their day-to-day issues with transporting gas cylinders during or after delivery. Questions were asked to get an idea about the various MSDs they face and about what existing products from the market they use to accomplish the task.

Using Ovako Work Assessment System (OWAS), photos of working postures of lifting and transporting cylinders under various loads (filled and empty) were analyzed and results were obtained to classify whether postures need immediate corrective measures or not.

A frequency distribution table is made to study the frequency of occurrence of specific postures (shown in annexure).

Results

The survey was conducted on 63 people which includes 60 home users and 4 deliverymen of LPG cylinders. The percentages of age groups of the respondents were: 15-20 years (38.1%), 21–40 years (30.2%), 41–60 years (28.6%), 60+ years (3.2%). The respondents were of almost equal percentages of males and females. The regions from where the respondents come from are southern (Karnataka: 38.1%, Kerala: 20.6%, and Tamil Nadu: 20.6%), western (Maharashtra: 15.9%, Gujarat: 1.6% and Daman and Diu: 1.6%) and north-eastern (Assam: 1.6%) parts of India. More than half of the home users said that they faced difficulty with moving around the LPG cylinder. Most of the users (49.2%), said that they use a circular trolley stand to move the LPG cylinder after receiving it. 40.7% of the users said that they don't use a trolley. 42.4% of the users said that they live in a flat or apartment while 57.6% said they live in an independent house. When asked about how many cylinders changed in a month, 69.5% of the users said they changed once a month, 27.1% said twice a month and 3.4% said more than two in a month.

Measures taken by users to reduce pain during/after transporting the cylinders include: taking help from family members, asking the deliveryman to place it in the kitchen, dragging or rolling the cylinder, using a trolley, taking rest in between, moving slowly without sudden movement and shaking or stretching the arms. Only half of the respondents agree that their products or methods are comfortable.

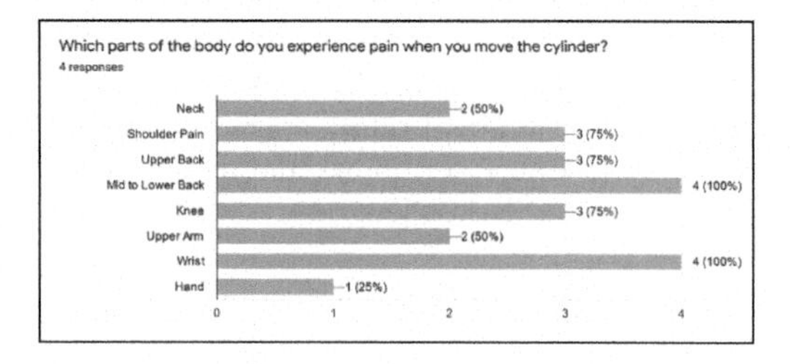

Fig. 1. Distribution of occurrence of MSDs in LPG deliverymen when delivering LPG cylinders

All four deliverymen interviewed said that they faced difficulty delivering LPG cylinders to their customers. Almost all of them said that they do not use a trolley and deliver to both apartments as well as independent homes. They deliver about 15-30 cylinders in a day, about 22 being average. Users were asked about in which parts of the body they felt pain after carrying cylinders; their responses are illustrated in figure 1.

Measures taken by deliverymen to reduce pain during/after transporting the cylinders include: taking rest after a whole day of work or taking break in small intervals. When asked to rate how comfortable are they with existing products or methods they use on a scale of 1-5 (1-least, 5-most comfortable), three of the four deliverymen gave a rating of 2 and one deliveryman gave 3 out of 5. Most of them said that they have faced injuries when transporting cylinders.

Analysis of 10 different postures of handling LPG cylinders (by deliverymen and home users) was done using OWAS, where they carry an empty or a full cylinder (weighing 15.6 kg and 29.8 kg respectively). It was found that only 2 out of 10 postures scored "no action required" (category 1), 7 scored "corrective actions should be done as soon as possible" (category 3), while one posture scored "corrective actions for improvement required immediately" (category 4) according to OWAS. None scored in category 2. Hence, postural analysis has revealed the possible reasons why LPG handling results in body pain or MSDs in the long run. The frequency distribution table shows that the leading cause of MSDs was due to postures leaning forward, inappropriate way of holding cylinder with hands, and handling of heavy weights repeatedly.

Proposals

Based on our research, we would like to propose the following methods to alleviate the said problems: using a trolley with a caterpillar track (belt) instead of wheels, so that it is able to climb on staircases as well as on a variety of surfaces; providing wearable massage devices or heating pads that are periodically activated to increase blood flow in order to relax the muscles after work; using extra people to deliver LPG cylinders to homes; hence, the load carried by a single person is reduced and the deliverymen get enough time to relax; restructuring work shift timings so that people are working in an alternating manner between periods of work and rest.

This research was done during the ongoing COVID-19 pandemic at the time, hence face-to-face interviews were limited and data collection was done through an online form with a sample size of about 60 people. Direct, in-person observations of existing methods used in homes or warehouses were not possible. This research, therefore, reflects the opinions of only a limited section of the society only from a few specific states of India. Therefore, this can be considered only as a preliminary study.

Conclusion

Our research finds that users are not satisfied with the existing solutions for transporting LPG cylinders. Safety features or securing mechanisms are inadequate or absent in the products available in the market and components like wheels used are not of good quality, which results in less user comfort. It was also found that some users were also not aware of various products available in the market. This prompts users in homes or workers in storage areas to manually handle gas cylinders without the help of any equipment because they find it easy for picking up and transporting them. People also need to be educated about the need of maintaining correct work postures, taking rest, and using relaxation techniques like massages.

References

Chen, Y. L., and Chiang, H. T. (2014). Atilt rolling movement of a gas cylinder: A case study. Work, 49(3), pp.473–481.

Kim, J. N., Jeong, B. Y., and Park, M. H. (2017). Accident analysis of gas cylinder handling work based on occupational injuries data. Human Factors and Ergonomics in Manufacturing & Service Industries, 27(6), pp. 280–288.

Mohd Tajwi, N. A. B. (2015). Design and development of cylinder LPG trolley for domestic use. Project for Bachelor of Manufacturing Engineering (Manufacturing Design) with Honours. Universiti Teknikal Malaysia Melaka.

Mack, K., Haslegrave, C. M., and Gray, M. I. (1995). Usability of manual handling aids for transporting materials. Applied ergonomics, 26(5), pp. 353–364.

Okunribido, O. O., and Haslegrave, C. M. (1999). Effect of handle design for cylinder trolleys. Applied Ergonomics, 30(5), pp. 407–419.

Okunribido, O. O., and Haslegrave, C.M. (2003). Postures adopted when using a two-wheeled cylinder trolley. Applied ergonomics, 34(4), pp. 339–353.

Roaslni, M. A. B. (2019). Climb Gas Tank Trolley with Brake. Project for diploma in mechanical engineering (packaging). Mechanical engineering department, Polytechnic of Ministry of higher education, Malaysia.

Talapatra, S., Mohsin, N., and Murshed, M. (2019). An Ergonomic Approach for Designing of an Industrial Trolley with Workers Anthropometry. American Journal of Industrial and Business Management, 9(12), pp. 2156–2167.

Wan Zainul Abidin, W. N. S. A. (2017). Design and Develop A Stair Climber Trolley. Project for Bachelor of Mechanical Engineering Technology (Automotive) with Honours. Universiti Teknikal Malaysia Melaka.

Malek, MohammedShakil S., and Bhatt, Viral (2022). Examine the comparison of CSFs for public and private sector's stakeholders: a SEM approach towards PPP in Indian road sector, International Journal of Construction Management, DOI: 10.1080/15623599.2022.2049490

Proposed Concept for Mysore Pak Packaging

Vaishnavi, S.,[a] Shruti Dinesh,[b] V. Pranav,[c] Siddhant Kawari,[d] and Arunachalam Muthiah[e]

[a]School of Industrial Design, Unitedworld Institute of Design, Karnavati University, Gandhinagar, Gujarat – 382422
E-mail: arunachalam@karnavatiuniversity.edu.in

Abstract

Mysore Pak, an authentic South Indian sweet from Mysore city has been selling for the last 80 years and yet has not been acknowledged by folks of other parts of India. A sweet that has not yet found its perfect style of packaging. Generally, the external packaging design of a product does play a major role in attracting customers. Hence, understanding the roots of Mysore Pak and its existing packaging has been performed to get a firm grip on generating a new packaging design. Topics such as Mysore Pak's origin, its specialties, and its ultimate taste were researched. Characteristics of its origin (Karnataka), the residing population, and the uniting features were also researched through surveys and interviews. Problems with the Mysore Pak packaging were also recognized to be rectified in the newly proposed solution. A final redesigned Mysore Pak packaging solution is also proposed.

Keywords: Mysore Pak, Packaging, Redesigning Packaging, Bangalore Culture

Introduction

Mysore Pak is the most popular sweet made with pure Ghee in Karnataka. It is generally made using just four main ingredients. It is a sweet famous for its flavorful combination of ghee and oil with a blend of sugar and chickpea flour otherwise known as Besan. It is generally found in brick shape, which is slightly hard from outside but smooth & textural from the inside. This is the one main sweet dish in Karnataka that melts in one's mouth and leaves behind the rich taste of ghee which is lingering that makes one want more of it. The ghee enhances the taste hence making it so mouth-watering (Ms. Priyanka Joshi, 2008). The Mysore Pak has an ochre color to it, and it varies from yellows to browns. It is dense, soft, and has a honeycomb texture in it. The fat plays a crucial role; it contributes to the texture, mouthfeel, and aroma too.

Mysore Pak is a sweet not only famous in Karnataka, but it is a well-known sweet throughout South India (Dr. Chidambara Swamy and Dr B. Shankar, 2012). It is known as the "King of Sweets" in the south. ("Traditional Sweets of Tamil Nadu I Famous Sweets ITamil...") During Dussehra, the people are supposed to make fifty-one traditional foods and it is said that without Mysore Pak it is incomplete. It is one of the signature sweets of Karnataka traditionally served during meals at weddings and other festivals of South India as well as during baby showers. The dessert Mysore-Pak and the second half word "Pak" in it is a true contemplation of India and the thought behind is what the country stands for: a sweet concoction of a variety of cultures, practices, traditions, religions, ethnic cities, beliefs, and a lot more. This delightful experience of the sweet hasn't yet reached the people of other Indian states. Bangalore is the old capital of Mysore, Karnataka, and has a population of over 12 million, hence making it the third most populous city in India. Bangalore has the most floating population, which has resulted in being culturally diverse. It comprises a spirited blend of people belonging to a variety of religions, castes, and communities with close to 65% of its migrants coming from outside Karnataka. Being the IT hub of the country, it attracts 35% of India's pond of 1 million IT professionals. Lal Bagh and Cubbon Park are some of the parks for which the city is called the "Garden city of India." ("Culture of Bangalore I Bangalore Culture Iangalore...") Bangaloreans call the city "Namma Bengaluru" and are quite proud to be from the city. The word Namma which means "our" is observed throughout the city, for e.g.: Namma metro, Namma city.

Packaging is essential and plays a key role which can be explaied as a socio-scientific method that delivers the goods to the final consumer. It is a crucial factor that is influenced by varied factors and is changing but at the same time, it conveys the benefits of the product to the consumer.

Indian sweets are generally prepared using brown sugar or white sugar, such as the Muktagachha Monda sweets (Like Pinni, Rewi, Mysore Pak). They sweets are popular for their perfect taste and unique texture but get spoiled rapidly when relative humidity is higher than 86%. For these types of sweets - the product deterioration property, the moisture adsorption/desorption properties, the shelf-life requirements, etc. has to be taken into account. Hence, functional and economic packaging is yet to be designed. (Kumar, K.R., 2009)

Dairy food products are food products that get spoiled easily as they are highly prone to oxidation as they need packaging with low-oxygen absorbability to assure shelf life (Gerschenson et al., 2017).

In food packaging, the material of the packaging plays a crucial role in the industry. It is believed that "The apt choice of package highly depends on the efficiency of the packaging material used." This is because - food is a product that is directly consumed by users and packaging acts as a container that will be in constant contact with the food from the post-cooking to pre-consumption period. Hence, there is a higher probability by the inner packaging layer to impact the food inside in positive & negative ways through both physical and chemical means. The first impression of a product is given using the packaging of the product. The purchasing decision of a product is highly dependent on the product's packaging. "According to research, packaging of a product controls one-third of customer's buying decisions." In a market governed by rivalry, and with a consumer under considerable time constraints, the product is what draws the consumer in. Color, visuals, size, proportion, design, and substance are all important aspects of packaging, as is innovation. "They have a critical role in seizing the attention of the consumer/buyer." ("[Solved] Identify and describe a product you feel has ...") "If a product has different and eye-catchy packing it automatically gathers the customer's attention.

The current packaging system: The packaging system of products generally comes with 3 layers of individual packaging: Primary-packaging, secondary packaging, and tertiary packaging for increasing the products' shelf life and simultaneously increasing its safety while transportation. Primary packaging material is the packaging that comes in direct contact with the food, and it plays a key role in preventing the food from getting spoiled by oxidation. Secondary packaging ensures physical protection for the product, like a surrounding carton. Finally, tertiary packaging is used for transportation purposes facilitating storage and handling (Gerschenson et al., 2017). Handling packages is another major aspect of packaging design. Consumers prefer to have packaging which is easy to grasp and handle such as bags/ containers which can be opened without difficulty or hassle while working or driving, microwavable-boxes and ready-to-eat dishes in grease-free foil are much in demand as compared to packaging sizes for a single person, large families, or the elderly. ("Requirements of Food Packaging -- interpack Packaging Fair") The storage life of the product depends on how it is handled by the user and also by the packaging of the product. Packaging consisting of more than one layer provides mechanical stability and easier handling of packages by retailers.

Sustainability in the packaging domain is receiving increasing attention. The Food-industry is increasingly interested in producing efficient and innovative solutions to guarantee quality & distribution while considering environmental sustainability. Considering current days, plastics take up to 50% of the primary packaging of food, and around 30% of plastics wastes is landfilled. The industry of packaging in specific has been under immense pressure for years to reduce the waste of packaging, considering over-packaging & to improve recyclability. Many organizations have defined "sustainable packaging" by setting a set of principles and strategies which could guide decision-making. ("Case study of TetraPak India Ltd - sustainable development ...") The research identified that packaging would support sustainable development if the packaging were effective, efficient, cyclic & clean.

Packaging helps in identifying products easily. (Underwood, Klein, and Burke, 2001, Silayoi and Speece, 2004) said that packaging helps in product identification and distinguishing similar products. This in-turn aids buyers to pick a product from a vast range of related products, hence stimulating the consumer's buying nature. Therefore, packaging plays a vital role in consumer communication in terms of promoting and advertising and can act as an important factor that affects consumers' choice of products and affects their buying behavior.

Some studies indicate that shapes and colors used on the packaging acts as a huge role in attracting consumers' attention.(Garber et al., 2000; Schoormans and Robben, 1997). Underwood et al. (2001) mentioned that pictorial representation on packaging entices customers of less familiar brands. ("Packaging features and consumer buying behavior towards ...") Studies have further mentioned that single signs like colors (Gordon et al., 1994), illustrations, brand names (Rigaux-Bricmont, 1981), and materials used (McDaniel and Baker, 1997) deliver brand meaning apart from pictorial representation ("role-of-packaging-on-consumer-buying-behavior").

Packaging is meant to be functional: meaning it must follow these main functions to be a good packaging.

such as: Protection- to protect the contents from external influences like environmental influences, Convenience- to package the product in a desirable consumer size, Communication- the product packaging is to act like a "silent salesperson"

A product can help to identify distinct brands (Klimchuk and Krasovec, 2007) and simultaneously also helps in brand identification and promotion of the product. Packaging design is the commonly used element to promote products by directly giving product information, and indirectly affecting the consumer's choice to make a final purchase (Well, Moriarty and Burnett, 2006).

Packaging design has three aspects: creating front visual graphic design, forming the structure; design and providing the infographics. Graphic design includes brand name and details, the typography of information, pictures, and colors used. Perceptible colors have unique themes. Most people are drawn toward products with matching personalities. (Nilsson and Ostrom, 2005; Ampuero and Vila, 2006; Klimchuk and Krasovec, 2007)

Mysore Pak is a sweet that has existed for more than 80 years but still, this sweet has not found its ideal packaging style. It is a sweet that neither comes in the dry sweets category nor in the wet sweet one. This is because the sweet has a special tendency of releasing ghee in small quantities that brings its peculiar flavor. Also, from the research conducted above, we can conclude that Mysore Pak is a sweet that requires a high amount of recognition to subsequently improve its sales all over Karnataka. Hence, the packaging of Mysore Pak must be redesigned to an efficient design that will not only improve the sales of Mysore Pak but will solve the problems with the existing one.

Methodology

The research methodology selected is a questionnaire survey method and interviewing method.

A survey of 20 different questions was circulated which was targeted at Mysore Pak consumers who reside in both Karnataka and other states of India. People from varied age groups were asked to participate in the survey to understand the differences in the opinion of people regarding Mysore Pak and its origin, Karnataka. The survey majorly comprised questions about people's views regarding Bangalore as a city as well as about the features of Mysore Pak sweet (such as its popularity, packaging issues, serving techniques, etc.). Questions like which color or location can be associated with Karnataka were asked to trigger their thoughts and express their opinion about the points that unite the people of Bangalore and their thoughts. The Mysore Pak-related questions were intended to know the customer's feelings, what qualities they prefer in their Mysore Pak packaging, and what packaging issues they undergo. Apart from the online survey, interviews were taken amongst several types of sweet shop employees to understand in depth. The interview mainly targeted knowing details regarding the average quantity of Mysore Pak sold by shops to understand the consumer demographics which in turn tells us the amount of packaging sold.

Employees of different professional experiences were interviewed to understand the difference of opinions on the types of packaging to understand the existing products in the market Questions regarding existing packaging problems if any and material of the packaging was enquired, telling us about the scope for future packaging and to keep sustainability in mind.

Results

From the survey data, it is shown that a significant number of people consider buying the sweet due to the aesthetic appeal of Mysore Pak packaging and it is a huge factor influencing the sales of Mysore Pak, but these people also think that packaging should be improved because the existing packaging does not give due credit to this South Indian delicacy (R^2 = 0.0649). Data also shows that packaging can be significantly improved by changing or improving the existing packaging material to more sustainable packaging that also takes into account factors like grease control and pieces of each sweet sticking to each other. These include materials like plastic boxes with biodegradable film or sustainable paper cartons.

A study was conducted in India comprising 208 people coming from diverse backgrounds. It was to understand the variables affecting the Mysore Pak packaging and the respective consumer's thoughts. Amongst these people, an average of 41% of the participants belonged to the age group between 18 - 25 years old and 58.4% of them resided in Karnataka. The survey confirms that 40.5% of Karnataka residents could closely associate Vidhana Soudha with Bangalore. Also, 40.5% of the participants felt that green colors is the ideal color that can be easily linked with Bangalore city, maybe because Bangalore is known as a Garden city. On the other hand, 42.6% of the participants that belonged to the other parts of India felt that Brigade Road and Vidhana Soudha were the two locations that represent Bangalore with 23% and 25.3% respectively. The green and blue palettes displayed voting values of 31% and 39% respectively, being the highly chosen categories. From the survey conducted, we understood that the majority of the audience is highly fond of consuming Mysore Pak and believes it to be an ideal gifting option. Out of 208 respondents, 74 people also felt that Mysore Pak is relatively a good sweet but isn't largely popular over the nation. They consider that the popularity of this sweet can be enhanced by improving the packaging of the sweet. Supporting this hypothesis, 150 participants agreed to the fact that the exterior packaging does influence their purchase of sweets. From the survey, we understood that 56.2% of the people who answered preferred to serve the Mysore Pak on a plate rather than directly from the box for various reasons such as It is not convenient, it may not look nice, may look petty and they did not like the existing packaging as well. Out of the 208 responses, we have noticed as shown in Fig. 1 that 65.9% of people felt that the packaging of sweets does influence their purchase and from this group, 92.6% of people felt that the existing packaging of the Mysore Pak can be improved. When asked which packaging, they would prefer their sweets to be put in, 54.2% of the people preferred paper cartons, 23.4% preferred metal boxes and 19.8% preferred plastic boxes as shown in Fig.2. The survey also gave us insight on what people, in general, think about the packaging of Mysore Pak and how it influences the way they serve their sweets to their guests at home.

The interview was conducted in the famous location Jayanagar, Bangalore. It is one of the food hubs in Bangalore where famous sweet shops are present. Three popular Mysore Pak sweet stores were selected and interviews among the sweet shop vendors were conducted. The first store was Adyar Anand Bhavan where Sri Kumar age 29 was interviewed who was relatively new to the field with only two years of experience. Their shop has a sales of 10 kg per day selling dry, soft and badam flavored Mysore Pak's daily. The average sales per day were 10kg. They sealed their plastic boxes with rubber bands to prevent the delicious sweet from getting soggy from outside. With this kind of packaging the sweet's shelf life is up to 10–15 days. The smallest packaging box of 250gms can accommodate five pieces of Mysore Pak. The second store was Pramila sweets where Nanjappa was the interviewee who was comparatively much more experienced with 10–15 years of work in this field. They basically sell the sweet in cube shape in both soft and dry textures. On average, their customers bought 250-500gms of Mysore Pak which leads to the shop selling 40kgs of Mysore Pak per day. They sell their sweets in cardboard boxes based on the customer's requirements and the shelf life of their sweets is up to 10 days. The third shop interviewed was the famous - Shri Krishna Sweets. He has 22 years of experience in this field. Their shop sells Mysore Pak in long cuboidal shapes in both dry and soft variants. They sell up to 20kgs in plastic boxes. Their largest box can accommodate up to 15 pieces which is about ¾ kg. Their sweet's shelf life is 15 days.

The above survey was conducted to understand how the marketing of Mysore Pak can be improved by redesigning the existing packaging design. Majority of the respondents have accepted the fact that the external packaging design does affect their choice of purchase. The fact of Bangalore is known as the Garden city has influenced most of the participants to link green color shades and also the famous Vidhana Soudha building with Bangalore city. Hence, to improve the exterior look of the packaging, elements of Vidhana Soudha blended with shades of green can attract many consumers and at the same time can keep the sweet more rooted towards its origin.

On the other hand, looking at the internal packaging design it is observed that the sweet boxes getting greasy from the inside and sweet pieces sticking to each other are the two major problems to be handled with the existing packaging. The improper closing system of paper carton lids has to be solved to satisfy the user's need of food freshness. Additionally, the participants have also expressed their willingness towards having packaging that looks more user-friendly and can be comfortably used for serving guests.

From the above interviews conducted one can observe that each shop sells different quantities of Mysore Pak based on various factors such as their shop's location, respective popularity, taste, etc. Mysore Pak is available in various types such as dry one, soft one, and one flavored with Badams. They are also sold in various basic shapes to attract customers. On average, 500g - 1.5 kg of Mysore Pak is bought by a customer.

Looking from the packaging point of view, it can be understood that most customers prefer plastic boxes. As it will prevent the ghee from oozing out of the container. Currently, 250gm, 500gm, and 1Kg capacity sweet boxes are available which can accommodate sweets from 5 pieces to 15 pieces.

Proposed Concept

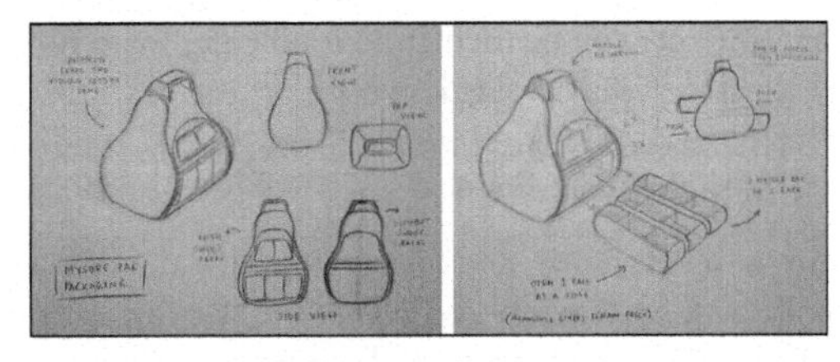

Fig. 1. Final concept in ideation phase

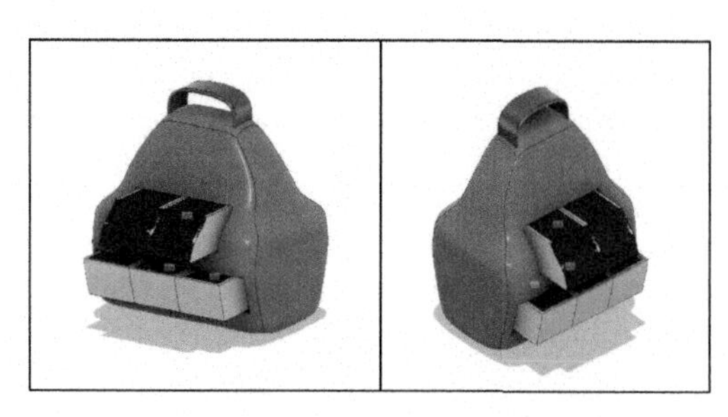

Fig. 2. Isometric drawing of the final concept from the CAD software.

Considering the points discussed above, new packaging designs exclusively for Mysore packs were ideated which will solve the flaws with the existing packaging design. From the survey conducted it can be observed that people can relate Vidhana Soudha to Mysore, Karnataka. Bangalore is also known as a green city; hence people associate the location with green color.

So, the proposed concept is a packaging design, whose external shape is inspired by the dome of Vidhana Soudha. The external illustration has traditional Mysore patterns with a blend of green colors. The internal packaging is done with the concept of individual Mysore Pak packaging. The sweet pieces are stored in racks placed in the packaging box. At a time only one rack has to be unsealed and the sweets inside can be consumed whereas the other racks remain unsealed leading to an increase in the food's freshness.

The bottom half will have three racks with each having the capacity to store three pieces of Mysore Pak and the top will have 2 racks that can hold two pieces of Mysore Pak each.

The following materials are used in this packaging design: For the external packaging and the sweet racks, carton paperboard (i.e., cardboard) can be used. For inner lining, layers of butter paper can be used., and food cling wrap for sealing the sweets (plastic wrap).

Conclusion

After researching and analyzing all the data collected, one can conclude that the packaging of Mysore Pak needs to be redesigned to adhere to the taste of the consumers.

From the survey conducted it is understood that the packaging's illustration and design must be changed by incorporating elements of the legislative house "Vidhana Soudha" along with the blend of green colors. Increasing food freshness, and preventing sweet pieces from sticking to each other has to be performed in the design phase. Individual sweet packaging concepts can also be considered to solve the above criteria.

Hence, covering all the above conclusion points - a solution has been proposed above.

A redesign packaging design for Mysore Pak, adhering to the taste of its audience. An individual sweets packaging Inspired from the Bangalore Culture. Once this concept is tried and tested by experts it can be progressively launched in the market.

Acknowledgment

A special mention to our mentor Mr. Arunachalam Muthiah for guiding us throughout the paper.

References

Swamy, C., and Shankar, B. Architecture and Heritage Resources of Mysore City.

Swamy, C., and Shankar, B. (2012). Architecture and Heritage Resources of Mysore City. International Journal of Modern Engineering Research, 2, pp. 139–43.

Sampurna, Saddi and Reddy, Sunki, Reddy, Yella. Preparation and Application of Trans-free Vanaspati Substitute in Selected Indian Traditional Foods.

Sampurna, S., and Reddy, S. R. Y. (2011). Preparation and application of trans-free vanaspati substitute in selected Indian traditional foods. Food science and technology research, 17(3), pp. 219–226.

Gowri, K. S. Kingdom of Mysore.

Lee, S. Y., Lee, S. J., Choi, D. S., and Hur, S. J. (2015). Current topics in activeand intelligent food packaging for preservation of fresh foods. ("Current topics in active and intelligent food packaging ...") Journal of the Science of Food and Agriculture, 95(14), 2799-2810

Kumar, K. R. (2003). Design and development of packages for traditional food products. Journal of Rural Technology, 1(1), 33–36

Leia Noemi, Rasa and Carolina Gerschenson, L. N., Jagus, R., Resa, C. P. O., and Patricia, C. (2017). Biodegradable packaging applied to dairy products. Advances in Dairy Products, 328.

Piergiovanni, L., and Limbo, S., 2016. Food packaging materials (pp. 33–49). Basel, Switzerland: Springer.

Malek, MohammedShakil, S., and Bhatt, Viral (2022). Examine the comparison of CSFs for public and private sector's stakeholders: a SEM approach towards PPP in Indian road sector, International Journal of Construction Management, DOI: 10.1080/15623599.2022.2049490

Livelihood Experiences of Working Women with Disability during COVID-19: Predicament and Prospect

Twinkal,[*,a] *Kshama R. Kaushik,*[b] *and Jagdish Jadhav*[c]

[a]Ph.D. Scholar, Department of Social Work, Central University of Rajasthan, India
[b]Secretary & Chief Functionary, Rajasthan Mahila Kalyan Mandal Sansthan, Rajasthan, India
[c]Professor, Head of Department, Department of Social Work, Central University of Rajasthan, India
E-mail: [*]twinkleleads@gmail.com

Abstract

The world faced an unprecedented crisis that required global responses for far-reaching consequences to our lives. They are a severely marginalized group of people who have difficulties in the public and private spheres and daily harassment, exploitation, exclusion from society, stigmatization, and prejudice. This study also examined the difficulties and obstacles they encountered when looking for an opportunity after the lockdown. The data was collected using a qualitative approach. Ten women with disability aged 18 - 40 were selected through purposive sampling from the rural region of Ajmer, Rajasthan. This study involved families from low socioeconomic backgrounds, disabled women, and workers who have been employed for more than three years. The data was gathered using a face-to-face interview and observation technique. The research presented in this paper is anticipated to shed light on several issues, including a lack of social support, a lack of awareness, a lack of resources, an economic crisis, and poor socioeconomic conditions.

Keywords: Barrier, Challenges, COVID-19, Livelihood experiences, Rajasthan, Women with disability

Introduction

The ripple effects of limiting people's rights grew wider and wider as the virus spread. The COVID-19 pandemic also contributed to the highest unemployment rate due to forced and voluntary business closures. The loss of jobs, prolonged unemployment, low pay, migration, business losses, and other hardships affecting necessities made employment one of people's most significant difficulties (Wong et al., 2021). Pandemic has impacted the vulnerable population of society, especially the person with a disability. The study shows that 84.2% person with a disability stated that their lives had been affected due to the lockdown, around 45.7% of persons with disabilities were forced to borrow money during the lockdown mainly for livelihood, and 84.7% had to borrow or request support for food to cope with financial crisis (GVS, et al., 2020). People with disabilities are defined as having "long-term physical, mental, intellectual, or sensory impairments that, when combined with other factors, may prevent them from fully and effectively participating in society in a manner that is comparable to that of others" (Mactaggart, et al., 2018). The Convention on the Rights of Persons with Disabilities recognized the significance of the right to work early on by adopting article 23 of the Universal Declaration of Human Rights, and article 27 reiterates the right of people with disabilities to work on an equal basis with others. Additionally, the Sustainable Development Agenda (SDA.2030), which the United Nations General Assembly also endorsed in September 2015, declares that it would leave no one behind in its global effort to promote social and economic development for the benefit of people with disabilities.

However, livelihood is the central well-being of a person with a disability. Chambers and Conway (1992) defined *livelihoods* as how individuals or households can meet their basic needs. It encompasses remunerated not only labor but also an individual's capabilities (e.g. level of education, skills), assets, and participation in other productive activities (e.g. farming for direct consumption) (Mactaggart et al., 2018). Persons with disabilities in the Northern countries perform worse economically than people without impairments, as seen by lower employment rates, income, and earnings (Hanass-Hancock and Mitra, 2016). Disability-related exclusion from employment opportunities is pervasive because of a lack of educational background and occupational training. The lack of physically accessible infrastructure, extravagant and inaccessible transportation, inaccessible buildings and facilities, a lack of employment opportunities and housing, inconvenient (rough) roads, and social discrimination limits the ability of disabled people to support themselves (Mulubiran, 2021). Research conducted by Verma and Chandarkanta (2016) showed that people with a disability are ready to work outside. Still, there is no accommodation for them, so their parents and relatives do not permit them to work. The accessibility to entry into public services is also a measure of livelihood strategy. The disabled-friendly public services symbolize social progress, remove the barrier, and help reduce social stigma. Chacko and Anooja (2015) study show that only 31.67% have access to public services.

It becomes more difficult for women with disability to improve their livelihood and social status through developmental work than the man doing similar work. In India, women with disability were twice as likely not to work as men with disabilities, which showed double discrimination against women (Mactaggart et al., 2018). Women with disability face multiple forms of prejudice as being a woman, disabled, and poor. The problem can occasionally be compounded by age, geography, and additional vulnerability related to ethnicity and minority status. According to research by Halder (2008), 80% of girls have taken various vocational courses at different institutions; however, the results are unsatisfactory regarding job chances. According to a survey, people with intellectual disabilities are seen as being unable to work outside and being a risk to society. Girls and women come from an underprivileged population and often face sexual assault or other forms of violence. Still, there is a lack of study on the struggle faced by women with disability in improving their livelihood patterns. This paper emphasized the livelihood circumstances and experiences of women with disabilities during the pandemic, which was worsted in the vulnerable sections of society. This study also examined the difficulties and obstacles they encountered when looking for an opportunity after the lockdown.

Methodology

The qualitative approach was used to explore the livelihood experiences of working women with a disability during the COVID-19 period. This approach was used to understand the challenges and difficulties faced by women while pursuing any occupation.

Sample Design

Table 1. Socio-Demographic Characteristics of Respondents

The participants were ten working women with disability from the rural region of Ajmer, Rajasthan. Informed consent was taken, and participants were told about the aim and objectives of the study. The interviews were recorded for analysis purposes. All the women's age ranged from eighteen to thirty-seven years. The study showed that 60 percent of women had secondary and 20 percent had primary education. Fifty percent of women with disability were self-employed; tailoring and running shop was common among these women. A maximum number of women earn below 6k per month, and some women earn below 2k.

Process of Data Collection

Participants were selected with the help of the Rajasthan Mahila Kalyan Mandal, Rajasthan. The organization is inspired by the Gandhian philosophy and strives to improve the lives of people with disability. The organization helped in identifying the women with disability who are actively participating in the livelihood practices. The term "Women with Disability" in this context refers to women who have any kind of disability that prevents them from working like non-disabled people and supporting their families through their means of subsistence. Women who have worked for more than three years experienced COVID-19, and solely reside in Ajmer's rural areas were eligible to participate.

The data were collected from women with disability. Researcher took help from the Community-based Rehabilitation (CBR) workers who are providing services in Rajasthan Mahila Kalyan Mandal. The researcher along with CBR workers visited the residence of participants and collected the data.

Tools for Data Collection

Interview Schedule

The information was gathered through a semi-structured interviewing process that included questions about the obstacles that women with disabilities had to overcome during the COVID period. The face-to-face interviews were conducted at the residence of the respondents according to their convenience and availability. The participants shared their views and made the interview insightful through its content. The interview lasted from twenty minutes to at least forty-five minutes.

Process of Data Analysis

All of the interviews were transcribed verbatim. Thematic analysis was used for identifying themes. It is used to identify, analyze, and understand the challenges and difficulties faced by women with disability in the difficult time of COVID-19.

Findings and Discussion

Major four themes about the livelihood experience of working women with disability were analyzed. The themes explored the challenges and obstacles faced by women with disability. The major themes are the challenges and barriers during the lockdown, challenges, and barriers after the lockdown, social support, and domestic violence.

Challenges and Barrier faced during the Lockdown

Regardless of disability, women with disability engage themselves in work and struggle to enhance their self-esteem and autonomy. In this study, four women with disabilities were employed in the private sector (teaching, tailoring, housekeeping, and working in a nursery), five women were self-employed and running their businesses in this study, and one woman with a disability was working as an unskilled worker (working in NREGA). Employment is a significant factor that depends on knowledge, abilities, and overall health. It permits the individual to support their family and society while earning an income. People with disabilities rarely get the chance to improve their well-being. They relied on their spouse and other family members heavily. This study shows that women with disability lack self-esteem and financial autonomy. Seven out of ten disabled women lost their employment during the first lockdown. After losing their job, they had no other means of support. In addition, they had obligations to other family members that were neglected during the lockdown. The financial crisis was the main problem they were dealing with. During the lockdown, transportation facility was unable, so the demand and supply of necessary product were affected. Families used their savings in this dire circumstance and went into debt.

Before Covid, my shop was running well, and I was able to work. During the lockdown, groceries got finished, and we had a financial crisis. I had to close my shop. Till now, my shop is closed. I am struggling to arrange the money to restart the shop and to pay the debt (statement by a 28-year-old woman with a loco motor disability).

Most families relied on loans from friends and family to make ends meet. They utilized this cash on necessities. In the lockdown, the majority of the households experienced debt. Their aspiration to be independent was dashed. The findings (GVS, et al., 2020) demonstrated that lockdown had a detrimental psychological impact on people with disabilities, with financial constraints being the most frequent cause. People claimed that COVID-19 had undermined their resolve and self-assurance.

Socio-demographic Characteristics of Respondents	Number of Respondents	Corresponding Percentage
Age (N=10)		
18-22	1	10
23-27	3	30
28-32	5	50
32-37	1	10
Above 37		
Education (N=10)		
Primary	2	20
Secondary	6	60
Intermediate	1	10
Graduate		
Post Graduation	1	10
Illiterate		
Marital Status		
Unmarried	6	60
Married	3	30
Separated/ Divorced	1	10
Occupation		
Skilled Worker	4	40
Unskilled Worker	1	10
Self-employed	5	50
Income		
Below 1k	1	10
1k-3k	3	30
4k-6k	6	60
Above 6k	0	
Socio-economic Status		
6Lower	4	40
Middle	6	60
Higher	0	

Allah showed us the worst days during the lockdown. I took a loan before lockdown; during the lockdown, I was not able to pay now, I will have to pay double amount of the loan (Statement by 28-year-old woman with loco motor disability).

Some women with disability were self-employed and were getting work. However, things became worse for them as well. After the lockdown, they were not getting any work.

Everyone is a tailor in their home; I used to get few orders for making suits. However, in lockdown, I was getting only 2-3 suits in a month (Statement by 30-year-old woman with loco motor disability).

One of the biggest challenges that persons with disabilities encounter on a daily basis is poverty. Most disabled ladies claimed that the village's water facilities were inadequate. Even during the summer, water is delivered to rural communities every eight-ten days. Even getting water to drink was difficult for them. Some homes lacked water connections; they had to fill water with buckets. The challenges got worse during the lockdown. They had to carry drinking water from outside. According to study in the same area of concern (Hasan et al., 2021) stated that COVID-19 had a detrimental effect on the lives of individuals with disabilities and the level of food security in their households. They were forced into extreme poverty due to their drastic income loss, and they had to turn to friends, neighbors, and relatives for financial assistance. Millions of individuals, especially disabled women, were affected by the lockdown, impacting their daily lives. The most vulnerable members of society suffered the most. During the lockdown, women with disabilities faced challenging circumstances.

Challenges and barriers faced after the lockdown

After the lockdown, the majority of the women returned to their jobs. Due to their lack of social support, commencing work again was difficult for them. The majority of them do not have enough work to support themselves. The majority of disabled women received the same pay after the lockdown as they had before.

I was hired as a teacher at a private school while also managing my father's business. I was dismissed from the school during the lockdown, and the store was shut down. Daily needs were a struggle for us. I considered starting an E-Mitra in the village after lockdown. I frequently went to the Sarpanch to ask him to provide me a place to set my shop. I have also made numerous visits to government officers" offices. I did not get any support from them. They are giving place only to wealthy people (Statement by the 28-year-old woman with locomotor disability).

This study shows that disabled women's income is insufficient to sustain their way of life. Although they rely on their family for support, their choice of self-employment as a means of subsistence is affected by outside forces. Women with disability are not getting opportunity and government services are not reaching to the rural region. The same worries were raised in the study, which revealed that women with disabilities have a greater unemployment rate and, even when they find jobs, typically earn low pay. Impaired mobility, a lack of support from their families and formal groups, and unfavorable physical surroundings are additional challenges they confront regarding access to public services.

Social Support

The role of social support in the lives of people with disabilities is significant. Without social support, it becomes challenging to establish social resilience and harmony in daily life. Families were crucial throughout this challenging time. This study discovered that most family members supported disabled woman's independence and accepted their status. Family members assisted them with household chores and provided financial support during the lockdown. The findings are consistent with the research that person with a disability has more reliance on their immediate family member and staff within their social circle. The neighborhood and friends showed no financial support but only moral support.

Everyone was suffering from financial issues. My neighbor did not help by giving money, but they gave moral support (statement by the 25-year-old woman with a locomotor disability).

The majority of the women with disability received help from the government and village sarpanch. They received food grain, but the quality was not good. They did not get any financial help from the government. All were receiving the disability pension on time during and after the lockdown. They also received help from non-profitable organizations regarding food kits during the lockdown.

The food quality was not good. Grain used to come with fungi. We did not have any choice of leaving that food. We washed them and had food (statement by a 32-year-old woman with locomotor disability).

Some women with a disability did not receive any help from the government and non-profitable organizations regarding food and financial help. Lockdown proved to be more difficult for them. The findings are consistent with (Mathias, et al., 2020) that social support was reduced during the pandemic due to long-term isolation and financial crisis. Government and non-profit organizations provided services to women with disabilities in terms of grain kits; some families received financial help too.

They were giving food only to the wealthy people. No one helps us (statement by 30-year-old woman with locomotor disability)

Social support is crucial for the development of the individual. Women with disability were required to receive formal and informal social support. Still, when discussing legal assistance, the efficiency is low.

Domestic Violence during Covid Period

Early marriage has a long tradition in Indian culture. In the country's rural areas, it has grown widely and evolved into the societal norm. The prevalence of early marriage was found to be relatively high among disabled women. Early marriage has a negative impact on a child's right to protection, education, and health (UNICEF). According to UNICEF research, at least 1.5 million girls under 18 get married in India every year, making it the country with the most significant proportion of child brides worldwide—about a third of all child brides globally. Any discussion of domestic violence can be upsetting for anybody. When we discuss women with disabilities, it is a delicate subject. In this study, out of ten women, six married at an early age, 12-15 years. Early marriages affect not only the girl child but also the family and community. It can be contributing factor to domestic violence. The threat of domestic violence increased during the lockdown. Most women with a disability experienced no violence at home during the lockdown. Some women claimed that their partners assisted them with housework.

By God's grace, my husband understands me and accepts by disability. He helps me with domestic work such as washing and cutting vegetables, preparing tea, and taking care of children (statement by a 32-year-old woman with a locomotors disability).

However, several women were unaware that abuse is a kind of violence as well. They experience abuse regularly, so it has become the norm.

I got married at the age of 14. In starting of my marriage, my husband and in-laws used to beat me. But now he does not beat me; sometimes he abuses.... It is normal to abuse here (Statement by a 35-year-old woman with locomotor disability).

My brother teases and calls me "langdi." He is short-tempered. Whenever he got angry, he started beating me. Its brother-sister fights (statement by a 24-year-old woman with a locomotor disability).

Awareness regarding domestic violence is less in the rural region. Verbal abuse is considered usual and accepted in society. Women with disabilities experienced numerous family disputes during COVID-19, which disrupted the peace in the home. The same worries were reinforced by (Afifah, 2021), who said that the double strain of the pandemic and declining income have led to a rise in family disputes.

Overall, the research highlighted the difficulties and obstacles that disabled women must overcome to support themselves, underscoring the need for intervention. With the assistance of the government and non-profit organizations, the practitioner may offer a source of income. Social workers can assist them in improving their well-being or gaining confidence in their particular line of employment.

Limitations of the Study

The study is restricted by its small sample size and the inclusion of participants in livelihood services. In addition, families confronted the same issues concerning socioeconomic status, domestic circumstances, and issues affecting women with disabilities. It was difficult for researchers to search the women with disability who were practicing livelihood services in a rural region. The women were not ready to talk about domestic violence issues because all family members were sitting together while discussing. Families found it challenging to discuss their experiences with their means of subsistence. The researcher did not face any language barrier.

Conclusion

This study presents a thorough knowledge of the difficulties and constraints that disabled women experienced regarding pandemic livelihood patterns. The rights of women with disabilities, including their freedom of speech and expression, right to health, right to employment, and right to movement, have been violated by Covid-19. The key to the nation's economic development is empowering women, particularly those who had to live with social identity obstacles, disabilities, class, and sexual orientation. Action must be taken to address the hurdles in addressing the livelihood practices among women with disabilities.

Acknowledgment

We wish to thank the parents who participated in this research. We also wish to thank Rajasthan Mahila Kalyan Mandal, Rajasthan for helping with data collection.

Disclosure Statement

The authors reported no potential conflict of interest.

References

Afifah, Q. A. (2021). Analysis of the Impact of the Covid-19 Pandemic on Family Harmony: Case Studies on Family with and without People with Special Needs. Indonesian Journal of Community and Special Needs Education .

GVS, M., S, K., MG, L., S, S., and S, T. (2020). A Strategic Analysis of Impact of COVID-19 on persons with disabilities in India. Hyderabad, India: Funded by CBM India Trust, and Humanity & Inclusion (HI).

Hanass-Hancock, J., and Mitra, S. (2016). Livelihoods and Disability: The Complexities of Work in the Global South. International Perspectives on Social Policy, Administration, and Practice .

Hasan, M. T., Das, A. S., Ahmed, A. I., Chowdhury, A. M., and Rashid, S. F. (2021). COVID-19 in Bangladesh: an especially difficult time for an invisible population. Disability and Society, 1362-1367.

Mactaggart, I., Banks, L. M., Kuper, H., Murthy, G. V., Sagar, J., Oye, J., et al. (2018). Livelihood opportunities amongst adults with and without disabilities in Cameroon and India: A case control study. PLOS ONE .

Mathias, K., Rawat, M., Philip, S., and Grills, N. (2020). "We've got through hard times before": acute mental distress and coping among disadvantaged groups during COVID-19 lockdown in North India - a qualitative study. International Journal for Equity in Health .

Mulubiran, T. (2021). Livelihood Assets and Strategies of People with Disabilities in Urban Areas of Ethiopia. Scandinavian Journal of Disability, 63-73.

National Conference on State Legislatures. (2020, 03 05). COVID-19: Impact on Employment and Labor. Retrieved 07 12, 2022, from https://www.ncsl.org/research/labor-and-employment/covid-19-impact-on-employment-and-labor.aspx

Verma, S. K., and Chandarkanta. (2016). Livelihood Condition of Disabled People: A Case Study of Tiruchirappalli District of Tamil Nadu. Bangladesh rural Development Studies .

Wong, J., Ezeife, N., Kudla, A., Crown, D., Trierweiler, R., Capraro, P., et al. (2021). Employment Consequences of COVID19 for People with Disabilities. Journal of Occupational Rehabilitation

Developing Storytelling as a Method of the Design process in Bachelors of Interior Design Education

Kriti Malkani,[*,a] *and Dr. Rikimi Madhukaillya* [b]

[a]Unitedworld Institute of Design, Karnavati University, Gandhinagar, India
[b]Unitedworld Institute of Design, Karnavati University, Gandhinagar, India
E-mail: *kriti@karnavatiuniversity.edu.in

Abstract

Stories are a great method to start understanding what difficulties exist once designers have a basic understanding of the environment in which they are working. Designers can build new stories based on their user research, or they can just utilize it as a springboard to create a larger narrative. In the current study, storytelling and visual narratives are emphasized as crucial components of the design-thinking process. At various phases, stories are utilized to both comprehend and construct theory-based narratives. Additionally, they aid in enlarging perspectives on narrative techniques and the design-thinking process in design education. Studying the perspective of the learner while designing is crucial to comprehending the teaching process. The study sheds light on how designers collaborate and work alone to develop new ideas, and how using storytelling as a medium for cognitive processes might make those ideas more original. Both a user's expectations and a designer's experiences can be matched by stories.

Keywords: Interior design narratives, Storytelling, Design process, Education, Design pedagogy

Introduction

An excellent approach to connecting and communicating is via stories. We have been sharing stories with one another for aeons to pass on knowledge and instil moral ideals. Since excellent stories also require a lot of profound imagination, storytellers have always been valued and revered for their abilities and creative understanding. Storytelling is a form of narration, expression of imagination, and ideas with a blend of experience depicted into an interesting framework of sequences connected in harmony. A good story comes with capturing the imagination, creating dynamism and surprise elements (Dahlström, 2019, 11). Any narration including all these elements makes the story meaningful and interesting for the audience to value it at every level. As designers, telling stories is our way to engage in the design process. Essentially, stories can be used in four ways during the design process (Hunsucker and Siegel, 2015, 3)

Research: Designers must comprehend the issues, experiences, and pain points of the user as part of the original design research. They can create knowledge in a way that makes it simpler and more comprehensible by using storytelling techniques. This can aid in identifying the actual issue and direct a designer toward spatial design problems.

Ideation: Inquiring is an essential aspect of the design process. Designers recreate stories from research to better grasp the challenge. Research can be used to kick off this step of the design process. Designers can build tales based on the data given and extend the perspective to gain a complete picture in order to develop real concepts.

Prototyping: An explanation of the answers is required when applying the concepts to comprehend the area. As a result of creating prototypes, these can be expressed using the storytelling technique. A story that uses the user as the main character in consecutive frames can be built using the solutions that can be sketched. Storyboarding and videography are two examples of visual storytelling tools that might be useful at this stage of the procedure. (Hunsucker and Siegel, 2015, 5)

Presentation: The last stage in the process is presented. This is the final step where stories can work as a great way of selling design. As stories are persuasive, this stage helps designers sell their ideas to users and clients.

This framework aids in keeping users interested in design. However, design tales must be distinct from movies, novels, and films. While films are more concerned with sound, music, and human emotions, design tales must be practical and result in tangible effects.

Designers must be genuine. Designers can learn narrative abilities to help them turn their ideas into design solutions. The designer's primary duty to its users is to keep them in the spotlight as the main protagonists of their story. This is comparable to how they are the focal point of the designs.

Storytelling in Design

Every creative process in design goes through a series of steps including framing the problem, idea generation, designing, and then finally testing it. A point of view helps in discovering the source to figure out problems by understanding life experiences. As part of the storytelling process, it's important to first write the story. These include a beginning (the launch), a middle, and an ending (the landing part). Further, it needs to be spoken or at least converted to a dialogue for conversation. This is part of oral communication in storytelling. Ultimately once both these are put together, the visual storytelling comes into action. The brain develops visual images of the storyline which can be put to advantage and capitalized using Vitruvian aspects of Attractiveness (novelty and appeal), soundness (quality and reliability), and Utility (useful and function) (Huber, 2017).

When the stories of fiction are told, imagination gets utilized to conjure up situations that teach something about the world that analysis would not otherwise reveal. Because of this attribute, the story is a method that designers are inclined to adopt when creating their designs. (Parrish, 2006)

Writing design stories can help to improve the designer's viewpoint at every stage in the design process. Following certain guidelines for writing design stories can help to make the user experience better and more tangible. These guidelines can include: Writing design stories for clients in order to communicate the design. This might be incorporated into the project plan. Analysis details can be used as backstories. Outcomes can be surprising elements for improvising designs. A single idea connecting the story elements can help to maintain the design language. Learners' responses can be recorded for formative assessments of design stories for assessment processes. Every aspect of a user's experience can be recorded and translated into a setting and set of frames. In addition, imagination plays a crucial role while writing a story, prototyping the results, and designing a formative assessment technique to evaluate the processes. Stories are exploratory devices to narrate user participation in design. Also, empathy is an important element in enhancing the storytelling process in design. This helps to anticipate an authentic experience for the user. Empathy is a process of putting oneself in place of the user, and re-iterating the entire experience. It is only when one can feel what the user experiences, can one design the story accordingly. For this, we need to know our users, collect surveys or do some research, become learners, adapt to create design stories, and further learn acting skills. Students can also be taught to defend their design decisions by experiencing a user's perspective in design. (Parrish, 2006, 79,80)

Mediums of Stories (History of Storytelling)

Looking at our history, Aristotle pointed out the interrelation between the way stories are told and the human experience. He gave the world seven golden rules of storytelling such as: (Dahlström, 2019)

1. The arranging of events
2. The story's main character
3. The main idea that connects the frames/scenes/incidents is referred to as the idea.
4. Conversation, discourse, and speech
5. Song - Sound to accompany the story
6. The setting for the character, speech, and events.
7. A powerful aspect/ creating impact

People tell stories in a number of mediums:

Oral medium - The medium of telling stories by means of talking is one of the oldest versions in storytelling. It is the most basic way of passing on information. Elderly people used this medium to communicate and share their experiences. Storytelling began with recording experiences.

Written medium: Scriptwriting is a very important part of storytelling. People write their experiences in form of both personal and public notes. In these modern times, people take to blogging and recording experience through digital media. History has seen a lot of holy scriptures telling the tales from the past. This has given rise to a lot of moral values which exist well in the present.

Visual narratives: Rock art and fresco are some very popular examples of visual stories from ancient times. Our ancestors used the most local day-to-day materials in the purest and most creative form to express their experiences and record them. These were later also found on leaves, wood, and a number of daily materials used. Visual narratives exist in a variety of forms and offer inspiration in the field of storytelling expressions.

Interior Design Education

In order to create an interior environment that is practical and reasonable, comfortable and beautiful, and meets the needs of people's material and spiritual lives, interior design is based on the nature of the use of the building, the environment, and the corresponding standards. It also uses material and technical means and architectural design principles. (Wu, 2016, 68)

Education in interior design has been crucial in recent years as we learn, explore, create, serve, and transform the world. Interior design education contributes to the identification of this approach, which has sparked problem solutions across numerous industries like architecture, planning, furniture design, light design, exhibition design, and even set design. Humans identify their wants in daily life and strive to regularly discover solutions to these demands.

Worldwide, there are many design colleges pertaining to numerous design courses as part of the undergraduate, postgraduate, and doctoral programs. As part of the undergraduate curriculum in interior design, the courses vary from a variety of verticals ranging from furniture details to space design of all kinds. Students are given exercises to make their design thinking skills powerful and help them translate ideas into prototypes. Interior design education includes the teaching of the space environment. Space environment uses values related to functional requirements, historical contexts, architectural elements and styles, atmosphere, and many other psychological factors. Furthermore, functional principles, economic principles, security principles, and feasibility principles are features of interior design education. It is critical to recognise that interior design is more than just a production vertical; it also addresses social, cultural, and artistic aspects of user experiences. Interior design affects the impact of events in space. Spatial design is related to understanding user behaviour which further adds to the element of memory in the viewer of space (Wu, 2016, 68). Interior design education includes a wide variety of narrative features ranging from diversity, humanities, experiences, historical heritage, and spatial arrangements to minute details of elements with materials, colors, lighting, and finishes. It is a diverse field with a lot of inputs in exploring a wide range of narrative elements.

Visual Narratives as part of Storytelling

Design narratives can be used to enhance design processes and add meaning to a lot of design products. The importance of narratives in design depends on how, when, and where it is being used with respect to the context, purpose, and user. Narratives can be theoretical as well as visual. Theoretical narratives are supported by oral and written scripts. Visual narratives are a combination of three parameters: oral, written, and visual parameters. "Thinking like a designer can transform the way you develop products, services, processes—and even strategy." (Brown, 2008) Design thinking is a standard process that leads to innovative ideas and problem-solving. Designers work to solve problems based on the perspective of the target users. This involves a sequential process of understanding the problems, creating ideas, visualizing the solutions, and eventually designing prototypes (Muller-Roterberg, 2020, 9,10). While storytelling is also a sequential framework of incidents with a beginning, body, and ending, design thinking blended with it helps to visualize and communicate better.

Interior Design student's Journey and Thinking process

The field of interior design is very demanding and necessitates a great deal of knowledge. Although it offers you a bright career, it also necessitates a lot of labor. It requires a lot of hard work, practice, and ingenuity. It is entertaining while also involving numerous design thinking and design process-related processes. The most crucial aspect of a student's path through education and practice is design thinking. A design student in this particular field explores all forms and processes in design thinking right from the beginning of interior design education. Thus, a student's journey is quite challenging and demands creative solutions. Students in the design field are expected to come up with innovative solutions in design.

Existing Courses on Storytelling Worldwide

A lot of courses online and in different universities talk about storytelling as part of UX design and Interaction design. Storytelling has been widely used as an expression in various fields, but interior design specifically. It is important to look into these courses in order to understand the structure of teaching, the content, learning outcomes, and evaluation.

1. Coursera has the following course listed:

Course name: Powerful Tools for Teaching and Learning: Digital Storytelling

Content: The course is divided into 5 weeks. Each week concentrates on different exercises and activities related to forming a good story. It begins with choosing a random topic, but with a purpose for digital story formation. This is followed with scriptwriting and supported with storyboards. Students are allowed to click pictures, create diagrams and add audio recording to narrate the story. This is followed with assembling all the parts together and creating a combined sequence of framework. It ends with a proper assessment of the entire framework of ideas and revising it if required.

In addition to this there are more courses which shall be listed for a wider study of how such workshops and courses can be conducted.

Storytelling as a pedagogic tool

Term "Pedagogy" is derived from Greek paidagōgikos. Pedagogy is the discipline that is about the theory and practice of teaching. Pedagogic is an adjective and it refers to one of various teaching methods. For the present study, since storytelling is considered a teaching method to produce the desired effect, it is called a pedagogic device (Sethi, 2018)

Conclusion

Storytelling is one of the oldest and most creative forms of expression, according to all the assessments. This technique for transferring one's experience into media has been extensively studied. Over time, people have chosen from a range of media to capture their experiences. With time, this technique has evolved into a means of bringing dreams to life as well. This approach now serves as both a tool for expression and for exploring and decoding processes.

The technique aids in the conversion of concepts into scripts and visual storytelling. A tale can aid in giving a goal significance. Organizing ideas into a string or sequence can be beneficial. It offers designers numerous chances for interaction and communication. As a result, there are many different verticals within the design that can be used to solve issues relating to humans, their environments, and their experiences. Using stories to create better design solutions and make the journey more exciting.

Specific to interior design, stories can help to make meaning with the long-sum processes. As interior design has a lot of subdivisions in terms of categories and possibilities, the students can explore a range of design processes using narratives. Thus, narratives can become essential elements of the interior design process.

References

Brown, T. (2008, June). Design Thinking. Harvard Business Review, June 2008 (2008), 11. https://readings.design/PDF/Tim%20Brown,%20Design%20Thinking.pdf

Dahlström, A. (2019). Storytelling in Design: Defining, Designing, and Selling Multidevice Products. O'Reilly Media. 1491959371

Huber, A. (2017). Telling the Design Story: Effective and Engaging Communication. Taylor & Francis.

Hunsucker, A., and Siegel, M. (2015). Once Upon a Time: Storytelling in the Design Process. Cumulus Conference proceedings Learn x Design: The 3rd International Conference for Design Education Researchers.

Lawson, B. (2018). The Design Student's Journey: Understanding how Designers Think. Routledge.

Mcquillan, V. (n.d.). T1 - Design by Narrative. Telling Brand Stories with Interior Design and Architecture. Inner Magazine, 2018(2), 10.

Muller-Roterberg, C. (2020). Design Thinking for Dummies. Wiley.

Parrish, P. (2006, August). Design as Storytelling, 50(4), pp. 72–82. https://eric.ed.gov/?id=EJ774598:79,80.

Sethi, J. (2018). Storytelling as a pedagogic device Developing a manual for primary school teachers. Shodh ganga. Retrieved 12 11, 2021, from http://hdl.handle.net/10603/312629

Wu, Y. (2016). Research on Interior Design Education Based on Narrative Theory (Vol. Proceedings of 4th IEESASM conference). Atlantis Press. 10.2991/ieesasm-16.2016.15: 68.

Review on Incorporating Visual Storytelling as a Method of the Design Process in Design Education

Kriti Malkani[*,a] *and Dr. Rikimi Madhukaillya* [b]

[a]Unitedworld Institute of Design, Karnavati University, Gandhinagar, India
[b]Unitedworld Institute of Design, Karnavati University, Gandhinagar, India
E-mail: [*]kriti@karnavatiuniversity.edu.in

Abstract

In order to arrive at novel conclusions, design thinkers employ a wide variety of methods. The use of narrative in the design process is becoming increasingly common. The process of designing with a story in mind. As a component of the larger design thinking process, the storytelling technique has been investigated at varying depths in virtually all design disciplines. Global student populations are unconsciously employing this method to generate fresh ideas for product design. The term "interior design" refers to the art of planning and creating aesthetically pleasing and functional interior environments. Each area has its backstory from which the user's context and actions may be gleaned. One approach that might help with this is using fiction writing while planning an interior space. In the context of design education, this investigation may lead to the development of more effective strategies for reaching acceptable interior design solutions. This study aims to investigate and experiment with the design process as a narrative for the development of original answers to problems.

Keywords: Narrative, Design cycle, Visual, Design instruction, Technique

Introduction

As soon as designers have a firm grasp of the environment in which they are tasked with solving issues, they may learn from narratives. Storylines uncovered through user research can be re-created by designers, or they might serve as inspiration for an expanded narrative. Hunsucker and Siegel (2015). In this research, we highlight the importance of storytelling and visual narratives in the design-thinking procedure. Theories are understood and new theoretical narratives are created through the use of stories at many different phases. In addition, they aid in broadening perspectives on the role of storytelling techniques and design thinking processes in the classroom. In order to fully grasp the educational process, it is crucial to examine the viewpoint of the learner while the product is being developed. The study provides fresh insight into how designers collaborate and think creatively to develop new approaches to old problems, and demonstrates how the cognitive process inherent in storytelling may help designers come up with novel, effective solutions. Narratives can be tailored to either the needs of the user or the designer's background.

Literature Review

Can you define storytelling?

We can see that stories have always played an important part in society and that their worth has only increased as time has passed. In 2019, (Dahlström) As Dahlström (2019) puts it, "stories have a compelling influence on us because they connect with us emotionally."

One of the best ways to bond and share ideas is via shared stories. Humans have been using tales as a means of communication and moral instruction for centuries. Because excellent stories need a lot of in-depth imagination, storytellers have traditionally been held in high esteem for their abilities and insight (Dahlström, 2019, 1, 2). In storytelling, the teller narrates events, describes settings, and introduces characters through a series of vignettes that flow smoothly into one another. For a tale to be successful, it must be able to "capture the imagination," as well as "create dynamic and surprise components" (Dahlström, 2019, 11). Any telling that incorporates all these components into its fabric is bound to be appreciated on several levels by its listeners. To participate in the creative process, we designers rely on narratives. Briefly, there are four stages of the design process in which narratives can be employed: For example: Hunsucker and Siegel, 2015, p.

First and foremost, designers must do research to gain an understanding of the user's issues, experiences, and pain points. They can employ narrative techniques to make complex ideas more accessible and comprehensible. This can aid in isolating the root cause and leading the designer toward spatial problems.

A key component of the creative process is generating ideas by questioning assumptions. To better grasp the issue at hand, designers often act out scenarios drawn from the background study. This step of the design process can be kicked off with some research. Concrete concepts may be generated by designers by weaving together narratives from the available data and expanding their view to see the big picture.

The prototyping process requires a description of potential solutions in order to effectively apply the concepts and get an understanding of the environment. As a byproduct of prototyping, they might be represented through narrative. Sketching out potential responses can help construct a narrative in which the user takes center stage. Storyboarding and videography, which are visual storytelling approaches, can also be useful at this point. (Hunsucker and Siegel 2015, p.)

In the end, the procedure culminates in the presentation. Telling compelling tales at this stage may be an effective means of marketing the final product. This is a crucial step in the design process because of the persuasive power of tales to win over target audiences (in this case, users and customers).

Maintaining consumers' interest in the design is made easier by this approach. However, design narratives cannot and should not be compared to movies, books, or television shows. Unlike movies, which tend to concentrate on narrative elements like ambiance and character development, design stories should have a clear purpose and an obvious outcome.

Realism is essential for designers. Designers can be educated in the art of narrative construction in order to better articulate and implement their ideas. It is the designer's primary responsibility to ensure that the users remain the focus of the narrative at all times. This is analogous to how they play a starring role in the artwork. In their study (Hunsucker and Siegel, 2015, pp.)

Storytelling in Design

Problem definition, ideation, design, and testing are the standard phases of any design process. By gaining perspective, one can better grasp the circumstances around an issue and, in turn, find its solution. To narrate a narrative, one must first put pen to paper. There is the start (the launch), the main body, and the conclusion (the landing part). Moreover, it must be delivered or at least transformed into a dialogue for use in conversation. This is an integral aspect of telling a narrative orally. When these two elements are combined, only then can visual storytelling begin. Using the Vitruvian principles of Attractiveness (novelty and appeal), Soundness (quality and dependability), and Utility (useful and function), the mind creates visual representations of the plot. (Huber, 2017).

Fiction writers use their creative faculties to make up scenarios that illuminate truths about the world that cold analysis would miss. Because of this quality, designers frequently utilize narrative while conceptualizing new ideas (Parrish, 2006).

When a designer writes down their thought process, it can allow them to see things more clearly at every level of the process. The user experience may be improved and made more concrete by adhering to specific criteria while developing design stories. The following are examples of what these rules may cover: Crafting client-facing design narratives that effectively convey design intent. The project plan might be updated to include this. Backgrounds can be constructed from analysis information. Unanticipated results may be a welcome source of inspiration when it comes to making last-minute adjustments to plans. When there is a central theme that ties together the many parts of the tale, it's easier to keep the visual style consistent. It is possible to use students' answers to design stories as formative assessment data. A user's whole session can be captured and turned into an environment and sequence of images. Additionally, creative thinking is essential for coming up with plot points, developing prototypes, and coming up with a formative evaluation strategy to analyze the procedures. The use of stories as an investigative tool to describe user involvement in design is common. Empathy is also crucial in the design process while delivering a tale. This aids in preparing for a genuine user experience. By placing oneself in the user's shoes, one might develop empathy for their experience. Only by empathizing with the player can one craft a narrative that truly engages them. To achieve this goal, it is necessary to understand our target audience, conduct appropriate surveys or research, adopt a growth mindset, iteratively generate design stories, and hone our acting abilities. Students can learn to justify their design choices by putting themselves in the shoes of a user during the design process. According to (Parrish, 2006, pp. 79,80)

Mediums of Stories (History of Storytelling)

Aristotle saw the connection between narrative form and human experience as he examined our past. He outlined seven guidelines for effective tale-telling, including: As reported by Dahlström (2019).

1. Plot - arrangement of incidents
2. Character – the main character in the story
3. Idea – the main idea connecting the frames/ scenes /incidents
4. Speech – dialogue, and conversation
5. Song - Sound to support the story
6. Decor - the space where the character, dialogue, and events take place
7. Spectacle – impact-making element

There are several ways that people share stories:

One of the earliest forms of storytelling was transmitted orally, through the act of repeating tales from one person to another. It's the simplest approach to sharing knowledge with others. This platform was utilized by the elderly to interact and share their life stories. The first steps in telling stories were made through the documentation of events.

When telling a tale, scriptwriting is crucial. People document their lives in a variety of journals, some of which are accessible to the public. People in the current day frequently use digital media to document their lives, such as through blogs. Many sacred texts have been written throughout history, each detailing a different period of time. This has resulted in many long-standing moral principles.

Examples of ancient visual tales include rock art and fresco. Our forebears recorded and communicated their lives using the most basic, locally available resources in the most original ways possible. These were subsequently discovered on leaves, wood, and other commonly-used objects. There is a wide range of visual narrative styles that may be mined for ideas in the art of storytelling.

Interior Design Education

According to the American Society of Interior Designers (ASID), "Interior Design is based on the nature of the use of the building, the environment, and the corresponding standards, the use of material and technical means and architectural design principles to create a functional and reasonable, comfortable, and beautiful, indoor environment that meets the material and spiritual life needs of the people." (Wu, 2016, 68)

Education in design has grown more vital in recent decades as a means to learn about, investigate, shape, create, serve in, and alter the world. The design process has led to innovative approaches to solving problems in a wide range of fields, and design education play a key role in branding this method. Every day, we humans assess our situations, identify problems, and actively seek answers. Bothra (2019). There are several schools of design across the world that provide a wide variety of design-related undergraduate, graduate, and doctorate degree programs. Courses in interior design at the undergraduate level cover a wide range of specializations, from the finer points of furniture construction to the conceptualization of whole new sorts of environments. The tasks are meant to strengthen the student's design thinking abilities and aid in the translation of concepts into prototypes. The environment is a topic covered in interior design courses. Values from practical requirements, historical contexts, architectural components and styles, atmosphere, and many other psychological variables go into creating a space's environment. As an added bonus, students of interior design learn to apply the concepts of function, economy, security, and feasibility. It's important to keep in mind that interior design encompasses more than simply the aesthetics and functionality of a space; it also takes into account the social, cultural, and artistic needs of its end users. The significance of events occurring in space is modified by the architecture of the surrounding interior. Understanding human behavior is essential to spatial design, which further enhances the viewer's ability to recall past experiences in the place. (Wu, 2016, 68) Diversity, the humanities, experiences, historical heritage, spatial layouts, and the finer points of components with materials, colors, lighting, and finishes are all part of the narrative characteristics covered in interior design curricula. It's a wide-ranging area where many people have contributed to the investigation of various story components.

Visual Narratives as part of Storytelling

Lots of different design outputs can benefit from design tales, and such narratives can also enrich the design process. When, why, and where a story is told in a design project determines how significant it will be. Both theoretical and visual narratives have value. Theory is backed up by both written and spoken stories. Oral, textual, and visual elements all come together to form a visual narrative.

Design Thinking and Storytelling

"Product, service, process, and even strategy development may all benefit from "design thinking" (Brown, 2008). It is common knowledge that the design-thinking process produces novel insights and answers to challenges. It is the job of designers to find solutions to challenges from the consumers' point of view. The steps here include analysis, ideation, visualization, and prototyping (Muller-Roterberg, 2020, 9,10). Combining narrative with design thinking improves visualization and communication, despite the similarities between the two.

Design student's Journey and Thinking process

As a career option, design is difficult since it requires attention to detail and extensive expertise. Career success is possible, but only after putting in a lot of effort. It calls for a lot of study, effort, practice, and creativity. It's a lot of work, but it's also a lot of fun, and it requires a lot of design-related processes and ways of thinking. Students' educational and professional experiences benefit most from exposure to design thinking. From the very first days of studying interior design, students are encouraged to investigate various methods and approaches to design thinking. As a result, the path of a student is fairly difficult and calls for innovative approaches. Students pursuing a degree in design should be prepared to think creatively about design problems.

Discussion and Gaps in Research

It is clear from all of these considerations that storytelling is one of the earliest and most original forms of self-expression. This approach to adapting personal experience into a creative form has been thoroughly investigated. People have used many different formats throughout history to document their lives and culture. Over time, this process has also been used as a means of materializing fantasies. This strategy is now also used for discovering new information and unraveling hidden meanings. This process facilitates the development of screenplays and visual storytelling from the ground up. A narrative may provide depth to a goal. It's useful for stringing together related thoughts. It opens up vast possibilities for designers to engage in conversation and collaboration. Therefore, many other user-centered verticals are incorporated into the design to address issues that arise from the interplay between users, their surroundings, and their activities. Better design solutions may be conceived through the use of storytelling, and the process itself can be enlivened.

Since there is currently no dedicated design tool that prioritizes visual storytelling as a core methodology, there are still significant knowledge gaps. In addition, tales have been incorporated into pedagogy in a wide variety of contexts, albeit not in any way unique to design education or visual storytelling. This opens the door for further research into the method's applicability in the wider context of design education.

References

Austin, T. (2020). Narrative Environments and Experience Design: Space as a Medium of Communication. Taylor & Francis.

Bothra, S. (2019). Design education in India its role in imparting values of socially responsible design. Shodh ganga. http://hdl.handle.net/10603/279437

Brown, T. (2008, June). Design Thinking. Harvard Business Review, June 2008(2008), 11. https://readings.design/PDF/Tim%20Brown,%20Design%20Thinking.pdf

Cross, N. (2011). Design Thinking: Understanding How Designers Think and Work. Bloomsbury Academic.

Dahlström, A. (2019). Storytelling in Design: Defining, Designing, and Selling Multidevice Products. O'Reilly Media. 1491959371

Huber, A. (2017). Telling the Design Story: Effective and Engaging Communication. Taylor & Francis.

Malek, M. S., and Gundaliya, P. J. (2020). Negative factors in implementing public–private partnership in Indian road projects. International Journal of Construction Management, DOI:10.1080/15623599.2020.1857672.

Borikar, H., Bhatt, V., and Vora, H. (2022). Investigating The Mediating Role Of Perceived Culture, Role Ambiguity, And Workload On Workplace Stress With Moderating Role Of Education In A Financial Services Organization. Journal of Positive School of Psychology, ISSN, (9233-9246)

Malek, MohammedShakil S., and Bhatt, Viral (2022). Examine the comparison of CSFs for public and private sector's stakeholders: a SEM approach towards PPP in Indian road sector, International Journal of Construction Management, DOI: 10.1080/15623599.2022.2049490

Borikar, M. H., and Bhatt, V. (2020). Measuring impact of factors influencing workplace stress with respect to financial services. Alochana Chakra Journal, ISSN, (2231-3990).

Does Musically Responsive School Curriculum enhance Reasoning Abilities and Helps in Cognitive Development of School Students?

Ashraf Alam[*,a] *and Atasi Mohanty*[a]

[a]Rekhi Centre of Excellence for the Science of Happiness, Indian Institute of Technology Kharagpur, India

E-mail: *ashraf_alam@kgpian.iitkgp.ac.in

Abstract

Employing music as a pedagogical tool is a relatively new approach to classroom teaching that offers a variety of methods and strategies for teaching and learning. In the current investigation, research papers were systematically reviewed, and subsequently, evidence was summarized. The authors started by framing questions, followed by summarizing the findings, and finally interpreted the results, and concluded. Summarily putting the findings together, it is concluded that listening to or playing music has a direct bearing on students' intelligence, cognitive abilities, and brain functioning. Education involving the application of music as a pedagogical tool facilitates students' intellectual development. Furthermore, making music actively engages the brain and increases its capacity by increasing the strength of connections among neurons, thereby leading to enhanced cortical and cognitive development. Authors have also critiqued the limitations of existing literature to improve future pedagogical designs.

Keywords: Academic Achievement, Curriculum, Learning, Music, Pedagogy, School, Education, Teaching

Introduction

Humans have been aware of natural rhythms and music since the beginning of their existence. Rhythm is one of the most prevalent childhood and adolescent pleasures; in fact, it begins in the cradle and continues throughout life. From youngsters jumping rope to the football field, where the school scream lends rhythm to both voices and bodies, we see it everywhere. Rhythm is primarily physical and emotional; it is an inherent bodily response to rhythm's emotional appeal (Alam, 2021). The pure joy of rhythm causes one to dance. If the environment is favorable, the entire school sings with various rhythms. To teach the meanings of words, we continue to play out the rhythms of nature. Perhaps it is the motion of the waves, the flying of the birds, the swinging of the trees, and other natural phenomena that take on greater significance when we play them out. Creative dramatics is recognized as an educational and socializing force. The creative spirit is a mysterious power that incorporates feelings of self-expression. If a student's background of values is to be widened, every student with some degree of skill and a desire to create possibilities must be provided with opportunities to fulfill that ambition. Poetry is a good fit for rhythms. Children should be encouraged to sing and dance their rhymes and poetry. A collection of children's poetry is always brimming with joyful spirits ready to break out and dance. In poetry and literature, the pattern of emotional rhythms should be appropriate for young students. We archetypally must implant an appetite to appreciate and comprehend poetry. If an educator has to make poetry germane and relevant to children, it is she who must first like it. We must endeavor to view and Interpret poetry through the eyes of the poets. What matters is not what we see on the written page, but what revelations and discoveries it makes (Alam, 2022a).

Music and Reasoning

Music's positive impact is from producing music rather than from merely listening to it mindlessly. Rauscher and colleagues in1997 investigated the effects of music keyboard instruction in preschoolers (3- and 4-year-olds). A matched group received private computer training on an equal basis. Before and after training, students were given four conventional, age-calibrated spatial reasoning exams, one evaluating spatial-temporal thinking and three testing spatial recognition. The keyboard group improved significantly on the spatial-temporal exam alone, according to the results. The lack of progress in the computer group suggests that keyboard training has had a very particular influence on this form of thinking. Music keyboard instruction, according to Rauscher and her colleagues, can help students study traditional courses like mathematics and science, which need spatial-temporal reasoning (Alam, 2022b).

Music and Students' Cognitive Development

According to music instructors, students' engagement in school music has a favorable influence on other aspects of their lives as well. Students' self-discipline, dexterity, coordination, self-esteem, cognitive skills, listening skills, creative ability, and personal expression, according to music instructors, are all improved by musical engagement. Most music instructors, however, are unaware of studies that back up their views and observations.

In the 1980s, the Gemeinhardt Company performed two large studies on the school band movement. They spoke with band directors, music merchants, band and non-band parents, and students (band and non-band). The majority of the individuals asked in all categories identified many of the benefits a student may obtain from being part of a musical band, according to the preliminary findings of Gemeinhardt's research. Achievement, recognition, discipline, pleasure, active engagement, and expanding relationships are some of the advantages. According to a poll, 96 percent of band parents think that many individuals do not realize or comprehend the benefits of the band. In fact, 95% of band parents agreed with the proposition that music band delivers educational benefits that are not available in other classes and 78% believed that band is more educational than extracurricular activities. The benefits of a band are discussed in broad terms by the directors of music bands. Discipline, teamwork, coordination, skill development, pride, lifetime skills, accomplishment, cooperation, self-confidence, sense of belonging, responsibility, self-expression, creativity, performance, companionship, building character and personality, self-esteem, social development, and enjoyment are just a few of the benefits listed by these directors (Alam, 2022c).

Non-band parents, non-band students, drop-out band parents, and drop-out band students were polled in the second Gemeinhardt research. There is widespread agreement across these four groups that the band improves self-esteem, self-confidence, and a sense of achievement. When offered the options "Agree a lot," "Agree a little," or "Don't agree," 91 percent of non-band parents, 90 percent of drop-out band parents, 79 percent of non-band students, and 82 percent of drop-out band students answered, "Agree a lot." There is thus an urgent need that we become more outspoken about the advantages of musical engagement and communicate this knowledge with school boards, principals, parents, and students. Combining this type of data with scientific evidence of music's positive influence is a compelling case that needs to be communicated to the general population. According to research findings, "music offers something for everyone" or "everyone needs music."

Music and Academic Achievement

A lot of research has been conducted on the impact of music on students' academic achievements. High school music students have been demonstrated to have higher grade point averages than non-music students in the same school. In 1981, when this experiment was carried out, the overall grade point average for music students at Mission Viejo High School in Southern California was 3.59, while the overall grade point average for non-music students was 2.91. In the same survey, 16 percent of music students had a 4.0 overall grade point average, whereas just 5% of non-music students received a 4.0 overall grade point average. Ninety percent of alumni of the New York City School of Performing Arts went to college post the completion of school. Participation in high school music programs, according to research, helps kids develop the abilities needed for a range of contemporary jobs (Alam, 2022d).

Hurwitz and his colleagues in 1975 looked at whether music instruction helped first-graders enhance their reading skills. For seven months, the experimental group practiced listening to folk tunes and recognizing melodic and rhythmic aspects for 40 minutes each day. No treatment was given to the control group, which was similar in age, IQ, and socioeconomic level. The experimental group had considerably higher reading scores than the control group after getting music listening training, with the former scoring in the 8^{8th} percentile and the latter scoring in the 7^{2nd} percentile. Because both groups were taught by the same instructor, the disparities in scores were not due to greater/lesser reading instruction. These findings pose two issues. First, was the increase in reading ability due to music or because of the diversity in educational experiences? Second, how could music training possibly help reading when the experimental group did not learn to read music but rather learned to listen to it and detect musical ideas?

Successful music students have attributes and talents that are typically deemed vital by employers in business, education, and service organizations. Music education helps pupils enhance their writing and communication skills, as well as their analytical abilities. Being successful in music requires a lot of self-control, and music majors have the highest SAT scores in all disciplines. "We seek students who have participated in orchestra, symphonic band, chorus, and theatre. Such students demonstrate the requisite enthusiasm and ability that is required to manage the time that our university is looking for. They indubitably demonstrate the dynamic handling of a full academic load while still learning something new. These kids already know how to get active, which is exactly the kind of campus environment Stanford is looking for.," remarked Fred Hargadon, former Dean of Admissions for Stanford University, in an interview.

According to Christensen, research investigations have repeatedly proven that kids benefit from participating in extracurricular activities. Individual successes in student activities (theatre, music, etc.) can predict college success more accurately than SAT scores, class rank, and grades in high school. Dropout students, on the other hand, have the lowest level of engagement in school activities, according to surveys. Participation in a musical ensemble was the most common co-curricular activity in American high schools, according to the Mode of American Youth.

Concluding Remarks

Music educators and researchers must take the initiative in disseminating this knowledge to those who can make a difference in the future, such as school boards, educational administrators, parents, and legislators. They must speak up in support of the incorporation of music as a pedagogical tool for effective teaching-learning. Music educators must take an active role in promoting this art. We could see an increase in enrolment in music classes as more people become aware of the studies in this area. Music integration into the curriculum as a tool for enhanced learning is an area that will grow in importance. With such high dropout rates, educators must pool their resources and make use of the instruments at their disposal in order to provide more effective education.

We now have enough evidence to form a fresh perspective on music in classrooms. Music has profound biological foundations at its most fundamental level. As a result, kids have a high level of perceptual and cognitive processing ability for the fundamental components of music. Even though newborns cannot comprehend words, parents, and caregivers automatically musically connect with them since melodic stimulation always captures their attention. Young children certainly like music, as seen by their spontaneous musical behavior. Furthermore, the human brain has musical building blocks that may be identified.

This biological inheritance broadens our understanding of human endowment and emphasizes the value of music as a teaching tool. The most extensive workout for brain cells and their synaptic linkages appears to be music-making. Pupils' intellectual growth is aided by music education, which includes both listening to music and producing music. It also aids students in learning other fundamental topics such as reading.

Parents sometimes believe that music has a disadvantage since it is so entertaining. Is it possible that something so fun can also be so vital in education? Absolutely. Music provides excellent chances for communication, expression, creativity, and group collaboration, as well as being beneficial to the brain and enhancing learning and intellectual growth. Instead of asking, "Why music?" we should question, "Why not music?" and "How can teachers utilize music to help students achieve educational goals?."

We shall use all the available resources to help students develop their cognitive capabilities. According to the current research, music appears to be a key player in achieving that aim. Humans have sensed the rhythms and sounds of nature since the dawn of time. One of the most prevalent pleasures of human life has always been rhythmic music. It begins in the cradle and lasts the whole of one's life. Rhythm is mostly physical and emotional in nature. It is a natural bodily reaction to a person's emotional and mental state of mind. Every person is born with the talent and capacity to create in some way. This may be cultivated using rhythmic music. Because of this, youngsters in their early learning stages are encouraged to sing rhymes and poetry, as well as dance to them. This may be demonstrated in later stages of growth and cognition, all the way up to pre-college education and learning.

The findings demonstrate unequivocally that music education provides kids with the basic ability to study, succeed in other major academic disciplines, and acquire the capabilities, skills, and knowledge necessary for lifelong success. There have been several studies that show that music education increases cognitive abilities, helps develop superior working memory, improves attendance rates, develops fine motor skills, prepares the brain for accomplishment, and prepares pupils to learn. Students' academic progress is also aided by music education, which enhances memory and retention of verbal material, improves standardized test results, advances reading abilities, and promotes mathematics achievement.

Additionally, this instructional philosophy improves students' creativity, study habits, behavioral conduct, and self-esteem. Musical participation increases students' personal expression, listening abilities, cognitive skills, coordination, dexterity, and self-discipline, among other things. It also has a favorable influence on other aspects of students' lives. Despite all these benefits, rhythmic music instruction in schools is frequently disregarded, and as a result, its true potential stays untapped. It can make a significant difference in pupils' academic progress when used in school. Music also has profound biological foundations for brain development.

This research tried to unravel and investigate the impact of music on the learning outcomes of children. As a result of the use of rhythmic music instructional strategies, students' rate of cognitive development improves. To summarise the findings, it can be concluded that: (a) music is not only enjoyable, but it also improves brain development and enhances abilities such as reading and solving mathematical problems; (b) listening to or playing music has a direct impact on a child's intelligence, cognitive abilities, and brain functioning; (c) education involving the use of rhythmic music pedagogical tools facilitates students' intellectual development; (d) learners who choose music as a form of expression benefit intellectually; and (e) music help with language acquisition, readiness to read, and general intellectual development. It also encourages good behavioral attitudes and reduces absenteeism in middle and high schools; boosts creativity; and helps with social development, personality adjustment, and self-worth.

Acknowledgments

We are highly indebted to Prof. Priyadarshi Patnaik of IIT Kharagpur for his ceaseless patronage and valuable feedback.

Declaration of Competing Interest

The authors declare that they have no known competing financial interests or personal relationships that could have appeared to influence the work reported in this paper. No external agency in any capacity influenced the design of this research work, data collection, and subsequent analysis.

Funding

The reported work is not funded by any external agency, and the costs incurred in carrying out this research are borne solely by the authors.

References

Alam, A. (2022a). Investigating Sustainable Education and Positive Psychology Interventions in Schools Towards Achievement of Sustainable Happiness and Wellbeing for 21st Century Pedagogy and Curriculum. ECS Transactions, 107(1), 19481.

Alam, A. (2022b). Positive Psychology Goes to School: Conceptualizing Students' Happiness in 21st Century Schools While "Minding the Mind!'Are We There Yet? Evidence-Backed, School-Based Positive Psychology Interventions. ECS Transactions, 107(1), 11199.

Alam, A. (2021). Possibilities and Apprehensions in the Landscape of Artificial Intelligence in Education. In 2021 International Conference on Computational Intelligence and Computing Applications (ICCICA) (pp. 1–8). IEEE.

Alam, A. (2022c). Mapping a Sustainable Future Through Conceptualization of Transformative Learning Framework, Education for Sustainable Development, Critical Reflection, and Responsible Citizenship: An Exploration of Pedagogies for Twenty-First Century Learning. ECS Transactions, 107(1), 9827.

Alam, A. (2022d). Social Robots in Education for Long-Term Human-Robot Interaction: Socially Supportive Behaviour of Robotic Tutor for Creating Robo-Tangible Learning Environment in a Guided Discovery Learning Interaction. ECS Transactions, 107(1), 12389.

Music and Its Effect on Mathematical and Reading Abilities of Students: Pedagogy for Twenty-First Century Schools

Ashraf Alam[*,a] *and Atasi Mohanty*[a]

[a]Rekhi Centre of Excellence for the Science of Happiness, Indian Institute of Technology Kharagpur, India
E-mail: *ashraf_alam@kgpian.iitkgp.ac.in

Abstract

Music-integrated pedagogy is a relatively new approach to classroom teaching that offers a variety of methods and strategies for teaching and learning, employing music as a pedagogical tool. In the current investigation, research papers were systematically reviewed, and subsequently, evidence was summarized. The authors started by framing questions, followed by summarizing the findings, interpreting the results, and eventually reaching a conclusion. Summarily putting the findings together, it is concluded that music is not only fun but also improves brain development and enhances capabilities such as reading and solving mathematical problems. Learners opting for music as one of their courses have been found to perform better in mathematics, since both the subjects, i.e., music and mathematics, although appear to be different when viewed superficially, share profound structural similarities at the deeper level. Authors have also critiqued the limitations of existing literature to improve future pedagogical designs.

Keywords: Mathematics, Reading, Academic Achievement, Curriculum, Learning, Music, Pedagogy, School, Education, Teaching

Introduction

According to brain research, music not only makes us happy, but also helps us develop, build, and shape our brains and boosts our skills in other academic disciplines like arithmetic and reading (Alam, 2022a). Music research is generating a lot of interest in the academic and scientific community. New results on the impact of listening to or playing music on children's intellect, cognitive capacities, and brain functions are reported in popular newspapers almost regularly. Georgia's governor thinks music is important enough to risk political embarrassment. He argued that the state shall cover the cost of sending a classical music CD to each new mother in Georgia. Even though he did not get the money sanctioned, Sony provided him with the requisite number of CDs. In Florida, a bill is being considered that would urge preschools to play music for at least 30 minutes each day. What exactly is going on?

When it comes to making profound yet difficult decisions concerning curricular and extracurricular activities, many school administrators and instructors need to be mindfully aware of the most recent and relevant research findings in the field of music and education. Should music education be eliminated, abridged, maintained, or expanded in schools? We will outline some of the most important discoveries in music research concerning its impact on students' academic achievement. However, regardless of research findings, we believe that music and arts are vital and not optional components of education. We shall not have to defend music in the classroom only based on its non-musical benefits. However, because the advantages are becoming more obvious, educators want up-to-date knowledge in order to make educated judgments regarding the role of music in classrooms.

Music and Mathematics

Gordon Shaw, a physicist, would not have gotten involved in a heated political dispute, or any debate for that matter had his work revolved only around arcane scientific publications and particle accelerator experiments. As a professor emeritus, Shaw became interested in brain functioning. With this, Mozart came into the picture, leading to a massive controversy. Shaw and his colleagues discovered Mozart's effect, i.e., the capability of a Mozart sonata, under precise settings, to advance listeners' reasoning and mathematical capabilities. Shaw has more than a dozen papers on this theme, most of which have been published in "A" category journals such as Neurological Research. The findings are however quite divisive. Mathematics and music have a long and symbiotic relationship. Both disciplines have a history of producing kid prodigies. Einstein (who played the violin) and Saint-Sans (who dabbled in mathematics) are examples of cross-over skills. The musical scale is a near approximation of a tidy logarithmic frequency progression.

The question Iow is whether the link between these two areas of human ability can be used for educational purposes. Shaw and his colleagues felt it to be possible. According to him, listening to Mozart's sonata prepares the brain's hardware for mathematical activities. Shaw presses the button on a battered boom box for the hundredth time in his messy office at the University of California, Irvine. For a few minutes, Mozart's Sonata for Two Pianos in D Major (K. 448) trills through the room. The succession of rising and falling notes in quick, symmetrical patterns is astonishing even for Shaw, who does not play an instrument. The same may be said of his group's results. Students who listened carefully to the composition improved their arithmetic and spatial-temporal thinking scores. The Mozart effect was popularized by author Don Campbell in his book of the same name, which advices listening to classical music as a cure-all for all physical and mental ailments (Alam, 2021).

Is there a connection between the way brain cells operate and the effect? Vernon Mountcastle, a renowned neuroscientist, produced a model of the neuronal anatomy of the brain's cortex in 1978. It displays the firing patterns of groupings of neurons and resembles a thick belt of colorful American Indian beading. Shaw and his colleagues created their own model based on Mountcastle's model. Shaws' colleague at the time, Xiaodan Leng, utilized a synthesizer to convert these patterns into music on the spur of the moment. It was not precisely toe-tapping music that blasted out of the speakers, but it was music nevertheless. Music and brain wave activity, according to Shaw and Leng, is constructed on the same patterns. In 1991, the two co-authored a study in which they hypothesized that music may help in cognitive development (Alam, 2022b).

There were significant responses from neuroscientists and psychologists to Shaw's findings. They have not denounced it, but they neither have recognized it completely. Other academics' attempts to replicate his findings have been sporadic. Ellen Winner, the director of the Harvard Graduate School of Education, undertook a study wherein they analyzed hundreds of papers relating to the effects of music education on academic achievement and found a definitive correlation. The studies of Shaw and Rauscher are fascinating, albeit they should be taken with a grain of salt. Their first study, published in 1993, demonstrated that listening to Mozart helped college students improve their performance on a spatial-temporal exam. However, the impact only lasted 10 to 15 minutes. In the next trial, 3-year-olds who took piano lessons outperformed a control group by 35% on equivalent test scores. This impact remained for several days. Most children who find it difficult to understand proportional math, according to Shaw, believe that music, when paired with adequate teaching, can assist them immensely.

The Mozart sonata has been demonstrated to diminish abnormal brain-wave activity in comatose epileptics and enhance spatial-temporal cognition in Alzheimer's patients in studies conducted by Shaw and some of his colleagues. After being reared on repeated Mozart sonata recordings, rats, whose brains are thought to have basic neuronal processes comparable to those of humans, sped through mazes quicker, according to Rauscher's research. According to Shaw, it works with rats and epileptics in comas. The brain's reaction to this music acts as a Rosetta Stone for decoding the brain's neurophysiological structure.

Later, a team led by Shaw stated that young children who had spent months listening to Mozart sonatas and studying the piano had increased their "ratios and proportions" scores by 27%. They were compared to a control group that attended similar enrichment classes but without music. Shaw states that the Mozart-trained children, who come from a variety of ethnic and socioeconomic backgrounds, are now three grade levels ahead of their peers in math. Finally, he says that if school boards happen to remove music from the curriculum, they will be committing a major blunder.

Similar studies in the field of mathematics have shown that music students outperform non-music students in math exam results. Enhanced music education, according to Maltster, can lead to increased arithmetic learning. A comparison of all sections of the California Test of Basic Skills was made in a study conducted in Albuquerque, New Mexico public schools. Music students who had been in an instrumental class for two or more years performed much better than non-music students. Grace Nash, a music instructor in Arizona, discovered that introducing music into math courses made it easier for pupils to grasp multiplication tables and arithmetic formulae.

Music and Reading

To see how music education can help with reading, we looked at the three stages of learning to read: visually recognizing words, learning the correspondences between visual parts of words (graphemes) and their spoken sounds (phonemes), and achieving visual recognition of words without going through the earlier stages. The second, or phonemic, step is the most crucial. Music aids reading by enhancing the second step of hearing out words. The association between musical sound discrimination and reading ability was discovered by Lamb and Gregory in 1993. In addition to performing normal reading exams, 1st graders were assessed on their ability to sound out meaningless syllables that they saw on cards, as well as pitch awareness, which required them to listen to pairs of musical notes or chords in order and indicate if they sounded the same. They were also put to test in terms of timbres. Finally, researchers tested students' phonemic awareness by asking them to listen to spoken words and determine whether they started or finished with the same sound. After that, the researchers looked at the correlations between the various test results. They discovered a strong link between children's ability to read both standard and phonic material and their ability to differentiate pitch. Awareness of timbre was not a factor.

Data shows that strong pitch discrimination helps with the second step of learning, i.e., the phonemic stage. The most significant component in delivering word information is changing the pitch of words. Because music training requires improved pitch discrimination, the connection between reading and music instruction is clear. Timbre awareness, on the other hand, is unrelated to reading abilities, indicating that the gains are due to specialized pitch training rather than an enhanced variety of students' educational experiences. Without a doubt, further research will be required to deepen our knowledge. However, the outcomes of both types of research are consistent, indicating that specialized music exposure aids reading.

By including rhythms of music into the curriculum, a reading program in New York was able to significantly enhance reading achievement results (Alam, 2022c). Learning to read music improves a student's capacity to accomplish abilities such as reading, listening, predicting, forecasting, memory training, recall skills, concentration strategies, and speed reading. Music students have also been found to outperform non-music pupils on achievement exams in reading and arithmetic. Researchers have strongly emphasized the importance of music training in the development of perceptual and cognitive skills.

Concluding Remarks

Music educators and researchers must take the initiative in disseminating this knowledge to those who can make a difference in the future, such as school boards, educational administrators, parents, and legislators. They must speak up in support of the incorporation of music as a pedagogical tool for effective teaching-learning. Music educators must take an active role in promoting this art. We could see an increase in enrolment in music classes as more people become aware of the studies in this area. Music integration into the curriculum as a tool for enhanced learning is an area that will grow in importance. With such high dropout rates, educators must pool their resources and make use of the instruments at their disposal to provide more effective education.

We now have enough evidence to form a fresh perspective on music in classrooms. Music has profound biological foundations at its most fundamental level. As a result, kids have a high level of perceptual and cognitive processing ability for the fundamental components of music. Even though newborns cannot comprehend words, parents, and caregivers automatically musically connect with them since melodic stimulation always captures their attention. Young children certainly like music, as seen by their spontaneous musical behavior. Furthermore, the human brain has musical building blocks that may be identified.

This biological inheritance broadens our understanding of human endowment and emphasizes the value of music as a teaching tool. The most extensive workout for brain cells and their synaptic linkages appears to be music-making. Pupils' intellectual growth is aided by music education, which includes both listening to music and producing music. It also aids students in learning other fundamental topics such as reading.

Parents sometimes believe that music has a disadvantage since it is so entertaining. Is it possible that something so fun can also be so vital in education? Absolutely. Music provides excellent chances for communication, expression, creativity, and group collaboration, as well as being beneficial to the brain and enhancing learning and intellectual growth. Instead of asking, "Why music?" we should question, "Why not music?" and "How can teachers utilize music to help students achieve educational goals?."

We shall use all the available resources to help students develop their cognitive capabilities. According to the current research, music appears to be a key player in achieving that aim. Humans have sensed the rhythms and sounds of nature since the dawn of time. One of the most prevalent pleasures of human life has always been rhythmic music. It begins in the cradle and lasts the whole of one's life. Rhythm is mostly physical and emotional in nature. It is a natural bodily reaction to a person's emotional and mental state of mind. Every person is born with the talent and capacity to create in some way. This may be cultivated using rhythmic music. Because of this, youngsters in their early learning stages are encouraged to sing rhymes and poetry, as well as dance to them. This may be demonstrated in later stages of growth and cognition, all the way up to pre-college education and learning.

The findings demonstrate unequivocally that music education provides kids with the basic ability to study, succeed in other major academic disciplines, and acquire the capabilities, skills, and knowledge necessary for lifelong success. There have been several studies that show that music education increases cognitive abilities, helps develop superior working memory, improves attendance rates, develops fine motor skills, prepares the brain for accomplishment, and prepares pupils to learn. Student's academic progress is also aided by music education, which enhances memory and retention of verbal material, improves standardized test results, advances reading abilities, and promotes mathematics achievement.

Additionally, this instructional philosophy improves students' creativity, study habits, behavioral conduct, and self-esteem. Musical participation increases students' personal expression, listening abilities, cognitive skills, coordination, dexterity, and self-discipline, among other things. It also has a favourable influence on other aspects of students' lives. Despite all these benefits, rhythmic music instruction in schools is frequently disregarded, and as a result, its true potential stays untapped. It can make a significant difference in pupils' academic progress when used in school. Music also has profound biological foundations for brain development (Alam, 2022d).

This research tried to unravel and investigate the impact of music on the learning outcomes of children. As a result of the use of rhythmic music instructional strategies, students' rate of cognitive development improves. To summarise the findings, it can be concluded that: (a) music is not only enjoyable, but it also improves brain development and enhances abilities such as reading and solving mathematical problems; (b) listening to or playing music has a direct impact on a child's intelligence, cognitive abilities, and brain functioning; (c) education involving the use of rhythmic music pedagogical tools facilitates students' intellectual development; (d) learners who choose music as a form of expression benefit intellectually; and (e) music help with language acquisition, readiness to reading, and general intellectual development. It also encourages good behavioral attitudes and reduces absenteeism in middle and high schools; boosts creativity; and helps with social development, personality adjustment, and self-worth.

References

Alam, A. (2022a). Investigating Sustainable Education and Positive Psychology Interventions in Schools Towards Achievement of Sustainable Happiness and Wellbeing for 21st Century Pedagogy and Curriculum. ECS Transactions, 107(1), 19481.

Alam, A. (2022b). Positive Psychology Goes to School: Conceptualizing Students' Happiness in 21st Century Schools While "Minding the Mind!'Are We There Yet? Evidence-Backed, School-Based Positive Psychology Interventions. ECS Transactions, 107(1), 11199.

Alam, A. (2021). Possibilities and Apprehensions in the Landscape of Artificial Intelligence in Education. In 2021 International Conference on Computational Intelligence and Computing Applications (ICCICA) (pp. 1–8). IEEE.

Alam, A. (2022c). Mapping a Sustainable Future Through Conceptualization of Transformative Learning Framework, Education for Sustainable Development, Critical Reflection, and Responsible Citizenship: An Exploration of Pedagogies for Twenty-First Century Learning. ECS Transactions, 107(1), 9827.

Alam, A. (2022d). Social Robots in Education for Long-Term Human-Robot Interaction: Socially Supportive Behaviour of Robotic Tutor for Creating Robo-Tangible Learning Environment in a Guided Discovery Learning Interaction. ECS Transactions, 107(1), 12389.

An Exploration of the Complexities Involved in the Regulation of Green Buildings

Akash K. Prajapati,[*a] *Dhiraj Bachwani,*[b] *and MohammedShakil Malek*[c]

[a]M. Tech (Construction Project Management-Purusing, Parul University, Vadodara, India
[b]Assistant Professor - Department of Civil Engineering, Parul University, Vadodara, India
[c]F. D. (Mubin) Institute of Engineering and Technology
E-mail[*]: akashprajapati0110@gmail.com

Abstract

Every country's economic development depends on the construction industry. In terms of GDP, the construction sector contributes significantly to the country's economy. The flexible and effective concept of green building has been embraced by the modern eco-friendly construction industry. The purpose of this paper is to investigate the complexity of green building construction and attempt to provide a solution. The author has reviewed numerous research papers in the field of green building and identified several issues. The problems that were found were separated into three groups: Driver related issues, Rating System-related Issues, and other Issues. To overcome these issues, A framework is necessary in the context to all the construction parameters. This paper provides future direction for further research work.

Keywords: Green Building, Rating System, Complexity in Green Building Construction, Challenges Faced in Green Building, Challenge Faced in Management of Green Building, Issues in Green Building.

Introduction

It is estimated that by 2050, the rise in CO_2 will raise global temperatures by 4.5 degrees. To prevent global warming, the building industry must be carbon neutral, or zero, by 2050. Energy-related CO_2 emissions are expected to rise from 32.3 billion metric tonnes in 2012 to 43.4 billion metric tonnes in 2040 (Zhang et al., 2017). As a result, green buildings are being prioritized in the construction industry over energy-saving issues such as site planning, waste management, material selection, and design (Babayev et al., 2017).

A green building reduces or eliminates climate and environmental impacts, according to the World Green Building Council. It incorporates siting, design, construction, operation, maintenance, renovation, and deconstruction to reduce the impact of energy, water, materials, and other natural resources. Green building grading tools, often termed certification, examine and recognize green structures.

Benefits of Green Building or Green Construction

- Building environmental
- Economic Benefits.
- Social Benefits

Why Green Building?

Positive Impact on the Environment for Every One Billion Square Feet of Green Building:

- Emissions of carbon dioxide were reduced by 12 million tons.

- Energy Cost Reduction by 15000 GW.
- Approximately 45,000,000 gallons of water are saved every year.

How much water and energy each type of building use is depicted in the graph below:

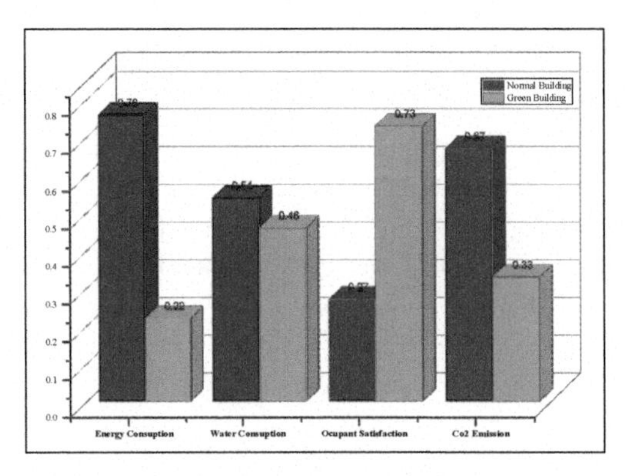

Fig. 1. Comparison of Normal & Green Building
Source: Compiled from LEED by Author

Green Building status in India

According to the most recent information made available by the Indian Green Building Council, India has successfully achieved 7.17 billion square feet of "Green Building Footprint." Across the country, approximately 5.77 million acres are currently being developed as part of nearly 6,000 large-scale construction projects and environmentally conscious programs totaling nearly 6,000.

Top 10 States in India for Green Building Projects

Green construction practices have been implemented by all Indian states. If you look at Certified Square Meter, Maharashtra is at the first spot. Increasing Green Buildings is a priority for the state of Gujarat, which is why they have launched the Panchamrut Yojna initiative.

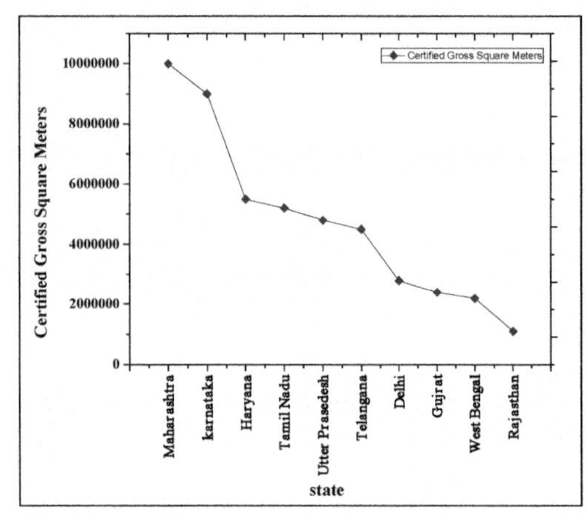

Fig. 2. Green Building projects in Top 10 States in India
Source: Compile from USGBC by Author

There are so many different rating systems adopted in different countries & nations. However, in India, there are primarily three types of Green Building Rating systems that are widely used.

LEED stands for Leadership in Energy and Environmental Design.

GRIHA stands for Green Rating for Integrated Habitat Assessment.

IGBC stands for Indian Green Building Council.

Current Scenario and Future targets of Green Building

The market for green buildings is anticipated to reach a value of USD 9,91,19.3 million by the year 2030, showing a 17.48% CAGR over the projected timeframe (2022 – 2030).

The market for green buildings in India grew during the years 2020 and 2021 by 5.1%, reaching a value of $21.0 billion in 2021. Throughout its entire existence, the market averaged an annual compound growth rate of 7.2% during the time period of 2017 and 2021.

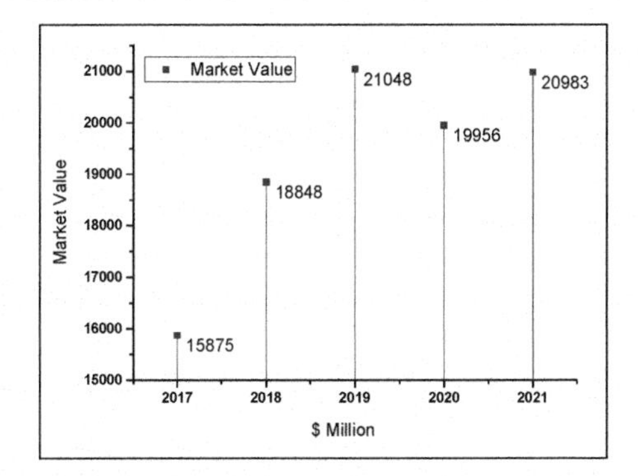

Fig. 3. Green Building Market Value

Source: Compile from Global Data by Author

Why do we need to overcome these issues?

Indian Government targeted to archive the market for green buildings is expected to be worth USD 20 billion by 2030. But the projects go time overrun due to the Complexity of the green building project. Also due to the Time overrun, we have to spend more money on certain projects and because of that project goes into Cost overrun. That's why we need to overcome these issues.

Literature Review

The number of papers from different publications like Elsevier, Springer, MDPI, Emerland, etc., were referred by the Author. The cited study was organized into the following sections: Rating system; Challenges, Barriers, and Issues; Case Study; Costs, and Time; and Other Related Categories. The literature review found that the majority of the research focus was on the Green Building Rating System and the Rating System Challenges.

Referred paper is divided into Different Category

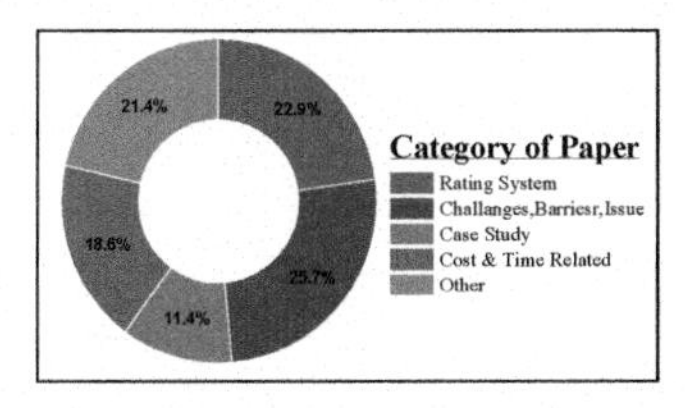

Fig. 4. The papers that have been cited are organized into various sections

Findings from the Literature Review

The author has compiled a list of issues that have been faced all around the world in Green Building initiatives. The author then divided the issues into its parts, as stated below: Driver Related issues, Rating System issues, and other issues. The Author also made use of official documents and reports from the Indian Green Building Council. Various other issues based upon scrutiny of reports are also highlighted as a part of the findings of literature review.

Driver Related Issues

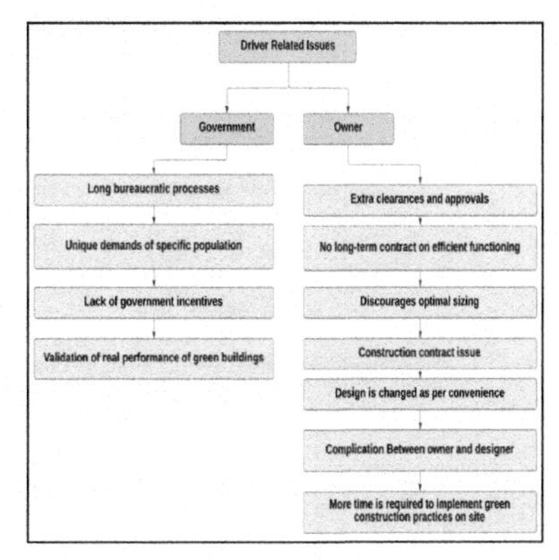

Fig. 5. Issues Associated with Drivers

Rating-System Related Issues

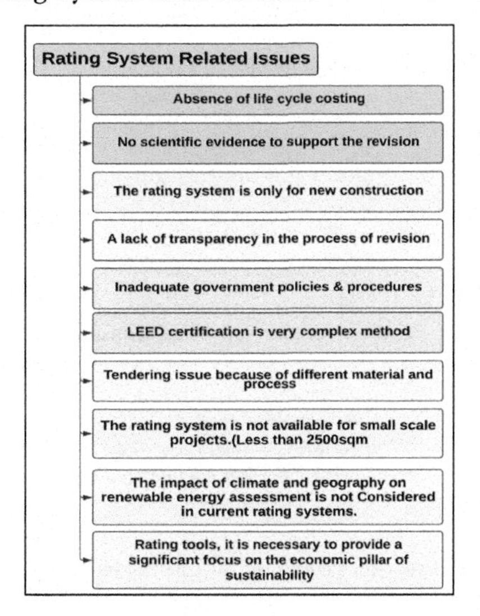

Fig. 6. List of issues Associated with Rating Systems

Other Issues

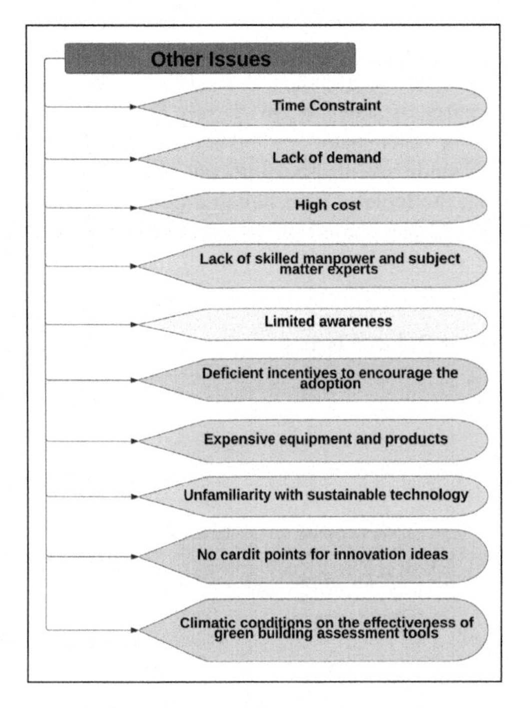

Fig. 7. Additional Issues List

Issues find in IGBC & Gov. reports

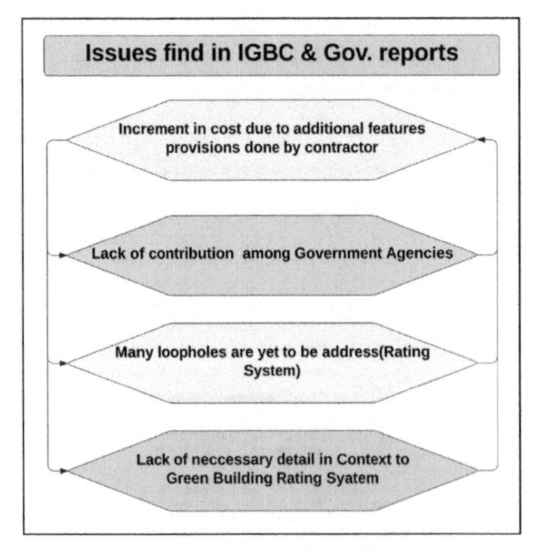

Fig. 8. Issues highlighted by the IGBC and government reports

Conclusion

The value in the treemap hierarchy is presented in rectangular form because Rectangles are simple to understand because they accurately represent the value of the measure due to their size and color. The majority of the problems are related to drives and the rating system, as illustrated in the following treemap. The majority of which have an impact on green building design and construction. The Issues listed under "other issues" are linked to the "Driver related issue" and the "Rating System relate" issue, both of which have an impact directly or indirectly on green building construction.

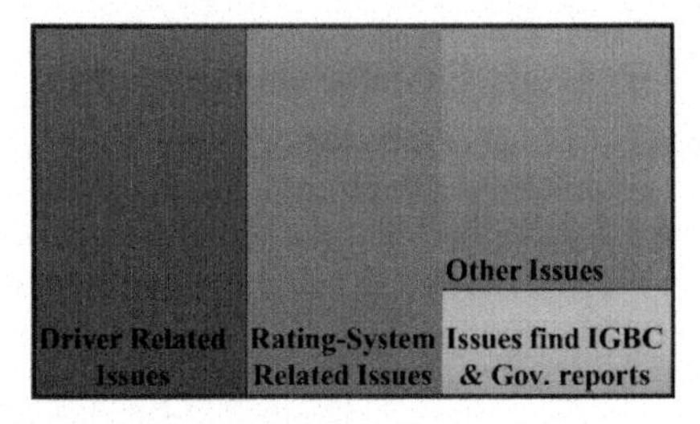

Fig. 6. Hierarchy Map of Issues

Recommendations

The author recommends having a Standard framework in order to address the problems (Mentioned above) that are interconnected with each other directly or indirectly.

The building criteria should also be to the above-mentioned framework in order to reduce the complexities of green building development.

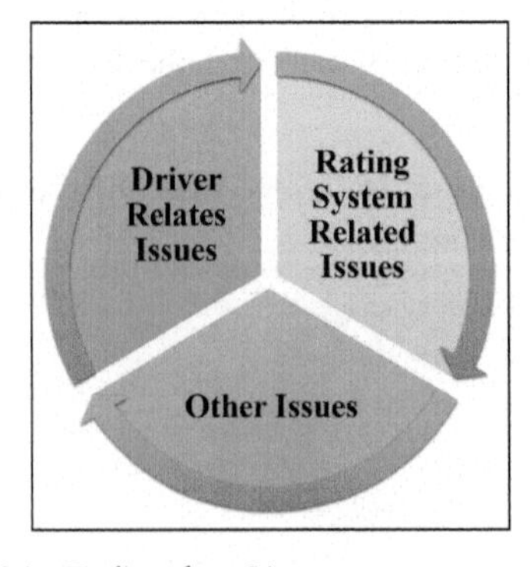

Fig. 9. Major Findings from Ligature survey

Future Direction

It is suggested that the research can be extended with the help of various research tools to find out the impact ratio of the issues. After, finding that the result can be validated through the case study approach.

References

Javed, N., Thaheem, M. J., Bakhtawar, B., Nasir, A. R., Khan, K. I. A., and Gabriel, H. F. (2019). Managing risk in green building projects: toward a dedicated framework. Smart and Sustainable Built Environment, 9(2), pp. 156–173.

Javed, N., Thaheem, M. J., Bakhtawar, B., Nasir, A. R., Khan, K. I. A., and Gabriel, H. F. (2019). Managing risk in green building projects: toward a dedicated framework. Smart and Sustainable Built Environment, 9(2), pp. 156–173.

Malek, MohammedShakil S., and Bhatt, Viral (2022). Examine the comparison of CSFs for public and private sector's stakeholders: a SEM approach towards PPP in Indian road sector, International Journal of Construction Management, DOI: 10.1080/15623599.2022.2049490

Malek, MohammedShakil S., and Gundaliya, P. J. (2020). Negative factors in implementing public–private partnership in Indian road projects, International Journal of Construction Management, DOI: 10.1080/15623599.2020.1857672

Malek, MohammedShakil S., and Gundaliya, Pradip (2021). Value for money factors in Indian public-private partnership road projects: An exploratory approach, Journal of Project Management, 6(1), pp. 23–32. DOI: 10.5267/j.jpm.2020.10.002

Malek, MohammedShakil S., and Zala, Laxmansinh (2021). The attractiveness of public-private partnership for road projects in India, Journal of Project Management, 7(2), pp. 23–32. DOI: 10.5267/j.jpm.2021.10.001

Malek, MohammedShakil S., Saiyed, Farhana M., and Bachwani, Dhiraj (2021). Identification, evaluation and allotment of critical risk factors (CRFs) in real estate projects: India as a case study, Journal of Project Management, 6(2), pp. 83–92. DOI: 10.5267/j.jpm.2021.1.002

Mir-Babayev, R., Gulaliyev, M., Shikhaliyeva, S., Azizova, R., and Ok, N. (2017). The impact of cultural diversity on innovation performance: Evidence from construction industry of Azerbaijan. Economics & Sociology, 10(1), p.78

Nguyen, H. D., and Macchion, L. (2022). Risk management in green building: a review of the current state of research and future directions. Environment, Development and Sustainability, pp. 1–37.

Wu, Z., Jiang, M., Cai, Y., Wang, H., and Li, S. (2019). What hinders the development of green building? An investigation of China. International journal of environmental research and public health, 16(17), p. 3140.

Zhang, Y., Chen, W., and Gao, W., 2017. A survey on the development status and challenges of smart grids in main driver countries. Renewable and Sustainable Energy Reviews, 79, pp. 137–147.

The Makers and Users of Fashion, a Study of Contrast

Pritam Saha[*,a]

[a]Karnavati University, Gandhinagar, India
E-mail: [*]pritamsaha@karnavatiuniversity.edu.in

Abstract

Aari embroidery has been a predominant craft in traditional Indian textiles since the Mughal period of 16th Century India and currently, this forms a major technique of surface design in Indian Couture. However, quite often the technique is not credited and is generalized as hand embroidery. This study looks at the couture collections of 5 of the most prominent Indian Couturiers and maps the usage of *aari* embroidery in their collections. The methodology adopted is mainly observational from secondary data. This is a review paper that mainly analyses images of high-end Indian couture garments and visually examines the content of the same with respect to *aari* embroidery. The constructs of the study will be measured using indicators like the percentage of embroidery in the entire product which will be visually observed and noted by the researcher, followed by observations and discussions on the data.

Keywords: Artisans, Class-divide, Consumers, Handicraft, Sustainability

Introduction

The hands that make and the hands that use them are drifting further apart, is the starting point for this paper. This paper will identify gaps, through research on the lifestyles of the artisans and put them forth in contrast with that of the consumers of Indian high fashion.

This is a narrative that we know has been existing in all aspects of society but gets more vivid when portrayed through the high fashion industry because of the extreme contrast.

Thus this brings forth a narrative of the contrast that is existent between the hands that work in long shifts with the needle, thread, and other hand tools, to the plush parties where high fashion is worn. What strikes as most interesting was a talk the researcher once had with an embroidery artisan working with a designer based out of Kolkata. It was about the amount of time they spend in making an Indian handcrafted garment, which is so much in contrast with the amount of time the end user wears it. Something that has taken 2 months to craft may be worn for 2 hours, yet the emotions it might generate will be attached for a lifetime. A comparison of this sort often makes us wonder about the frivolity of this all.

What are the thoughts the makers of these garments have while embroidering them to the fullest, will they ever be able to see what their creation gets to see when they travel all over the world and find their space in fancy walk-in closets of their consumers? At times these closets are bigger than the size of the artisan's own home. Should the lifestyle of makers be as contrasting as that of the users which I would be touching on in this research?

The paper aims to study the contrast in lifestyle between the artisans and the consumers of high fashion in India and to bring forward the unsung narrative of artisans employed by the fashion industry in India, studying the contrast that exists between them and the end users of high fashion, offering a unique anecdotal point of view of the same.

The review of literature spanned through many topics for this paper, the key ones being understanding the lifestyle choices of the artisans from their perception and analyzing the same using available literature.

In an interview with Gudda Prasad, Artisan, ARDY 2M Design Studio, Mumbai, he says that - "we feel sad that karigar work might end but the responsibility lies with big people like my boss and the merchants. They should realize that if karigars are leaving this profession, why and what are their problems? If the payment is like what is offered here, that's one thing, but if we are working for 12, 13, 14 hours and without timely payment nobody will want to continue. Our work involves a lot of imagination – our ideas depend on colors we see from photos that people bring" (Border and Fall, 2017). This interview with an actual artisan enlightens us a lot about their perceptions, and how they are dependent on designers/consumers for both financial stability as well as inspiration/direction on what they want to work on.

A thesis called - Artisan Voice: an investigation of the collaborations between skilled, traditional textile artisans in India and foreign textile and fashion designers from the artisans shows artisans' perspective and the viewpoints of their craft industry in contemporary times (Emmett, 2015).

At this point, the researcher would agree with what Aravind Adiga (2008) mentions in his novel "The White Tiger" as majority of rural underprivileged India being half baked, where he says - "Me, and thousands of others in this country like me, are half-baked because we were never allowed to complete our schooling. Open our skulls, look in with a penlight, and you'll find an odd museum of ideas: sentences of history or mathematics remembered from school textbooks." and how this is relevant to a majority of artisans in today's digital age as well (Adiga, 2008).

Emmet (2015) talks about Amartya Sen, an economist who emphasizes social values in economic development, providing a capabilities model that appeared useful in distinguishing capability acquisition from entrepreneurial application. For Sen, an important objective of development is maximizing people's capabilities, and the freedom to pursue the kinds of lives they value. Sen defined capability as a combination of "functionings" or the various "doings and beings" that leads to well-being. Functionings range from elementary (nourishment, health) to complex (self-respect, social integration) and vary within different local cultures.'

Tereza Kuldova (2016) investigates the Indian luxury fashion market's dependence on the production of thousands of artisans all over India, revealing a complex system of hierarchies and exploitation, and helps us understand the perspective of elite consumers of high fashion as well (Kuldova, 2016). This gives numerous leads to the research and points out the existence of lack of awareness, education, and organizational structure as a major reason such contrast has existed for so long, and through this paper, I aim to identify this further as per the scope of this research.

The key objectives are to identify and study the points of contrast in lifestyle in terms of clothes, spaces, food, living conditions, etc. between the artisans and consumers of high fashion in India.

Methodology: The methodology mainly followed 3 phases, post the review, which was deciding on the Sampling Design, making the Final questionnaire based on that, and referencing and noting the data down to come up with results and conclusions.

Artisans employed by the fashion industry, both directly and on contract, based in Ahmedabad India, were chosen as a sample and interviewed via telephone/face-to-face to gain information about the points that would help identify the contrast.

The secondary data indicated that the karigars are either employed by designers directly or on contract, where they take the work from design studios and work in their homes. Both were approached for interviewing. A purposive sampling design was followed to select the respondents and the locale of the study. The respondents for the case study were selected to gather in-depth information about the persons engaged in this field and were personally interviewed via telephone to collect authentic first-hand information.

In total 6 artisans were interviewed. The data to identify contrast points from the point of view of the consumers were taken from luxury fashion bloggers, mainly identified as the top 10 influencers in India by various journals and articles.

Results & Discussion: The results and discussions on the research conducted are manifold with interesting findings about a lot of things that we have known as a society but have been choosing to ignore for the longest time. Especially while interviewing the artisans for this project the apprehension from their end was like why are these questions being asked, is this about a scheme, what will be the direct benefits of this, how come we are being asked about how we are, how have we been, what do we eat, what do you care about this? The first challenge was to get over the above apprehensions and make them comprehend the need for sensitization and awareness. How the outcome of such research can spread awareness across the various communities who are consumers of high fashion and start a conversation that people have been avoiding, how one can get aware of one's privilege when the same is shown through visual contrast? These were the higher aims of the project and this led to many conversations on broader subjects starting from fashion to a larger aspect of society.

The various aspects of the questionnaire led to the various results and discussions on identifying the points of contrast between them. A few key contrasting factors were identified as results and discussions on them.

The main categories wherein this contrast was studied are titled Food, Clothing, Work & Leisure, and Travel, wherein each aspect is studied from the perspective of artisans and consumers of high fashion.

The Food: What we eat has always defined who we are in various social scenarios. The food that is consumed both daily and while dining out at restaurants is one of the most important factors defining lifestyle for both the consumers and artisans. Food does not cover only the items consumed, but also the way it is taken, the setup, the utensils, and even the company points out the contrasts. One of the first results of this research points to most of the artisans have been having food cooked from a local restaurant daily as they have been deprived of cooking facilities by the owners of the units that they work in, as it is the same place where they embroider as well as stay, so there are chances of the fabrics getting damaged by a cooking setup in the same room, which also points that they do not have access to a kitchen daily. Whereas on the other hand not just the consumers of high fashion, even people belonging to the same socio-economic status as the artisans have access to making their own food daily, which is a basic right that the artisans have been deprived of for such a long time.

The second observation that follows is in most of the photographs it has been seen that the artisans eat on the floor, and there is no provision for a table and chair unless it is a very large factory. Even though the traditional Indian way is to sit on the floor and eat, in today's society a dining table is not even considered a luxury but an amenity that can be found in most households, as the artisans live at the place of their work, they have been deprived of that as well. On the other hand, it has been observed that most of them have 3 to 4 meals a day which is at par with an average Indian household and is a good sign. Eating out in a restaurant is not that common, although it occurs once a month for a few of them, and according to one of them, the most expensive meal that they have at a restaurant was a Biryani (an Indian rice dish) worth Rs. 80, which is a shocking revelation as that price for a biryani would not even be imagined to be by far expensive. Tea consumption is also a great way to contrast the 2 groups, an average quantity of tea the karigars consume in one go is in a small 50ml paper cup, on the other hand, the average ceramic cup used for serving tea in an Indian household range from 200ml to 350ml.

The clothing: There were 2 major aspects discussed in the clothing segment, one was the clothing they wear and the other was the clothing they make. As makers of high fashion products, it was always a known fact that the artisans would be the ones who will never be able to afford their products, Average price of what they wear at festivals or occasions is about Rs. 800, on the other hand, the festive wear they embroider could range anywhere from Rs. 30,000 to Rs. 3,00,000. It was also observed that most of them are not aware of the cost of making a high fashion garment other than their own labor cost, which was also a surprising find as they did not know the costs of fabrics in the market, etc. as they have never been exposed to a wholesome idea of that. What was also surprising to find was when asked how much profit they would keep if they had to sell products they make on their own, the average answer was about 30% to 40% whereas in reality most high fashion garments retail at least 500% of the making cost.

Work & Leisure: Work and leisure did not seem to have a well-defined boundary here because of the lack of existence of the latter. Most of them usually work for 10 to 12 hrs per day and are very keen to do overtime as they are paid by the hour. When asked what they do for leisure, most of them said they rest, mostly on Sundays, and tells us a lot about how the perception among them is that whatever you are doing while you are not working is considered leisure. Some said they used to play soccer in their village, and some used to watch films on mobile, but now it has been concentrated only on the kind of mechanical work they are expected to do.

Travel: Travel is another key pointer of lifestyle and it was found 5 out 6 people interviewed have never taken a flight in their life, which in itself talks about the stark contrast that exists. Most of the travel that happens is when they are visiting their hometowns/villages once a year to meet their family, and holidays are taken very rarely, that too in nearby spots from their home or work. The above results identify the contrasts that the paper aimed to study and are summarised below in "Table 1."

Table 1. A summary of results and discussions on the contrast of Lifestyle between artisans and consumers in Indian High Fashion

Lifestyle Indicators	Food	Clothing	Work & Leisure	Travel
Sample ↓	Avg. Cost of 1 Daily Meal	Avg. Cost of Daily Clothes.	Avg. Working hours per day	Avg. Frequency of travel for leisure
Artisans	INR 50	INR 500	12 to 14	Only for Work
Consumers	INR 500	INR 5000	5 to 6	Once a month

Significance and Conclusion

The experience of writing this paper was an enriching and extremely sensitizing one and has rekindled the curious learner in me and opened doors to a lot of further research and other projects.

This has enabled me to know the artisans in a new light and understand the contrast from a first-hand point of view. While going through the images and conversation it has taken me to know in depth about how interdependent our fashion industry is and yet the joy and experience of the same are only restricted to a very specific segment of the people.

This has also enabled me to understand the large-scale problems in society, reasons ranging from exploitation, superstitions, lack of awareness, and many more that have been the perils for a long time.

In conclusion, the need for smaller initiatives starting with sensitization and awareness about the hands that make fashion is needed more than ever before and I have tried to take a small first step through this project, looking forward to many more.

Information on Author Affiliation: Pritam Saha, Assistant Professor, Unitedworld Institute of Design, Karnavati University, Gandhinagar, India, pritamsaha@karnavatiuniversity.edu.in

References

Adiga, A. (2008). The White Tiger. London: Atlantic books.

Emmett, D. (2015). Artisan Voice: An investigation of the collaborations between skilled, traditional textile artisans in India and foreign textile and fashion designers from the artisans' perspective and the viewpoints of their craft industry in contemporary times. (Unpublished master's thesis). University of South Wales. doi:https://doi.org/10.26190/unsworks/18384

Kuldova, T. (2014). Fashion exhibition as a critique of Contemporary museum exhibitions: The case of "fashion India: Spectacular capitalism." Critical Studies in Fashion & Beauty, 5(2), 313–336. doi:10.1386/csfb.5.2.313_1

Kuldova, T. (2017). Luxury Indian fashion - a social critique. Bloomsbury Publishing Plc.

Malek, MohammedShakil S., and Bhatt, Viral (2022). Examine the comparison of CSFs for public and private sector's stakeholders: a SEM approach towards PPP in Indian road sector, International Journal of Construction Management, DOI: 10.1080/15623599.2022.2049490

Malek, MohammedShakil S., and Gundaliya, P. J. (2020). Negative factors in implementing public–private partnership in Indian road projects, International Journal of Construction Management, DOI: 10.1080/15623599.2020.1857672

Malek, MohammedShakil S., and Gundaliya, Pradip (2021). Value for money factors in Indian public-private partnership road projects: An exploratory approach, Journal of Project Management, 6(1), pp.23-32. DOI: 10.5267/j.jpm.2020.10.002

Gohil, P., Malek, M., Bachwani, D., Patel, D., Upadhyay, D., and Hathiwala, A. (2022). Application of 5D Building Information Modeling for Construction Management. ECS Transaction, 107(1), 2637–2649.

Malek, M. S., Gohil, P., Pandya, S., Shivam, A., and Limbachiya, K. (2022). A novel smart aging approach for monitoring the lifestyle of elderlies and identifying anomalies. Lecture Notes in Electrical Engineering, 875, 165–182.

Malek, M. S., Dhiraj, B., Upadhyay, D., and Patel, D. (2022). A review of precision agriculture methodologies, challenges and applications. Lecture Notes in Electrical Engineering, 875, 329–346.

Beyond Classroom: Impact of Covid-19 on Education System

Dr. Arun Kumar Singh [a,*] *and Simran Patawari* [b]

Unitedworld School of Law, Karnavati University, Gandhinagar, Gujarat, India
E-mail: *arun@karnavatiuniversity.edu.in

Abstract

The pandemic has helped us to better understand the existing injustices and the subsequent actions required to address the education of the more than 1.5 billion children whose learning has been hindered. The COVID-19 pandemic has primarily been a public health disaster having a significant influence on educational systems all over the world. This paper examines the impact of school closures brought on by the pandemic on children in India. The paper will emphasize, how the transition to distance learning has caused a "regression in learning" that, despite affecting people from all socioeconomic backgrounds, has disproportionately impacted the disadvantaged furthermore it will also discuss how it has brought people closer too. The schooling gap and learning loss brought on by the pandemic will cascade through an entire generation of students. All of the aforementioned difficulties will be covered in this essay, with a particular emphasis on remedies.

Keywords: Covid 19, Education, Significant Influence, NEP 2020.

Introduction

Education has a direct or indirect impact on all elements of a person's life. Historical research makes it clear that education has been crucial in creating morally upright people and a powerful country. The 21st century is said to be the age of science and technology and the world has become a small family due to rapid advancement in science and technology. People have connected virtually with the use of the internet with a click. Science and technology is used in all areas be it hospital, bank, industries, etc. Education is no exception. Science and technology have had a significant impact recently, particularly during Covid, but despite this, an entire generation of children has been impacted by this unprecedented upheaval, having possible implications that extended beyond the short and medium term, beyond the field of education. The potential impact could affect one's mental health, general well-being, socialization, and chances of becoming an active member of society, including in the job market. Unprecedented changes in the pace and scope of education run the risk of escalating the current learning crisis.

The Indian government has implemented several measures to combat COVID-19, including the first Janta Curfew, which was declared on March 22, 2020. Later, a 21-day lockdown was declared to contain the COVID-19 instances. The Indian government then decided to extend the strike till May 3, 2020, on April 14. This decision has affected the nation's various industries. The nation's education system, which is crucial to the country's economic future, is also affected by the lockdown. The government ordered all schools, colleges, and institutions to be closed. The COVID-19 outbreak disturbed the whole educational system. The strong basis of a face-to-face physical learning environment served as the cornerstone for the educational scenario. The educator and the educand engaged in active and integrative learning while sharing the same learning environment offline.

Numerous students from different states, classes, castes, genders, and regions have been impacted by COVID-19. The choice to close schools and convert traditional classrooms to digital platforms has increased learning disparities among kids, but it has also driven a lot of kids out of school because of the digital gap. Classrooms that were technology-enabled eventually became technology-driven. Textbooks and notebooks have mostly been replaced by electronic devices including phones, laptops, desktop computers, and tablets. The government has been taking action to make education digitization accessible and affordable for everybody as e-learning becomes the "new normal."

The national emergency prioritized protecting the lives of citizens of their country and focusing on health. For a very long time, education was put last in the process. Universities and institutions were unable to come up with an alternative plan of action right away. It was better understood as a "trial and error technique" for education. An evaluation was nearly overshadowed by how badly the process had gone wrong.

During the first few days of the educational conversation, only a few PowerPoint presentations were shared with the WhatsApp group or notes were distributed by email or WhatsApp. Many educational institutions lacked even a standard online collaboration tool for students and faculty. Only a small number of colleges used institutional Google accounts for daily curriculum transactions. A very little amount of technology was incorporated into lessons through presentations interspersed with a few activities or by having students work remotely on projects during class time.

India is a populous country, so education must be available even in the most isolated regions. However, everyone hopes to see this dream through to the end. More children were living there than there were gadgets. Students used the gadget for a very little amount of time. Having trouble connecting to the internet to access classroom instructions was another major barrier. Overall, the number of people receiving an education has decreased. Only a few groups of fortunate students had access to different online learning resources and uninterrupted classroom instruction.

Teaching has shifted to digital platforms due to the closing of schools to guarantee the health and safety of children, whether through online teaching methods, government portals, Direct-to-Home (DTH) channels, or other means. The wide disparities in access to fundamental digital infrastructure, such as electricity, gadgets like smartphones and computers, and internet connectivity, made remote learning difficult for many Indian students.

It is important to remember that there are numerous reports, publications, and polls whose findings imply that everyone can't profit equally from an online education system. Lack of equitable technological advancement across the country is the cause of the same.

Effects of Covid 19 on Educational Sector

Access To and Participation in Learning - In March 2020, while the majority of schools were completing the 2019–20 academic year, all educational institutions in India were temporarily closed. May would have been the start of the new school year, but due to the nationwide increase in COVID-19 cases, it was not able to restart in-school instruction. Closures have influenced millions of pupils in pre-primary through secondary schools. The significant disparities in the educational systems of states have come to light as a result of the shift from face-to-face to distant learning. There are disparities in the ability of teachers, the results of learning, the government-provided digital infrastructure, and access to technology. There is no data on how much digital information reaches kids, whether they are engaging with it, and what effects it is having, although a lot of it has been produced and distributed to help kids continue to learn from home. Many people who previously faced obstacles in getting an education—children with impairments, pupils in distant areas, children of migrant workers, immigrants, and asylum seekers—have had minimal or no schooling as a result of the COVID-19 problem, or they have fallen further behind their peers. This may drive many kids to give up on their education even when "normalcy" has been restored. (The Wire, 2022). For kids, going to school involves much more than just taking courses. Children miss out on the fun of play during the lockdown and interactions, sports, and friendly conversation. These extracurricular activities help shape a child's development just as much as academic study and are not taken into account through distance education a school also offers fundamental services. Services that are challenging to replace. Timing will tell the effects on non-academic areas of year-long school closure will have on a generation's growth and development of learners.

Health, Well-being, and Protection - More than 1.26 million schools in India have 120 million students participating in the midday food program. For many kids, this lunch serves as a major motivator. To go to school, but there is a lockdown and a COVID-19 issue In numerous states, disturbances have stopped the initiative, united territories, depriving kids of what is frequently the healthiest lunch of the day. Additionally, schools frequently provide amenities like feminine hygiene items. Given the significant obstacles that adolescent girls experience, these programs are crucial for preserving fundamental menstrual hygiene. The prevalence of taboos, particularly in rural India, has increased censorship and restrictions on girls' mobility, worries of early marriage, sexual assault, and grave health hazards Additionally eliminating this provision have increased the difficulties. teenage girls due to the closing of schools.

It is commendable that the government has increased social support through the Pradhan Mantri Garib Kalyan Yojana, which offers a combination of monetary and non-financial social aid to safeguard poor and vulnerable households. Even before the epidemic began, just 39% of eligible mothers with babies received maternity benefits under the Pradhan Mantri Matru Vandana Yojana, even though India's social security programs cover significant life phases (PMMVY). Additionally, most of India's employment is unorganized and unprotected by social security. The COVID-19 epidemic brought to light the fundamental problems with India's social protection system, which is disjointed and primarily carried out by intricate centrally sponsored initiatives. Investing in integrated social protection can assist people's managing shocks and other risks, as well as multidimensional vulnerabilities (both social and economic) (UN India, 2020).

Violence against children, sexual minorities, and women is more likely to occur during times of crisis. The risk of gender-based violence is enhanced by physical separation, movement constraints, infection fear, and confinement at home, as well as by rising tensions and economic stress, particularly in situations when the risk is already high. In the week after the lockdown, the National Commission for Women saw a more than two-fold increase in reports of violence against women and girls (Outlook India, 2020). The decline in calls received by helplines run by non-governmental organizations (NGOs) has been observed during the same period, which may be partially attributable to the challenges women have in seeking support owing to mobility issues. Nearly 30% of the contacts to Childline India Helpline were from people looking for safety against child abuse and violence (The Hindu, 2022).

Finances - India's economy has been severely impacted by the pandemic, and unemployment has increased. Due to the combined effects of the demand shock and supply disruptions as a result of closure (in May 2020), the growth rate projections for the Indian economy for 2020–21 have regularly decreased from 1.9% (in April 2020) to 1.2%. The growth rate forecasts for the Indian economy for 2020–21 have frequently dropped from 1.9% (in April 2020)15 to 1.2% as a result of the combined impact of the demand shock and supply interruptions following the shutdown (in May 2020) (United Nations Department of Economic and Social Affairs,

2022). The Indian economy was predicted by the IMF to decline by 10.3% in 2020 in October. For the first time in 40 years, India's growth rate was negative. Since the first cases of the coronavirus were discovered in January 2020, there are now more people without jobs than before. According to the Centre for Monitoring Indian Economy, the unemployment rate in India reached a high of 24% in May 2020 (26.4% in urban areas and 23.4% in rural regions), before falling to less than 7% by November 2020. About 122 million Indians lost their employment in just a month of April. Amongst these, 1.3 million of them were employed as laborers or small-business entrepreneurs. There are also a significant number of self-employed people (18.2 million) and paid workers (17.8 million) (Relief Web, 2022).

Challenges

Future economic, educational, and digitalization strategies must address long-standing challenges of inequality and a digital divide as a result of the pandemic's hasty shift to e-learning. Although the desire to increase e-learning is admirable, the majority of teachers and students in government schools lack the resources, equipment, and skills necessary to participate in this digitalization process. Various challenges faced are:

- Very few teachers had any prior training or experience delivering high-quality classes through a blended approach employing digital tools and online platforms, indicating a low capacity for digital and e-learning skills. This will further exacerbate learning quality disparities between students with teachers with higher capacity vs those with lower capacity.
- Poor structural support - teachers thought the government was not supporting them enough. They were not given the technological resources (mobile phones, laptops, dependable internet connectivity, etc.) they required to continue teaching from a distance. They also did not receive any guidance on how to navigate the dense content on e-learning platforms like DIKSHA, and there was little information on how to reach all children. Some states didn't even pay instructors their wages.
- Lack of capacity and access to cutting-edge technology made it difficult for teachers to reach every student, especially those who lived in distant locations or were at risk of dropping out.
- Preference for in-person instruction: Teachers and union officials agreed that no matter how visually or aurally arresting, engagement with children cannot be replaced by dense content.

Statistics

While there is an obvious problem with the power connection in nearly 99.9% of households in India, the quality/voltage of the electrical supply is relatively low and notably slow, especially in rural India. Only 47% of homes in rural areas have access to electricity for more than 12 hours (Protiva Kundu, Scroll, 2022). Additionally, only 11% of Indian families have any kind of computer, and only 24% of Indian households have access to the internet, even though 24% of Indians own smartphones (Ministry of Statistics and Programme Implementation 2016). Unfortunately, just 8% of families with kids between the ages of 5 and 24 have access to both a computer and the internet. Rural and urban India differ significantly from one another as well. In contrast to 23.4% of urban Indian households, only 4.4% of rural Indian families own a computer. Urban households in India have greater access to the internet as compared to rural households in India which are evidenced by the fact that only 15% of rural Indian households have access to the internet, as compared to 42% of urban Indian households. Therefore, regularly offering classes online had a financial impact because students must pay for internet access. The internet has been crucial for the advancement of education and learning, but it has also paved the door for numerous cybercrimes like hacking and cyberbullying and exposed children to graphic and violent content that can incite, defile, and corrupt young minds.

Table 1. Disparity between Urban and Rural India in terms of computer owners and internet access (in percentage)

	Computer owners	Access to internet
Rural India	4.4	15
Urban India	23.4	42

Preschools are essential for setting the groundwork for a child's healthy psychological, physical, and social development in addition to early learning. The potential of early children's holistic development continues to be severely threatened by the closing of schools and other institutions that offer early care and education. In this pandemic, there is also an increasing need to pay attention to children's psychosocial and mental health issues. Likely, the financial difficulties brought on by the loss of their jobs and income will have a negative impact on their mental health. As their neurological development is compromised, children who experience an acute nutritional shortage, a lack of protection or stimulation, or continuous exposure to toxic stress are more likely to experience lifelong difficulties (Centre for Early Childhood Education and Development).

Impact of school closures on marginalized pupils disproportionately: Even while the goal of remote learning solutions is to ensure that all children continue to study, it is generally known that the most disadvantaged kids may not be able to take advantage of these changes. When schools are closed, teaching is more difficult: Since the majority of teachers now instruct online students from their homes, distance learning has an impact on them. According to an ASSOCHAM and Primus Partners poll, to conduct proper online classes, teachers at government schools comprising 17% only were provided the efficient training, against the teachers in private schools which comprised a healthy number of 43.8% (ASSOCHAM, 2020).

There is already a teacher shortage in India. Regarding professional qualifications, there were around 14.6% of teachers in government schools, 9.7% in government-funded schools, 25.4% in private non-funded schools, and 58.7% in fifteen other categories of schools (NIEPA, 2017).

In addition to teaching online, teachers are also required to do COVID duties, which have had a negative influence on their health and well-being. Many teachers are experiencing financial insecurity right now, particularly those who work as contractual teachers in government schools and low-fee private schools. Many of them are also dealing with irregular paychecks, pay reductions, or even losing their jobs as a result of the pandemic. Through NISHTHA, MHRD, NCERT, and current policies and budgetary initiatives are strengthening the capacities of teachers and school administrators at the elementary 16 level throughout the nation. The technology has been modified to offer online training to the 24 lakh existing unskilled teachers and school administrators. The State of Kerala trained 81,000 primary teachers in 11,274 schools across the state since online education has replaced traditional means of education as a consequence of the lockdown. The apprenticeship was delivered by the Kerala Infrastructure and Technology for Education (KITE) organization. KITE also launched KOOL, the Kerala government's e-learning platform for teachers, from which 12,000 instructors have benefited.

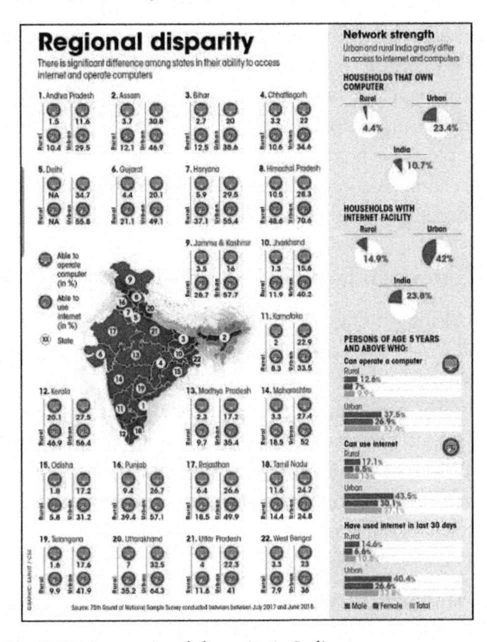

Fig. 1. Depicts regional disparity in India

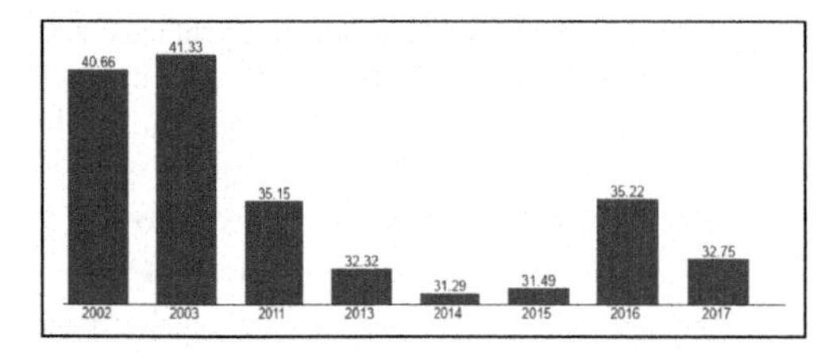

Fig. 2. Depicts Teacher-Student Ratio
Source: The Global Economy

Online education has become popular due to the Covid-19 pandemic, but an analysis by the global education network Quacquarelli Symonds (QS) indicates that the Indian internet infrastructure is still not prepared to sustain the change. A 2019 government poll shows that only 24% of homes have internet connectivity. The percentages are much lower in rural India, where only 4% of households have access. A 2018 NITI Aayog review found that 55,000 Indian villages lacked coverage related to mobile networks. A 2017–18 survey performed under the aegis of the ministry of rural development found that more than 36% of schools in India lack proper means and operated without electricity.

Therefore, it is evident that the online mode of education has been disadvantageous to the entire nation however it has had a different proportion impact on a different classes of people. The next section of the paper puts forth a certain recommendation to help mitigate the consequences of the unprecedented pandemic

Recommendations

Working on Digital Divide - Access to technology will increase through working with mobile networks to provide free or significantly reduced data for educational purposes. Make considerable financial efforts to offer a consistent electricity supply to even the most remote locations, adapt current educational software to cope with slower internet connections, and add digital services to social protection programs.

Training Faculties - Invest in teachers and give them ongoing support so they can utilize a hybrid type of instruction to offer all of its pupils' e-learning and distance education both during and after the closure of the school.

Strategies Digital Learning for the Marginalized - To ensure that the most marginalized students do not drop out and can finish their education, contextualized methods to promote distance learning are needed. This will make them more resilient in the case of another systemic shock. Finding out why specific students are more likely to drop out or fall behind, creating context-specific solutions to extend the to reach these students, teaching them the fundamentals of language and mathematical intricacies, addressing the underlying causes of why students are at risk, and offering a safety net against dropping out should be the main goals of the strategy.

Appropriate Data Collection - In-depth monitoring is required to get real-time data. This should contain information on school infrastructure, particularly the caliber of the electrical supply and internet connection, compliance with safe school policies, and water, sanitation, and hygiene (WASH) facilities in schools.

Analysis

The COVID-19 pandemic, which began in April of last year, resulted in over one lakh children becoming orphans, losing a parent, or being abandoned, according to the National Commission for Protection of Child Rights (NCPCR), which has informed the Supreme Court that these children need care and protection. Measures of supervision and safety should be taken in favor of 1,01,032 children who were in dire need of care as per the national data. The statistics show that 396 children have been deserted, both the parents of 92,475 children died, and 816 children have been orphaned.

One commendable measure taken by the national commission was to devise an online tracking application called Bal Swaraj. This application is used by the commission to track the number of children who need care and protection. The main purpose was to monitor and track children digitally and also track children who have lost their parents in the Covid. This system very well tracks the child right from the beginning of the process that is the production before the child welfare committee until the end which is restoration to their parents or guardian. One way that this can also help prevent the infringement of the right to education is that during the process of monitoring all the data regarding the child is collected by the commission. This Data can be used to monitor the number of children that have lost access to their education. As a result, the number of children who do not have access to education can be monitored and hence education can be restored using different technical and physical means for the children. The main aim however was to prepare the children for the upcoming Covid waves and therefore there are emergency care and protection were a priority however it is important to notice that the opportunity to use the collected data to enforce the right to education can be effectively implemented by correct measures and new protocols

One main action under the Bal Swaraj scheme was that the National Commission launched CiSS application whereby all the children in the street situations could be tracked down these children were then rehabilitated and could be provided with open shelters counseling services other services like De-addiction services the same mechanism could be used to provide them educational opportunities. Various non-profit organizations and institutions could be given this data so that they can effectively provide the children with proper educational facilities.

The Government of India has added a few late amendments to its flagship National Education Policy 2020 (NEP) that concentrate on the integration of technology in education in response to these difficulties. India must systematically incorporate technology into its educational system, according to the NEP, or else it runs the risk of being "at a perilous disadvantage in an increasingly competitive world" (Ministry of Human Resource Development, National Education Policy, 2020). To "facilitate decision making on the induction, deployment, and use of technology, by providing to the leadership of education institutions, State and Central governments, and other stakeholders, the latest knowledge and research as well as the opportunity to consult and share best practices," it has proposed the creation of the National Educational Technology Forum (NETF). These suggested changes are necessary, but they will need to be complemented by significant investments in digital infrastructure, teacher training, and education.

Conclusion

In sum, due to the pandemic, millions of children in India have had limited or no access to education. To address the immediate effects on these affected pupils as well as longer-term structural changes to the national system, including access to and integration of technology, education policy reform is essential. Up until now, India's institutions have primarily been responsible for filling the gaps in education policy. Although the emphasis on public health was vital, it has led to a national education deficit, especially for kids from lower socioeconomic backgrounds. To make sure that no student is left behind in school or higher education, India will prioritize public policy change and more investments in the education system going forward.

References

COVID-19 Crisis Will Push Millions of Vulnerable Children into Child Labour," The Wire, as in August 2022.

UN India (2020). COVID-19: Immediate Socio-Economic Response Plan, May 2020

Domestic Violence Cases Across India Swell Since Coronavirus Lockdown," Outlook India, as in December 2020.

COVID-19 lockdown - Rise in domestic violence, police apathy: NCW," The Hindu, as in August 2022.

World Economic Situation and Prospects 2020 Multimedia Library - United Nations Department of Economic and Social Affairs, as in August 2022.

Joint un socio economic response plan," Relief Web, as in September 2022

Indian Education can't go online – only 8% of homes with young members have computer with net link, Protiva Kundu, Scroll, as on 24-08-2022.

Ministry of Statistics and Programme Implementation (2016): "Disabled Persons in India: A statistical pro le 2016.

Digitalisation of Education: A Readiness Survey, ASSOCHAM (2020)

Government of India, Ministry of Human Resource Development, National Education Policy 2020, New Delhi, 2020, p. 57.

Implementation of Rawls Theory of Justice in the Present Indian Reservation System

Arpit Vihan[*,a]

[a]Assistant Professor, UWSL, Karnavati University, Gandhinagar, India
E-mail: [*]arpit@karnavatiuniversity.edu.in

Abstract

The researcher in this piece would try to understand the working of Rawls's theory of justice in the present Indian reservation system as this notion has been used to support a variety of "affirmative acts," including the fair distribution of assets and preferential treatment of minorities in America. Therefore, the researcher believes that it would be interesting to apply it to the present Indian Reservation system. The author further discusses provisions and policies regarding the concept of reservation in India. The role of the judiciary in interpreting and implementing Rawls's theory is another aspect discussed in the paper. To understand the problems relating to Rawls's theory and its implementation in India, a brief comparison of the theory and its application in other jurisdictions has also been discussed in detail.

Keywords: Rawls, Theory of Justice, Reservation, Affirmative action

Introduction

In the 20th century, John Rawls significantly influenced the resurgence of natural law theory. According to him, "justice and fairness are not a mere moral doctrine but a political conception which applies to all political and social institutions of the society." Numerous "affirmative measures," (Rosenfeld, 1991) like the fair distribution of assets and preferential treatment of minorities in America, have been justified using John Rawls theory. According to his idea, "economic and social disparities should be designed with everyone's best interests in mind and should call for respect for the rights and demands of others." To ensure equality of opportunity, the initial social and cultural handicaps of an individual have to be taken into consideration to evolve a just and equitable society (Paranjape, 2019). The primary goal according to Rawlsian theory is to let go of the disadvantages in society and culture that a person is born into and later experiences in life. A person belonging to lower strata suffers by the virtue of being born into a particular social stratum. It must not just be done but it must be seen to be done that everyone is provided with equal opportunities which would ultimately result in the betterment of a person's social standing. This would furthermore remove all hindrances that a person from a vulnerable section may face in achieving the highest goals.

Overview of Rawls Theory of Justice

> *"In order to strengthen a chain, we should start with strengthening its weakest link..."*
>
> – *John Rawls*

In "A theory of justice," Rawls says-

> *"Whatever their starting position in the social system—that is, regardless of the income class they are born into—those who possess the same level of aptitude and ability and the same willingness to use them should have the same chances of success. Everyone who is similarly motivated and endowed should have about equal opportunities*

for success in all spheres of society. Social class shouldn't have an impact on the expectations of those with the same skills and goals."

Equal distribution among equals implies that, in accordance with a certain standard of discrimination, unequal circumstances have to be treated differently, which raises the question of whether choosing that particular standard is ethically appropriate (Gauba, 2013). Again, there must be differential treatment under the law in order to account for differences, therefore the question of when the law should stray from equality must be decided on a principle other than equality.

He propounded the basic principles of justice, namely,

1. Everyone should be entitled to the most extensive, basic freedom possible while still allowing for the freedom of others.
2. Positions and offices that are open to everyone should be associated with social and economic inequalities that may be expected to benefit everyone.

He also pointed out that in a formal system of equality, those with substantial social and educational resources will reap the benefit of being placed in a better position. In contrast, those with meager initial resources will continue to receive only meager returns. Thus, the distribution which takes place must be must provide an initial and equal start to all the members of the society (Dias RWM 1994). The distribution of opportunities, liberties, wealth, and resources must be available to all.

Based on the first principle, all citizens are given access to all essential rights. Every citizen must be guaranteed basic freedoms, including the right to vote, freedom of conscience, and freedom of thinking, in a just society (Dev A 2013). It can at times get difficult to understand and imply the second principle because of the word "reasonably" in the policy as it is understandable that anything "reasonable" in principle may not be reasonable for a considerable number of people but we cannot fault Rawls for using the word carelessly or casually. There are countless interpretations of what justice is.

We can point out here that Aristotle's views are also on a similar line, he believes that "the legislator must in accordance to merit, exercise the distribution of wealth, honor and other divisible assets of the community which may be allotted among the members in equal and unequal shares. The initial position of equality among individuals is where Rawls contends that "judgments concerning the principles of justice are most likely to be reasonable and impartial if they are formed in reflective equilibrium" and are not influenced by passing or shifting circumstances.

According to the justice as fairness doctrine, creating the greatest amount of happiness for the greatest number of people is not as important as the ideal of equal rights for all citizens (Campagnolo, 2011). However, in utilitarian theory, the goal of bringing the greatest possible amount of happiness to the greatest possible number of people takes precedence above the idea of universal citizenship and equal rights.

It is reflected in Rawl's work that he was not completely in line with the concept of utilitarianism. He argues that the utility principle may force those who are already at a disadvantage in comparison to others in terms of their ability to obtain basic social goods (such as rights, opportunities, income, and wealth) to endure even greater disadvantages if this redistribution of rights and opportunities results in greater happiness for a larger number of people (Clayton, 2001). This is one of the main flaws in the utilitarian theory, according to him. Furthermore, if this redistribution of rights and opportunities results in more satisfaction for a broader number of people, "a large number of individuals who currently enjoy advantages over others in their capacity to obtain basic social goods may do so in greater measure."

It would thus be seen that the Rawlsian theory of equality and justice takes into account the realities of social structure and the need for formulating a uniform policy whereby those who have remained backward and neglected due to socio-economic disabilities stand at par with those who have better resources and social status.

Indian Reservation System and Policy

At the time of framing the Indian Constitution, the constitution-makers consciously applied what is now called "Rawls Theory of Justice." The major reason behind applying this theory was to create a social order backed by justice. By formulating such social order, the constitution-makers made sure that socioeconomic equality was very much present. They also made sure that such social order must be subject to some exceptions. They allowed inequality in certain cases wherein it was for the betterment and upliftment of those who were least well-off. Inequality was permitted to produce the greatest possible benefit for the downtrodden section of society.

John Rawls, in his work, has emphasized a world in which persons belonging to the vulnerable section of society are allowed to rise. It is emancipated in his work that he intended to remove every hindrance in the way towards the betterment of the people belonging to lower strata and give them equal opportunity to develop. This law should take upon itself the task of forced redistribution of wealth so as to achieve a fair division of material resources among the member of societies and the same is easily connected to Articles 14, 15, 16, and 17 of the Indian Constitution, which largely accomplishes the same task of removing the initial barrier that a person of a lower caste or minority religion must encounter.

Article 46 seeks to protect the weaker sections from social injustice, whereas Article 39-A contains a directive for legal aid to the poor ensuring that the poor have access to justice and courts. Also, the directive principles contained in Article 39(a) to (g) further require the state "to remove inequalities based on wealth and ensure distributive justice to all alike and ensure fair distribution of material wealth to eliminate the disparity between the people of India."

The courts have ruled that for the underprivileged and minorities to receive the equal opportunity, the state must engage in affirmative action by providing them with "preferential treatment" or "protective discrimination" and by taking steps to lessen inequality. As per Dr. Basu, "the fundamental idea of equality requires using legitimate classifications to give disadvantaged classes of citizens privileges in order to better their circumstances and move up to positions of equality with some more fortunate classes of citizens" (Basu, 2016).

Here we can see a certain resemblance to Rawls's theory. The court here is of the opinion that equality does not merely means a medium to provide justice, but justice must be equal. There can be no equality between people living in two different strata. Preference to a class of persons whether based on caste, creed, religion; place of birth, domicile, or residence is embedded in our constitution and *equality of unequal is secured by treating them unequally*.

To get a better understanding of the Indian policies relating to reservation, we need to focus on specific provisions under the Indian Constitution. Article 15(4) is one such provision that explicitly provides reservation to the backward classes. The first amendment Act, of 1951 added it. "Article 15(4) is read as an enabling provision under which the state may make special provisions, but it is not obliged to make such a provision." The two most contentious issues with Article 15(4), which is now also the case with Article 15(5) and Article 16(4), have been to determine the backward classes and the extent or quantum of reservation (Shukla, 2013).

Article 15(5) was added as a result of the 93rd Amendment Act of 2006. Because of the Supreme Court's ruling in P. A. Inamdar v. the State of Maharashtra, its acquisition "became imminent." The Supreme Court further held that a quota of State seats in unassisted professional institutions is a significant infringement on the freedom and autonomy of independent, professional education institutions, and it cannot be justified as a restriction under Article 19 of the Constitution (6). Article 16(4) specifically states that "any backward class of people who, in the State's opinion, is not adequately represented in the services under the State, shall be given preference in appointments or posts."

Critical Analysis and Findings

After examining both, the Rawls theory of justice which must preferably be called in this scenario distributive justice and the present Indian Reservation system, the researcher would like to conclude by saying that there are certain loopholes in the current reservation system. Now the critical question here is whether these loopholes are the result of John Rawls's theory of justice or are being wrongly interpreted and implemented by the court.

There are a few weaknesses in the theory of justice as provided by John Rawls. The use of the term "reasonable" by Rawls raises the question of what exactly is reasonable. The Indian courts have tried to interpret it in their way, but at times it can get troublesome and ambiguous. But it is just not possible and reasonable to blame it all on Rawls. The provisions dealing with equality in India, for example, Article 15(4) of the constitution run contrary to Article 15(1). Proper implementation of Article 15(3) has not yet been done. Judiciary has been reluctant in providing quotas to various classes and also derives certain doctrines from curbing this problem. Still, the unruly behavior of the political parties has been a factor in ruining the work of the court, as is the case with Article 15(5). After *P.A. Inamdar v. the State of Maharashtra* the government curb the judicial pronouncement by introducing an amendment. The political interference in these matters, solely for political gains, has resulted in the depletion of the whole idea behind the purpose of reservation.

The court clearly stated it in the *Mandal Commission case* that the reservation should not exceed the 50 percent limit. But we can see that at present, after providing reservations to the Marathas in Maharashtra, the reservation now in educational institutes is around 68% of the total seats in the state which is in clear violation of the *Mandal* judgment. This decision by the Maharashtra government is to gain the support of Marathas in the upcoming Lok Sabha election. Such political stunts are getting common in India with each passing election but such intervention by political parties is contrary to the numerous observations of the apex court of India and to the principle of distributive justice as laid down by John Rawls must be addressed and criticized.

Another sphere where we can see the wrong implementation of the theory is in the case of reservation provided by the government in promotions. There are express provisions in the constitution providing blanket protection to such reservations. This kind of equality mechanism runs directly in conflict with Rawl's theory of justice as it is rather enhancing discrimination and inequality in the workplace. We utterly overlooked the innocent, qualified, and likely economically backward upper-caste citizens while applying Rawls' theory to the welfare of the socially and economically disadvantaged. Reservation overstepped its bounds and brought in too much "justice," which is now riding in the road of social injustice and inequality. Reservation was meant to secure social fairness. To construct a new social order, social justice encompasses much more than just giving preferential treatment to OBCs and SCs/STs.

To analyze Rawls's theory closely, we must look at its implementation in other jurisdictions. Let us try to compare Rawls's theory in India and in the USA. A significant difference between U.S., and India is that in the U.S., universities have taken upon themselves the onus of increasing the representation of disadvantaged groups unlike in India. Here, we have governing bodies and organizations working under the government or the concerned authority.

Statement of Conclusion

Based on the above-mentioned observations it could be said that the problem in India is not Rawls's theory but its implementation and reservation. The role of the judiciary is essential, and the judiciary has been quite vocal about its stand which is in accordance and in line with that of Rawls and U.S. courts but the legislature, for its benefits creates a hindrance in the proper implementation of Rawls theory of justice.

References

Gauba, O. P. (2013). An Introduction to Political Theory, (5th ed, MacMillan, 2013), 429.

Matthew Clayton, Rawls and Natural Aristocracy, (2001). Croatian Journal of Philosophy, I(3) https://warwick.ac.uk/fac/soc/pais/people/clayton/research/rna.pdf

Paranjape, N. V. (2019). Studies in Jurisprudence and Legal Theory, (8th ed CLA, 2019), 163.

Rosenfeld, M. (1991). Affirmative Action and Justice: A Philosophical and Constitutional Inquiry. Yale University Press. http://www.jstor.org/stable/j.ctt1dr3888

Ashwin Dev, A. (2013). Caste based reservation in India: A development issue, http://its-a-wonderful-lyf.blogspot.com/2013/09/caste-based-reservation-social-justice.html

Gilles Campagnolo, Criticisms of Classical Political Economy. Menger, Austrian economics and the German Historical School, (2011) Routledge studies in History and Economics

Basu, Durga Das (2016). Commentary on the Constitution of India, (8th ed. vol 1 Wadhwa and Company, 2016), 980.

Shukla, V. N. (2013). Constitution of India, (12th Ed, EBC), 94.

Dias, R. W. M. (1994). Jurisprudence, Fifth Edn., P-65

Malek, MohammedShakil S., and Bhatt, Viral (2022). Examine the comparison of CSFs for public and private sector's stakeholders: A SEM approach towards PPP in Indian road sector, International Journal of Construction Management, DOI: 10.1080/15623599.2022.2049490

Financial Inclusion: Conceptual Understanding to Indian Report Card

*Kishor Bhanushali**

*Karnavati University, India
E-mail: *kishor@karnavatiuniversity.edu.in

Abstract

Financial inclusion is one of the critical factors in reducing poverty, increasing and sustaining inclusive economic growth, providing an opportunity for participation of all the sections of the society, in main stream financial system and active participation in economic growth. Present study is an attempt to present comprehensive picture of conceptual framework of financial inclusion and present status of India in global scenario in general and South Asian scenario in particular. Such study provides empirical pieces of evidence of progress made by India, results of efforts of Indian action. Such study will provide important insight to policymakers towards the results of their effort and direction for the future.

Keywords: Financial Inclusion, Banking, Findex, World Bank, Economic Growth, Poverty

Introduction

Though it looks simple and easy to understand, the term financial inclusion has different meanings based on the context and coverage. In its simple term, it means including all the sections of society in mainstream financial markets and institutions. It has been defined differently by different researchers, It has been defined as the timely delivery of financial services to the less privileged and disadvantaged sections of society, through the process of financial accounts, to allow them to access financial services ranging from savings, payments, and transfer, to credit and insurance (Sarma and Pais 2011; Lenka and Barik, 2018; Asare Vitenu-Sackey and Hongli, 2020). Financial inclusion provides financial services to individuals and households at affordable prices, at a convenient place and time without any discrimination. Most talked about model of financial inclusion, Grameen model is based on philosophy that access to credit is fundamental human right, which emphasizes lending to poor on terms that are suitable to them, teaching them financial principles (Yunus, M 2003;). According to UNDESA and UNCDF (2006), the concept of financial inclusion encompasses two primary dimensions: (a) that financial inclusion refers to a customer having access to a range of formal financial services, from simple credit and savings services to more complex ones such as insurance and pensions; and (b) that financial inclusion implies that customers have access to more than one financial services provider, which ensures a variety of competitive options. In the light of above meaning, understanding and identifying the factors which affect the level of financial inclusion is important to policymakers and the academic field as well (Nyoka 2019). The process of financial inclusion starts with identification of those groups who are not covered under traditional financial system and designing of programs and policies for their inclusion. The main focus of financial inclusion process is to reach out to those disadvantaged groups who tend to be excluded (Olarewaju Afolabi, 2020). The process not only talks about providing services, but it also covers the quality of services. According to former RBI Governor Rangarajan financial Inclusion is "the process of ensuring access to financial services and timely and adequate credit where needed by vulnerable groups such as weaker sections and low-income groups at an affordable cost" (Simon, 2020, Son et al., 2019). Financial Inclusion is also defined as the process that marks improvement in quantity, quality, and efficiency of financial intermediary services which helps improve lives, foster opportunities, and strengthen economies (Babajide et al., 2015). Local savings are promoted through financial inclusion, leading to increased productive investments in local businesses (Babajide et al., 2015). The process of financial inclusion is also linked to social inclusion. The European Commission has defined financial exclusion as, a process whereby people encounter difficulties accessing of using financial services and products in the mainstream market that are appropriate to their needs and enable them to lead a formal social life in the society to which they belong (Timbula et al., 2019). Definition of financial inclusion by World Bank is considered to be most comprehensive in terms of coverage and policy action. The World Bank (2018) defines financial inclusion as "Individuals and businesses have access to useful and affordable financial products and services that meet their needs like transactions, payments, savings, credit, insurance which are delivered responsibly and sustainably." There are research studies where instead of defining financial inclusion some authors have considered it better to define financial exclusion. Financial exclusion is process whereby people are struggling for accessing for the usage of financial products and services in market, forcing them to leave their ordinary social life (Sinclair, 2001; Asare Vitenu-Sackey and Hongli 2020). Government of India's "'Committee on Financial Inclusion in India" has defined financial inclusion as the process of ensuring access to financial services and timely adequate credit where needed by vulnerable groups such as the weaker sections and low-income groups at an affordable cost (Rangrajan Committee, 2008). At some places, financial inclusion has been presented in terms of access to various financial services. Access to financial services has been measured by World Bank in the context of having an account at bank of another type of financial institution such as a credit union, microfinance institution, cooperative, or the post office; has a debit card connected to an account at a financial institution with their name on it; received wages, government transfers, public sector pension, or payments for agricultural products directly into an account at a financial institution in the past year; or personally paid utility bills or school fees from an account at a financial institution in the past year.

Financial Inclusion: Indian Report Card

As stated above government of India and Reserve Bank of India have been working actively in the direction of increasing financial inclusion in India. Various policies are in place to bring those un-served into mainstream financial services. Financial Inclusion in the Indian context in the post-liberalization period has witnessed a significant increase in the number of bank branches in urban, semi-urban, and metropolitan regions. Yet, a large section of population is deprived of financial products and services in India at present (Peachy and Roe, 2004;). Out of 19.9 crore households in India, only 6.82 crore households have access to banking services. As far as rural areas are concerned, out of 13.83 crore rural households in India, only 4.16 crore rural households have access to basic banking services (Madurai, 2011). Share of adults with an account has more than doubled since 2011, to 80 percent. An important factor driving this increase was a government policy launched in 2014 to boost account ownership among unbanked adults through biometric identification cards. This policy benefited traditionally excluded groups and helped ensure inclusive growth in account ownership. Between 2014 and 2017 account ownership in India rose by more than 30 percentage points among women as well as among adults in the poorest 40 percent of households. Among men and adults in the wealthiest 60 percent of households it increased by about 20 percentage points due to a strong government push through biometric identification process (Simon 2020; Demirgüç-

Kunt et al., 2018). Still there is a gap between the need and the actual situation. "This fact has been confirmed by NSSO survey which shows that over 80 million poor people are living in the cities and towns of India and they lack access to the most basic banking services" (Madurai, 2011). Need for more financial inclusion can also be accessed in the fact that the share of non-institutional sources in total credit to cultivator households increased from 34 percent in 1991 to 39 percent in 2002, and further to 41 percent in 2012. The increase in the combined share of agricultural and professional moneylenders (from 18 percent in 1991 to 31 percent in 2012), outperforming commercial banks as a source for rural credit, is a worrisome phenomenon" (Simon, 2020).

Having an account with financial institutions is first step toward financial inclusion. Access to financial institutions leads to habits of savings, investment, borrowing etc. Global Findex database is world's most comprehensive data set on how adults save, borrow, make payments, and manage risk. Drawing on survey data collected in collaboration with Gallup, Inc., the Global Findex database covers more than 140 economies around the world. Compiled using nationally representative surveys of more than 150,000 adults age 15 and above in over 140 economies, the 2017 Global Findex database includes updated indicators on access to and use of formal and informal financial services. It has additional data on the use of financial technology (or fintech), including the use of mobile phones and the internet to conduct financial transactions. By collecting detailed indicators about how adults around the world manage their day-to-day finances, the Global Findex allows policymakers, researchers, businesses, and development practitioners to track how the use of financial services has changed over time. The database can also be used to identify gaps in access to the formal financial system and design policies to expand financial inclusion. For Indian report card, we will use World Bank Findex database. The details about access to financial institutions in India as compared to global picture shows that nearly universal access to financial institutions has been achieved for countries like Denmark, Finland, Norway, Sweden, Netherland, Australia, Canada, New Zealand, Belgium, and Germany where more than 99 percent people have access to financial institutions. On the other hand countries in the bottom of the pyramid include South Sudan, Madagascar, Niger, Sierra Leone, Chad, Afghanistan, Central African Republic, Guinea, Cambodia, and Pakistan where less than 20 percent population has an account with financial institutions. The data about India shows that nearly 79.30 percent populations have an account with financial institutions. Access to financial institutions in India is very high when compared to global average of nearly 60 percent and South Asia average of around 53.6 percent, but seems lower when compared to 95.5 percent and 85.8 percent in the case of high-income OECD and non-OECD regions (Fig. 1).

Financial Inclusion: Gender Gap

Gender gap in access to financial institutions is most widely discussed in literature. Female normally have less access to financial institutions as compared to their counterpart. The data at the global level indicates that nearly 64 percent of male have an account with financial institutions as compared to 56.5 percent among female. For India, these numbers are 84.4 percent and 76.4 percent for male and female respectively, which is relatively higher than many regions except high-income regions, both OCED and Non-OECD.

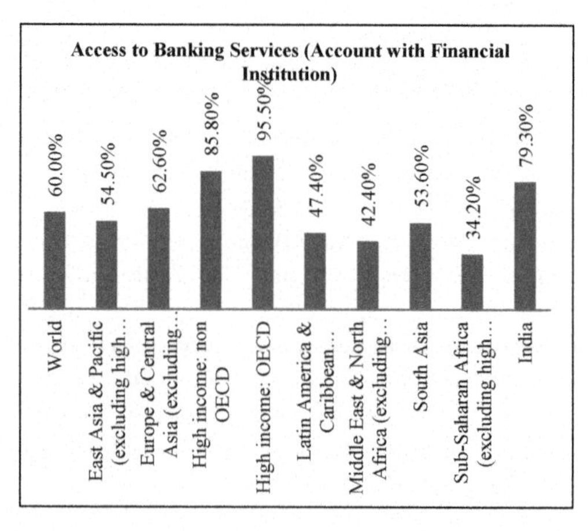

Fig. 1. Access to Banking Services (Account with Financial Institution)

The financial inclusion in India is much higher as compared to South Asia region of 58.3 and 49.5 percent respectively for male and female. The gap between financial inclusion of male and female is around 8 percent in India which is much higher than 1.5 percent and 4.30 percent respectively in the case of high-income OECD region and Europe and Central Asia region, but slightly lower than 8.8 percent in the case of South Asian region (Fig. 2). To probe further to check the association between access to a financial institution and gender in India, Chi-square test of independence has been conducted. The results of chi-square test of independence show that their account with financial institution is gender are not independent (Chi-square 16.297, df 1, N 3000, $p < 0.05$).

Access to Financial Institution and Education

The level of education is one of the most important factors affecting financial inclusion. At global level, only around 36.8 percent population having completed primary or less level of education have bank accounts as compared to 89.6 percent among those who completed education level of tertiary or more. Corresponding data about India are 76.0 percent and 96.20 percent respectively. When we compare India with South Asian Region the corresponding value in the region is 45.6 percent and 84.5 percent respectively.

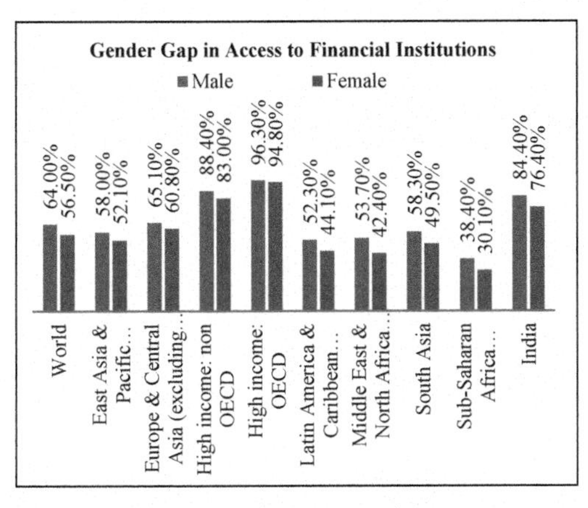

Fig. 2. Gender Gap in Access to Financial Institutions

Thus it is evident that financial inclusion in India is much better than global as well as regional scenarios. India is performing relatively better than high-income non-OEDC group (Figure 3). Further analysis has been undertaken to check the association between the level of education and account with financial institutions in India. The results of Chi-square test reject the null hypothesis that account with financial institution and education are independent (Chi-square 54.954, df 2, N 2991, $p < 0.05$). There is significant difference in the level of education across various levels of education. Proportion of population, having an account with financial institution increase with their level of education. Thus government needs to focus on people with lower levels of education to bring them to mainstream financial system.

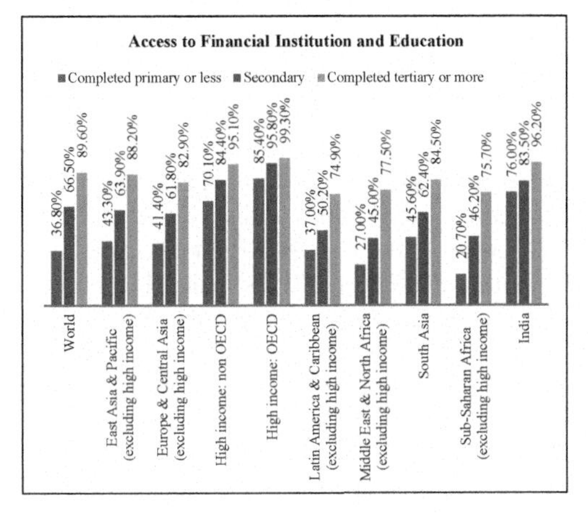

Fig. 3. Access to Financial Institution and Education

Financial Inclusion and Income

Main objective of financial inclusion is to bring economically lower strata of society in financial system. The comparative picture of financial inclusion across different economic group shows that 75.6 percent of poorest 20 percent have access to financial institutions as compared to 48.6 percent and 48.4 percent respectively for World and South Asian region. On the other extreme, 84.5 percent richest 20 percent in India have access to account when compared to 71.2 percent and 61.2 percent respectively for World and South Asian region. Thus Indian performance in terms of access to bank account is much better compared to global average and regional average but there is long way to go when compared to high income OECD and Non OECD region (Table 1).

Table 1. Access to Financial Institution and Income

Region/India	Poorest 20%	Second 20%	Middle 20%	Fourth 20%	Richest 20%
World	48.6%	54.0%	58.2%	62.8%	71.2%
East Asia & Pacific (excluding high-income)	44.3%	47.7%	52.4%	58.6%	67.7%
Europe & Central Asia (excluding high-income)	47.6%	54.8%	61.4%	67.4%	74.8%
High income: non-OECD	77.4%	82.0%	84.9%	89.2%	92.7%
High income: OECD	92.0%	94.0%	95.3%	96.4%	98.3%
Latin America & Caribbean (excluding high-income)	31.1%	37.8%	44.9%	51.9%	65.1%
Middle East & North Africa (excluding high-income)	27.7%	32.3%	40.0%	44.4%	59.9%
South Asia	48.4%	50.7%	51.0%	53.9%	61.2%
Sub-Saharan Africa (excluding high-income)	20.7%	25.9%	29.1%	36.0%	50.1%
India	**75.6%**	**77.6%**	**79.6%**	**78.3%**	**84.5%**
Source: Computed from Word Bank Database					

Chi-square test has been undertaken to understand the association between access to financial institutions and level of Income. The results of the chi-square test reject the null hypothesis that account with financial institution and level of income are independent (Chi-square 16.988, df 4, N 3000, $p < 0.05$). Thus we conclude that the proportion of people having an account with financial institutions depends on their level of income. Government needs to divert its efforts towards covering those with low income in terms of having an account with financial institutions.

Financial Inclusion and Work Status

Financial inclusion is directly related to working status, though it is important both for working and non-working populations. The data shows that the global average data about access to financial institutions among non-working and working is respectively 51.8 percent and 65 percent. As compared to global average data about India shows much better results of 74.3 percent and 83.2 percent respectively. Indian performance is much better as compared to South Asian average of 44.5 percent and 61.6 percent respectively. There is still scope for improvement for India when compared with the performance of high-income economies where the percentages are 91.5 percent and 97.8 percent respectively for non-working and working populations (Fig. 4).

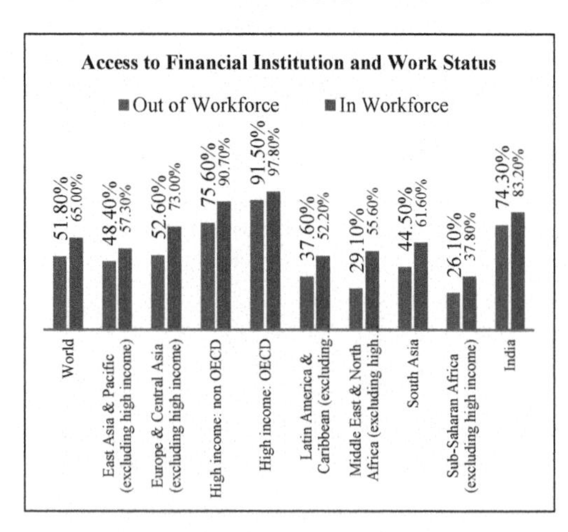

Fig. 4. Access to Financial Institution and Work Status

Chi-square test conducted to check the association between access to an account with financial institutions and working status in India shows that both the variables are not independent, null hypothesis is rejected (Chi-square 35.084, df

1, N 3000, $p < 0.05$). Thus there is difference in the proportion of the working and non-working population in terms of having an account with financial institution.

Conclusion

Indian report card reveals that the Indian financial sector has been performing much better as compared to the South Asian region when measured in terms of access to banking accounts across gender, education, income and working status, which reflects on the success of efforts. The access to banking services in India is lower when compared to high-income regions. Gender disparity in access to banks is area where special attention is required. In terms of income, India needs to work towards providing banking services to poorest of poor and those with low literacy rates. Overall picture shows that India has performed relatively better in terms of accounts with financial institutions. Now it is time to put efforts make sure that financial services attached to bank accounts should be used by the population. Efforts are required towards financial inclusion of female, lower level of education, low income, and non-working population in India.

References

Asare Vitenu-Sackey, P., and Hongli, J. (2020). Financial Inclusion and Poverty Alleviation: The Contribution of Commercial Banks in West Africa. International Journal of Business, Economics and Management, 7(1), 57–70. https://doi.org/10.18488/journal.62.2020.71.57.70

Babajide, A. A., and Adegboye, F. B. (2015). Financial inclusion and economic growth in Nigeria.

Madurai, K. K. (2011). Financial Inclusion in India. SSRN Electronic Journal.

Nyoka, C. (2019). Education Level and Income Disparities: Implications for Financial Inclusion through Mobile Money Adoption in South Africa. Comparative Economic Research, 22(4), 129–142. https://doi.org/10.2478/cer-2019-0036

Olarewaju Afolabi, J. (2020). Impact of Financial Inclusion on Inclusive Growth: An Empirical Study of Nigeria. Asian Journal of Economics and Empirical Research, 7(1), 8–14. https://doi.org/10.20448/journal.501.2020.71.8.14

Peachy, S., and Roe, E. (2006). Access to finance: What does it mean and how do savings banks foster access. Bruxelas, Belgium: World Savings Banks Institute (WSBI).

Sarma, M., and Pais, J. (2011). Financial inclusion and development. Journal of International Development, 23(5), 613–628.

Simon, T. D. (2020). Determinants of Financial Inclusion in India. Journal of Management, 7(1).

Timbula, M. A., Mengesha, T., and Mekonnen, Y. (2019). Financial Inclusion And Its Determinants Among Households In Jimma Zone Of Oromia Regional State, Ethiopia. International Journal of Commerce and Finance, 5(2).

Yunus, M. (2003). Banker to The Poor: Micro-Lending and the Battle Against World Poverty. New York, NY: PublicAffairs, 2003

Contemplating the Problems and Issues Related to Corporate Social Responsibility in India

Hiral Borikar Dhal[*a] *and Dr. Kishor Bhanushali*[b]

[a,b]Karnavati University, Gandhinagar, India
E-mail: [*]hiraldhal@karnavatiuniversity.edu.in

Abstract

The idea of corporations having a social duty has received much attention lately. Corporate Social Responsibility is an emerging idea that reflects the changing nature of businesses' approaches to social marketing. Many of the world's most successful businesses have come to understand the promotional value of aligning themselves with important social concerns. Companies now have a new competitive edge thanks to CSR and other forms of cause-related marketing. "Corporate Social Responsibility" means a company's duty to its many constituencies. It's based on the idea that businesses have a responsibility to society as well as a drive to succeed for their happiness. Similar to sales promotion, corporate philanthropy, corporate sponsorship, corporate Samaritan deeds, and public relations, this might be a smart marketing action that helps a company succeed by helping others.

Keywords: Corporate Social Responsibility, CSR Practices, Economic, Environmental, and Social imperatives, Companies Act, Corona

Introduction

The shift towards building more welcoming companies is a new and emerging strategic problem on the CSR agenda. In this article, we'll take a closer look at how creative economic and social development ideas might help marginalized communities become integral links in a business's value chain. As a part of their business plan, many corporations are looking to better establish inclusive business techniques to increase market access, which are linked to community investment initiatives and poverty reduction. Although these strategies can open up new sales channels, they also present chances to "create shared value' via product, service, and company innovations where CSR can play a pivotal role. Similarly, executives of multinational corporations with several divisions are increasingly interested in the causes and effects of corporate social responsibility (CSR). These business tycoons are well aware of the reality that business customs, regulatory structures, and stakeholder pressure on CSR may differ widely among geographies and industries. In addition, they are well aware of the pressure exerted on division managers by workers, suppliers, community organizations, NGOs, and Government to expand their CSR efforts.

It was observed that the humanitarian and humanitarian approach lasted after the 1947 independence era, but was characterized by attentive professional participation in social prosperity and transformation. When it comes to corporate social responsibility (CSR), Indian businesses had chosen a more progressive approach since the 1990s, when liberalization took hold in the country and globalization began to have an impact on the level of competition and the nature of the economic environment. As a result of the globalization and economic business environment, Indian industries and organizations have improved their knowledge and preparation of CSR, and they are now using creative approaches to addressing social and environmental concerns within their own, long-established organizational framework. Since the year 2000, CSR in India has experienced rapid growth as more and more businesses embrace the concept and work to incorporate it into their operations. Though less in number, such businesses are still abundantly present. The Indian Government is just beginning to address the issue of corporate social responsibility. Instead of being integrated into day-to-day operations, CSR in India is still managed by company founders and overseen by top executives.

Corporate social responsibility is largely seen as an "add-on" obligation by Indian enterprises, although a select few see it as integral to their operations. Over the past decade, we've seen corporate social responsibility (CSR) grow among a select group of businesses, with those organizations increasingly devoting resources to strategic innovations that contribute to the betterment of society at large. As a result, many corporations have shifted their attention to need-based creativity that aligns with national goals including public welfare, education, economic security, water conservation, and resource management. The landscape of CSR in India has mostly stayed the same. Still, it has continued to include national-level conferences focused on the potential role and obligation of the business sector in giving or contributing towards alleviating societal challenges. The future of corporate social responsibility (CSR) in India may be deduced from existing models.

Meaning/Concept/Evolution CSR

A consensus on a single definition of CSR may take time to achieve. Many different ideas overlap with the term "corporate social responsibility," making it a difficult subject to grasp. However, all existing definitions of the term include the effect that businesses have on society, including but not limited to issues of social welfare, poverty alleviation, and environmental impact, such as climate change, sustainable development, etc.

CSR is "the responsibility of companies for their impacts on society," as defined by the European Commission (EC)., and it foresees that businesses "should have in place a method to incorporate social, environmental, ethical, human rights, and consumer issues into their business operations and core strategy in close partnership with their stakeholders" if they are to fully fulfill their social responsibilities Malek, MohammedShakil S., and Gundaliya, P. J. (2020)

Corporate social responsibility (CSR) is "the limiting commitment by business to contribute to economic development while improving the quality of life of the workforce and their families, as well as community and society at large," according to the World Business Council for Sustainable Development (WBCSD).

"Corporate social responsibility is a management concept whereby companies integrate social and environmental concerns in their business operations and interactions with their stakeholders," as defined by the United Nations Industrial Development Organization (UNIDO).

CSR "urges business to embrace the triple bottom line approach whereby its financial performance can be harmonized with the expectations of society, the environment, and the many stakeholders it interfaces within a sustainable manner," according to the Ministry of Corporate Affairs (MCA), Government of India (2011).

S-135 of the Companies Act, 2013 defines corporate social responsibility (CSR) to include, but not be limited to, projects or programs related to activities specified in the schedule (specify which schedule) or such activities undertaken by the board (specify which board) in accordance with the recommendations of the CSR committee. (Applicable portions of the Act should be consulted)

Concept

As we have shown, companies are no longer expected to serve the obsolete function of pure profit generators. On the other hand, businesses are subject to substantial pressure to adopt policies compatible with their economic, social, and environmental missions. More and more people are beginning to understand that corporations have far-reaching effects on society and have a unique opportunity to make constructive changes. This is the driving force behind the concept of corporate social responsibility (CSR), which is now about integrating ethical concerns, social systems, and environmental factors into commercial activities rather than stopping at acts of charity or philanthropy. In the modern era, corporate social responsibility (CSR) is business-driven, strategic, and embedded in all aspects of operations. While this may be true, defining CSR within a rigid framework proves to be a formidable challenge.

Though it has been done informally for centuries, the tradition is still widely observed. Religions and ideologies from all over the world stress the need to do business morally and charitably, particularly when dealing with society's most vulnerable members Borikar, M. H., and Bhatt, V. (2020) Traditional primary drivers of CSR have been philosophy, religion, and compassion. Religious practices like Seva, daan, zakat, and darshans have influenced the rich and the poor. CSR actions have been driven and shaped by religious beliefs, cultural norms, and family traditions. Before 1850, when the industrial revolution began, the wealthy built hospitals, schools, and churches for the public's benefit. They also fed the hungry during times of crisis, like natural catastrophes and pandemics. Giving back to the community was also a means to raise one's social standing.

Evolution

With the rise of colonialism, there was a noticeable shift in CSR strategy (1850 onwards). Tata, Birla, Godrej, Singhania, Modi, and Bajaj were just a few of the notable industrialists of the period. They helped pioneer the notion of CSR through their philanthropic organizations, healthcare and education institutions, and other community development initiatives (Sharma, 2013). Even in semi-urban and rural locations, business owners invested some of their earnings in improving their communities through infrastructure projects like building new schools.

In the years immediately after independence, the Government, not the private sector, was the driving force behind progress. Time may be accurately defined as an age of regulations, controls, and commands due to the strict laws and policies controlling private industry, such as industrial licensing, heavy taxes, and legislations covering labor and environmental concerns and corporate governance. Unfortunately, many of the PSUs the state set up at the time had weak management and hence failed in their mission to help the poor. As a result, private businesses had higher hopes for the country's economic and social progress. However, before 1980, CSR efforts typically fell below the mark set by the public.

As in the 1980s, Indian businesses began incorporating CSR into their overall strategy (Gahlot, 2013). Companies could expand quickly because of the economic boost provided by globalization and economic liberalization, both of which were initiated in the 1990s. It gave businesses a chance to demonstrate their commitment to helping society. Indian businesses have become more socially responsible in response to Western customers' pressure about labor and environmental standards in business partner nations.

The last 15 years have seen a dramatic shift in corporate social responsibility (CSR), with more and more businesses developing CSR strategies that reflect national goals in areas such as public health, education, skill development, the environment, and more. The Government has made concerted attempts to encourage businesses to implement CSR. The "Guidelines on CSR and Sustainability for Central Public Sector Enterprises" were issued by the Department of Public Enterprise (DPE), Ministry of Heavy Industries, and Public Enterprises to serve as a benchmark for CSR spending by government agencies. The Board of Directors approved a change to these policies in April 2013 that set aside funds for CSR and environmental initiatives. These rules also mandated publicizing and reporting on CSR initiatives (Malek, and Bhatt, 2022).

Companies Act/CSR Spending

The Lok Sabha and the Rajya Sabha approved the Companies, Bill in 2012 and 2013, respectively. With the President's signature in August 2013, it became law and is now known as the Companies Act, 2013. Section 135 of the Companies Act 2013 became effective with the notification of CSR guidelines by the Ministry of Corporate Affairs in February 2014. As of April 1, 2014, the Act and CSR regulations are in effect.

This Act's groundbreaking provision was the need for CSR for a certain group of businesses. The law became India the first and, to our knowledge, only country to give CSR a legally binding mandate. Companies with an annual turnover of Rs. 1000 crores or more, or a net worth of Rs. 500 crores or more, or a net profit before taxes of Rs. 5 crores or more in any financial year (any of the three preceding financial years as laid out in Rule 3(2) of Companies CSR Rule, 2014) are required by section 135(1) of this Act to devote at least 2% of their average net profit of the three immediately preceding financial years to CSR activities.

Section 21 of the Firms (Amendment) Act 2019 clarifies how companies that haven't been in business for three years but otherwise fulfil the conditions outlined in Section 135 calculate their CSR spending. Those businesses will allocate two percent (2%) of their net income from the prior fiscal year to CSR initiatives. Companies that have been in operation for less than three years may need help to meet their required CSR obligations, especially if they are young businesses in desperate need of capital to grow.

If any portion of the CSR budget remains unspent after the fiscal year, the Unspent Corporate Social Responsibility Account must be established and funded within thirty (30) days. If this money isn't used in accordance with the CSR policy within three (3) fiscal years of the date of such transfer, it should be given to a fund listed in Schedule VII of the Act. The fundamental goal of making a difference in the community they serve may be undermined if they set up an account to deposit the leftover funds. Such a redistribution of unused CSR funding looks to be a tax in disguise. By establishing the National CSR Fund, the Government has implied that CSR contributions are necessary for funding national initiatives. Moreover, it gives corporations who could be more enthusiastic about implementing the system a way to still meet the letter of the law while undercutting the spirit of CSR.

This new era of corporate social responsibility (CSR) represents a paradigm change from shareholder value maximization to stakeholder value optimization. The fact that it is required by law, and has the inherent capacity to achieve commercial goals, has given it increased prominence. It will go to the next evolutionary stage when corporations begin to see it as serving their own self-interest (profit maximization) by proactively resolving the concerns of their stakeholders rather than simply complying with it because it is mandated.

Literature Review

The conventional justification for CSR comes from ethical and moral concerns. When it comes to assessing the contributions of businesses to society at large, Bowen (1953) was likely one of the first to do so. According to Steiner (1971), there is a direct correlation between the size of a company and the degree to which it is expected to contribute to the promotion of social welfare. Freeman's (1984) observation that profit maximization for shareholders should not be the main business purpose summed up this concept from the stakeholders' perspective. In addition to protecting, preserving, and nurturing human values and promoting socio-economic welfare (Prabhakar and Mishra, 2013; Nitish Desai Associates, 2017), it is now widely acknowledged that adopting principles of socially responsible behavior by the organization elevates the organization's position in the society (Prabhakar and Mishra, 2013). There are, however, dissenting, albeit few and weak, views based on economic considerations. According to Fredman (1970), putting social interests ahead of commercial ones might be considered stealing from shareholders.

CSR has long been recognised in India as a significant type of informal charity. Many worldviews condemn exploitative corporate practices.

Companies' approaches to Corporate Social Responsibility (CSR) challenges in their supply chains are diverse, as are the management systems and tools they employ, as reported by Francesco et al. (2008). (SCs). Kabir (2011) argues that firms engage in CSR practises in order to establish and maintain a positive corporate image in which they are seen as social organisations and that community participation ranks highest among CSR initiatives. According to Brooks (2012), corporate responsibility (CR) is about how a company considers and acts upon the concerns of its many stakeholder groups. Shrivastava (1995) argues that the industrial and environmental problems that are currently occurring are among the greatest dangers to humankind. The origins of these disasters lie in the operations, goods, and methods of manufacturing of large corporations. We need a more robust understanding of corporate social responsibility if businesses are to respond effectively to such emergencies (CSR).

American economist Milton Friedman questioned the idea of the economic bottom line as the sole basis for corporate responsibility in 1958. Several businessmen reached an agreement through discussion and debate on the idea that society can function adequately if businesses simply obey the law and generate prosperity, and then use taxation, private charity, and other options to channel more money and resources toward achieving social ends.

Currently, CSR is at an intriguing developmental stage that calls for a meeting of opposing perspectives. Creating maximum shareholder profit while being fair to all stakeholders (including employees, customers, suppliers, the local community, Government, and the environment) is the definition of social responsibility (Paramashivaiah and Puttaswamy, 2013).

Aupperleet al. (1985) found a neutral relationship between CSR and profitability, while Russo (1997) found a positive relationship between environmental performance and financial performance. McWilliams (2002) provides support for the idea that CSR strategies, when paired with political strategies, can be used to create sustainable competitive advantage firms. Strategic CSR, in the sense that corporations deliver a public benefit in combination with their marketing/business plan, is what attracts socially responsible customers, as defined by Baron (2001). According to Friedman (1970), CSR is a sign of managers' self-interest and consequently diminishes shareholder wealth.

Companies are given adequate time to conduct an accurate calculation of the opportunity cost of CSR spending since the Act's CSR standards adhere to a "comply or explain" philosophy rather than a "comply or otherwise" principle. An examination of the 500 largest firms in the world reveals that their CSR budgets will total less than 5 crores on average. This is hardly excessive when compared to the roughly 97,000 crores that these businesses earn each year (Sarkar and Sarkar, 2015). It is, therefore, absurd to argue that CSR expenditure significantly harms the economic interests of shareholders.

Measuring and Evaluating CSR Carroll (2000) investigated the question of whether or not corporate social performance should be monitored and if so, why. His research led him to the conclusion that CSP should be assessed since it has been recognized that it is a significant problem for both businesses and societies.

Challenges

The Act's net profit application criterion of Rs 5 crores looks arbitrary in comparison to the net worth and turnover thresholds of Rs 500 crores and Rs 5,000 crores, respectively. Too many successful businesses are left out of these restrictions.

Corporate social responsibility (CSR) is controversial because some people believe it should be optional for businesses rather than mandated by the Government. As a result, the 2019 changes have made room for a variety of options, like postponed CSR expenditure. The unused balance can now be carried over to the following year. Since mandatory compliance may result in ineffective use, this clause needs to be softened.

The Act does not include any language dealing with the consequences of failing to comply with CSR requirements. Violations of the Act that were not otherwise expressly allowed for in any other part of the Act were to be punished in accordance with Section 450. The 2019 change adds a new punishment provision for violating Section 135, which should improve compliance. Yet, the breach shouldn't be treated as a crime under the penal rules for at least the first five years.

Investing in community service projects close to home makes perfect sense. - Since the local population has ceded a great deal of power in the form of land and other resources, the company's operations must have an effect on the community as a whole in order to alleviate the suffering of its residents and enhance their standard of life.

A provision for required community contribution, no matter how little - say, 0.2% of the CSR plan outlay of the local region - is also recommended. Sweat equity is a common method of payment in rural communities.

We applaud the media's efforts to raise awareness of CSR efforts by publicizing positive examples of firms doing good in their communities. The apparent power of exposure and branding exercise causes many NGOs to engage in event-based programming; as a result, they lose out on important grassroots interventions.

Effect of Corona Pandemonium On CSR

In 2020, the Government expanded the concept of corporate social responsibility to include aid to former members of the Central Armed Police Forces (CAPF) and Central Para Military Forces (CPMF) and their families (CSR). The MCA's Circular from March 23, 2020, designating the COVID pandemic as a declared catastrophe and making the use of CSR money for COVID-19 an authorized CSR activity, would help businesses to fulfill both their legislative obligations and their social responsibility to the community. Spending on preventative health care, sanitation, and providing quarantine facilities, among other things, has been notified as qualifying CSR expenditures, as have contributions to the PM CARES Fund and Disaster Management Authorities. Government actions against COVID-19 have received massive funding from CSR funds.

It's safe to say that the media's focus on CSR and sustainability in 2020 will be remembered as a watershed year for the movement to raise public awareness of the importance of the idea. The importance of community involvement and fair treatment has grown amid the current chaos. Businesses from all around the world have stepped up to aid communities and individuals in need by doing things like donating money to improve healthcare facilities, equipping themselves to combat the virus, and redirecting resources to expand healthcare product manufacturing. Companies have been known to help their workers out in many ways, like allowing them to work remotely, offering financial aid, etc. The Covid-19 epidemic was also said to have pushed corporations in a new direction, one that prioritized true and honest CSR (Hongwei and Harris, 2020).

Significance of the Study

CRS programs work on making the world a better place by providing direct benefits to society, nature, and the community in which a company operates. In addition, a corporation may reap advantages for the organization as a whole as a result of the projects. Employees may feel more satisfied with their jobs and be more motivated to stay with the firm if they are aware that the business supports charitable organizations. Additionally, people of society may be more inclined to prefer to do business with firms that are aiming to create a more conscious good influence that extends beyond the limits of the company's business.

Conclusion

It will be necessary to put in place adequate procedures to address the inevitable problems and difficulties that will arise when the new CSR Policies are put into effect. In India, the business sector has to travel a long road from being an

"Economic Citizen" to a "Responsible Citizen." While there is more clarity on the execution of CSR, a little more improvement and engagement is necessary with respect to CSR policies and procedures across diverse firms. In India, the business sector has to travel a long road from being an "Economic Citizen" to a "Responsible Citizen," and while there is more clarity on the execution of CSR, little more improvement and engagement with regard to CSR policies and procedures across diverse firms is necessary.

References

Gahlot, S. (2013). Corporate Social Responsibility: Current Scenario. Research Journal of Management Sciences. 2 (12) http://www.isca.in/IJMS/Archive/v2/ii2/3.ISCA-RJMS-2013- 105.pdf

Malek, MohammedShakil S., and Bhatt, Viral (2022). Examine the comparison of CSFs for public and private sector's stakeholders: a SEM approach towards PPP in Indian road sector, International Journal of Construction Management, DOI: 10.1080/15623599.2022.2049490

Malek, MohammedShakil S., and Gundaliya, P. J. (2020). Negative factors in implementing public–private partnership in Indian road projects, International Journal of Construction Management, DOI: 10.1080/15623599.2020.1857672

Prabhakar, R., and Mishra, S ((2013). A Study of Corporate Social Responsibility in Indian Organization: an Introspection. Proceedings of 21st International Business Research Conference 10–11 June 2013, Ryerson University, Toronto, Canada, www.wbiworldconpro.com/uploads/canada-conference/2013/management/1370168444-430-

Ciliberti, Francesco, Pontrandolfo, Pierpaolo, Scozzi, Barbara (2008). Investigating corporate social responsibility in supply chains: an SME perspective Journal of Cleaner Production, 16(15), 1579–1588.

Brooks, L. J., Corporate Responsibility, Encyclopedia of Applied Ethics (Second Edition),2012, Pages 645–655

Paul Shrivastava, Industrial/environmental crises and corporate social responsibility, The Journal of Socio-Economics (1995), 24(1), 211–227.

Paramashivaiah and Puttaswamy (2013). CSR for Sustainable Development: An Empirical Study on the Corporate Social Responsibility-Focus of Top Listed Companies of BSE, AL- SHODHANA, 1(2) http://staloysius.edu.in/c/document_library/get_file?uuid=7f9338fb-b64c-4d23-b793-2b694ICI099c&groupId=11530

Borikar, M. H., and Bhatt, V. (2020). Measuring impact of factors influencing workplace stress with respect to financial services. *Alochana Chakra Journal, ISSN*, (2231-3990), Volume IX, Issue VI, June/2020, Page No. 1122.

Visser, W. (2008). Corporate Social Responsibility in Developing Countries in Andrew Crane, Dirk Matten, Abagail McWilliams, Jeremy Moon and Donald Siegel (eds.) The Oxford Handbook of Corporate social responsibility, Oxford University Press. www.oxfordhandbooks.com/view/10.1093/...001.../oxfordhb-9780199211593-e-021 Available at www.oxfordhandbooks.com/

Untapped Power of Music-Integrated Pedagogy: Its Role in Enhancement of "Behaviour and Self-Confidence" among School Students

Ashraf Alam[*,a] *and Atasi Mohanty*[b]

[a]Rekhi Centre of Excellence for the Science of Happiness, Indian Institute of Technology Kharagpur, India
E-mail: [*]ashraf_alam@kgpian.iitkgp.ac.in

Abstract

Music-integrated pedagogy is a relatively new approach to classroom teaching that offers a variety of methods and strategies for teaching and learning, employing music as a pedagogical tool. In the current investigation, research papers were systematically reviewed, and subsequently, evidence was summarized. The authors started by framing questions, followed by summarizing the findings, and finally interpreted the results, and concluded. Summarily putting the findings together, it is concluded that employing music as a pedagogical tool facilitates students' behavior and self-confidence. Listening to or playing music has a direct bearing on students" intelligence, cognitive abilities, and brain functioning, thereby leading to academic achievement. Furthermore, making music actively engages the brain and increases its capacity by increasing the strengths of connections among neurons, thus leading to enhanced cortical and cognitive development. Authors have also critiqued the limitations of existing literature to improve future pedagogical designs.

Keywords: Academic Achievement, Curriculum, Learning, Music, Pedagogy, School, Education, Teaching

Introduction

Rhythmic music pedagogy includes several tactical pedagogical approaches for teaching and learning that use music as a pedagogical instrument, particularly in the realm of rhythmic music. It is more crucial than ever to sustain the notion of music as a collaborative creative process that transcends boundaries in a multicultural and fluid world when there are inclinations toward apathy and separation. Rhythmic music pedagogy can play an important role in a child's social and intellectual development in a rhythmic ecosystem.

Scientists have mixed opinions on whether musical competencies and language learning abilities are linked or separate. In normally developing children, there is a link between receptive grammar and rhythm perception (Alam, 2021). These findings support the theory that music and language both use the same brain functionalities for rule-based temporal processing. The major goal of incorporating rhythmic music instruction into schools is for students to be able to play, construct symbols for, and execute rhythms based on those symbols. Many music programs in schools aim to assist students to become skilled in reading, writing, and playing musical notation.

Throughout the third decade of the twentieth century, the cautious acceptance of rhythmic music as a teaching tool expanded and bloomed. Changing educational philosophies, from 19th-century conservatism to the New School's progressive perspectives, as well as advancements in curricular areas and a rising knowledge of Dalcroze's concept of Eurhythmics - pushed rhythmic instruction to the forefront of today's basic music curriculum. From the late nineteenth century through the third decade of the twentieth century, music was gradually included in early-childhood school curricula. The transition from a conventional curriculum that is based on singing and note-reading to a more progressive perspective of rhythmic movement as an outlet for the holistic development of the child's physical, social, cultural, and expressive requirements was prompted by transforming educational ideas.

The progressive New School (a group of philosophers whose goal was to achieve educational goals through social activism) persuaded music educators to develop practices for a comprehensive elementary curriculum with free and creative rhythmic expressions as a result of growing awareness of Jaques- eurhythmics, Dalcroze's techniques, and acceptance of John Dewey's child-centered curriculum. While conservative thinkers such as T. P. Giddings, Will Earhart, Carl E. Seashore, and Jacob Kwalwasser believed that rhythm was an inherited trait that could not be changed through training, Charles H. Farnsworth, Karl K. Gehrkens, Mabelle Glenn, and James L. Mursell championed the cause and tailored curricular materials to eventually include it as a key component in the elementary school curriculum.

Influence of Music on Self-Confidence and Behaviour

The present flurry of interest in music research is not attributable to a single, or even a collection of new studies on music and behavior. Instead, three things have occurred. First and foremost, the general public is aware that music and behavior research exists and is as serious a topic of study as any other. Second, an increase in the quantity and range of research conducted throughout the world has rekindled interest in previously published but sometimes forgotten discoveries. And third, academics, educators, and policymakers are increasingly willing to consider the significance of music in students" lives in a larger and much broader context.

The spirit of a transdisciplinary approach to music and behavior underpins these three characteristics. This method promotes mutual respect, communication, and collaboration within disciplines such as psychology, biology, medicine, education, computer science, and music therapy, rather than a compartmentalized perspective of academic fields. Neurobiology and cognitive neuroscience, two multidisciplinary sciences, have greatly benefited our understanding of this issue. Experiments such as examining the functional activity of the human brain during music cognition or mental rehearsal of music are gaining traction and will soon be widespread. We could see the beginnings of transdisciplinary science of music if we go far enough into the future (Alam, 2022a).

In the vast field of music study, these are exciting times. But what does this mean for educators in grades K–12? Should we play Mozart in the cafeteria to boost SAT/IIT-JEE results, as suggested by some wag, and accepted seriously by many? Most instructors understand that "learning pills" in the shape of Mozart do not exist. Learning to listen to, understand, enjoy, and perform music has a lot of value, but it is best accomplished with help from experts and with practice. Only a few types of research captured our attention because our need for knowledge has overtaken the field's ability to offer it. More relevant investigations are much needed to produce numerous, convergent conclusions. Fortunately, such research is starting to appear.

The Norwegian Research Council for Science and Humanities discovered a link between musical ability and strong motivation, in kids who were more likely to succeed in school. They concluded that positive self-perception, high cognitive competency scores, self-esteem, and interest and engagement in school music all had a strong link. Whitwell arrived at a similar result, claiming that creative music engagement increases self-image, self-awareness, and favorable views toward oneself. Marshall discovered that participation and accomplishment in school music fosters a good self-image, which in turn motivates urban black middle school children to excel academically. Individuals learn to encourage one another, retain commitment, and connect for collective goals through participation in group music activities at the high school level. Musical experiences instill: 1) a good attitude; 2) a positive self-image; 3) a drive to attain greatness; 4) cooperation; 5) group cohesion; and 6) the capacity to establish objectives (Alam, 2022b).

Roots in Biology

Rhythmic music has always been seen because of culture and social interactions rather than as a biological gift. Scientific evidence, on the other hand, suggests differently. Several shreds of evidence back up the theory that music has deep biological foundations. To begin with, if music had a major biological component, animals would have had the fundamental musical ability, which appears to be the case. Monkeys, for example, can reason in terms of musical abstractions and can calculate the fundamental frequency of a harmonic series. Music is everywhere, even though its applications differ from culture to culture. Lullabies and musical "baby talk" are the most effective ways for parents and caregivers to interact with newborns across cultures. Biological traits are frequently shown early in life, before cultural influences and changes in behavior. The considerable capacity of young infants to comprehend musical components and behave in musical ways has been well demonstrated by past studies. Toddlers, for example, spontaneously display musical behaviors. In addition, new research has indicated that neonates have musical talents. They can, for example, discriminate between two notes; remember the contour or pitches of melodies, mentally segment or "chunk" extended melodies into smaller phrases, and can comprehend rhythm as efficiently as adults do. They can even utilize music to recall events from the previous day.

We may thus anticipate human brains to be structured in such a way that it can process musical activities. Basic musical building blocks exist in the brain, which is specialized to process fundamental musical aspects. Pure tone, pitch, complicated harmonic connections, rhythm, and melodic contour, for example, are all very sensitive to neurons. Melody is processed by the right hemisphere of the brain, while language is processed by the left hemisphere. Music is honored in the same way as language is honored in the brain's functional design. As a result, children's brains are well-equipped to comprehend music, and consequently, they appreciate and participate in music long before they enter kindergarten. Children would develop this natural channel of communication, expression, and cognition if parents and caregivers encouraged children's spontaneous musical activities as much as they encouraged linguistic behaviors. Teachers may feel more at ease when it comes to teaching music. Fortunately, given children's insatiable curiosity and the fact that instructors do not need to be music experts to effectively teach using it, there exist lots of possibilities to include music in classroom teaching-learning.

Music, Brain, and Its Synapses

Schools will continue to embrace a take-it-or-leave-it mindset as long as educators and parents regard music as relatively unimportant. We should not, however, continue to dismiss musical ability as a frill when we know it is part of our biological heritage. Furthermore, when we consider the benefits of music for brain development, it becomes impossible to justify reducing or eliminating music from the curriculum. Perceptual systems include auditory, visual, sensory, tactile, and kinesthetic; cognitive systems include symbolic, linguistic, and reading; planning systems include fine and gross muscle movements action and coordination; feedback systems include evaluation of actions; motivational or hedonic systems include pleasure. A student engages in all these systems when s/he participates in creating or playing music (Alam, 2022c). This is evident from the stages required in reading a music sheet with many complicated symbols, playing the piece, then analyzing the results before adjusting the performance.

There have been a lot of studies that demonstrate a link between music and brain development. Dr. Frank Wilson is a professor of clinical neurology at the University of California, San Francisco School of Medicine. His research claims that practicing with musical instruments improves coordination, focus, and memory, as well as eyesight and hearing. He goes on to say that learning to play an instrument refines the brain's growth as well as the complete neurological system. Dr. Wilson stated in a lecture at the California Music Educators Association State Convention, that he has discovered that people become active participants in their physiological development through music. People may find themselves and a sense of self in community, he claims, through participating in music. His study has revealed that music participation links and develops the brain's motor systems in a manner that no other activity can.

Dr. Wilson backed this up with data from UCLA brain scan tests, which reveal that music involves more brain functions (both left and right hemispheres) than any other activity. These discoveries are so profound that they will lead to a global recognition that music is an absolute requirement not only for the brain but also for an individual's holistic development. A different study found that performing music improves cognitive abilities. Musical engagements improve students" perceptual, creative, and visual talents as well as their aesthetic literacy. Whitwell in 1977 addressed the left-brain/right-brain dichotomy. He argues that musical intelligence is an autonomous, distinct kind of intelligence, a type of intellect through which an individual may directly communicate in his own intrinsic form.

This appears to back Wilson's claim that music has a developmental effect on the brain. Whitwell criticizes the school system for educating just half of a person's brain. He claims that the key to paying attention or daydreaming is to engage the right side of the brain in the learning process. To be a complete human, one must have equal access to both domains of understanding (left and right brain), and this access must include a creative activity such as dynamic engagement with musical performances. In his address "A Neurologist Looks at Musical Behavior" at a conference on the Biology of Music-Making, Tedd Judd concludes that music involvement includes multiple sections of the linked brain. Children who do not have access to engagement with musical programs, according to Dr. Jean Houston of the Foundation for Mind Research, are harming their brains. They are not exposed to non-verbal modalities, which can help them acquire reading, writing, and math much more quickly.

Influence of Music on Mind

Trehub and scores of other scientists have testified on the biological roots of music at the New York Academy of Sciences meeting, interspersing their PET scans and MRIs with snippets of Celine Dion and Stravinsky. These lines of evidence indicate that the human brain is wired for music and that music enhances various kinds of intellect. The fact that humans can recall scores of songs and recognize hundreds more is perhaps the most compelling indication that music has a specific home in the brain's "grey matter". Music has a significant effect on the mind, it not only evokes emotions such as passion, belligerence, tranquility, or dread, but it does so even in those who have no prior knowledge of what a certain crescendo implies, such as when the killer is about to appear on the movie screen. According to

University of Montreal scientist Isabelle Peretz, the brain appears to be adapted for music. The music center is in the temporal lobes of the brain, immediately below the ears. Patients have reported hearing songs so clearly when neurosurgeons poke these areas with a probe that they wonder, "Why is there a phonograph in the operating room?". The temporal lobes are also where epileptic convulsions usually begin, and "the power of music" is no cliche for some epilepsy patients.

The corpus callosum is a large trunk line that connects the left and right hemispheres of the brain. Researchers led by Dr. Gottfried Schlaug of Beth Israel Deaconess Medical Center in Boston compared the corpus callosum in 30 non-musicians to the corpus callosum in 30 professional string and piano players and discovered significant differences. In musicians, the front half of this thick wire of neurons is bigger, especially if they started practicing before the age of seven. The corpus callosum joins the two sides of the prefrontal cortex, which is responsible for planning and foresight. It also links the two sides of the premotor cortex, which maps out actions before they are carried out. According to Schlaug, "these linkages are important for coordinating quick, bi-manual motions" like those performed by a pianist's hands in an allegro movement. Something else might be explained by the cerebral highway that connects the right and left brains. Emotion is related to the right brain, whereas intellect is linked to the left (Alam, 2022d).

Concluding Remarks

Music educators and researchers must take the initiative in disseminating this knowledge to those who can make a difference in the future, such as school boards, educational administrators, parents, and legislators. They must speak up in support of the incorporation of music as a pedagogical tool for effective teaching-learning. Music educators must take an active role in promoting this art. We would see an increase in enrolment in music classes as more people become aware of the studies in this area. Music integration into the curriculum as a tool for enhanced learning is an area that will grow in importance. With such high dropout rates, educators must pool their resources and make use of the instruments at their disposal to provide more effective education.

References

Alam, A. (2022a). Investigating Sustainable Education and Positive Psychology Interventions in Schools Towards Achievement of Sustainable Happiness and Well-being for 21st Century Pedagogy and Curriculum. ECS Transactions, 107(1), 19481.

Alam, A. (2022b). Positive Psychology Goes to School: Conceptualizing Students" Happiness in 21st Century Schools While 'Minding the Mind!'Are We There Yet? Evidence-Backed, School-Based Positive Psychology Interventions. ECS Transactions, 107(1), 11199.

Alam, A. (2021). Possibilities and Apprehensions in the Landscape of Artificial Intelligence in Education. In 2021 International Conference on Computational Intelligence and Computing Applications (ICCICA) (pp. 1-8). IEEE.

Alam, A. (2022c). Mapping a Sustainable Future Through Conceptualization of Transformative Learning Framework, Education for Sustainable Development, Critical Reflection, and Responsible Citizenship: An Exploration of Pedagogies for Twenty-First Century Learning. ECS Transactions, 107(1), 9827.

Alam, A. (2022d). Social Robots in Education for Long-Term Human-Robot Interaction: Socially Supportive Behaviour of Robotic Tutor for Creating Robo-Tangible Learning Environment in a Guided Discovery Learning Interaction. ECS Transactions, 107(1), 12389.

Casualty of Dignity and Other Rights of Children Born Out of Casual Relationship: A Legal Conundrum

*Dr. P. Lakshmi**

Professor and Dean, Unitedworld School of Law, Karnavati University, Gandhinagar, India
E-mail: *deanuwsl@karnavatiuniversity.edu.in

Abstract

No person shall be deprived of his life and personal liberty except under procedure established by law enshrined in Article-21 of the Indian Constitution does not refer to the physical act of breathing, but it has a much wider meaning. It is one right that has been expanded by the Courts to include every meaningful right needed to live as a human. One such right which has been recognized by Courts in recent times is, the right to choose their partner which includes the right to decide about the nature of the relationship with the partner in the way they choose. The object of this paper is to discuss whether this right of liberty given to parties to choose their partner under Article 21 conflicts with the right to live with dignity, right of maintenance, and other related rights to those children born out of this kind of casual relationship.

Keywords: Live-in-relationship, walk-in walk-out relationship, Rights of children, Right to life

Introduction

A drunken man was walking in the road. He was slinging his hands left and right. One of his arms came close to the nose of a person who was walking near him. He instinctively clenched the fist of the drunkard and nailed him down. He said, "Is this not a land of liberty? " while getting up, feeling the spot where he had been struck. Do I not possess the freedom to raise my arms? My arms made it. You obviously have the right to freedom, but you must realize that it ends where my nose starts, the other man added.

This story was shared as a joke by Rev. A. C. Dixon in 1894 at the "Thirteenth International Christian Endeavor Convention."This is not just a story. It has a deeper meaning and is an important principle of the rights of a person. Right and duty are correlated. You have the right to swing your arm, but you must see it does not hurt anyone. It also explains, when there are conflicting rights, where the right of one person ends and where the right of another begins. In the early period when humans were living as nomads, almost like animals, several of the rights which are being recognized now as human rights were not realized by them as their rights. When human civilization advanced and they started settling as a society, they found the need for certain "rights" for themselves. They also found that if they need those rights for themselves, the rights of all members of society should be protected and reasonably restricted. The concept of right and duty started developing.

In the very early period of civilization, the right to have sexual relationships was free and the result was only maternity of the child was known and paternity was not. The nature of the human child is that it must be protected and nurtured for a longer period. When two people were responsible for a bringing child into existence, the duty should also be on both, to protect the child.

Hence knowing the paternity became important. That was possible only if there is an exclusive union of a male and female. This is how in a slow graduation process, the institution of marriage developed. It recognized the right and duties of the parties to the relationship and the offspring of the relationship which was eventually legitimated by law. The object of this paper is to analyze whether the recently recognized or developed institution of "Live-in relationship" in alternative to the institution of marriage maintains the co-relation between the right and duties of all the members of the institution, particularly the children born out of the relationship.

Legal Status of Living Together Relationships in India vis-à-vis Live in Relationship

In India, the relationship between a man and woman becomes legitimate and the offspring born out of the relationship are considered legitimate when there is a valid marriage between the two. How they can get into this legally valid union called marriage, what are their rights and duties after getting into the union, how they can get out of the union, what are their rights and duties after getting out of the union towards each other and children of the union is governed by personal laws. Under these personal laws, getting into a relationship and getting out of the relationship is a right that comes with restrictions and duties. Recently some are of the view that there should not be too many restrictions in choosing their life partner and the nature of the relationship they would have with their partner. Those who subscribed to that view felt they have the liberty to enter the relationship and get out of the relationship as per their wish and started getting into a walk-in walk-out relationship popularly called live-in relationship. It resembles marriage and parties cohabiting, but without any legal obligations. Currently, there are no laws governing this kind of relationship but in some recently decided cases, the Indian Courts have recognized this relationship.

Court's View About the Validity of Live-in Relationship

In a live-in relationship, he or she can any time decide to get into or move out of the relationship. Though morally it is not widely accepted by a society like India, no law expressly validates or prohibits it. In fact, in some of the cases, the Indian Courts have said it may be morally wrong to have such a relationship but legally there is no prohibition. In the case of Payal Sharma v. Nari Niketan, the Allahabad High Court acknowledged this idea. Justice M. Katju and Justice R.B. Misra were on the bench. while deciding this case observed that "In our opinion, a man and a woman, even without getting married, can live together if they wish to. This may be regarded as immoral by society, but it is not illegal." Generally, this was the stand of the Courts when the question of validity of live-in relationships came before them.

Though the Courts repeatedly pointed out that there is no law to approve or to disapprove of Live-in relationships, they always tried to do justice by interpreting the existing laws, whenever the parties to the relationship approached the Court seeking any right or remedy from the Court. Velusamy vs. D. Patchaiammal 2010 (10) SCC 469 is the case in which the Supreme Court interpreted the term "through a relationship in the nature of marriage" Live in relationship under section-2(f) of the domestic violence act 2005 includes extending protection given under that section to a woman of live-in relationship. In this case, the Court specified that weekend relationships or occasional relationships or one-time relationships will not be covered under this term. They must have cohabited voluntarily and must have shown themselves as akin to spouse for a considerable period. Through this case and number of other cases, Courts have made it clear that, parties entering a relationship knowing well the non-committal nature of the relationship, cannot expect any remedy from the Courts as there are no laws to provide remedy.

Recently the increasing number of cases that came before various Courts and the varied decisions of the Courts have highlighted the fact that there is a huge uncertainty. One of the early cases in which the Supreme Court conversed about live-in relationships is S. Khushboo v. Kanniammal and Anr. It was in September 2005, "India Today conducted a study. It was about the sexual habits of people in big cities with an objective to raise awareness related to sexually transmitted diseases. Kushboo being a popular cine actress, her opinion on the subject was asked for and published. While expressing her opinion she mentioned that pre-marital sex and live-in relationships are on the rise. She called for acceptance of such relationships by society with a caution that women of such relationships should protect themselves from unwanted pregnancies and sexually transmitted diseases. This was distorted and published in some local newspapers which resulted in the filing of many criminal cases against actress Kushboo. While deciding this case the Court mentioned that "India is a country where marriage is considered as an important social institute." Casual relationships between man and woman may not be acceptable to everyone in society and may be viewed as an attack on the sanctity of the institution of marriage, but there can be individuals or groups who see nothing wrong in these kinds of relationships. As the notions of social morality are subjective, sex outside marital settings and live-in relationships may be morally or ethically wrong but not legally.

In the case of Alok Kumar v. State, Alok Kumar was married and had a child. When this marriage was subsisting, he started living with another woman who had a child through another marriage. She filed criminal charges against him. She stated that she had been living with him for almost five years because he had promised to wed her after his wife filed for divorce. She accused him of betraying their marriage vows and physically abusing her. The Delhi High Court described the association as having the characteristics of a very casual walk-in, walk-out relationship. It is just like a contract that is to be renewed every day. It can be ended at the will of either party without the consent of the other. There is no legal liability or any entitlement. People who need protection of their rights in a relationship should choose marriage. After choosing live-in relationship they cannot be complaining of infidelity, betrayal, or immorality.

In the case of Chanmuniya v. Chanmuniya Kumar Singh Kushwaha the High Court of Andhra Pradesh and the Supreme Court had different opinion on the rights of the parties of live-in relationship. Chanmuniya claimed that she was married to Ram saran and after his death started living with his brother Virendra Kumar Singh as per their community custom. She alleged that he failed to maintain her and claimed maintenance. The High Court declared that she is not eligible to get maintenance under Section 125 Cr.P.C as she could not prove that there was a marriage between them. The Protection of Women from Domestic Violence Act of 2005 states that women in live-in relationships are entitled to the same protections as wives. This was upheld by the Supreme Court on appeal.

Aysha vs Ozir Hassan, (2007) is a case in which the Madras High Court and Supreme Court of India had very different views on the status of the relationship between the parties, their children, and their rights. Aysha claimed to have married Ozir Hassan and they are blessed with two girl children. She has filed a case for maintenance for herself and her children against Ozir Hassan. Ozir states that there was no marriage. As per Muslim custom, all Muslim marriages are recorded in the Nikah book and there is no such record available. The Madras High Court observed that Aysha and Ozir Hassan has lived together and have begotten two children. She was a spinster, and he was a bachelor before getting into this relationship and living together. If a girl above 18 years has sexual connection with a man above 21 years and during such a relationship, children are born her status will be that of wife and he will be elevated to the status of husband and the children will be legitimate children of the couple. The main legal aspect of marriage is consummation. If consummation has taken place between a spinster and bachelor from that time they will be treated as husband and wife.

The observation of the Madras High Court that "a valid marriage does not necessarily mean that all the customary rights pertaining to the married couple are to be followed and subsequently solemnized" was challenged before the Supreme Court of India by an Advocate named Uday Gupta. He sought for deletion of the Madras High Court's observation on the solemnization of marriage as the remarks of the High Court may demolish the very institution of marriage. The Supreme Court clarified that living together for a long time without any marriage, may give presumption of marriage and they will be conferred with the status of husband and wife. The observation of the Madras High Court has no universal application, and it is restricted only to the facts of the specific case.

Defending women against domestic abuse The first piece of law to recognize a live-in relationship is the Act of 2005. It grants civic rights and protection to females who live with a man but are not wed to him legally. Under the Act, the phrase "live-in relationship" is neither used nor defined. It makes use of the phrase "connection" in the "nature of marriage," which has been interpreted by the courts to cover live-in relationships. In the case of Deepak V Soniya Deepak , Soniya was working in the company of Mr. Deepak. She was appointed as a salesgirl in the company by the wife of Deepak who was the Managing Director of the Company. Deepak and Soniya developed relationship and were living together for nearly 15 years and had two children through the relationship. Soniya requested remedy. under the Protection of Women from Domestic Violence Act, 2005, and Additional Sessions Judge Thane recognized the situation as a live-in relationship and approved Soniya's request. In the Bombay High Court, a revision petition was brought against the Additional Sessions Judge's ruling. She had. been cohabitating with a man who, according to the court, was already married. According to the Protection of Women from Domestic Violence Act, a relationship of this type cannot be considered to have the "nature of marriage." There is therefore no way to help her or her kids.

A writ of Habeaus corpus was filed by father of a 19-year-old girl Rifana Riyad, in High Court of Kerala. He alleged that Rifana is under illegal custody of an 18-year-old boy Hanize, and she should be set at liberty. The girl and the boy appeared before the court and submitted that they are deeply in love with each other from school days and have the right to decide about their life as both are majors. The Court dismissed the petition stating that the girl is a major have every right to live with anyone of her choice, even outside wedlock as live-in relationship has been statutorily recognized by the legislature itself.

In the case of Nandakumar v. State of Kerala also the views of the High Court of Kerala and the Apex Court of India was different on the issue of two consensual adults living together without the consent of their parents. Ms. Thushara got married to Nandakumar in a temple without the consent of her parents. At the time of marriage, she was 19 years old and Nandakumar was less than 21. Her father Binu Kumar filed a Habeas Corpus petition in the Highest Court of Kerala. The Judiciary handed over the custody of Thushara to her parents as it considered the marriage invalid and did not recognize the right of two consenting adults living together in a relationship. The Supreme Court held that the marriage is only voidable and not void. Being adults, they have the right to live together as live-in association is recognized and a major girl (above 18) has the right to choose whom she wants to live with.

In the most recent judgment, the Rajasthan High Court refused to grant police protection to the parties who were living together as they were already married to some other person. It said that a married person living with the spouse of another is an immoral act and giving protection to them will be endorsing an immoral act and refusing to recognize their relationship.

In a very recent case heard through video conference by the High Court of Punjab and Haryana at Chandigarh, a girl who is a major and boy who is a major but not of marriageable age entered into an agreement to live together and started living together in live-in relationship. It was strongly opposed by their parents and the couple sought for protection. The right to choose partner is included in the Article 21 of the constitution of India. The boy and girl who have sought protection under Article 21 of the Constitution are major and have every right to live their lives as they desire. If they have decided to live together in a live-in relationship, there cannot be any justifiable reason for anyone to object to it.

The High Court of Punjab and Haryana at Chandigarh was approached by another couple seeking protection for their life. A Hindu boy and a Muslim girl got married in a shiv Mandir against the wishes of their parents. The Court said the marriage is invalid, but their relationship can be treated as live-in relationship and granted protection to them.

The decision of the Punjab and Haryana High Courts was different in the next two immediate cases decided by it. When Gulza Kumari and Gurwinder Singh who were in a live-in relationship approached the Court seeking protection the Court said that the object of their approaching the Court is to get a seal of approval for their live-in relationship. Live-in relationship is socially and morally not acceptable and protection cannot be granted. On 6th June 2021, the Supreme Court decided this case and said, "It concerns life and liberty:" and granted protection to them.

The Punjab and Haryana High Court turned down a couple's request for protection in the following instance. An 18-year-old girl and a 21-year-old man who claimed to be living together sought protection from the girl's parents. The Court refused to say "if such protection as claimed is granted, the entire social fabric of the society would get disturbed."

Within few days of this decision the same Court rendered another decision in the case of the couple who approached it seeking protection. Ms.Soniya aged about 22 years and Mr.Anil aged about 19 years were in a live-in relationship as they had to wait for Mr.Anil to attain the marriageable age to get married. It was opposed by the girl's parents, and they sought for protection. The Court observed that living together without the sacrosanctity of marriage is not an offense.

Within a week the Punjab and Haryana High Court alone have given several contradicting judgments on the issue of live-in relationships. This highlights the fact of high uncertainty in decision of the Courts on the issues related to live-in relationships. The various issues that arose out of this non-committal relationship not only affected the parties but the innocent children related to the parties of the relationship. They might be children born out of the relationship or children of the parties with some other partner.

Legal Status Children Born Out of Live-in Relationship

Indra Sarma, an unmarried lady, began living with Mr. Sarma, a married man with two children, in the case of Indra Sarma v. V.K.V. Sarma. She quit her career and spent 18 years with her partner in a shared home. She sought remedies from the court under several sections of The Protection of Women from Domestic Violence Act, 2005, including maintenance and other redresses, when the relationship ended. She claimed that as a result of their relationship, she contracted three pregnancies, all of which she aborted. She claimed that Mr. Sarma used to make her use contraceptives to prevent pregnancy.

This case highlights the fact that the rights of the children of the relationship is affected even before their birth. It affects their social status, right to property, right to be loved and protected by parents. From the time of the Privy Council Courts have always tried to protect the interest of children within the existing framework of laws. Under Section 114 of the Indian Evidence Act, the Court may presume presence of certain facts. This provision has been used by Courts to declare the relationship between a man and a woman as a valid marriage if they have been cohabiting for a lengthy period. It makes the children of the relationship legitimate, and their rights are protected. But the issue is, the Courts have said the presumption cannot be applied if the cohabitation period is short. The Supreme Court ruled in the case of Tulsa v. Durghatiya that in order for society to recognize a child born out of such a relationship as a legitimate child, the parents must have lived together and shared a home for a sizable amount of time. It also cannot be a "walk-in and walk-out" relationship.

What length of time of cohabitation is considered lengthy or short is unclear based on court rulings? Once more, this presumption is inapplicable if one of the parties to the connection is a validly subsisting spouse. The relationship's children's rights may not be safeguarded in certain circumstances.

Rights of Parties to Live-in Relationship Vs Rights of Children of Live-in Relationship Under Article 21 of the Constitution

If the parties desire to cohabit in a live-in relationship, it is their right under Article 21 of the Constitution, according to the Supreme Court's interpretation of the S. Khushboo v. Kanniamma case. In related judgments, the courts have upheld the argument that parties have the freedom to select their partner and determine the nature of their relationship under Article 21 of the Indian Constitution, which guarantees the right to life and personal liberty.

But after entering this relationship, they cannot expect to have mutual rights and duties as in the case of a valid marriage. The Delhi High Court in the case of Alok Kumar v. State said, people who decide to have "live-in relationships" cannot later complain of infidelity or immorality. It is a contract of living together "which is renewed everyday by the parties and can be terminated by either of the parties without consent of the other party.

The missed fact here is Article 21 is equally applicable to the child of the relationship. If the casual relationship of his parents is going to brand him with illegitimacy and will not give him rights as enjoyed by a child born in a valid marriage it affects his right to live with dignity guaranteed under Article 21 of the Constitution. The right to swing your arm is only to the extent it does not touch the nose of the other person.

The Supreme Court ruled that children born into a live-in relationship can inherit the parent's property in the case of Bharatha Matha v. R. Vijaya Renganathan. The Special Bench of the Supreme Court of India stated that the nature of the relationship between the parents should not be the determining factor of a child's rights in Revanasiddappa v. Mallikarjun. A child born into a live-in relationship who is innocent is entitled to all the same rights and privileges as children born into legally binding unions.

From these cases, it is obvious that even in the absence of specific legislation the Courts have been giving broader interpretations to the existing law for protecting the rights of innocent children. But in the absence of specific legislation, there is no certainty about the rights of such children.

Conclusion

Without any special or specific law governing live-in relationships, if it is recognized under Article 21 of the Constitution, the rights of the children born out of the relationship should also be recognized under the same Article 21 of the Constitution. To protect the rights of the children of such a relationship the following legislative actions are required.

1. A child born out of a live-in relationship should be treated as a legitimate child born out of a valid marriage.
2. From the moment a child has conceived the parties to the relationship should be considered married under the Special Marriage Act. The moment a child is conceived it is no more a relationship between two people. If the parties to the relationship are in a valid subsisting marriage with another partner, the live-in relationship should be treated as a bigamous marriage and the consequence should follow. To this effect, suitable amendments must be made to the Special Marriage Act.

References

Daily Kos. 2022. Dear right-wing protesters: your "liberty' to. swing your fist ends just where my nose begins. [online] Available at: <https://www.dailykos.com/stories/2020/4/16/1937928/-Dear-right-wing-protesters-your-liberty-to-swing-your-fist-ends-just-where-my-nose-begins> [Accessed 6 October 2022].

(2010) 5 SCC 600

2010 SCC OnLine Del 2645

(2011) 1 SCC 141

Criminal Revision Application No.341 0f 2014

WP(Crl) No.178 of 2018

Vakeela and another v. State of Rajasthan and others, [S.B. Criminal Miscellaneous (Petition) No. 4271/2020]

Mafi and another Vs. State of Haryana and others CRWP No.691 of 2021, Date of Decision: 25.01.2021

Nasima and another vs. State of Haryana and others CRWP-2148-2021, Date of decision: 03.03.2021

Ujjawal And Another Vs. State of Haryana and Others CRWP-4268-2021(O&M) Date of decision: 12.05.2021

CRWP No.4533 of 2021 (O&M), Date of Decision: 18.5.2021

Women in Civil Engineering Profession: Career Profile of Indian Women

Rameshwari Rao[a,] and MohammedShakil Malek[b]*

[a]PhD Research Scholar, Gujarat Technological University, Gandhinagar, India
[b]Principal, F. D. (Mubin) Institute of Engineering and Technology, Gandhinagar, India
E-mail: [*]rameshwarirao37@gmail.com

Abstract

Women in engineering are vital to national growth. It improves women's quality of life. Men and women choose various engineering careers. Daily terminology, in which women are soft and technology is rigorous, matches the conflict between being a woman and an engineer. Men dominate civil engineering, whereas women are underrepresented. Current literature discusses women's problems in this profession. Long and inflexible work hours, limited networking opportunities, and harassment and discrimination are problems. These concerns cause women limited job prospects and stress. These issues frequently lead to low career prospects and high levels of stress for women. Women's responsibilities in engineering, particularly in civil engineering, are presented and discussed in this study with a focus on the obstacles, challenges, and opportunities. This will help women have a better grasp of civil engineering careers and educational options.

Keywords: Gender equality, women working in civil engineering, construction, and education

Introduction

All modern cultures recognize that employment and education are essential to a woman's rights as well as a country's ability to advance economically and socially. For the development of the country, women's involvement in the engineering profession is essential. Additionally, it has a big role in raising women's quality of life. Despite this acknowledgment, women continue to have a very small role in many industries and make very little contribution. Despite what is believed, there are few women enrolled in undergrad programs in several disciplines in India. Although there has been a steady rise in the number of women studying engineering over the years, there are still far fewer of them than there are in the sciences, arts, and medicine at the bachelor's level. In engineering, there were around 2 women for every 100 men in the early 1970s, but by 2000, that ratio had increased to 30 women for every 100 men. Even though the first woman obtained a degree in engineering in 1892, there has traditionally been little female involvement in this crucial sector, and this continues to be a cause for concern around the world.

The number of women enrolled in engineering colleges and universities has increased dramatically in recent years. The annual output of female engineers has multiplied several times when you consider the remarkable growth in the number of engineering institutes. As a consequence, there are now many more female engineers than there were in past years. While it is relatively low in certain programs, the proportion of women is considerable in others. The professional paths taken by men and women in engineering are also different. The cultural divide between being a woman and being an engineer is also reflected in common speech, which portrays women as soft while portraying technology as harsh. Who the woman engineer is and what she will contribute have been hot topics in public conversation as well as feminist critiques of the development of science and technology.

In this article, statistics on female engineers' career characteristics, graduation rates, and enrollment and graduation rates by year are collected and analysed. Most jobs are still sex-segregated, and women are still disproportionately represented in certain fields. In addition to giving women equal access to well-paying, high-status engineering careers, equitable participation would also provide new perspectives to scientific and technical advancement.

Methodology

Detailed literature was studied to perform this research and tried to identify the exact scenario of women's role in civil engineering field.

Engineering Participation of Women in Developing Countries

Data and information about gender are abundant, especially in the area of diversity engineering. Only 6% of qualified engineers and technicians and 9% of the engineering workforce are women. Less than 10% of engineers in Europe are female, with England having the lowest percentage. The highest percentages are in Latvia, Bulgaria, and Cyprus, at over 30%. 15.8% of engineering and technology undergraduates in the UK are female, which is lower than the over 30% female engineering enrollment rate in India. Since 2012, a very steady proportion of young women have chosen to major in engineering and physics. In the UK, there must be a significant rise in the number of engineers. The annual shortage of qualified engineers was estimated to reach 55000 as of 2015, necessitating an increase in the enrollment of engineering students in UK universities. While both male and female engineering and technology students express an equal degree of interest in a career in the field, in 2011 66.2% of male graduates and 47.4% of female graduates found employment in the field. Engineering students are second only to physicians in terms of finding full-time positions and earning decent pay, with 84% of 300 female engineers responding that they were either pleased or very happy with their career choice. By 2025, the GDP might increase by up to $28 trillion if women are allowed to realize their full potential at work.

According to the Bureau of Labor Statistics, women make up around 13% of the engineering workforce and 13% of the civil engineering workforce (BLS). With a 24.3% rise by 2018 and the greatest growth anticipated in the Engineering Services industry, civil engineering is likely to develop faster than almost all other engineering disciplines.

Women got 19.7% of all civil engineering degrees in 2010, compared to 18.2% of all engineering degrees, according to the Engineering Workforce Commission (EWC). According to the EWC, during the last five years, civil engineering has regularly granted 26% of all Master's and Doctoral degrees to women, more than any other engineering area. Master's degree awards in civil engineering are 3% higher than those in all other engineering areas, while doctoral degree awards are 4% higher.

Engineering For Women

The percentage of women seeking B.Tech. degrees at IIT Bombay increased from 1.8% in 1972 to 7.9% in 2005, according to prior studies. This pattern seems to indicate that more women are enrolling in engineering programs, maybe on the assumption that job opportunities would be numerous after they graduate. Women got 23.9% of all engineering degrees in 1998 compared to 1.5% in 1980, demonstrating a steady and significant increase in women's engineering enrolment in India. Despite the significant increase in female enrolment, the reality for female engineers after graduation remains dire, with companies openly or covertly rejecting them. Improvements in employment have not followed increases in enrolment and graduation rates. Women's involvement in the workforce decreased from 34% in 1980 to 32% in 1998. Kerala has a 30% enrolment rate and has the highest female literacy percentage of any state. The unemployment rate for female engineers was significantly higher, at 35.9% in 1990 and 41.5% in 1998. The unemployment rate for women with engineering degrees is five times higher than for men.

Civil Engineer Women

One of the oldest and most varied engineering specialties, civil engineering is largely concerned with the planning, building, and administration of infrastructure in both our built and natural environments. It involves creating an infrastructure for roads, railways, docks, harbors, and airports as well as a sustainable society with access to clean water, power, and waste treatment. The care profession of the built environment, it also includes environmental protection. The topic includes railway lines, roads, tunnels, and bridges, as well as sewage and water systems, coastal development, and geotechnical engineering. Buildings are constructed under the direction of civil engineers. They use computer technology and contemporary materials to construct buildings that meet the requirements of an expanding population while protecting the environment, reducing storm-related risks, and meeting the community's long-term needs. Careers in civil engineering may be diverse and fascinating, including anything from geology to risk management to surveying. In this line of work, women have the potential to positively impact the lives of others. The satisfaction of seeing a project through from conception to design and completion is another benefit. As a civil engineer, your daily duties can involve handling budgets, reviewing project drawings and reports, monitoring orders and equipment deliveries, responding to client requests for revisions, resolving issues, and making sure that work is finished on schedule and under budget.

Construction Sector Difficulties

There is no denying that during the last 30 years, several rising countries have seen a phenomenal increase in construction employment and output. The building represents a significant portion of investment, hence a rise in construction activity is closely related to economic expansion. Numerous studies have shown that when countries create their core infrastructure in the early stages of growth, construction output climbs extremely fast, often exceeding the pace of expansion of the economy as a whole.

Civil Engineering Students' Declining Trend

In various countries, fewer students are choosing to major in civil engineering. Success patterns have changed in the modern world, and many prospective students believe that working in engineering will be more challenging than other professions. This mindset may result from out-of-date study programs, a perception of a high workload, perceived low salaries, a dearth of opportunity for research, or the idea that civil engineers are just technicians who do not get to the top like, for instance, business or management graduates. However, some of this might be attributed to the fact that, recently, the public has not been sufficiently informed about civil engineering in contrast to science or other sectors of engineering and technology. The difficulty of studying civil engineering, which has a higher mathematical component than other study programs like the social sciences, and the low starting salary compared to other professions are other reasons for such perceptions; the new Bachelor degrees in civil engineering may make this problem worse. Firms that specialize in civil engineering and other built-environment fields don't encourage ongoing professional development; instead, they recruit engineers as needed and let them go when their contracts are up. Even though they are the resources that such enterprises depend on, civil engineers are frequently seen as disposable and less important than other professions in the construction industry hierarchy. Critical phases of construction and supervision at construction sites, which are often situated far from corporate headquarters and require additional time for travel or working away from home, are particularly time- and work-constrained.

India With Several Opportunities

The field of civil engineering, which may be challenging to work in, is no longer dominated by men. In actuality, it gives women a number of lucrative possibilities. It is a prevalent misconception that males predominate in the field of civil engineering and that women would find it difficult to work on exterior projects. Civil engineering is a specialized field at this time. In the field of civil engineering, it is now possible to operate remotely or on computers in air-conditioned offices. In many institutions nowadays, the ratio of girls to boys is higher than 20:80. Urbanization is increasing, which has a significant impact on this profession's prospective growth. One is not required to do on-site labor. The development of new technologies has made estimation work easier. All that is required for girls to succeed in the tough field of civil engineering is their commitment to work there. Working in the public sector offers chances and benefits in fields including infrastructure building, maintenance and repair, design and planning, and unconventional technology usage. In the foreseeable future, civil engineers would be in great demand, according to the infrastructure projects of the Indian government. The sector offers engineers the chance to work on large-scale projects while also giving them career security, in addition to good pay. In India, entry-level freshmen may easily earn between $15,000 and $35,000 per month, and after three years of experience, the salary can exceed $50,000.

All of the vocations have a diverse group of people from different fields of employment. You work together on project design with professionals from practically every technological area, as well as with lawyers, real estate agents, locals, planners, architects, contractors, and surveyors, to name a few. It is a really nice piece of work. Seeing something you built get constructed and function properly is exhilarating. It's also gratifying to work through design challenges, and trying to come up with the best design that costs the least amount of money while still looking good is almost like solving a puzzle or playing a game., and women are good at it; they typically plan ahead and have the ability to think outside the box to find original solutions. Additionally, they pay close attention to the details, which is important in this market. There are a lot of women working as civil engineers nowadays.

Why Study Civil Engineering?

To construct enough homes, offices, transportation, clean water supplies, and energy sources for the future as well as to provide sustainable constructed solutions to safeguard the environment, more engineers are required. Engineering jobs pay well, have good long-term prospects, and provide you the freedom to go anywhere, even internationally, if you so choose. Many ladies who exhibit curiosity, originality, or practicality seek degrees in science and mathematics and work in civil engineering, where they enjoy resolving construction-related problems.

Women in the Construction Industry Confront Challenges

Every industry is predominated by men. Women have struggled with several problems relating to income disparity and gender bias for a long time without coming to a resolution. Due to the nature of the labor, civil engineering is mostly a male field. It is crucial to highlight that female laborers are working on the same construction site. In actuality, they are almost as numerous as their male counterparts. This is one of the finest arguments in favor of female civil engineers letting go of their concerns about male dominance in the field. The macho culture of the construction business is evident in the argumentative, combative, and unstable nature of its interpersonal connections. As a consequence, both male and female workers experience a very unpleasant work atmosphere. Women are a significant minority in society, therefore it might be scary for a new female employee to arrive at an all-male workplace for the first time. According to Professionals Australia (2010), there are still problems in the workplace, particularly with relation to wage disparity, workplace culture (including harassment), and views of women's capacity to fill the same jobs as men.

According to Watts, working long hours, which often included working on the weekends, was a major concern for many women. Standard ten-hour days in the construction industry can lead to longer workdays during peak periods, with no advance notice provided to workers regarding overtime. Working a lot of hours demonstrates a devotion to your job, according to industry lore. Engineers often work on the weekends in addition to the weekdays during the busiest times of a building project. For women who are just entering the job, these problems do not provide ideal working circumstances. However, it seems that women are less ready than men to put up with such antiquated, rigid, and ultimately unproductive working circumstances, particularly when they have children and need to make a delayed return to work.

Amaratunga et al. noted that the perception of the industry, career knowledge, culture, and working environment, family obligations, male-dominated training programs, and recruitment practices have been identified as the main barriers to women entering and remaining in the civil construction workforce. According to Dainty and Bagilhole's poll, women have risen at a rate of one hierarchical level less than their male counterparts who are similarly aged and experienced. In most businesses where women are either underrepresented on decision-making committees or are not selected for positions where they have decision-making authority, women find it challenging to advance from junior to middle management. In addition, male-dominated informal networks and cultures can serve as obstacles to the advancement of women. Most crucially, corporate disparities in compensation and opportunity for growth deter women from applying for senior management roles.

Due to the flexibility offered by the employment, women find it simple to work on many technical specialties. Flexible work hours must be better arranged in the construction sector. Flexibility is now a major barrier to women succeeding in the area of civil engineering. Options like flexible work schedules, working from home, nurseries, and daycare facilities on the workplace site or close by are often preferred by women with children. Few organizations are interested in creating infrastructure that will support women's desire to work after having children.

The prevalence of sexual harassment against women is quite high, but it is not exclusive to the construction business. There is a substantial body of research supporting various forms of sexual violence against women in the construction business. This is because males often see female employees as invaders and resort to violence to keep them out of what they perceive to be a male domain. In addition to sexual assaults, the construction business often sees physical and verbal harassment of women. One of the main issues with the construction sector that prevents women from choosing it is workplace harassment.

Significant Result:

After the detailed study, to encourage more women to study engineering and pursue career opportunities the following issues needed to be sorted out:

- Making training settings more gender-sensitive
- Women's appropriate training items
- Using counseling and mentorship to increase chances for women in high-tech fields
- Positive role models are needed.
- Encourage People to Participate
- Determine Indicators and Specific Goals
- Encourage Recognition and Highlight Achievements

Conclusion

In engineering, and notably in civil engineering, women are underrepresented. On the other hand, many women who major in engineering go on to have very lucrative, interesting careers. All disciplines of business provide a variety of choices for women, and civil engineering in particular offers a real chance to change the world and improve the quality of life in regional areas. Contrary to the past, there are now open/opening new disciplines of civil/structural engineering where women's "Patience' and naturally "Creative' attitude may be usefully exploited. Without regard to gender, sincerity, honesty, and attitude help individuals succeed or achieve their goals. Attracting, retaining, and developing talented women to carry out the expanding program of work is part of the solution. To encourage the hiring of women for higher positions in technology, administration, and management, employers, training providers, schools, and the general community all have a part to play in creating a supportive and inspirational environment.

References

Bachwani, D., and Malek, M, S. (2018). Parameters Indicating the significance of investment options in Real estate sector. International Journal of Scientific Research and Review, 7(12):301-311.

Gohil, P., Malek, M., Bachwani, D., Patel, D., Upadhyay, D., and Hathiwala, A. (2022). Application of 5D Building Information Modeling for Construction Management. ECS Transaction, 107(1), 2637–2649.

Malek, M. S., Dhiraj, B., Upadhyay, D., and Patel, D. (2022). A review of precision agriculture methodologies, challenges and applications. Lecture Notes in Electrical Engineering, 875:329-346.

Malek, M. S., Gohil, P., Pandya, S., Shivam, A., and Limbachiya, K. (2022). A novel smart aging approach for monitoring the lifestyle of elderlies and identifying anomalies. Lecture Notes in Electrical Engineering, 875:165-182.

Malek, MohammedShakil S., and Bhatt, Viral (2022). Examine the comparison of CSFs for public and private sector's stakeholders: a SEM approach towards PPP in Indian road sector, International Journal of Construction Management, DOI: 10.1080/15623599.2022.2049490

Malek, MohammedShakil S., and Gundaliya, P. J. (2020). Negative factors in implementing public–private partnership in Indian road projects, International Journal of Construction Management, DOI: 10.1080/15623599.2020.1857672

Malek, MohammedShakil S., and Gundaliya, Pradip (2021). Value for money factors in Indian public-private partnership road projects: An exploratory approach, Journal of Project Management, 6(1), 23–32. DOI: 10.5267/j.jpm.2020.10.002

Malek, MohammedShakil S., and Zala, Laxmansinh (2021). The attractiveness of public-private partnership for road projects in India, Journal of Project Management, 7(2), 23–32. DOI: 10.5267/j.jpm.2021.10.001

Malek, MohammedShakil S., Saiyed, Farhana M., and Bachwani, Dhiraj (2021). Identification, evaluation and allotment of critical risk factors (CRFs) in real estate projects: India as a case study, Journal of Project Management, 6(2), 83–92. DOI: 10.5267/j.jpm.2021.1.002

Parikh, R., Bachwani, D., and Malek, M. (2020). An analysis of earn value management and sustainability in project management in construction industry. Studies in Indian Place Names, 40(9), 195–200.

Factors That Make Public-Private Partnerships Appealing for Highway Projects in Gujarat

Bhatt Sarthak[*,a] *and MohammedShakil Malek*[b]

[a]PhD Scholar, Parul University, Vadodara, Gujarat, India
[b]Principal, F. D. (Mubin) Institute of Engineering and Technology, Gandhinagar, Gujarat, India
E-mail: [*]sarthakbhatt7@gmail.com

Abstract

The limited availability of public money to finance the necessary investment, as well as the necessity to use private sector expertise to enhance the quality and efficiency of public services, have prompted such a PPP agreement. PPP provides a framework for the public and private sectors to collaborate in order to enhance public service delivery via the provision of infrastructure and associated ancillary services for infrastructure project development. PPP uses open, competitive methods to generate ideas, skills, and talents from both sectors in order to satisfy the community's needs, expectations, and development objectives. Gujarat carried out an empirical questionnaire study (India). Respondents to the survey evaluated the fifteen elements that impact PPP efforts.

Keywords: Attractive factors, PPP, Highway projects, India

Introduction

Private-public infrastructure collaboration has increased. Due to limited public funds required to finance vital improvements and the need to depend on private sector expertise to improve public service quality and efficiency, PPP agreements have been pushed. PPP investigates alternate buying methods. PPP stands for private-sector delivery of public services. Public-private partnerships include the private sector in nation-building and accelerate the delivery of high-quality public goods and services via joint ventures without squandering scarce resources. Eleventh Five Year Plan forecasts considerable infrastructure spending to enable strong annual growth. The public sector can't fulfill big promises with minimal funding. At least 30% of the entire plan's infrastructure demands must be met by the private sector. PPP may be a solution.

Infrastructure requires PPPs. Deputy Planning Commission Chairman Montek Singh Ahluwalia said private investors manage risks better. Governments may generate cash in many ways, but they tend to underprice risks and collect low, inadequate, and unsustainable user fees for public services. Private investors are better positioned to analyse market risks, foresee changing demands, and react with suitable solutions. Better technology and private capital boost resource efficiency, freeing up public revenues. Head of the European Commission Delegation to India, Bhutan, and Nepal Francisco da Camara Gomes noted the need to learn from international experience for policy-making and highlighted the EU India Economic Cross-Cultural Program, which has allowed European and Indian civil society organizations to interact.

Due to budget shortfalls, governments worldwide seek public-private partnerships to develop roads. PPPs maximize public and private capacities and operations while sharing risks and benefits. The public sector owns the structure but gives the private sector development and project management rights for a specified time. Public sector social responsibility, environmental awareness, and local experience could benefit from private sector entrepreneurship, management abilities, and financial astuteness.

PPP projects are long-term concession contracts with specific quality of service and measurable performance objectives, implying direct user participation. These agreements enable infrastructure development on public land. Dealers may charge for public goods. User services are government-run. These projects require government oversight to enforce concession agreements and laws. The government should manage a PPP project to fulfill users' time, cost, quality, and quantity demands. A well-defined institutional structure is needed for contract supervision. Monitor and enforce each concession contract's terms.

Literature Review

Literature gives PPP suggestions. This section covers PPP successes. In PPP contracts, the government should guarantee assets are acquired on time and that services meet pre-agreed requirements. The government shouldn't care "how" goals are accomplished and shouldn't prohibit private sector involvement. The government should regulate businesses and services, embrace innovations and new technology, and help and reward the private sector. In case of default, the government should resume service.

Effective procurement decreases transaction costs, negotiation, and deal-closing time. Project briefs and client demands should be included in bids. Price-only bidding may not assist form a strong private consortium or save the state money. The government needs a partner.

Successful PPP implementation requires a stable and competent host government. Corporate interests aren't political or social. Private-sector victims should be reimbursed.

Private sector engagement in public infrastructure requires project money, experts say. A robust financial market with low borrowing costs and a range of financial instruments would stimulate private-sector PPP ventures.

Methodology

Design Questionnaire

This section discusses whether Gujarat should use PPP for public works projects. A Gujarati questionnaire compared PPP to Hong Kong (H.K.) and Australia. Three jurisdictions reached different results. Each jurisdiction has different practices, cultures, geographies, traditions, etc. Contrasts are fascinating.

Gujarati public, private, and researcher respondents took the survey. Private-sector respondents made up 48%. 27% public sector, 25% researchers. In Gujarat, public sector departments are common, but private companies handle road PPP.

Data Collection

Gujarat's empirical questionnaire research on PPP project success was compared to H.K., and Australia. The questionnaire's intended respondents were public and private industry practitioners and researchers.

60 Gujarati, 20 H.K., and 20 Australian respondents received surveys. Some respondents were handed blank survey forms in case they had PPP-savvy colleagues or friends who wanted to participate. Gujarat, H.K., and Australia each returned 42 completed questionnaires, with 70%, 40%, and 30% response rates. Since the questionnaire was sent from Gujarat, perceived geographical limits led to a reduced response rate in H.K., and Australia. Since this essay is largely about Gujarat, H.K., and Australia were used as references.

For a better result, we focused on respondents with direct or indirect PPP road project expertise. 42% of respondents had 20+ years' experience. 41% have 11–20 years of experience. Only 17% have less than 10 years experience. Half of the respondents in H.K., and Australia have 21+ years of industry experience. All three sites had experienced responders, confirming the findings' authenticity and reliability.

Data Analysis

This study's quantitative analysis includes mean score ranking, Chi-square, and ANOVA. The five-point Likert scale was adopted to rate each element's importance. Esther Cheung's survey respondents from India/Gujarat, H.K., and Australia triangulated the criteria's importance (2009).

Economists, biologists, managers, and other academics use ANOVA. This method needs multiple examples. Gujarat, H.K., and Australia were covered.

Tables 1 and 2 exhibit the validation output before one-tail ANOVA analysis.

Table 1. PPP Attractive Factors Anova Output-1

Factors	Avg	Variance
Resolve the issue of public spending constraint	4.1875	1.0119
Offer a unified service platform for municipal infrastructure needs.	3.3281	1.5573
Lessen the percentage of public funds dedicated to capital expenditures	3.7656	1.3251
Costs for the completed service will be capped at this point.	2.8750	1.0952
Enable fresh ideas and methods to flourish	3.6250	1.7302
Lessen the overall price tag of the job.	2.6875	2.0913
Get the job done faster and with fewer hassles.	3.8438	1.3085
Devise a plan to shift the burden of risk onto the private investor	3.3750	1.3174
Spend less on bureaucracy in the public sector.	3.1875	1.2024
Help for the local economy	3.5938	1.6101
Up the ante on construction and upkeep	3.6406	1.4084
The Transfer of Technology to Domestic Business	3.7500	1.2698
Reliance on public funds is prohibited or strictly regulated.	2.8437	1.4037
Speed up the creation of your projects.	2.7500	1.4920
Resolve the issue of public spending constraint	3.6093	1.4164

Table 2. PPP Attractive Factors ANOVA OUTPUT-2

SoV	SS	df	MS	F	P	F
Between Groups	179.06	14	12.93	8.18	0.00	1.9
Within Groups	1297.50	934	1.4			
Total	1519.1830	959				

Table 3. Chi-square analysis of PPP success factors

CHI – SQ.	GUJ. H.K.	GUJ. AUS.
No. of survey respondents	42 & 8	42 & 7
Chi-square value	119.1	134.13
Critical Chi-square value (From Table)	22.741	22.741
Degree of freedom (dF)	14	14
Level of Sig.	0.05	0.05

Table 2 shows that for 14 degrees of freedom (df), F value is greater than F critical value, showing that respondents rated PPP's enticing attributes consistently. This allows survey analysis. 42 Gujarat, 8 H.K., and 7 Australia respondents are shown in Table 3. Chi-square critical value for both groups was 22.741 at 14 degrees of freedom and 0.05 significance (119.1 and 134.13 respectively). The data is valid since respondents rated PPP's beauty consistently.

Discussions

Gujarat, H.K. and Australia scored better in several categories. Respondents didn't rank financial motives as the primary reason for PPP.

H.K's financial reserves and budget surplus have funded public works projects. Government officials seldom borrowed when they could grant inexpensive cash. H.K. adopted PPP for efficiency. The Easther Chaung research was five years before this one, so PPP uptake and practitioner views may have changed. Easther Chaung's research is a reference.

"Reduce public sector spending" rated first in Gujarat and third in H.K. Both administrative systems regarded this as a PPP draw. Financing public sector projects drive worldwide PPP implementation. Many experienced PPP practitioners believe financing isn't the primary motivation to choose PPP. Budget-conscious administrative systems commonly use PPP. This financial attraction attracts governments worldwide, especially when spending public money on competing goals. Both sets of responders rated this highly, however with different emphases. This factor ranked 12th among Australians. This may not restrict Australia's public sector budget. Before verifying this conclusion, more Australians are required since just 11 participated. Gujarat's second-most-attractive rationale was "save time and deliver the project." Australia placed 3rd, H.K. 12th. Geographically, this aspect was equally important to Gujarat and Australia. Reducing public money for capital investment was Gujarat's third positive, ranked sixth by H.K., and last by Australia. Gujaratians rated attractive aspects 2.69 to 4.20. 1.51 is Gujarat's variance. H.K. averaged 2.94-3.79 and Australia 2.36-4.45. 0.85 and 2.09 Australia has bigger mean differences than Gujarat and H.K. H.K. respondents consistently rated the 15 attractive traits higher than Gujarat and Australia. Over "3" on a Likert scale from 1 to 5 (1 = least important, 5 = most important) denotes importance. Gujarat received four below-par grades. Cap final service cost, Transfer technology to local businesses, Non-recourse or restricted public financing, and Reduce total project cost scored less than 3 but more than 2.5. This is because the immediate advantages of this attractive aspect were not obvious, therefore the other eleven were better. In H.K. and Australia, these enticing traits scored low, showing that the first group disagreed but the second agreed with Gujarat. 11 Gujarat components scored between 3 and 4. Respondents may have contributed more but didn't.

Conclusion

This article analyses the perspectives of respondents from Gujarat (India), H.K., and Australia on fifteen PPP project success determinants. Gujarat's top five attractions were:

1. Restrict public spending
2. Reduce public money related to capital investments
3. Maintainability
4. Enhance construction

Australia's poll had substantially larger mean disparities than Gujarat and H.K's. H.K. respondents scored the 15 appealing features significantly more consistently than Gujarat and Australia.

References

Malek, MohammedShakil S., and Bhatt, Viral (2022). Examine the comparison of CSFs for public and private sector's stakeholders: a SEM approach towards PPP in Indian road sector, International Journal of Construction Management, DOI: 10.1080/15623599.2022.2049490

Malek, MohammedShakil S., and Gundaliya, P. J. (2020). Negative factors in implementing public–private partnership in Indian road projects, International Journal of Construction Management, DOI: 10.1080/15623599.2020.1857672

Malek, MohammedShakil S., and Gundaliya, Pradip (2021). Value for money factors in Indian public-private partnership road projects: An exploratory approach, Journal of Project Management, 6(1), 23–32. DOI: 10.5267/j.jpm.2020.10.002

Malek, MohammedShakil S., Saiyed, Farhana M., and Bachwani, Dhiraj (2021). Identification, evaluation and allotment of critical risk factors (CRFs) in real estate projects: India as a case study, Journal of Project Management, 6(2), 83–92. DOI: 10.5267/j.jpm.2021.1.002

Malek, MohammedShakil S., and Zala, Laxmansinh (2021). The attractiveness of public-private partnership for road projects in India, Journal of Project Management, 7(2), 23–32. DOI: 10.5267/j.jpm.2021.10.001

Gohil, P., Malek, M., Bachwani, D., Patel, D., Upadhyay, D., and Hathiwala, A. (2022). Application of 5D Building Information Modeling for Construction Management. ECS Transaction, 107(1), 2637–2649.

Malek, M. S., Gohil, P., Pandya, S., Shivam, A., and Limbachiya, K. (2022). A novel smart aging approach for monitoring the lifestyle of elderlies and identifying anomalies. Lecture Notes in Electrical Engineering, 875, 165–182.

Malek, M. S., Dhiraj, B., Upadhyay, D., and Patel, D. (2022). A review of precision agriculture methodologies, challenges and applications. Lecture Notes in Electrical Engineering, 875, 329–346.

Bachwani, D., and Malek, M, S. (2018). Parameters Indicating the significance of investment options in Real estate sector. International Journal of Scientific Research and Review, 7(12), 301–311.

Parikh, R., Bachwani, D., and Malek, M. (2020). An analysis of earn value management and sustainability in project management in construction industry. Studies in Indian Place Names, 40(9), 195–200.

Risk Management in Public–Private Partnership-Based Infrastructure Projects: A Critical Analysis

Dixit N Patel[*,a] *and MohammedShakil S. Malek*[b]

[a]Ph.D. Scholar, Gujarat Technological University, Gujarat, India
[b]Principal, F. D. (Mubin) Institute of Engineering & Technology. Gandhinagar Gujarat, India
E-mail: *199999912517@gtu.edu.in

Abstract

Infrastructure has boosted India's economy. However, India's infrastructure is still developing. Indian firms lack risk management technology. Therefore, PPP infrastructure projects fail. The report seeks to identify India's biggest risk management concerns and propose remedies. This article gives a high-level overview of the framework for risk management in the PPP-based infrastructure sector, defining the major risks associated with infrastructure projects, which decision-makers must do when selecting how to manage infrastructure project risks. India's infrastructure drives economic growth. However, India's infrastructure is developing. Indian companies lack risk management technologies. Hence PPP infrastructure projects fail. A comprehensive literature review of the risk management framework in the PPP-based infrastructure sector identifies the main infrastructure project hazards in this article.

Keywords: Risk Management, PPP, Infrastructure Projects.

Introduction

There are always potential dangers with any undertaking. However, the sophistication of the risk is a direct consequence of the magnitude of the effort and resources invested in the infrastructure improvements. Furthermore, risk management must be meticulously executed throughout the whole of public-private partnership infrastructure projects, beginning with the bid evaluation and ending with the contract signing., due to the intrinsic uncertainties and volatility of each step. Risk is inevitable, yet it can be mitigated, distributed, transferred, or accepted. Each project manager should make it a priority to mitigate the exposure implicit in their organization's construction projects, which are highly complicated and frequently require considerable financial commitments. The proposed effort incorporates risk management all across the entirety of infrastructure project life cycles.

There is a lot of unpredictability in the construction industry due to environmental factors; circumstances can be very rigid due to the complexity of each project, delays caused by a lack of resources, supervision, and poor planning, technical issues, and the difficulty of dealing with the personalities of stakeholders in PPP mode projects. PPP projects depend on government policies and public interest. Without that project, success requires risk management. Thus, risk management is proactive. Infrastructure developers and managers will profit from identified hazards. These infrastructure projects will be better planned, scheduled, organized, and controlled with known hazards. Infrastructure projects address these risks.

Need and Objective of Research:

As a result of the diversity of its constituents—from homeowners to builders to design professionals to government regulators—the construction sector presents unique challenges. Many elements and causes contribute to the variety of issues that affect PPP projects. The ability to recognize risk variables and assess their impact on Indian infrastructure projects is, thus, crucial. To identify the most significant technical challenges with project performance and to articulate solutions, we will assess current practices in relation to time, cost, and owner satisfaction.

Many studies are conducted on infrastructure projects that use the public-private partnership model. The primary goal of this assessment is to determine which risk factors are most significant in Indian infrastructure projects and assess their relative importance for future research.

Methodology

Primary and secondary literature reviews can be split. Journals, articles, theses, periodicals, and research papers make up the primary literature evaluation on Indian infrastructure initiatives. Factors of danger in Indian infrastructure projects were compared in a study of secondary sources. As indicated below, we conducted a primary literature assessment of articles discussing Indian infrastructure initiatives.

Recent economic disasters, as surveyed by Craciun (2011), have prompted fresh thought about the threats to unsecured infrastructure investments. Through an examination of the current financial chaos on global capital markets (2008–2010) from the vantage points of project finance funding methods and more traditional direct investment framework, a new category of risk associated with financing has been identified by drawing on a variety of typologies established over the past thirty years by a wide range of researchers and writers.

Governments at the federal, state, and municipal levels in India have reportedly launched massive infrastructure-building programs in response to the expanding economy, as reported by Visvanathan et al. (2012). Recent transportation infrastructure projects in India went significantly over budget and behind schedule because of the extensive list of statutory and non-statutory permits and permissions that each project needed before construction could begin.

Public-private partnerships (PPPs) have been employed by Zambia as a method of delivering projects, as stated by Ngoma et al. (2014). Conservation and new infrastructure are both challenging due to limited resources.

According to Jayasudha et al. (2014), time and budget restrictions exacerbated initial bridge project hesitance and risk due to the complexity and fluidity of the problems involved.

Large multidimensional projects, as discovered by Chris Harty et al. (2014), lessen anxiety and boost confidence. By inquiring into the role that knowledge plays in contemporary risk assessment, they cast doubt on both presumptions. Risk management in big infrastructure and construction projects has been studied extensively, revealing the knowledge needs and repercussions for project and construction management.

According to research by M. S. Khan et al. (2015), PPP helps governments at both the federal and state levels get their initiatives off the ground. Although PPP has great promise for India's economic future, it confronts obstacles such as a lack of contemporary technology and equipment, project cost overruns, and wasteful material use. Which may make it tough to gauge PPP quality.

Non-standard financing in the nation is cited as a major cause of the technical difficulties seen in infrastructure projects by Yadav et al. (2015). The project's proposed plan for managing potential threats is straightforward in comparison to similar plans. It synthesizes the results of previous studies and lays out detailed instructions for international businesses interested in funding India's infrastructure development. Government management and communication should be strengthened.

Infrastructure inadequacy may be managed with the use of PPPs, as stated by Ruchi Sharma (2015). The finance minister and the Deputy Chairperson of India's Planning Commission have both stressed the importance of investing in infrastructure to develop road networks into a world-class service for the country's rapidly expanding economy and business.

The public-private partnership model was singled out by Jagdale (2016) as a cutting-edge, effective method of constructing and renovating infrastructure in both developed and developing nations. Inherently flawed, PPPs can never succeed. They thus developed a high-level plan for handling risks, detailing potential threats and how to counteract them.

To develop a list of significant risks in PPP highway projects and to construct a risk allocation framework to determine the optimal mitigation techniques, Carbonara et al. performed a Delphi poll among experts.

Private sector participation and investment are central tenets of India's infrastructure policy, as stated by Gupta et al. (2013). Infrastructure PPPs are crucial to the success of the 12th five-year plan. Main threats to current PPP highway projects were identified using questionnaires (under NHAI and MPRDC). Based on the results, it is recommended that a Regulator for PPP Road projects be set up to keep an eye on the dynamic socioeconomic landscape, propose measures to reduce risk, and create a win-win scenario for everyone involved.

According to Ezeldin et al. (2013), public-private partnerships (PPP) have been adopted in Egypt, and several projects are currently being evaluated for potential implementation. When risks are properly allocated and a suitable contract is in place, the best risk manager can be responsible for mitigating those risks. The survey revealed the top 26 threats. Dangers are ranked and compared to those in China, India, and Singapore.

In order to meet project goals for time, money, quality, safety, and environmental sustainability, managing risks is a crucial part of the management process. The feasibility study is second in terms of riskiness, behind construction. Contractors and subcontractors with strong construction and management skills should be recruited early to guarantee safe, efficient, and high-quality construction, and clients, designers, and government authorities should work collaboratively from the feasibility phase to address any hazards.

Rashed argued that infrastructure is essential for long-term economic expansion, as inadequate facilities might stifle development. Private infrastructure projects (PIP) were encouraged by governments across the world as a means of improving productivity and, in some cases, cutting costs. It was important for developing nations to ensure an equitable distribution of risks associated with infrastructure projects while keeping private agent losses to a minimum.

Major Findings

As it stands, there is no tried-and-true method for managing the various threats that face entrepreneurs and investors; without a clear answer to this problem, infrastructure funding will dwindle. The best timing for making investments will be thrown off by the chaos. The infrastructure sector in India is undergoing significant transformation, which introduces numerous new business risks. For this reason, Indian infrastructure projects need to adopt a more systematic approach to risk management. To maintain a competitive edge and grow during this moment of transition, the Indian infrastructure sector must learn to effectively incorporate risk management into its operations. Overruns in land acquisition and operational costs are also common, as are changes in legislation and delays in project approval and permits. A radical rethinking of PPP is required if it is to outperform the more conventional methods in terms of cost-effectiveness. The capacity of governments to manage and control risk will be considerably enhanced by the adoption of integrated risk management systems.

Conclusions

The infrastructure industry in India is crucial. The infrastructure industry in India is a generous contributor to both the public and private sectors, whether measured by the number of jobs created, the scope of activities conducted, or the impact on national development. More study is needed to find ways to successfully control the detected episodes. In order for the Indian infrastructure sector to maintain its competitive edge and persevere during this moment of profound change, it is crucial to determine how to effectively incorporate risk management into Indian infrastructure projects.

References

Gupta, Anil Kumar, Trivedi, Dr. M.K., and Kansal, Dr. R. (2013). Risk Variation Assessment of Indian Road PPP Projects. International Journal of Science, Environment, And Technology. 2(5), 1017–1026.

Khan, Dr. M. S. and Ojha, V. V. Neeraj Kumar (2015). A Study of Public Private Partnership (PPP) with Special Reference to Infrastructures. Indian journal of applied research. 5(1), 241–243.

S. B. Jagdale, (2016). Risk: Awareness, Identification and Mitigation in PPP Projects. International Journal of Engineering Research. 5(1), 85–89.

Malek, MohammedShakil S., and Bhatt, Viral (2022). Examine the comparison of CSFs for public and private sector's stakeholders: a SEM approach towards PPP in Indian road sector, International Journal of Construction Management, DOI: 10.1080/15623599.2022.2049490

Malek, MohammedShakil S., and Gundaliya, P. J. (2020). Negative factors in implementing public–private partnership in Indian road projects, International Journal of Construction Management, DOI: 10.1080/15623599.2020.1857672

Malek, MohammedShakil S., and Gundaliya, Pradip (2021). Value for money factors in Indian public-private partnership road projects: An exploratory approach, Journal of Project Management, 6(1), 23–32. DOI: 10.5267/j.jpm.2020.10.002

Malek, MohammedShakil S., Saiyed, Farhana M., and Bachwani, Dhiraj (2021). Identification, evaluation and allotment of critical risk factors (CRFs) in real estate projects: India as a case study, Journal of Project Management, 6(2), 83–92. DOI: 10.5267/j.jpm.2021.1.002

Malek, MohammedShakil S., and Zala, Laxmansinh (2021). The attractiveness of public-private partnership for road projects in India, Journal of Project Management, 7(2):23-32. DOI: 10.5267/j.jpm.2021.10.001

Gohil, P., Malek, M., Bachwani, D., Patel, D., Upadhyay, D., and Hathiwala, A. (2022). Application of 5D Building Information Modeling for Construction Management. ECS Transaction, 107(1), 2637–2649.

Malek, M. S., Dhiraj, B., Upadhyay, D., and Patel, D. (2022). A review of precision agriculture methodologies, challenges and applications. Lecture Notes in Electrical Engineering, 875, 329–346.

Construction Safety Practices: An Analysis

Kardam Tank,[a,] Jaydeep Pipaliya,[b] and MohammedShakil Malek[c]*

[a]PG student, Parul Institute of Engineering & Technology, Civil Engineering, Parul University, Vadodara, Gujarat

[b]Assistant Professor, Parul Institute of Engineering & Technology, Civil Engineering, Parul University, Vadodara, Gujarat

[c]Principal, F. D. (Mubin) Institute of Engineering & Technology, Gandhinagar, India

E-mail: [*]kardambtank@gmail.com

Abstract

All workers, whether full-time or part-time, are entitled to a safe and healthy workplace in accordance with the company's policy. A large number of people lose their lives in the building industry, countless disasters and permanent human harm occur, and the construction process is often interrupted. Injuries in the construction industry will continue to have far-reaching consequences, from the suffering of the wounded employees themselves to construction delays and lost productivity for the construction supplier to increased workers' compensation premiums. These deaths cannot be tolerated, and they rob the building industry of its credibility in modern society when so much progress has been made in this area. The industry as a whole suffers when workers are injured or killed. This literature study details the steps necessary to ensure workplace safety and shows how five main challenges or causes contribute to the high accident rate.

Keywords: Safety Management System (SMS), Construction Sector, Safety Outcomes, Accident, Health & Safety, Employer, Workplace, Contractor.

Introduction

The necessity to improve safety execution and well–being is a common concern among specialists and researchers because the construction sector is one of the riskiest activities among different other industries. However, this level of control is conceivable in other endeavors where gradually steady circumstances are substantially established, unlike the conditions that are normally considered from the perspective of safety and well-being in the building business. When compared to other industries, the construction sector is often enormous. As result, it is vulnerable to many health risks (lordsonmillar and Subramani, 2014). In the past few decades, the construction and development industries have advanced significantly, which has favourable impact on people lives. In any case, a similar development came after an increased risk to human prosperity.

Accidents at work are a serious problem for the industry as whole, not just the employers. Not only are the workers lives and physically work at risk, but there are also indirect costs associated with construction accidents, such as lost labor productivity, supervision cost, worker confidence, and equipment damage (Roth and smith, 1991). Construction workers' injury payments are roughly twice as much as those received by workers in other companies (Georgine ET, 1997). Although construction has a high rate of fatalities & injuries that to other industries (Leigh and Robin, 2004). Yet estimates for the number of fatalities and injuries are rarely available (Waehrer et al., 2000).

Literature Review

A hypothetical foundation for developing health and safety rules is provided by the written surveys. However, the Occupational Safety and Health Act (OSHA) of 1970 has resulted in major improvements, there are still more workplace injuries and deaths than in other sectors (Bureau of Labour Statistics, 2015). According to one estimate, as many as 60,000 people die each year while working in the construction industry throughout the world. "any unplanned incident that results in damage and suffering and misfortune," as defined by the UKHSE (United Kingdom's Health & Safety Executive HSE, 1999). "an unplanned, unforeseen, uncontrolled incident that results in damage to commodities and things and causes bodily harm, loss of life, or the prospect of such events" (Heinrich and Grannies, 1959) is the definition of an accident. Accidents occurring on a construction site might be very unforeseen and convoluted. The security of an economic sector may be gauged not simply by the absence of a single incident, but by the frequency with which such incidents occur over a specific period (Sanchez et al., 2017). Safety and health standards, such as those issued by OSHA, have a major impact on the construction business. The mental, emotional, and financial toll that workplace injuries, illnesses, and disabilities have on employees and their families should be enough to spur the executive's framework for safety into action. Avoiding the direct and indirect costs of employee injuries, illnesses, and property damage is another important driver for security management. Improvements in people's habits may have a major impact on the number of preventable deaths and injuries. The high death toll is inexcusable in the highly developed culture of Morden, where the development industry has seen exponential expansion at the expense of lost productivity. The results of the study show that the high number of accidents in the construction industry may be attributed to a small number of high-risk practices. To reduce the number of construction-related casualties, experts have developed a safety management system.

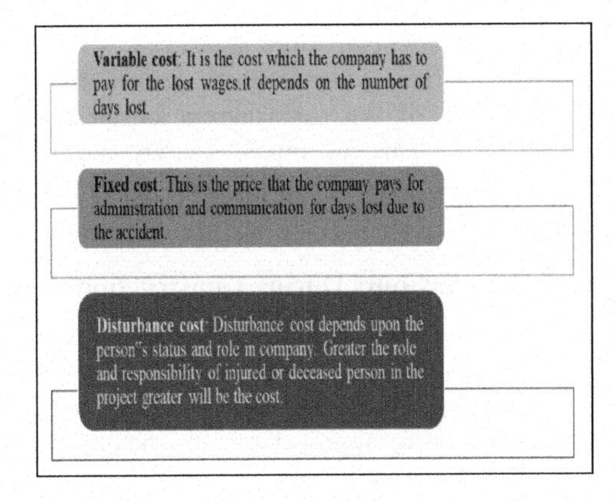

Fig. 1. Rikhardsson and impgaard (2004) say categorized accidents cost in three types

Safety Management System

SMIC defines a safety management system as "a progression of established, association-wide procedures that accommodates effective hazard-based basic leadership" Some models increase building safety and prevent accidents. Models describe a variety of accidents. Researchers have studied what causes accidents and how those causes are connected to environmental variables that lead to fatalities and injuries. Multiple factors boost the location's security. Human aspects matter. Environment, management, technology, and infrastructure are all external elements. Construction accident psychology has been extensively studied because of the human factor. Human mistake is a primary cause of deadly construction accidents. This suggests that personnel are either careless about their safety or find defensive riggings a nuisance. According to the assessment, site officials didn't emphasize worker safety. Choudhry and Fang (2008) researched hazardous behavior, which led to conferences in Hong Kong's development sector. Nobody wanted to study safety measures, they found.

Unsecure procedures and practices. People's inability to recognize risks, lack of competence or preparation, exposure to hazardous working conditions, and usage of protective gear were blamed on not following safety protocols (PPE). Employee behavior typically contributes to workplace accidents, but it may also be the main cause. Procrastination, weariness, a busy schedule, or poor judgment can induce risky behavior. Size of the construction business is another factor. This encompasses safety culture, policy, project coordination, and economic pressure.

Hazards Behaviours During Construction

1. No one is protecting their ears with earplugs (when noisy equipment and machinery is used i.e. Hilti, grinder, cutter).
2. Not dressing appropriately for the job and the local climate.
3. Guardrails and safety harness not being used on high platforms.
4. Not organizing various tools or other small equipment properly.
5. Major openings unattended or unprotected
6. Avoiding the use of air purifying respirators (in dusty environments).
7. There are no safety helmets at the scene.
8. Lack of hand protection when working with hot or sharp materials on the job site.
9. Absence of eye protection when utilizing cutting and welding equipment, such as goggles.
10. Not wearing safety traction boots at work
11. Nails were left in the wood, scattered around.

Numerous books, periodicals, and articles have been written on building safety, highlighting its flaws and suggesting that nothing has been done to reduce human suffering. But that's not all. Improved safety management. Europe and the U.S. turn to Britain's record for guidance. 2001 (Holt/Lampl) (Holt & Lampl). Construction workers' safety attitudes are evolving. "Zero injuries" is the result. This idea was created to explain the popular perception that construction mishaps are random and inevitable. ISSA promotes worldwide health and safety. Vision Zero seeks to eradicate avoidable deaths and injuries globally (Wustemann, 2017). Accidents were more common in the construction business before "Safety First" became the slogan. Prior cost forecasts included fatalities. One life for every 1,000 square feet of floor area, mile of tunnel excavated, or two levels added. Many believed construction accidents were inevitable and that nothing could be done to lessen them. Many contractors felt powerless at the time. Some contractors believed it was their ethical obligation to keep their employees safe and productive (Samelson and Levitt, 1993).

Construction Safety Measures Procedure

A. Effective Energy Strategy

A structured plan to ensure employee health. A project's safety strategy should be simple to modify. Includes a program for preventing injuries, the company's safety policy, PPE guidelines, a manual for using power tools and other tools, emergency procedures, safety responsibilities, general advice for avoiding unexpected and unpredictable hazards, and what to do, how to do it, who to call, etc. in the event of an accident. The policy should promise employees a safe workplace. Construction businesses often produce unified manuals explaining their activity-based safety regulations. Every employee who takes work photos needs access to this repository. The safety team's roles include developing a security plan, setting regulations and updating them as required, sharing information efficiently, and preserving safety. There have been more transactions with large organizations. Uncomplicated completion of crucial tasks. So, prepare. Need "emergency action plan."

B. Meeting and Training of Safety

Every welfare and wellness program includes security planning. Security and construction workers should be taught risk assessment, management, and safety awareness to work more cautiously and earn more. New reps must be informed about workplace security before coming. All reps should acquire PPE, be taught in its correct usage, and know when to use it. Encourage workers to raise safety concerns to management. Every mistake, however little, must be reported to the site manager. On-the-job training, peer-to-peer training, workplace demonstrations, and toolbox dialogues are also effective. Effectively manage the training duration. Once a week, keep meetings to 20 minutes; once a month, 45 minutes. Meetings last 10 to 15 minutes and are usually casual. Workers should be trained in their native language or a language they readily grasp.

C. Arranging for Medical & First Aid

In the event of an injury, first aid is given; it is the urgent medical response for treatment. First aider is a known somebody who administers first aid. A first responder is someone with a health and safety executive HSE certificate who has completed a safety course or training. For every 50 workers on a construction site, at least one first aider is needed. Foil blankets, hemostatic dressings, tourniquets, disposable aprons, and individually wrapped wet wipes are all examples of items that may be found in a first aid kit. Treatment areas, potable water, and a central health care hub are required for providing initial care. It may help keep an injury from worsening. It goes without saying how important it is to make the right first-aid judgments in emergencies. An employee's life might be spared as a result. It's important to note that pills and medicines are not considered first aid supplies. In cases of cardiac arrest, only aspirin may be used. The employer must ensure that the first aid kit is always in working order. No matter how big or small the project is, it must provide access to basic medical services.

For the sake of worker safety, the Health and Safety (First Aid) Regulations of 1981 mandate the following for all workplaces, including construction sites:

- A first aid kit sufficiently stocked to treat all employees present;
- A designated medical professional to take charge in the event of an accident.

D. Administrative Stance

An efficient safety and health strategy will lay forth strict standards that the company must follow. When it comes to avoiding mishaps on the job, nothing is more important than management's duty to provide a safe working environment. The construction project's leadership is responsible for creating a comprehensive safety plan and providing all workers with necessary PPE. Safety in the workplace may be enhanced via the use of incentives and rewards for workers who follow all necessary safety measures. Businesses' productivity and workers' safety will both improve as a result of these rules. Effective management structures and procedures are required for carrying out the policy's goals.

Research Methodology

Literature search of published publications was performed using search engines for general, engineering, and other related themes. A total of 40 articles were found, 20 of which were only concerned with safety, 15 with safety and quality, and 4 with construction quality.

Conclusion

According to the report, the construction sector is riskier than other sectors and has a higher risk of fatal accidents. These indications are the result of numerous important and elements that have an impact on construction industry safety. The study also shows that some risky behaviors, like as failing to use hand gloves, earplugs, safety boots or eye protection, among others, contribute to coincidences. Dedication is required to promote and increase the level of safety in this profession. Better awareness of safety culture and a greater emphasis on safe practices will help minimize the occurrence of such incidents. Stakeholders may play an important part in bolstering the management system's procedures for ensuring employee and customer safety. Safety precautions include things like a strategy, training, first aid kits, and management rules. According to the research, building accidents not only result in the loss of human life but also impose substantial monetary losses. This includes the cost of lost productivity due to injured workers, lost managerial time, diminished worker confidence and equipment loss expense. The amount of accident reduction can also be increased by teaching the workforce about safety and rewarding the worker for each month that goes by without ab injury.

References

Choudhry R. M., and D. Fang (2008). Why operatives engage in unsafe work behavior. Investigating factors on construction sites. Safety Science 46(4), 566–584.

Everett, J. G., and P. B. F. Jr (1996). Costs of accidents and injuries to the construction industry. Journal of construction engineering management. 122(2), 158–164.

Lordsonmillar, R and Subramani, T. (2014). Safety management analysis in construction industry. Journal of Engineering Research and Applications 4(5), 117–120. ISSN: 2248–9622.

Leigh, J., and Robbins, J. A. (2004). Occupational disease and workers compensation. Costs, coverage and consequences. Milbank Quarterly 82(4), 689–721.

Nancy, M. Samelson and Raymond E Levitt, (1993). How Safety Save Lives and Money, Construction safety management, 2nd Edition, john wiley & sons. inc, 1-5. ISBN 0-471-59933-6

Rikhardsson, P. M., and M. Impgaard (2004). Corporate cost of occupational accidents: an activity based Analysis. Accident Analysis & Prevention 36(2): 173-182.

Robson, L. S., et al. (2007). Effectiveness of occupational health & safety management system Interventions. Systematic review. Safety Science 45(3): 329-353.

Roth, R. D and G. R. Smith. (1991). Safety programs and construction manager, Journal of Construction Engineering and Management, 117(2), 360–371

Sanchez, F. A. S., Pelaez, G. I. C., and J. C. Alis, (2017). Occupational safety & health in construction: a review of applications and trends. Industrial Health, 55, 210–218.

Malek, MohammedShakil S., and Bhatt, Viral (2022). Examine the comparison of CSFs for public and private sector's stakeholders: a SEM approach towards PPP in Indian road sector, International Journal of Construction Management, DOI: 10.1080/15623599.2022.2049490

The Relevance of Kitchen Vastu Guidelines in Relation to Architecture

Zankruti Raval

Karnavati University, Gandhinagar, India
E-mail:

Abstract

'Vastu' is derived from the Sanskrit sound 'Vas,' which includes terms for garments, homes, and dwellings. Vastu Shastra describes 'Vastu' as immortal and mortal cohabitation. Vaastu Shastra is the positioning of items based on cosmic energy, which affects our life. Design, layout, groundwork, space organisation, and spatial geometry are explored. It sometimes combines Hindu and Buddhist beliefs. Vastu shastra depicts the link between space and form in a building (or collection of structures). The system combines navigation, astronomy, and astrology. Originating in Indian philosophy, mathematics, geology, geography, and religion, it examines topography, roads, and solar influences. Earth's magnetism, energy, and elements.

Keywords:

Introduction

The author of the samaraangana sutra-dhara argues that well made and planned constructions (attractive houses) will provide an abode of good health, wealth, wisdom, good offspring, tranquilly, and pleasure, as well as free one from debts. The disregard for architectural principles will lead to unnecessary travels, a bad reputation in society, tears, and disappointments. Villages, towns, cities, temples, and Raja Prasadas (government buildings) must all be erected according to vastu shastra. As a result, Vastu shilpa shastra is brought to light for the benefit of, satisfaction of, and overall welfare of the universe (Parashar, 2014). **Panchamahabhutas and principles of Vastu shastra:**

The words 'pancha','maha', and 'bhuta' form the word panchamahabhuta. 'Pancha' denotes five,'maha' denotes greatness, and 'bhuta' denotes existence. Panchamahabhuta is found in all living and non-living objects in the universe. As a result, panchamahabhuta refers to the five basic elements responsible for the creation of the universe, which includes people. Each person has a distinct panchabhautik constitution. In health, this constitution is in a state of balance, and any imbalance leads. There are five basic elements of nature including Vayu (Air), Jal (Water), Prithvi (Earth), Tejas (fire) and Ambar (sky) (Nandy, 2017).

Akasha (Space/sky): The term Akasha literally means "free or open space," "ether," "sky," or "atmosphere." (Williams et al., 2005). In human biology, akasha mahabhuta refers to the empty space in the body or cavity. Akasha is one of the universe's five basic constituents or primordial ingredients. All of the other elements are protected by the space(Aneesh & Deole, 2020).

Vayu (Air): The word 'vayu' literally translates to 'wind, air, or vital air.'(Williams et al., 2005). Panchamahabhuta is one of the panchamahabhuta. It is used in Ayurveda to denote mahabhuta and dosha in the body. Vayu is divided into two types: subtle (sukshma) and gross (vayu) (sthula). The subtle form exists as a vital force ('prana') in the vital force sheath (pranamaya kosha). Gross (sthula) vayu exists in the body as mahabhuta and dosha in the food sheath (annamaya kosha) (Joshi, 2018).

Jal (Water): The word 'jala' literally means 'water' or 'any fluid' (Williams et al., 2005). Rain, rivers, and seas are examples of water, which comes in three forms: solid, liquid, and gas. Water is essential for all life forms and eco-cultures, as it determines the type, form, and pattern of life in relation to the planet.

Agni(Fire): 'Agni' is a Sanskrit word that signifies 'fire.' The term "tejas" means "glow," "splendour," "light," "fire," "the bright appearance of the human body," "vital power," "essence," and so on. (Williams et al., 2005).The word teja also represents vitality or essence (oja).Day and night, seasons and energy, excitement, vigour, and passion are all represented by Agni, the god of heat and light. The thermonuclear reaction in the sun provides heat and light to the Earth.

Prithvi (Earth): 'Prithvi' is a Sanskrit term that signifies 'earth.' It also depicts the elemental form of the earth and can also refer to a location, a site, or the ground. (Williams et al., 2005) In charak Samhita aneesh suggests that Prithvi comes from the word 'pruthu,' which means 'having sthula guna (the property of bulkiness or thickness)' (Aneesh & Deole, 2020). Prath is also a word that means "to spread." Particles have another meaning. Prithvi is the one who is made up of many particles (Gupta, 2006). Prittvi is defined as having gandha (smell), rasa (taste), rupa (vision), and sparsha (touch). The North Pole and the South Pole are its two poles. This magnet's southern pole is in the northern hemisphere, and its northern pole is in the southern hemisphere. The distribution of load in Vastu Shastra is based on magnetic principles. (Tarkhedkar, 1995).

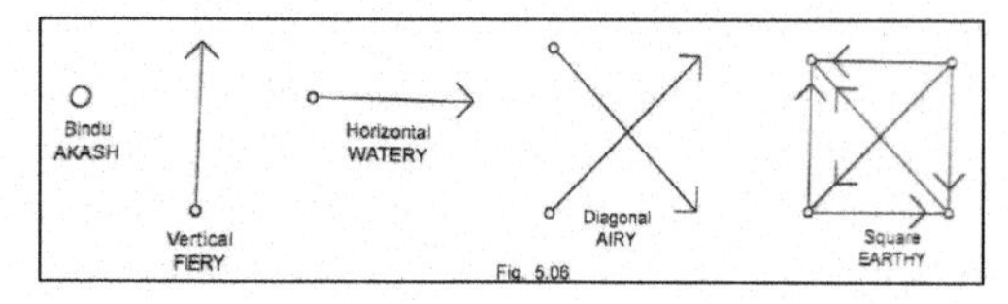

Fig 2. Representation of Five elements in form of Lines (Upadhyay & Mistry, 2006).

In vastu, the five elements are represented by distinct shapes, such as the square for earth, the diamond for ether, the triangle for fire, the crescent moon for air, and the circle for water.

(Upadhyay & Mistry, 2006)

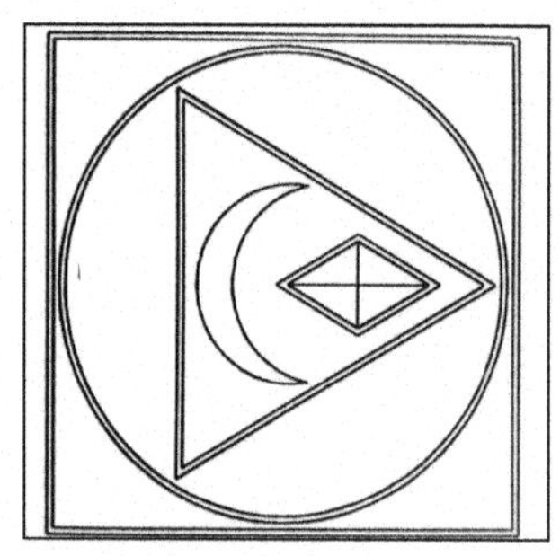

Fig 3. A diagram showing the shapes of Five elements (Upadhyay & Mistry, 2006).

These panchmahabhuta provide our bodies with internal energy such as proteins, carbs, and lipids, as well as external energies such as heat, light, sound, and air.

Fig 4: Co-relationship of elements, organs and the energy (Upadhyay & Upadhyay, 2006).

ELEMENT	FIRE	EARTH	AIR	WATER	ETHER
Energy form	Solar	Gravity	Wind	Rain	Space
Deity	Agni	Nairuti	Vayu	Isha	Sun
Direction	SE	SW	NW	NE	Center
Lord / Planet	Venus	Rahu	Moon	Jupiter	Ketu
Planet color	White	Blue	White	Yellow	Brown
Characteristic	Internal energy	Sleep demon	Wind speed	Immortal elixir	Creator
Represent	Vitality female issue	Fame longevity	Communication business	Knowledge spiritual	Balance creativity
Activity	Cooking	Sleeping	Storing	Meditation	Vacant
Metal	Silver	Iron	White metal	Copper / Brass	
Sub metal	Silver	Gold	Silver	Bronze	Cast iron
Gem	Diamond	Cats eye	Pearl	Topaz	Azurite
Symbol	Fire	Dagger	Deer	Kalash	Book
Shape	Triangle	Square	Crescent moon	Circle	Diamond
Body parts	Face	Feet	Breathe	Ocean	Naval
Sense	Sight	Smell	Touch	Taste	Sound
Organs	Eyes	Nose	Skin	Tongue	Ears
Swad	Hot / tikha	Sweet/madhur	Acidic/amla	Salty/turat	Bitter/kadu
Triguna	Rajas	Tamas	Rajas	Satwa	
Tridosha	Pitta		Vata	Kapha	Vata

Fig 5. Characteristics of the element in the tabular format (Upadhyay & Upadhyay, 2006).

Directions in Vastu shastra:

There are eight cardinal directions: north, south, east, and west, and the place where two directions meet is known as an intercardinal or ordinal point, such as North-East, North-West, South-East, and South-West (Sharma, 2020).

East: Lord Indra, sometimes known as the "King of Gods," rules this direction. In life, he bestows money, victory, and pleasure (Raman, 2000). Lord Indra is also said to bring rains, and when there isn't any, people pray to him and seek his blessings. It is customary to sit and pray to the sun in the east, believing that it brings happiness and wealth into our lives. Lord INDRA is the wealth-giver. To bring in plentiful prosperity, this direction should be left as open as possible.

South: Yama, the god of death, rules this direction. He is a Dharma manifestation who extinguishes evil forces and bestows positive things (Raman, 2000). It provides money, harvests, and enjoyment. On the contrary, it is believed that the south direction is inauspicious for any activity, particularly the entrance to the house, and it is recommended that the south direction be maintained heavy and laden at all times (Sharma, 2020).

South west: Niruti, the god who defends us from wicked opponents, directs us in this path. It's a source of personality, temperament, longevity, and death (Raman 2000 & Pegrum, 2000).

West: Lord Varun, the Lord of Rains, is in charge of this direction. He bestows his blessings in the form of natural rain, which brings life prosperity and pleasure. This direction's entrances and entry aren't very good. It ruins one's earnings possibilities.

North West: Lord Vayu directs this direction, and he brings us excellent health, vigour, and long life. It's a catalyst for change in business, friendship, and hostility. (Raman 2000 & Pegrum, 2000).

North: Kuber, the deity of wealth, is in charge of this direction. (Raman, 2000). Mirrors are considered lucky in this area because they are said to double your fortune. If there is considerable construction in the north, there will be a loss of wealth and prosperity. Lord Kubera's residence is in the north.

North east: Lord Eeshaan oversees the north east, which is a source of riches, health, and success. He bestows wisdom and understanding upon us, as well as deliverance from all sorrows and misfortunes (Raman 2000 & Pegrum, 2000).

Origin of Vastu Purusha and the Vastu Purusha Mandala

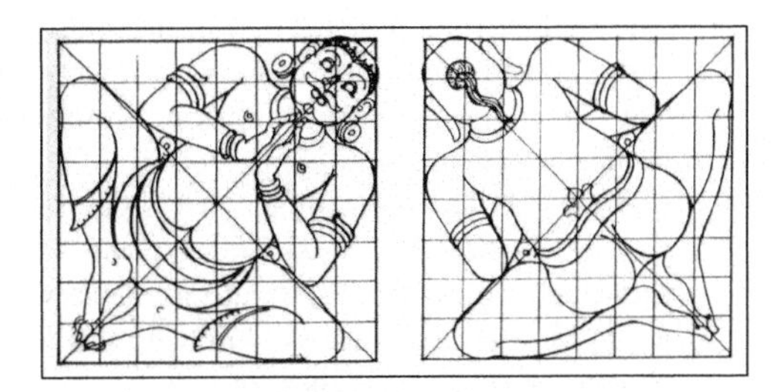

Fig 6. The vastu Purusha Mandala (Pegrum, 2000).

As a result, the shashtra provides prescriptive guidelines to avoid offending the Vastu purusha.A diagram known as the vastu purusha mandala is used to convey these principles.

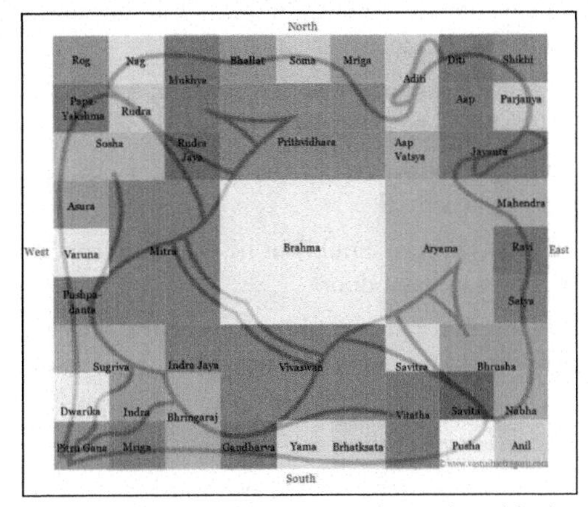

Fig 7. The vastu Purusha Mandala of 81 pada and leading gods (Pegrum, 2000).

Vastu Guidelines for Kitchen design

Kitchen:

Kitchen is the most affected area of the house where all kinds of energies prevail. A great deal of attention should be paid to the kitchen because it is where the lady of the house spends most of her time. For kitchen the most ideal place is south east corner zone which is governed by agni the god for fire and heat. The alternative choice for the placement should be in the North west corner. Kitchen should not be placed north-east, mid north, mid west, south west, mid south or the centre. Similarly it should not be placed below or above bed, pooja or toilet.

The kitchen's interior design should be as follows:

- The kitchen should be located in the south east (Agni) corner of the house or property.
- The kitchen's primary platform should be in the east and south east corners. The platform should not contact the kitchen's east or south walls. It should be one to three inches away from the walls at the very least. We can also build the kitchen with a protrusion of 1 to 2 inches.
- The stove or gas burner should be positioned in the south east corner, a few inches from the wall.

The platform's washbasin should be as far to the north east as practicable. The water pitcher and drinking utensils should be placed at the north east or north side.

Essential commodities such as grain, spices, and pulses should be arranged in the south or west.

Kitchen doors that open to the east, north, and west are beneficial.

The gas burner should not be placed in front of the kitchen's main door.

In the east and west directions of the kitchen, there should be one or two windows or air holes.

If the kitchen has a dining table, it should be on the north west or west side (Tarkhedkar, 1995).

Lightweight items should be stored in the east and north.

In the kitchen, the mezzanine level should face west or south.

Cooking should be done with the face to the east.

If you have a freezer in the kitchen, put it in the agni corner.

Empty or unwanted cylinders may be placed in the north west corner (Raman, 2000).

Electric equipment like microwave, oven, grinder etc should be placed in the southwest corner.

Allsort of waste should be collected in the Niruti corner.

'L' shaped platforms are more beneficial.

The colour of the kitchen area should be either of this: yellow, orange, pink, terracotta red. Black or violet colour should be avoided.

The slope of the kitchen or the placement of the nahni trap should be in the North east direction (Mehta, 2016).

According to Raman (2000) the various effects of planning the kitchen in different directions are as follows:

Location of Kitchen	Its effect on inmates
North east	Heavy expenditure and mental unrest
East	House wife will suffer from various health disorders and shall be unhappy despite all the comforts
South east	Ideal. Will bring health and happiness
South	Unrest and mental Tension
South west	Sickness and mental depression
West	Female related issues and bickering
North west	Heavy expenditure and tension
North	Intelligence, knowledge to children.

References

Aneesh, E. G., & Deole, Y. S. (2020). *Panchamahabhuta*. Charak Samhita Research, Training and Skill Development Centre (CSRTSDC). 10.47468/CSNE.2020.e01.s09.006

Bubbar, D. K. (2005). *The Spirit of Indian Architecture: Vedantic Wisdom of Architecture for Building Harmonious Spaces and Life*. Rupa & Company.

Dagens, B. (2007). *Mayamatam (2 Vols.): Treatise of Housing Architecture and Iconography* (I ed., Vol. II). Motilal Banarsidass.

Gupta, S. (2006, March 15). *Prithvi or Earth*. Sushmajee. Retrieved December 30, 2021, from http://www.sushmajee.com/reldictionary/dictionary/page-P-Q/prithvi.htm

Joshi, P. (2018, June 1). *Charaka Meemansa*. Vayuvivek. Retrieved 12, 2021, from https://charakameemansa.wordpress.com /2018/06/01/%e0%a4%b5%e0%

a4%be%e0%4%af%e0%a5%81%e0%a4%b5%e0%a4%bf%e0%a4%b5%e0%a5%87%e0%a4%95/

Kalra, H. (2017). *Is Vastu Shastra relevant to Architecture of the 21st century?* RTF | Rethinking The Future. Retrieved December 30, 2021, from https://www.re-thinkingthefuture.com/article/is-vastu-shastra-relevant-to-architecture-of-t he-21st-century/

Kumar, A. (2005). *Vaastu: The Art And Science Of Living* (reprint ed.). Sterling Publishers Private Limited.

Mehta, D. (2016). *Vastu Rachna* (Gujarati Edition ed.). N.M.Thakkar & Company.

Mitra, G. (2016, June 22). *Panchamahabhuta - The Concept and Explanation.* Aithein Healing. Retrieved December 30, 2021, from https://www.aitheinhealing.com/panchamahabhuta-concept-explanation/

Nampoothiri, S. (NA, September 12). *Vastu Shastra, Origin and its Principles - Sreejith Nampoothiri, World Famous Astrologer.* Mantrik yantra. Retrieved December 30, 2021, from https://www.mantrikyantras.org/vastu-shastra-origin-and-its-principles/

Ocean Health. (NA). *Chemistry of Seawater.* OCEAN HEALTH. Retrieved 12, 2021, from https://oceanplasma.org/documents/chemistry.html

Parashar, B. (2014, Jan). *Philosophy of Astrology and Vastu yoga.* Indian Astrology. Retrieved

12, 2021, from https://www.indianastrology.com/article/en/philosophy-of-astrology-and-vastu-yoga-5558

Pegrum, J. (2000). *Vastu Vidya: The Indian Art of Placement.* Gaia.

Raman, V. V. (2000). *Principles and Practice of Vastu Shastra* (Third ed.). Vidya Bhawan.

Sahasrabudhe, N. H., & Sahasrabudhe, G. N. (2005). *Cosmic Science of Vaastu.* Sterling

Drone Rules 2021: Analysis and Implications for India's UAV Programme

Arjun Sidharth,[a*] *Pratik Sinha,*[b] *and Dr. Sumaiya Sheikh*[c]

[a]Assistant Professor, Karnavati University, Gandhinagar, Gujarat
E-mail: *arjunsidharth15@gmail.com

Abstract

The deployment and use of Unmanned Aerial Vehicles (UAVs) for purposes of national security by sovereign governments, presents an entirely new dimension to statecraft. UAVs have several applications in a wide range of spheres. Increasingly, UAVs or drones as they are popularly called, are being used on a significant scale for both defensive and offensive purposes. In the Russia-Ukraine war, for instance, drones have been used widely by both entities for surveillance of enemy positions, and to target troops and critical infrastructure and assets. India too is not immune to challengers in the frontier regions, with drones being used by neighboring countries for strategic purposes. This paper aims to analyze the Drone Rules, 2021 published by the Union government to liberalize the rules and norms governing the production, deployment, and use of UAVs. The paper shall also delve into the various implications of these newly amended rules for India's UAV program.

Keywords: Unmanned Aerial Vehicles (UAVs), Production-Linked Incentives (PLI), Defence Research and Development Organization (DRDO), Ministry of Civil Aviation, Remote Pilot License

Introduction

The Drone Rules were published in the Ministry of Civil Aviation's gazette by gazette notification CG-DL-E-26082021-229221 on August 25, 2021. Additionally, the Ministry of Civil Aviation released the Drone Rules via gazette notification. The Certification Scheme for Unmanned Aircraft Systems was issued in a document with the number CG-DL-E26012022-232917 on January 26, 2022.

Drones are also referred to as Unmanned Aerial Vehicles (UAVs) or Remotely Piloted Aerial Systems (RPAS), and they can be remotely piloted by a pilot on the ground or controlled by technology alone.

The use of drones has the potential to bring enormous benefits to virtually every industry in the economy, including agriculture, mining, infrastructure, surveillance, emergency response, transportation, geospatial mapping, defense, and law enforcement, amongst others. Due to their reach, versatility, and ease of use, drones have the potential to be substantial creators of jobs and contributors to economic progress, particularly in India's more rural and inaccessible regions.

India has the potential to become the global drone hub by the year 2030 thanks to its historical advantages in the areas of innovation, information technology, thrifty engineering, and enormous domestic demand.

Drone

- It is a common term that refers to Unmanned Aircraft (UA).
- Drones were first created for the aerospace and military industries, but their greater safety and productivity have led to their widespread adoption in recent years.
- The autonomy degree of a drone can range from remotely controlled, in which a human being controls the movements of the drone, to advanced autonomy, in which the drone calculates its movement using a system of sensors and LiDAR (Light Detection and Ranging) detectors.[1]

Background

- Drone Rules: The Ministry of Civil Aviation issued a notification in 2021 that liberalized drone laws to promote research and development and to establish India as a center for drone manufacturing.
- It did away with a number of authorizations. The amount of paperwork that needs to be filled out has been reduced from 25 to 5, and the number of different sorts of fees has been lowered from 72 to 4.
- The use of micro and nano drones for non-commercial purposes does not require a remote pilot license, and there is no requirement for permission to fly drones in areas designated as green.
- Payloads of up to 500 kilograms have been authorized for drones, which paves the way for drones to be utilized as unmanned flying taxis.
- Additionally, the government has authorized foreign ownership of companies that operate drones.
- Production-Linked Incentive (PLI) Plan for Drones: The government has also approved a Production-Linked Incentive (PLI) scheme for drones and their components. This scheme will have an allocation of Rs. 120 crores for each of the next three financial years.[2]
- Strategic, tactical, and operational uses of this ground-breaking technology are all addressed in the PLI Scheme for the drone and drone component manufacturing sector.
- In September 2021, the Directorate General of Civil Aviation (DGCA) released an interactive airspace map to assist drone pilots in locating restricted airspace or areas in which they are required to complete specific pre-flight procedures before using their aircraft.
- The Indian government plans to create a comprehensive electronic property ledger as part of a program called SVAMITVA (Survey of Villages and Mapping with Improvised Technology in Village Areas). In order to help achieve this goal, the government has decided to use drones.

Classification of Unmanned Aircraft Systems

Based on the maximum combined weight, which takes into account the payload, the unmanned aircraft system is to be categorized as follows:

- With a weight of above 250 grams and less than or equal to two kilograms, a micro unmanned aircraft system falls into the following categories:
- To define a nano unmanned aircraft system, the craft must weigh less than or equal to 250 grams.
- Classifications of unmanned aircraft systems according to their maximum take-off weight are as follows:
- More than 2 kg but less than 25 kg is considered a small unmanned aircraft system.
- Larger than a 25-kilogram but smaller than a 150-kilogram UAV.[3]
- Those unmanned aircraft systems weighing more than 150 kilograms are considered "large."

Usage of Drones

- A surveillance system based on drones has been introduced for use in the railway industry.
- As part of the "SVAMITVA" program, the Survey of India plans to map the populated regions of villages using drone technology.

- India deployed drones to deliver COVID-19 vaccines. Manipur, Nagaland, and the Andamans & Nicobar Islands are currently participating in the pilot project that is being led by the ICMR.
- To ensure that lockdown procedures are followed to a tee, a drone is used to monitor potential hotspots for the COVID-19 virus and to monitor containment areas. They showed the device off to the Chandigarh police in a "containment zone."
- The Indian government's Ministry of Civil Aviation (MoCA) and Directorate General of Civil Aviation (DGCA) have given the Board of Control for Cricket in India (BCCI) a conditional exemption to allow the use of drones for live aerial filming during the India Cricket Season in 2021.

Drones in Defence Sector

Programs for Unmanned Aerial Vehicles (UAVs) and Unmanned Aircraft Systems (UASs) have been created by India's Defense Research and Development Organization (DRDO), which is the country's equivalent of the United States Department of Defense. The purpose of this project is to create a homegrown armament that will both supplement and replace the current fleet of unmanned vehicles. Some examples of this are as follows:

- DRDO Lakshya: This is a drone designed specifically to conduct stealthy aerial surveillance and the acquisition of targets. It is propelled into space by a rocket motor powered by solid propellant, and once it is airborne, it is kept aloft by a turbojet engine.[4]
- DRDO Nishant: It is primarily utilized for reconnaissance over hostile territory and is also employed for target designation, training, surveillance, damage assessment, and target designation. Nishant has successfully finished the stages of both development and user testing.
- DRDO Rustom: The Rustom unmanned aerial vehicle (UAV) is a Medium-Altitude Long-Endurance (MALE) system that was modeled after the American Predator UAV. In a manner analogous to that of the Predator, the Rustom is intended to be employed in both combat and reconnaissance roles. Although it is still in the prototype stage, it is anticipated that the Indian Air Force will use it to replace and supplement UAVs of the Israeli Heron kind.

Features of Drone Rules, 2021

On August 25, 2021, the Central Government issued a notification regarding the Drone Rules, 2021.

- Built on the tenets of trust, self-certification, and monitoring that does not interfere unnecessarily.
- Developed to usher in an era of supernormal growth while maintaining a balance between concerns over safety and security.
- Several clearances, such as the unique authorization number, the unique prototype identification number, the certificate of manufacturing and airworthiness, the certificate of conformity, the certificate of maintenance, the import clearance, the acceptance of existing drones, the operator permit, the authorization of R&D organizations, student remote pilot licenses, remote pilot instructor clearances, drone port clearances, and many more, have been deleted.
- The number of forms has been cut in half, from 25 to 5.
- The variety of fees has been cut from 72 to 4.
- The price was cut to almost nothing and no longer corresponded to the drone's size. There has been a significant reduction in the price of obtaining a remote pilot license; from INR 3000 (for a large drone) to INR 100 (for any type of drone), and it is valid for a whole decade.[5]
- The Digital Sky platform is now being developed as an online single-window solution that is user-friendly. There will be very little interaction with humans, and the vast majority of permissions will be generated automatically.

- An interactive map of the drone airspace, including both yellow and red zones, must be displayed.
- Flying drones in green zones do not require the obtaining of any permissions.
- The yellow zone, which is the area around the airport that requires clearance from ATC, has shrunk from 45 km to 12 km out from the airport perimeter.
- There is no requirement for a remote pilot license for micro drones (when used for non-commercial purposes) or nano drones.
- Entities engaged in R&D that operate drones in their own or rented premises inside a green zone are exempt from the requirements of obtaining a type certificate, a unique identification number, and a remote pilot license.
- The DGFT will be in charge of regulating the import of drones.
- Elimination of the prerequisite for obtaining import permission from the DGCA
- There is no prerequisite for obtaining a security clearance in order to obtain any kind of registration or license.
- The maximum weight limit for drones that are covered by the new drone rules, 2021, has been increased from 300 kg to 500 kg. This will also encompass taxi services provided by drones.
- The DGCA will be in charge of prescribing training standards for pilots of drones, supervising drone schools, and providing pilot licenses online.
- There are no limitations placed on the ownership of drone companies in India by international investors.
- The DGCA must issue the remote pilot license within 15 days of the pilot acquiring the remote pilot certificate from an authorized drone school via the digital sky platform. This must be done in order for the license to be valid.
- The Quality Council of India or other authorized testing institutions is responsible for carrying out the testing of drones in preparation for the granting of type certificates.
- The only time a Type Certificate is necessary is when a drone is going to be flown in India. There is no need to obtain a type of certification or a unique identification number in order to import or manufacture drones solely for export.
- Nano and model drones that are manufactured for study or pleasure are not required to have type certification.
- Using the self-certification path available on the digital sky platform, manufacturers and importers of drones can generate a unique identification number for each of their drones.
- The transfer and deregistration of drones may now be completed more quickly and easily thanks to enhancements made to the digital sky platform.
- If a drone is in India on or before November 30, 2021, it will be eligible to receive a one-of-a-kind identification number through the digital sky platform. This is provided that the drone already has a Device Acknowledgment Number (DAN), a GST-paid invoice, and is on the list of DGCA-approved drones.
- DGCA will mandate SOPs and TPMs for usage by users to self-monitor their compliance with digital sky's standards. There is no requirement for approvals unless there is a considerable variation from the procedures that have been defined.
- In the near future, notifications on safety and security features such as 'No Permission – No Takeoff' (NPNT), real-time tracking beacons, and geo-fencing will be provided. The sector will be given a head start of six months in order to comply with the regulations.
- The highest possible fine for infractions has been reduced to one lakh Indian Rupees. Earlier, it was in the multiple lakhs.
- To expedite the delivery of goods, drone lanes will be constructed.
- The government will establish a Drone Promotion Council, which will include involvement from academic institutions, start-up companies, and other stakeholders

to enable a growth-oriented regulatory framework.

Categories of Zones

According to the accompanying airspace maps for the Drone Rules 2021, the Indian airspace is divided into the following three zones: According to the accompanying airspace maps for the Drone Rules 2021, the Indian airspace is divided into the following three zones:

- The term "Red Zone" refers to the airspace with specific boundaries that are located above Indian territory or its territorial waters, as well as any installations or designated port limitations that are established by the Central Government outside of Indian territory. Only the federal government may provide the go-ahead for drone activities in this area.
- The "Yellow Zone" is the name given to the restricted airspace over Indian territory (including land and water) that has specific proportions. All UAS operations are forbidden and need prior authorization from the local air traffic control. Those areas of the sky above 400 feet (120 meters) in the green zone and those areas of the sky above 200 feet (60 meters) between 8 and 12 kilometers (km) from the outer boundary of an active airport will be labeled as the yellow zone.
- For unmanned aircraft system activities, the "Green Zone" is defined as the area up to 400 feet (120 meters) in altitude that is not located within a red zone or yellow zone on India's airspace map, and the area up to 200 feet (60 meters) in altitude that is located within a lateral distance of 8 kilometers (km) and 12 kilometers (km) from the boundary of an active airport.

Airspace Map of India for Drone Operations

On September 24, 2021, the Central Government of India, led by Prime Minister Shri Narendra Modi, announced the release of India's airspace map for drone operations.[6]

The newly relaxed Drone Rules, 2021 have led to the creation of the Drone Airspace Map 17 as a follow-up measure. The following is a list of the top ten features of drone airspace maps:

- The drone airspace map for India is an interactive tool that shows where the yellow and red zones are located around the country.
- The "green zone" refers to the undesignated area of the sky up to 400 feet in altitude. The area between 8 and 12 kilometers from the airport's boundary which is over 200 feet in altitude is also considered part of the green zone.
- Within green zones, the operation of drones with a combined weight of up to 500 kg is permitted without the need for any permission whatsoever.
- If you're flying above 400 feet in a green zone, you're in the "yellow zone." The yellow area also includes the airspace above ground level between 5 and 8 kilometers from the boundary of an active airport and the airspace above 200 feet from the perimeter of an active airport.
- Operations with drones in the yellow zone require approval from the relevant air traffic control body, which might be the AAI, the IAF, the Navy, or HAL, depending on the circumstances.
- The airport's yellow zone has shrunk from 45 kilometers to 12 kilometers in diameter.
- The "no-drone zone" is denoted by the color red, and inside its boundaries, remote-controlled aircraft can only be flown with authorization from the central government.
- Authorized entities can make alterations to the airspace map at any time they see fit.
- Before flying a drone, pilots must verify that the airspace map is the most up-to-date version and that there have been no changes to the zone limits.
- The map of the drone airspace is accessible to anybody and everyone on the digital sky platform without the necessity of requiring a login.

Registration of Unmanned Aircraft System

- Registration of Unmanned Aircraft Systems (UAS) on the digital sky platform and the acquisition of a unique identification number are prerequisites to the operation of UAS in the United Kingdom, as per the Drone Rules, 2021, unless an individual is exempt from the requirement of a unique identification number.
- The Director General is responsible for keeping a registration record of all unmanned aircraft systems that have been given a unique identification number in accordance with the Drone Rule, 2021.[7]
- Whoever operates an unmanned aircraft system is the one who is responsible for ensuring that the system complies with a current type certificate and operates in accordance with the requirements of the certificate.

Remote Pilot License

- General – An unmanned aircraft system may not be flown by anybody other than a person who is a current bearer of a valid remote pilot license and is enrolled on the digital sky platform.
- Classification – All unmanned aircraft systems, or subsets thereof, for which a remote pilot license is obtained must be included.

Eligibility for Pilot License

If a person meets all of the following requirements, then he or she is qualified to apply for a remote pilot license:

- ➢ The individual is least eighteen years of age but no more than sixty-five years of age at the time of publication.[8]
- ➢ possesses the equivalent of a tenth-grade education from a board that is recognized by the government.
- ➢ has completed the training required by the Director General and has met the standards of any remote pilot training organizations with which they are familiar.

Unmanned Aircraft System Operations for Research, Development, and Testing

In order to operate unmanned aircraft systems for research, development, and testing, the following individuals are exempt from the requirements of obtaining a type certificate, a unique identification number, previous approval, and a remote pilot license:

- any research and development institution that is either under the administrative supervision of the Central Government, a State Government, or the Union Territory Administration or that is recognized by one of these three levels of government;
- any educational institution that is either under the administrative supervision of the Central Government, a State Government, or the Union Territory Administration or that is recognized by one of these entities;
- any newly established business that has been acknowledged by the Department for the Promotion of Industry and Internal Trade;
- any officially recognized testing organization
- any manufacturer of unmanned aircraft systems in possession of a valid Goods and Services Tax Identification Number

Regulations of Import

Every shipment of unmanned aircraft must be approved by the Directorate General of Foreign Trade or another agency approved by the Central Government.

Production-Linked Incentive (Pli) Scheme for Drones and Drone Components

The Production-Linked Incentive (PLI) plan, which will apply to drones and drone components as of September 30, 2021, has been notified by the Central Government.

The newly liberalized Drone Rules, 2021 were issued by the Central Government on August 25, 2021. As a follow-through, the PLI scheme was launched on September 1, 2021. The PLI scheme and the new drone rules are both designed to stimulate above-average growth in the emerging drone industry.

As a result of the newly implemented regulations and the incentive program, the industry that manufactures drones and drone components might get an investment of more than 5000 crores of rupees over the course of the next three years. Revenue generated by the drone manufacturing sector might increase from an estimated Rs 60 crore in FY 2020–21 to well over Rs 900 crore by FY 2023–24. It is anticipated that the industry of manufacturing drones would result in the creation of more than 10,000 direct jobs within the next three years.

The extent of the drone services business (which includes operations, logistics, data processing, and traffic control, among other things) is far larger. Within the following three years, it is anticipated that it would rise to more than 30,000 crores. It is anticipated that the market of drone services will produce more than five lakh jobs during the next three years.

The PLI Scheme is characterized by its features, which are as follows:

1. First, over three fiscal years, Rs 120 crores have been set out for the PLI scheme for drones and drone components. This sum is almost twice as large as the total sales of all U.S. drone manufacturers in FY 2020–21.
2. An eligible drone and component producer can receive a bonus of up to 20% of the value she contributes to the final product.
3. The value addition must be determined by subtracting the annual purchase cost (net of GST) of drones and drone components from the annual sales revenue (net of GST) generated by sales of drones and drone components.
4. Only the drone industry has received such special treatment from the government, which has committed to maintaining the PLI rate at 20% for the full three years. The PLI rate decreases annually in other industries' PLI programs.
5. After the first year (FY) 2021-, the anticipated PLI scheme duration is three years. After assessing the program's impact and consulting with the industry, the PLI scheme will be extended or redrafted.
6. In another example of the special treatment afforded to the drone industry, the government has decided to lower the minimum value addition norm for drones and drone components from 50% to 40% of net sales. More people will be able to receive aid as a result of this.
7. PLI protects many parts of drones, including:
   Flight control module, ground station, and other components, as well as the Inertial Measurement Unit and Inertial Navigation System.
   Methods of Transmission (radio frequency, transponders, satellite-based, etc.)
   payload (cameras, sensors, spraying systems, etc.)
   Critical Elements for Safety and Security, e.g., "Detect and Avoid" Systems, Emergency Recovery Systems, Trackers, etc.
8. In order to keep up with the ever-improving drone industry, the government may periodically add new components to the list of those that are approved for use.
9. The government has decided to extend the incentive program to cover software engineers who work on drone-related goods.
10. The Government has maintained a low annual sales turnover eligibility requirement of Rs. 2 crores (for drones) and Rs 50 lakhs (for startups) for Micro, Small, and Medium Enterprises (for drone components). More people will be able to receive aid as a result of this.
11. The minimum annual sales revenue threshold for eligibility has been maintained at Rs. 4 crores (for drones) and Rs One crore (for other products) for non-MSME enterprises (for drone components).
12. For example, if you're a maker of drones or drone parts, you'll only be eligible for an incentive of 5% of your value creation.
13. The PLI premiums paid by a manufacturer cannot exceed 25% of the company's

annual revenue. More people will be able to receive aid as a result of this.

14. If a producer falls short of the required amount of acceptable value addition in a given fiscal year, she can make up the difference and still qualify for the incentive the following year.

References

Smith, Clint. (2018). The Drone. Poetry, vol. 213, no. 1, 2018, pp. 20–20. JSTOR, https://www.jstor.org/stable/26539629. Accessed 5 Nov. 2022.

Keene, Shima D. (2015). Lethal and legal?: the ethics of drone strikes. Strategic Studies Institute, US Army War College. JSTOR, http://www.jstor.org/stable/resrep11765. Accessed 5 Nov. 2022.

Watts, Adam C., et al. (2010). Small Unmanned Aircraft Systems for Low-Altitude Aerial Surveys." The Journal of Wildlife Management, vol. 74, no. 7, pp. 1614–19. JSTOR, http://www.jstor.org/stable/40801523. Accessed 5 Nov. 2022.

Upadhyay, Shreya (2019). India-US Defence Partnership: Challenges and Prospects. Indian Foreign Affairs Journal, vol. 14, no. 2, 2019, pp. 116–28. JSTOR, https://www.jstor.org/stable/48636718. Accessed 5 Nov. 2022.

Sandbrook, Chris (2015). The Social Implications of Using Drones for Biodiversity Conservation. Ambio, vol. 44, 2015, pp. S636–47. JSTOR, http://www.jstor.org/stable/24670995. Accessed 5 Nov. 2022.

Padmanabhan, Ananth (2017). Civilian Drones And India's Regulatory Response. Carnegie Endowment for International Peace, 2017. JSTOR, http://www.jstor.org/stable/resrep12772. Accessed 5 Nov. 2022.

Lightfoot, Thaddeus R. (2018). Bring on the Drones: Legal and Regulatory Issues in Using Unmanned Aircraft Systems. Natural Resources & Environment, 32(4), pp. 41–45. JSTOR, https://www.jstor.org/stable/26418852. Accessed 5 Nov. 2022.

Wojton, Heather M., et al. (2019). Pilot Training Next: Modeling Skill Transfer in a Military Learning Environment. Institute for Defense Analyses, 2019. JSTOR, http://www.jstor.org/stable/resrep22749. Accessed 5 Nov. 2022.

Toy Companies Using Unconventional Methods to Stay Relevant and Reach Evolving Minds of the Parents and Children

Deepikaa Raju,[*,a] *and Arunachalam Muthiah*[a]

[a]Karnavati University, Gandhinagar, Gujarat, India
E-mail: [*]deepikaraju09@gmail.com

Abstract

Toy Market has been one of the largest markets spread across the world with many familiar brands. These well-known brands are also known for some of iconic collaborations despite their steady positive growth. This research is dedicated to finding the need for such companies to collaborate. The collaboration of these companies varies based on the industry to which the other company belongs to. This paper presents a systematic review of the information available regarding such collaborations while keeping the statistical figures and objective responses of the customers, critics, and the company in consideration. Apart from the well-known reasons about the need to collaborate, the research also tends to focus on the light yet important aspects. Based on the research it can be concluded that companies find the need to collaborate irrespective of their sales and revenue generated.

Keywords: Toy Companies, Collaborations, Sales, Market & Trends

Introduction

Toys are objects designed for play, mainly for children but also used by adults in certain conditions. Recent times have opened many opportunities for the toy industry. They are not only seen as an item used by the children but are also a means to keep the craft & culture of a place or ethnicity alive. They are also considered to have a great social and educational impact on children and also create a sense of nostalgia among adults. In a fast-growing digital world, toy companies are making constant efforts to not be forgotten and to also reach new audiences.

In the age of digitization, toy companies need to keep growing and stay relevant. Today, play and toys are synonymous with imagining, planning, simulating, reacting, and communicating. Where most of the revenue gained by the toy industry is from young customers, digitization has got them leaning towards console gaming, browser gaming, and so on. The downside of digitization stays at the point that it has reduced the time spent by kids on physical activities. Many companies are finding various ways to be at par with the digitization; for example, companies have understood the fact that due to changing times, customers no longer shop in-store and have shifted to e-commerce. The following research paper highlights how toy companies constantly try to keep themselves relevant according to the trends and needs of the audience. One of the methods is shown with respect to the leading toy brand, Lego.

The global toys and games industry was worth USD 101.5 billion in 2018 and is expected to expand 4.6% from 2019 to 2025 ("Toys and Games Market Size"). In 2019, the U.S. toy sector generated over 623 thousand employment and had a 97.2 billion-dollar economic effect. The U.S. leads worldwide toy imports (Emma Bedford, 2020). In 2019, North American kids spent $300 on toys. The global toy business was worth around 90 billion U.S. dollars in 2019 ("Average amount spent per child on toys").

This paper aims to concentrate on the ways by which toy companies are reforming themselves to stay at par in the age of computers and video games. The paper discusses how the company can gain results in various forms and numbers.

Methodology

The review papers noted that several organizations have tried different ways to stay competitive in the increasing market. Companies have extended their perspectives and changed their beliefs to attract different clients. Digitalization has spread over the world in unexpected ways. Due to its many benefits, digitalization is hard to adjust to but necessary. The pandemic changed many activities and even altered some while the world was running on a group pattern that everyone had defined for themselves. COVID-19 halted society and business. During the crisis, a frightening mass migration of migrants on foot occurred. They worried about job loss, daily ration, and no Social Security (Chaudhary et al., 2020, p. 172). Some firms prospered after a financial shock. The pandemic altered consumer and business behavior. Low domestic growth and recession hit many countries.

The stock market demonstrates that the coronavirus has contributed to economic volatility, despite the presence of numerous other factors. Business Insider, n.d., Coronavirus Business & Economy Impact News The securities market was harmed by the Sensex's point decline. The virus prompted the exchange to fall by a record amount. The S&P 500 has increased by about 65% from its low in March and by nearly 16% overall. Nasdaq has increased by 44% year to date (Domm, 2020). Because certain firm shares appeared shaky, executives and investors became alarmed.

It was predicted that the market for dolls, toys, and games would shrink 1% from $102.6 billion in 2019 to $101.7 billion in 2020. The COVID-19 epidemic and subsequent containment measures have caused an economic slowdown in several nations, which is mostly to blame for the decline. After that, the market is expected to bounce back and grow at a CAGR of 9% starting in 2021 to reach $128.5 billion in 2023. Demand for smart toys is being driven by user engagement. Electronic games and toys that react to their surroundings are known as smart toys. Children are drawn to computer science, which lessens the risk of the smartphone industry. Released smart toys include Mattel's "smart" baby monitor, Spin Master's interactive Luvabella baby doll, Hasbro's bluetooth-connected Furreal motorised pet, Barbie's holographic future Barbie, and Lego's app-based coding kit (Markets, 2020).

The changing business environment and strategies needed the companies to come up with creative ways to boost growth. With all such activities happening in the background, some companies tried to find a way through the pandemic by collaborating with companies and products from different industries. Lego is one such toy company that has ruled the toy industry while being a common name heard in every household. The brand valued fun and creativity along with staying up with the trends and exploring in new directions. Apart from creativity the brand tries to impart learning and care. Along with their values, the Lego group showed exponential growth. Their growth surpassed the industry while setting new trends and milestones. Their bold and significant investments helped them achieve massive positive results. This also helped in increasing the consumers' sales due to the broader product range and also created a larger market. The Lego group was also able to create a larger market for itself in China.

Co-branding is effective regardless of sector or business type. A strategic marketing and advertising collaboration between two brands that benefit both brands. Co-branding can increase business, recognition, and market penetration, but it must be a win-win for all parties. Both audiences need value. Companies improve through collaborations. Collaboration success variables vary. Collaboration and interest rise with brand contrast.

Case Study: Lego Group

To study the impacts of teamwork, scientists will employ Lego, a widely recognized industrial brand. The Danish words “Leg godt,” meaning “play well,” are the inspiration for LEGO. We are dedicated to our ideals and our name. LEGO was invented by Ole Kirk Kristiansen in 1932. Kjeld Kirk Kristiansen, the grandson of the company’s founder, currently serves as CEO. It began as a little carpenter shop over 85 years ago and has since expanded to become a major toy manufacturer around the world (The LEGO Group History - The LEGO Group - About Us - LEGO.Com US, n.d.). Throughout the world, you can visit one of Lego Group’s many stores or one of the Legoland theme parks (Wikipedia contributors, 2021). Lego is committed to the principles of Creativity, Fun, Learning, Care, and Quality.

In terms of sales, income, and profits, Lego dominated its competition. In order to fuel its expansion, LEGO spent heavily on new products and infrastructure. Global retail sales increased in 2019 by 5.6%.

Earnings for the year were DKK 38.5 billion, an increase of 6%. Operating profit increased by 1% to DKK 10.8 billion. The bottom line increased by 3% to DKK 8.3bn despite substantial spending on long-term growth. With a positive operating cash flow of DKK 9.6 billion, the company is doing quite well. Every major market saw share growth. As predicted by Lego (2020b),

As a result of poor sales, Lego and Ikea worked together last year. The companies came up with the unique storage system BYGGLEK. The new BYGGLEK collection featured a LEGO brick kit and storage containers decorated with LEGO bricks to make tidying up the house more of a game. The BYGGLEK line debuted at IKEA shops across Europe (excluding Russia) and North America on October 1, 2020. Play, Display, and Replay: IKEA® and the LEGO Group Introduce BYGGLEK - a Resourceful Solution That Intertwines Play and Storage (IKEA® and the LEGO Group Introduce BYGGLEK - a resourceful Solution That Intertwines Play and Storage, 2020) details how IKEA of Sweden AB and the LEGO Group set out to eliminate obstacles to play in everyday life while designing a product that children and adults could enjoy together.

Children see creativity and inspiration whereas adults see disorder, and BYGGLEK will assist parents around the world confirm their children are engaging in more imaginative play at home. Your children’s artwork will look great alongside the BYGGLEK series and other IKEA products. (A Resourceful Solution That Combines Play and Storage; IKEA® and the LEGO Group Present BYGGLEK, 2020b).

Almost half of the children surveyed in the Ikea Play Report stated they wanted more time to play with their parents, and ninety percent of parents agreed that play is crucial to their children’s health, development, and happiness (Segran, 2020).

Rasmus Buch Lgstrup, a designer at LEGO Group, praised the creative approach, saying, “BYGGLEK is more than boxes.” Game and put away. The BYGGLEK brand of products is designed to bring more imagination and fun into people’s everyday lives. There are four different BYGGLEK goods available: a starter set of LEGO bricks, a set of three smaller boxes, and a set of larger boxes. The front studs and lid of BYGGLEK are “LEGO system live,” meaning that it is compatible with any LEGO element (Play, Display, and Replay: IKEA® and the LEGO Group Introduce BYGGLEK - a Unique Solution That Intertwines Play and Storage, 2020c). The Ikea Lego brick set with 201 pieces costs $14.99. In order for even the youngest child to be able to use this set, the designers used Lego parts that were less complex. The lack of instructions is intentional; the packaging for these parts is supposed to spark a child’s imagination, not limit it (Segran, 2020b).

After the partnership was announced, sales and Lego stock exploded. It is September 2, 2020, in Billund. Today, LEGO Group revealed its earnings for the first six months of the year. With a 7% increase over 2019, revenue reached DKK 15.7 billion. Global consumer sales for the brand climbed by 14% over the first half of 2019. Its operating profit of DKK 3.9 billion was up 11% from the previous year. This is according to research by The Lego Group (2020b). We saw results from our investments in long-term growth strategies like e-commerce and product innovation during the first half of the year. Building enthusiasts of all ages were drawn to our extensive product catalog, and we were able to successfully meet online orders thanks to our streamlined e-commerce platform and responsive global supply chain. We made sure our retail partners could continue their online sales by working with them. (2020).

Results

With sales rising, Lego partnered with Ikea again. Due to Lego's amazing sales, this was one of their most unexpected collaborations. This collaboration raised their stock price. The collaboration attracted curious clients of all ages and interests. The collaboration's nature confused reviewers. Containing mixed reviews, Lego and Ikea released BYGGLEK, a two-year-old product with three tiny boxes, two larger boxes, and a special Lego brick set. White plastic crates contain Lego studs on the bottom. You can build using the boxes while storing your bricks (Lindsay, 2020). The versatile bricks promoted creativity and learning while alleviating several housekeeping difficulties.

The product evoked a sense of nostalgia among the adults connecting them to their happy past and also providing a new range for the kids to explore. The versatility of the product played a huge role in its popularity of the product along with the brand value and trust of Ikea and Lego. The product turned out to be the stepping stone for many such versatile products in the world of interior design and toys. This also helped in changing the perspective and the preconceived notion held by many customers. The collaboration provided a platform for the company to maintain and increase its growth while opening the door to opportunities for future advancements. Along with reaching new audiences, the company managed to be fixed to the position of being every household's common name. While the pandemic could have been a severe blow for the not-so-digital section of the industry turned out to be a massive scope for the companies.

This collaboration did not only give Lego a financial boost but also increased its reach in the market by gaining new customers not only children but adults as well. The products launched were bought out of curiosity while some wanted to explore the new range. They were successful in breaking through with the partnership and established themselves as a versatile brand, changing the long and established image of a leading toy maker.

Conclusion

It can be concluded that the main reasons for the toy companies to collaborate irrespective of the sales include reasons like 'Nostalgia'. Toys have an extremely wonderful ability to provide customers with a feeling of nostalgia and familiarity. This also serves as a powerful tool to attract customers while they start to associate the brand with a sense of familiarity. Thus, increasing the likelihood of them purchasing products from the company instead of the competitors. (Use Interesting Promotional Toys in Marketing Campaigns, 2018).

Increasing Sales and collaborations are a definite way to increase sales as it changes the definite outlook of the company built over time. Changing perspectives of the customers and competitive brands. Partnerships with unexpected brands help in changing the perspective based on the outcome. Changing business & customer relations. Collaboration helps the customers to experience more from the brand. Reaching a new audience.

Hence, it can be concluded that toy companies are growing, developing their strategies, and considering collaboration to stay relevant and grow along with the fast-moving audience.

References

Chaudhary, M., Sodani, P. R., and Das, S. (2020). Effect of COVID-19 on Economy in India: Some Reflections for Policy and Programme. Journal of Health Management, 22(2), 169–180.

Cohen, S. G., and Mankin, D. (2002). Complex collaborations in the new global economy. Organizational Dynamics. 31(2), 117–133.

Gohil, P., Malek, M., Bachwani, D., Patel, D., Upadhyay, D., and Hathiwala, A. (2022). Application of 5D Building Information Modeling for Construction Management. ECS Transaction, 107(1), 2637–2649.

Malek, M. S., Gohil, P., Pandya, S., Shivam, A., and Limbachiya, K. (2022). A novel smart aging approach for monitoring the lifestyle of elderlies and identifying anomalies. Lecture Notes in Electrical Engineering, 875, 165–182.

Malek, MohammedShakil S., and Bhatt, Viral (2022). Examine the comparison of CSFs for public and private sector's stakeholders: a SEM approach towards PPP in Indian road sector, International Journal of Construction Management, DOI: 10.1080/15623599.2022.2049490

Muñoz, I., Gazulla, C., Bala, A., et al. (2009). LCA and ecodesign in the toy industry: case study of a teddy bear incorporating electric and electronic components. Int J Life Cycle Assess, 14, 64–72.

Airike, Peppi-Emilia, Rotter, Julia P., Mark-Herbert, Cecilia (2016). Corporate motives for multi-stakeholder collaboration– corporate social responsibility in the electronics supply chains, Journal of Cleaner Production. 131, 639–648.

Scott, I. (2016). Antitrust and Socially Responsible Collaboration: Chilling Combination. American Business Law Journal, 53(1), 97–144.

Yew Wong, C., Stentoft Arlbjørn, J., and Johansen, J. (2005). Supply chain management practices in toy supply chains. Supply Chain Management, 10(5), 367–378.

Zhu, H., Huang, X., He, Q., Li, J., and Ren, L. (2016). Sustaining Competitiveness: Moving Towards Upstream Manufacturing in Specialized-Market-Based Clusters in the Chinese Toy Industry. Sustainability, 8(2):176

Behavior of Speed Breaker in Urban Context

Prof. Varshin Vala,[*,a] *and Ali-Asgar Chunawala*[b]

[a]Professor, Product Design Department, UID Karnavati University, Ahmedabad
[b]Student, Anant National University, Ahmedabad
E-mail: *varshin@karnavatiuniversity.edu.in

Abstract

The roads are designed in such a way that they may suit the design needs as well as the transportation requirements. However, in some places, speed control is required. A speed-reducing measure must be implemented. This paper will cover the numerous speed-calming strategies that are used to reduce accidents/collisions or hazardous moments. Speed breaker is one of the most commonly utilized speed calming methods, along with speed bump, speed hump, and speed table. In this study, various types of speed breakers should be explored, as well as the problems associated with them, and appropriate remedial actions should be offered.

Keywords: Road Accidents, Speed Calming Measures, Speed Breakers, Behaviour

Introduction

To preserve the functionality of the roadway system, vehicles must maintain design speeds on various road types. The use of controls to maintain speeds. These techniques enhance both traffic flow and safety.

In a decade, traffic has changed significantly. More automobiles cause more accidents. Our nation's traffic safety is failing and requires action. Approximately one million three hundred thousand individuals each year are killed in motor vehicle accidents. Many of the 20–50 million additional injured persons become incapacitated. Road traffic injuries are costly to individuals, families, and governments. These losses include treatment expenditures and missed productivity for the deceased or disabled, as well as for family members who must miss work or school to provide care. Road traffic accidents cost most countries 3% of their GDP. Over ninety percent of road deaths occur in low- and middle-income countries. The biggest number of road traffic fatalities occur in India. Low-income motorists are more likely to be involved in an accident in high-income nations. Boys begin to experience road traffic crashes at an early age. Young males under the age of 25 accounts for 73% of road traffic fatalities, over three times as likely than young females.

Average speed increases the likelihood and severity of crashes. Every 1% increase in average speed raises the chance of fatal and major crashes by 4% and 3%, respectively. At 65 km/h, car fronts kill pedestrians 4.5 times faster than at 50 km/h. At 65 km/h, 85 percent of automobile occupants die in side collisions.

Drunk driving: Alcohol and psychotropic substances increase the likelihood of a deadly collision. At blood alcohol concentration (BAC) levels as low as 0.04 g/dl, drunk driving significantly increases the risk of a collision. Depending on the psychoactive substance ingested, driving under the influence of drugs increases the risk of road traffic collisions. Amphetamine users are five times more likely to be involved in a fatal crash.

Distracted driving: Numerous distractions impede safe driving. The use of mobile phones is becoming a traffic safety concern.

Mobile phone users are four times more likely to collapse than non-users. Using a phone while driving increases reaction times, especially braking and traffic signal reflexes, and makes it more difficult to maintain the proper lane position and following distance. Both texting and hands-free phone use are hazardous.

Designing roads can affect safety. The design of roads must prioritize road user safety. This requires amenities for pedestrians, cyclists, and motorcyclists. Footpaths, bike lanes, safe crossing locations, and other traffic calming measures contribute to the reduction of road user injuries.

Basics of Speed Breakers

Fast-moving traffic is controlled by speed bumps or speed breakers. Car traffic is slowed vertically for safety. These items come in a variety of forms, including speed tables, speed humps, speed cushions, and portable speed breakers.

Type of Speed Breaker

Driver speeds are supposed to be reduced to 10-15 mph over speed humps and 25–30 mph between them in a line of humps. They ought to be put in a position to prevent blocking off-street parking and bicycle lanes.

Wheel Stopper: Wheel chocks, also known as chocks, are wedges of durable material pressed firmly up against the wheels of a vehicle to stop unintentional movement. Chocks are positioned not only to set the brakes but also for safety. Rubber may occasionally be applied to the bottom surface to improve traction on the ground.

Cable Protector: Cable protector is a speed bumper that has a plate standing which is seen far by the driver and then the driver drove from it with the weight it goes down and it has gaps for cable flow also.

Corner Guard: Corner guards are used on the vertical wall of parking areas and the place where there is a sharp turn so oftentimes at night people face problems due to the dark, they can't see the turnings, and hazardous accidents take place.

Right Deployment of Speed Breakers

As per the Road Congress Guidelines, a speed breaker can be deployed on minor roads in below mentioned conditions. T-intersection of minor roads where the traffic is relatively low but there is an influx of high-speed vehicles with poor visibility. Slowing down vehicles at such intersections will reduce the chances of fatal accidents. Points where minor roads are intersecting with major roads; it is crucial to slow down the traffic to avoid fatal collisions Spot streets where schools, hospitals, colleges, shopping centers, police stations, or universities are located where the movement of traffic is faster. Speed breakers can lower the speed of ongoing traffic to reduce chances of accidents Spots on roads that are accident prone due to low visibility or intersection Long tangents where vehicles tend to over speed and create hazardous situations Points joining narrow or weak bridges to ensure safe movement of vehicles Before unmanned railway crossings to ensure vehicles cross after have a careful look of ascending train. Government construction sites including flyovers, underpasses, and any other civic development.

Required Specifications of Speed Breaker

As per the Indian Road Congress, a speed breaker should have a rounded hump of 3.7-meter width and 0.10-meter height to enable vehicles to cross at a maximum speed of 25 kmph. The distance between one breaker and the other one should be at least 100 to 120 meters.

Speed Breaker Sign

Drivers should be warned of the presence of speed breakers through warning sign boards. The warning sign should look like a hump or rough road. Signboard should read "SPEED BREAKER." Speed breakers should have a black and white alternate stripe to increase their visibility on the road.

Illegal or Unplanned Speed Breakers

Illegal or unplanned speed breakers could become a safety hazard and a reason for traffic chaos. Here is why:

Sudden braking causes traffic congestion; Reduces fuel economy and increases pollution levels; Vehicles face rapid wear and tear; Two-wheeler riders can skid due to sudden breaking; Patients in fast-moving ambulances can face severe health hazards due to sudden bumps; Slows down emergency vehicles including the fire brigade, police vehicle, and ambulance.

Behavior of Speed Breakers

Behavior of speed breakers shows how the speed breaker was designed for stopping people from fast driving in different areas like schools, housing, crossing, and public places often it also defines in different ways.

Literature

For a given set of safe speed limits for different vehicle kinds on that road length, Teja Tallam and Jyothi Makkea's (2017) research identifies the speed fluctuation of various vehicle classes at speed breakers and develops a model to forecast the optimal bump height for that road. Every 15 minutes, three locations counted the amount of traffic on the roads. These stations assessed the speed during peak hours both before and after speed humps. At 10 metres from the hump, cars slow down 52.69%.

According to Geetham Tiwari's analysis, Indian cities require traffic calming. By restricting cars from entering that space, provides common Traffic Calming measures like humps at intersection entrances and exits and raised walkways in all corners to make pedestrian approaches safe and comfortable.

Senthil Revathi Kumar (2020) Speed bumps served to slow down traffic. Bumpy roads slow down traffic. In locations with lots of pedestrians, speed bumps delay cars. A speed bump will shield pedestrians.

Sahoo Pakistan (2009). Most speed-control humps are sized by skilled design engineers or transportation authorities. Standards for speed control hump geometry are looked at. A hump's geometry is determined by the driver's hump-crossing speed and peak vertical acceleration.

Variety Vala The effects of chicanes, speed tables, and road narrowing on speed will be investigated by Aliasgar Chunawala in 2022. Urban traffic calming measures in Ahmedabad are examined via speed studies (India). Speed testing was done on traffic calming techniques. According to the survey, speed tables are the slowest. Speed tables and chicanes reduce it by 50%, while road narrowing reduces it by 35%. Accidents have decreased by 30% across all three traffic calming strategies. Tables of speed reduce crashes by 40%.

The main element of the non-Newtonian fluid speed breaker is LPPG OX50, one of three sizes of fumed silica nanoparticles. R972 has a specific surface area of 110 m2/g and primary spherical particles that are 16 nm in size. Primary spherical particles in R974 are 12 nm in size, and their specific surface area is 170 m2/g. Three average molecular weights of carrier fluid polypropylene glycol (H[OCH(CH3)CH2]nOH) are 400, 1000, and 3000 g/mol. Preparation Fumed silica particles are manually combined with a carrier fluid. After stirring, six three-roll mill passes enhanced STF dispersion. Shear force is produced by three horizontal rolls moving at different speeds in opposite directions to mix, refine, distribute, or homogenize viscous materials. bubbles from fully joined STFs are vacuumed. w/w STF of 7.5%, 10%, and 12.5%

Thermodynamics

Rheology was measured using stress-controlled Rheometric Scientific AR2000ex rheometers. Dynamic frequency testing was conducted using a 40 mm cone-plate device with a 4-degree cone angle and 0.4 mm Twitter gap. the link between carrier fluid viscosity and shear rate in steady conditions. At varying shear speeds, the Newtonian polypropylene glycol matrix viscosity does not change.

sheets: Organic compounds are present in carbamate-linked polyurethane (PUR or PU). While the majority of polyurethanes are thermosetting, some are thermoplastic. Polyurethane polymers are created using polyol and di- or thiocyanate. Two monomers are polymerized into alternating copolymers by polyurethanes. Isocyanates and polyurethane polyols have many functional groups per molecule. Polyurethane is used to make a variety of products, including high-performance adhesives, surface coatings, surface sealants, synthetic fibers (like Spandex), carpet underlay, rigid foam insulation panels, microcellular foam seals and gaskets, tough elastomeric wheels and tires (like those used on roller coasters, escalators, shopping carts, elevators, and skateboard wheels), automotive suspension bushings, electrical potting compounds, and more. Fluids are covered with sheets of tensile and abrasion-resistant polyurethane.

Its resistance to aging, chemicals, oils, solvents, and ripping can be used by speed breaker bodies.

Stainless steel plates Mild steel strips act as the system speed breaker (8mm thick). The speed breaker dome is riveted to the rectangular 3 mm angle iron frame. sheets of 6 mm mild steel. The speed breaker dome is riveted to the rectangular 3 mm angle iron frame.

The hump was measured by IRC-99-2018. Since fluids take on rounded forms, the hump is thought to be spherical. There are 3.7 m long, 0.2 m wide, and 0.1 m high one-way traffic humps. Highways in urban and rural areas employ this hump.

Research on rheometers: Rheometers gauge the dynamic viscosity and rate of shear-thickening fluids. Samples are tested using rotational rheometers. Samples of the load on two plates, a cone and plate, a cup and bob, or a vane system. The top plate's torque gauges the strain or shear rate. Viscometers are outperformed by rotational rheometers. Examples include the application of shear stress, oscillatory testing, and rotational testing with normal force control. Viscosity in motion is measured by a rheometer spinning. Plots of shear amplitude vs complex viscosity are shown at several angular frequencies. Viscosity can be used to compute shear rate.

Rheometer: Check to see that it is safely fastened to the lab stand, zeroed, leveled, and displaying 0% torque. Before connecting the sample cup, preheat the water bath to the test temperature. Let the bath warm up to the testing level. The gap setting on the cone spindle is electrical. Thread the cone spindle while the motor is off by gently pressing up on the rheometer coupling nut, gripping it securely with the wrench, and securing it with the spindle wrench. Thread a cone spindle by hand. Swing the tension bar under the cup to secure it to the mic ring without touching the cone. Polyethylene tubing is needed for tension bars.

separating Gap Setting is activated by turning the toggle switch to the right. The red pilot light turns on. no motor. The yellow contact light can be turned off by rotating the micrometer adjustment ring counterclockwise while gazing down at the instrument. Slowly turn the micrometer adjustment ring counterclockwise one or two scale divisions if the yellow contact light is not on. When the touch light becomes yellow, keep going. HIT. The nearest full scale division mark should be the rightmost or leftmost adjustment of the sliding reference marker. To align with the sliding reference marker line, turn the micrometer adjustment ring one scale division to the left. The yellow contact light is turned off. Decide on a measurement gap. Turn off the red pilot light (left). Carefully handle the sample cup.

Calibrating

Samples taken. Spindle sample volume is determined by the table. Determine your Brookfield Viscosity Standard fluid with a viscosity range of 10% to 100%. While the motor is off, remove the sample cup and fill it with viscosity standard fluid. Allow the rheometer, cone, and sample cup to cool. Begin. Choose a speed. Note the centipoise and percentage of torque viscosity (cP). Verify the viscosity reading's 1% deviation from the reference fluid.

Conclusions

STF viscosity and shear thickening increase with carrier fluid molecular weight. The Non-Newtonian fluid speed breaker boosts automotive fuel efficiency. Speed breakers don't require a complete stop, reducing traffic. Newtonian fluid speed breakers cost less to install and maintain.

Driving under the speed limit doesn't damage shock absorbers or steering systems. Installation takes an hour and is portable. constructed an IRC-compliant STF speed hump.

References

Bachwani, D. and Malek, M, S. (2018). Parameters Indicating the significance of investment options in Real estate sector. International Journal of Scientific Research and Review, 7(12):301-311.

Gohil, P., Malek, M., Bachwani, D., Patel, D., Upadhyay, D., and Hathiwala, A. (2022). Application of 5D Building Information Modeling for Construction Management. ECS Transaction, 107(1), 2637–2649.

Malek, M. S., Dhiraj, B., Upadhyay, D., and Patel, D. (2022). A review of precision agriculture methodologies, challenges and applications. Lecture Notes in Electrical Engineering, 875, 329–346.

Malek, M. S., Gohil, P., Pandya, S., Shivam, A., and Limbachiya, K. (2022). A novel smart aging approach for monitoring the lifestyle of elderlies and identifying anomalies. Lecture Notes in Electrical Engineering, 875, 165–182.

Malek, MohammedShakil S., and Bhatt, Viral (2022). Examine the comparison of CSFs for public and private sector's stakeholders: a SEM approach towards PPP in Indian road sector, International Journal of Construction Management, DOI: 10.1080/15623599.2022.2049490

Malek, MohammedShakil S., and Gundaliya, P. J. (2020). Negative factors in implementing public–private partnership in Indian road projects, International Journal of Construction Management, DOI: 10.1080/15623599.2020.1857672

Malek, MohammedShakil S., and Gundaliya, Pradip (2021). Value for money factors in Indian public-private partnership road projects: An exploratory approach, Journal of Project Management, 6(1), 23–32. DOI: 10.5267/j.jpm.2020.10.002

Malek, MohammedShakil S., and Zala, Laxmansinh (2021). The attractiveness of public-private partnership for road projects in India, Journal of Project Management, 7(2), 23–32. DOI: 10.5267/j.jpm.2021.10.001

Malek, MohammedShakil S., Saiyed, Farhana M., and Bachwani, Dhiraj (2021). Identification, evaluation and allotment of critical risk factors (CRFs) in real estate projects: India as a case study, Journal of Project Management, 6(2), 83–92. DOI: 10.5267/j.jpm.2021.1.002

Parikh, R., Bachwani, D., and Malek, M. (2020). An analysis of earn value management and sustainability in project management in construction industry. Studies in Indian Place Names, 40(9), 195–200.

Aesthetics of Distortion and the Absurd: Fusing Redemptive Existentialism and Berkeley's Metaphysics in Beckett's Plays

Rohit Majmudar*,a

[a]Karnavati University, Gandhinagar, Gujarat, India
E-mail: *rohitmajmudar@karnavatiuniversity.edu.in

Abstract

The conundrum between *chaos* and *order*, *distortion* and *congruity* have synchronicity of their own. Perceptual reality does not entertain binary opposites because there are none. It asks potent questions: Is the perception of the world around us a personal experience or is it impersonal? Is the *absurd* incongruous or another order that baffles the conservative mind? Is not every conformist structure a distortion of another, earlier structure? Art and literature, especially theatre and films are witnesses to life, a domino effect that has a mind of its own. The end is where the beginning is – a brilliantly used philosophy plays deeply within the microcosm of Beckett's paradoxical universe – *Waiting for Godot* and *Krapp's Last Tape* remain as the true testament of the distortion. To say that distortion is an order on its own, then, becomes the fulcrum, where absurd is but the *Other* that gives theatre, its own semiotic identity is in equilibrium.

Keywords: Aesthetics, Theatre of the Absurd, Berkeley, Metaphysics, Samuel Beckett, Existentialism

Introduction

Philosophy and science see transformation's domino effect. Dramatic creativity and absurdist theatre's attempts to reconcile paradoxes—from material to immaterial, real to virtual, and back to reality in cognitive disbelief—will be examined in this study. Joker's unpredictability is predictable. Everyone avoids it except in unique circumstances (a morphed word from physiology).

Art is fascinating but complicated, therefore contradicting statements are natural. Theatre and art are driven by emotion. Real emotions explode violently. But intense. Art expresses. Our emotions aren't expressed. The playwright's language is different from common speech and theatre.

Theater and art depict faraway emotions. Agony theatre imagines pain. Pain-sympathy? Pain and sympathy are unremarkable. Distant and immediate agony. Form and content may balance personal and impersonal art. Art is formed on human emotions and ideas, but once it has shape, it becomes its own, leaving the artist a spectator. Eric Clapton felt sad as a father when he lost his child, but his theatrical imagination converted it into an idealized experience in Tears in Heaven.

Logical Antinomies: Aesthetic Solutions Uniforms

Art is private and public. Even while they share ideas, all art has its own form. Fables are everlasting because they teach similar lessons in different ways. Cézanne developed distinctive self-portraits. Shakespeare excels at creating unique characters.

Readers and viewers cannot personify art if it is merely forming. Rethink this aesthetic affinity. How does the playwright engage audiences? Empathy—feeling something—is closer and further than compassion. Herbert Reed describes empathy by arguing that an open-minded viewer of Katsushika Hokusai's The Great Wave would be drawn to the massive wave's upswelling rather than the sad boaters and their peril. "When we feel sympathy for the unfortunate, we repeat in ourselves the sensations of others; when we gaze at a piece of art, we project ourselves into its form, and our experiences are characterized by what we find there, by the dimensions we occupy," he says. Even with empathy, art must be universal to the project. Since it captures the massive wave's sweep, Hokusai's painting will appeal more to sailors than landlubbers.

These are the core sentiments or archetypes. These emotions' deceptive expressions fascinate us because we feel them. We like real feelings. Othello is envious of Macbeth's ruthlessness. Why do fictitious characters appeal? Painting, albeit stressful, is alluring.

Art's symbolism overcomes this absurdity. Art places a concept on a separate level of reality, so it may abandon or modify everything incidental and just local and ephemeral and add much to depict the idea in all its distinctness and purity. Shakespeare made Hamlet fat because Burbage was. Burbage's corpulence contrasted with her second husband's crime. This lifts a single actor's individuality to universal forms.

Assimilation

Theatre or self-expression? Writers, painters, and musicians expect an audience and applause when they publish, exhibit, or play. Do these artists dream of the public or of self-expression? Dislocation, distortion, assimilation, and natural order distortion to generate new creative assimilation are the artist's most personal expressions. The artist retreats and seldom utilizes facts or laws to support his opinions. Without intimacy or concealment, art loses directness. Spiritual artists draw, paint, sing, carve, or create using suitable materials. Volitional action symbolises. Art ends with brain intuitions.

The aforementioned idea appears reasonable, or the artist's core preferences and the fact that others enjoy it or the artist gets affluent are coincidences. Artists must analyze their sentiments and ideas as a vision to succeed. Thus, all art is characterized by keen expectancy and calm detachment, alienation. Artists create by reading and watching their work. Audiences may hear an artist's worldly ideas.

Paradox

Existentialism and Waiting for Godot reject existence. Plays aren't bad. Berkeley's idealism—that nothing exists without observation—helps Beckett unveil essence. Beckett's plays mirror Berkeley's belief in seeing. Humans see meaning in a material-less universe. Waiting for Godot and Krapp's Last Tape show how we are viewed affects our self-worth and happiness. Self- and social acceptance empowers us. Beckett's tragedies highlight the necessity for contemplation and interdependence to live a meaningful life in a postmodern culture where ambition, technology, and upheaval may bring self-imposed loneliness.

Samuel Beckett's 1950s plays Krapp's Last Tape and Waiting for Godot examined human nothingness. Existentialists think his senseless repetitions make life monotonous. Beckett's plays are upbeat. According to a dialectical study of the plays, Beckett constantly warns his audience of Vladimir, Estragon, and Krapp's flaws and offers ways to overcome life's emptiness. Berkeley's idealism affected Beckett's existential dread response. Beckett brilliantly shows how this theory can be applied to the human mind and how, even if the cosmos may not have a purpose, people can find meaning in their relationships.

Samuel Beckett developed postmodernism after WWII. After seeing war, writers were confused and worried about the Cold War. Postmodernists predicted a bleak future due to history's seeming stagnation. Molière's Theatre of the Absurd highlighted human transience and incoherence. Shakespeare said, "Men and women were considered as mere performers in an absurd play, making their entrances and departures upon the stage of life and mouthing their stories 'full of sound and fury,' meaning nothing."

Krapp's Last Tape and Waiting for Godot are postmodern dramas. Existentialist Alain Badiou claims Beckett. Sartre's Waiting for Godot and Krapp's Last Tape show life's futility. Sartre, the originator of existentialism and Beckett's contemporary, felt that humans are worthless and that existence precedes essence. Existentialists advise taking full responsibility for their existence. Beckett showed that Sartre's views may cause despondency. Vladimir, Estragon, and Krapp are Sartreans trying to grasp life's meaninglessness.

Waiting for Godot begins with Sartre's existential crisis. rural road. Tree. Dead evening. While waiting for Godot, Estragon and Vladimir are squandering their lives.

Vladimir responds, "I'm beginning to agree with that" after Estragon declares "nothing to be done." Drama with two ordinary individuals waiting and talking makes no sense. Since the drama is aimless, Estragon and Vladimir's lives are idle chatter. Life challenges both characters.

Birth sins. Vladimir represents Sartre's "selects for himself and all men" universal person. "At this place, at this time, all humanity is us, whether we want it or not," Vladimir tells Estragon to help Lucky. Enjoy! Since everyone chooses the best choice, every decision improves humanity. Sartre emphasizes existential boredom and the idea that everyone's choices impact them in Waiting for Godot.

Krapp's Last Tape's tedium and unclear plot recall Sartrean existentialism. In the late twilight of the future, Krapp is oblivious of his past and unable to give his existence meaning. "Mother finally relaxes. Hmm. Black ball. Sees. Puzzled. Ball?" Krapp raises his head and stares into space to remember, but his existence has become meaningless. Since he no longer values his past learning, ideas, and philosophy, Krapp has not grown. He's too wounded to notice. "What I instantly recognized then was this, that the belief I had been holding all my life, namely—," says the younger Krapp. Krapp curses throughout video pauses. "I've been listening to that foolish bastard I took myself" for 30 years. Krapp proves Sartre's futility.

Vladimir, Estragon, and Krapp are depressed despite Sartre's claim that existentialism empowers by letting one define oneself. Beckett exploits Berkeley's idealism to ease existential suffering in Waiting for Godot and Krapp's Last Tape. Beckett's plays, which extend Sartre's universal person, represent Berkeley's belief that acceptance is fundamental for meaning. "Esse is precipi," stated eighteenth-century idealist Berkeley. According to his idea, the universe, including people, only exists when a seeing mind recognizes them, and everything in awareness, including form, color, texture, and taste, are sense perceptions. Berkeley believes only conscious minds can comprehend reality.

Waiting for Godot and Krapp's Last Tape show life's monotony, pointlessness, and fragmentation without Berkeleian perspective. The plays depict "an unimportant universe, its tension produced by the conflict between insignificance and man's desire to give himself meaning despite everything." Beckett believes humans are relational animals who can recognise essence but not meaning.

One scholar suggests Beckett is using Berkeley's theory of sensory object relativity and dependency. Estragon, Krapp, and Vladimir wait for Godot because they desire recognition and become irritated if they don't. Krapp only remembers faces. Beckett's insight counters existentialism's sadness. In Beckettian redemptive existentialism, we only know ourselves and are pleased when others see us.

Estragon and Vladimir seek fame in Waiting for Godot. "Why won't you ever let me sleep?" Estragon pleads Vladimir again. Vladimir says, "I felt lonely," as if he would die without attention. Pozzo, the play's strongest character, constantly asks, "Is everyone ready?" Is everyone watching me? Pig! Vladimir says the Boy Godot won't come that evening, combining sight and existence.

Vladimir and Estragon recognize Godot before he arrives, exhibiting Beckett's existentialism (appearance before essence) and idealism (to be is to be perceived). Vladimir concludes the speech when Godot doesn't show in Act II.

"Brains become dependent, delicate beings [in Beckett's plays]; they seek the support and comfort of being acknowledged," explains Berman. Vladimir wants attention. Beckett characters need exposure.

To be is to be perceived in Krapp's Last Tape. The drama shows Krapp's solitude. Gordon writes, "As the cassettes plainly illustrate, he has followed the life of the intellect, separating the 'grain' from the 'husks' to pursue his 'vision.'" "He believed would survive his illness. Krapp, who prefers alone," celebrated the sad occasion [of his birthday], like in previous years, silently at the Wine-house" with "not a person." Krapp wrote about his birthday vacation. Despite Earth's barrenness, I've never felt such serenity. "Seventeen copies [of his book] sold, of which eleven [were sold] at trade price to free circulating libraries across the seas," making his job unfulfilling. Krapp revisits the film's scenes he wishes he could relive, even though he's a detached artist. Krapp finds beauty in shared experiences rather than creative excellence, disappointing him. Berkeley said Krapp's best moments were recognized.

Like Vladimir and Estragon, Krapp obsesses with eyesight. His enormous yearning for approval and anguish over feeling alienated reveal that living a meaningful life needs the opinion of others. Krapp adores the sight of ladies. Despite calling his relationship with Bianca "hopeless," he remembers her "very warm" and "incomparable" eyes and "not much about her, other than an admiration to her eyes." Krapp loved nurses because she constantly watched him as he turned. Eyes! Krapp only remembers acknowledgment. Krapp often uses the boat scenario to demonstrate how women have shaped his self-image.

Finally, Beckett uses Krapp's refusal to accept himself to emphasize the need for self-perception for a meaningful existence. Beckett idealized self-awareness. Krapp no longer hears or sees himself. Krapp's "repeating past and projected future" complicates the play's futuristic setting. Recording distracts Krapp. Knowlson says Krapp's past comments "now constitute the only form of interaction that he can generate in a depleted, alone, and wretched existence." Krapp doesn't exist if space and time are relative.

Waiting for Godot and Krapp's Last Tape resolve existentialism's Berkeleian perception issues. Like Sartre, both works believe life is pointless. Berkeley's idealism eases Beckett's existential pain. Vladimir, Estragon, and Krapp's pointless existence defy Beckett. Beckett characters show us how to live meaninglessly.

Conclusion

In the postmodern world of ambition and technology, Beckett advises against living alone. Beckett's play emphasizes self-consciousness, trueness to the present self, and readiness to be understood despite life's meaningless repetitions and routines: Thus, only theatre can demonstrate human nature's absurdity. Theater must confront the dismal truths that most human endeavor is dumb and irrational, that connecting with people is difficult, and that the universe will always be a mystery. We may laugh and relax if we realize most of our goals are ridiculous.

Look inside and outward for purpose. Self-reliance is unattainable. Beckettian redemptive existentialism holds that being accepted and valued enhances life and gives purpose.

References

Bachwani, D. and Malek, M, S. (2018). Parameters Indicating the significance of investment options in Real estate sector. International Journal of Scientific Research and Review, 7(12), 301–311.

Gohil, P., Malek, M., Bachwani, D., Patel, D., Upadhyay, D., and Hathiwala, A. (2022). Application of 5D Building Information Modeling for Construction Management. ECS Transaction, 107(1), 2637–2649.

Malek, M. S., Dhiraj, B., Upadhyay, D., and Patel, D. (2022). A review of precision agriculture methodologies, challenges and applications. Lecture Notes in Electrical Engineering, 875, 329–346.

Malek, M. S., Gohil, P., Pandya, S., Shivam, A., and Limbachiya, K. (2022). A novel smart aging approach for monitoring the lifestyle of elderlies and identifying anomalies. Lecture Notes in Electrical Engineering, 875, 165–182.

Malek, MohammedShakil S., and Bhatt, Viral (2022). Examine the comparison of CSFs for public and private sector's stakeholders: a SEM approach towards PPP in Indian road sector, International Journal of Construction Management, DOI: 10.1080/15623599.2022.2049490

Malek, MohammedShakil S., and Gundaliya, P. J. (2020). Negative factors in implementing public–private partnership in Indian road projects, International Journal of Construction Management, DOI: 10.1080/15623599.2020.1857672

Malek, MohammedShakil S., and Gundaliya, Pradip (2021). Value for money factors in Indian public-private partnership road projects: An exploratory approach, Journal of Project Management, 6(1), 23–32. DOI: 10.5267/j.jpm.2020.10.002

Malek, MohammedShakil S., and Zala, Laxmansinh (2021). The attractiveness of public-private partnership for road projects in India, Journal of Project Management, 7(2), 23–32. DOI: 10.5267/j.jpm.2021.10.001

Malek, MohammedShakil S., Saiyed, Farhana M., and Bachwani, Dhiraj (2021). Identification, evaluation and allotment of critical risk factors (CRFs) in real estate projects: India as a case study, Journal of Project Management, 6(2), 83–92. DOI: 10.5267/j.jpm.2021.1.002

Martin Esslin, (1960). The Theatre of the Absurd. The Tulane Drama Review (United States: Cambridge MIT Press. 4(4), 5–6.

Explicit and Implicit Self-Esteem of Narcissists and Non-Narcissists

Tarundeep Kaur[*,a] *and Garima Garg*[a]

[a]Panjab University, Chandigarh, India
E-mail: tarun1deep1@gmail.com, garima23garg@gmail.com

Abstract

It has been observed that people with narcissistic personalities often display an exaggerated sense of self-regard and superiority. To understand this, the present research aims to study the explicit and implicit self-esteem of narcissists and non-narcissist. 90 male students of the age range 20-23 years were taken as the sample. NPI, Rosenberg Self-Esteem Scale, and Initials-preferences test were conducted on these students. A significant difference was found between the explicit self-esteem of narcissists M=(20.48, SD=5.73) and non-narcissists M=(17.28, SD=3.10). concluding that narcissists have high explicit self-esteem as compared to non-narcissists, which clarifies the reason behind their overt grandiosity. Whereas, there is no difference between the implicit self-esteem of narcissists and non-narcissists scores. Other factors besides self-esteem, can be studied to understand narcissistic behavior.

Keywords: Narcissist, Implicit self-esteem, Explicit self-esteem

Introduction

Narcissism is an inordinate fascination with oneself. Narcissists have an exaggerated or overly optimistic self-perception. The core processes of narcissism are attempts to protect, preserve, or reestablish a monumental view of self (Freud, 1986; Kohut, 1966, 1977; Kernberg, 1986). This is usually referred to as the "mask" concept, which is based on clinical findings that narcissistic people seek to perpetuate a grandiose façade while feeling threatened, inferior, weak, and/or vulnerable on the inside. According to these models, an inflated ego serves the practical aim of preserving oneself from emotions of inadequacy (Wright et al., 2021). The latest models show that people are organized similarly. When circumstances emerge that expose one's frailty, they contend that shifts in narcissistic expression are attempts to restore a desired superior state (Grubbs and Exline, 2016; Morf and Rhodewalt, 2001).

Among narcissists, exaggerated expectations or a strong sense of deservingness are frequent characteristics; this is known as a trait-level entitlement (Campbell et al., 2004). This aspect of narcissism is seen to be quite stable (Krizan and Herlache, 2018; Miller et al., 2017). Because of their high sentiments of deservingness, narcissists have unreasonable expectations and views about how others should see them. When narcissistic person notices a discrepancy between their self-expectations and the feedback they receive from people around them, they will employ techniques to restore their intended sense of superiority (Grubbs and Exline, 2016).

Narcissism is philosophically and experimentally linked to self-esteem. Indeed, it is implied in some recent research that several measures of narcissism and explicit self-esteem may represent the same underlying psychological construct (e.g., Rosenthal and Hooley, 2010). The majority of theorists and researchers view narcissism as a separate construct from self-esteem, describing it as a cluster of traits characterized by a grandiose sense of self-importance, egocentrism, and diminished empathy, despite the possibility of overlying between potentially flawed measures of these constructs. A craving for attention, as well as a manipulative and exploitative interpersonal style, are associated with high degrees of narcissism (Miller and Campbell, 2010). There is debate concerning the psychological causes of people with high levels of narcissism's exaggerated sense of self-worth. Some modern theoretical theories of narcissism believe that the individual's outer expression properly reflects their positive implicit and explicit evaluations of self, aggression, and social control (Miller et al., 2008).

Self Esteem

The amount of one's positive or negative self-evaluation is referred to as one's self-esteem (Zeigler-Hill and Jordan, 2010). There are several overt and covert ways to communicate self-esteem. Implicit self-appraisals happen instantly, don't require effort or focus, and could be difficult to obtain through introspection (Karpinsky and Steinberg, 2006). Explicit self-evaluations are purposeful and observable through reflection (Karpinsky and Steinberg, 2006). (Greenwald et al., 2002; Karpinsky and Steinberg, 2006).

Variable self-esteem is regarded to have its roots in low implicit self-esteem. To preserve a positive self-image, people both with high & low implicit self-esteem are more narcissistic, favor their group more, and use more defense mechanisms (Bosson et al., 2002). Narcissistic managers frequently associate their own admiration and respect with that of others, hence they are likely to have a high need for social acceptance. Due to their sensitivity to failure and criticism, they may become socially disengaged.

Objective of the Study

References from researchers who have studied narcissists in relation to their self-esteem are primarily from the western literature and relatively limited work is available in India. The objective of this research is to explore the implicit and explicit self-esteem of narcissists & non-narcissist. Based upon the literature review, the conceptualization of the term, and limited research evidence on the given topic in the Indian context, the following are the hypotheses that are put forth.

Hypotheses

The narcissists will show high explicit self-esteem, as compared to that of the non-narcissistic individuals.

There is no difference between narcissists and non-narcissists, in terms of implicit self-esteem.

Explicit self-esteem is negatively correlated with implicit self-esteem, in the case of individuals with Narcissism scores.

Explicit self-esteem is positively correlated with implicit self-esteem, in the case of individuals with non-narcissism scores.

Method

Sample

The sample of the present study consists of 90 male students, belonging to an age range of 20–23 years, from Thapar University, Patiala.

Measures used

Narcissism

It was measured using the Narcissistic Personality Inventory (Raskin and Hall, 1981). This version of NPI consists of 40 pairs of statements and the subject had to choose one out of the two statements, according to their preferences.

Explicit self-esteem

Participants completed the Rosenberg Self-Esteem Scale (RSES; Rosenberg, 1965), a well-validated measure of global self-regard (Blaskovich and Tomaka, 1991; Demo, 1985).

Implicit self-esteem

Initials-preferences: Based on the procedure developed by Nuttin (1985, 1987), participants evaluated each letter of the alphabet using response scales ranging from 1 (very unattractive) to 5 (very attractive).

Design

In this study, the dependent variable is explicit and implicit self-esteem and the independent variable is the narcissist.

t-test and correlation were used to compute the results.

Procedure

The consent of the participants was taken before conducting the test. The sample included 90 male students belonging to an age range of 20-23 years, studying at Thapar University, Patiala. The participants were given a set of three questionnaires. First was for Narcissism, there were two statements given for each question and the participants were asked to read each pair of statements and then choose the one that was closer to their feelings. They were to answer by writing the letter (A or B) in the space provided.

The next questionnaire was for the Explicit self-esteem by Rosenberg. This scale consisted of 10 items which were to be answered on a four-point scale – from strongly agree (SA) to strongly disagree (SD).

The last questionnaire was to measure Implicit self-esteem. Participants had to evaluate each letter of the alphabet using response scales ranging from 1 (very attractive) to 5 (very unattractive).

Result

After the data collection, the scores were compiled. Mean and standard deviation was applied to the NPI scores. The mean and standard deviation scores of NPI are M= (18.76, SD=5.52). To bifurcate the NPI scores into narcissist and non-narcissist, SD=5.52 was divided by 2, i.e. 2.76. The new SD=2.76 was used in the formula Mean + 1. SD (18.76+2.76=21.52) and Mean – 1. SD (18.76-2.76=16) to obtain the narcissist score and non-narcissist scores, respectively. Therefore, the scores above 21.52 fell in the category of a narcissist, whereas the scores below 16 were the non-narcissist scores. With respect to the NPI bifurcation, the explicit and implicit scores were also divided and a t-test was used to obtain the results.

The present research aims to study the explicit and implicit self-esteem of narcissists and non-narcissists. Three different measures were used to measure narcissism and explicit and implicit self-esteem. After the scores were obtained, a t-test and correlation were used to compute the results.

The above table shows the self-esteem scores of narcissists and non-narcissists, i.e., the mean and t-test scores. It has been found that the t value of implicit self-esteem is insignificant, in both first name ($t(39)=0.7$) as well as the last name ($t(47)=0.4$). This means that there is no difference between the implicit self-esteem of narcissists and non-narcissists. On the other hand, in the case of explicit self-esteem, the t value of narcissists and non-narcissists is found to be significant with $t\ (36)= 2.48, p<0.01$. Therefore, there is a significant difference between the explicit self-esteem of narcissists M=(20.48, SD=5.73) and non-narcissists M=(17.28, SD = 3.10). Also, narcissists have high explicit self-esteem as compared to non-narcissists.

A correlation was used to find the relation between the explicit and implicit self-esteem of narcissists and non-narcissists, respectively. After the results were computed, it was found that there is no correlation between the explicit self-esteem and implicit self-esteem (with first name $r = 0.02$ and last name $r = 0.03$) of narcissists. Similarly, no correlation was found between the explicit self-esteem and implicit self-esteem (with first name $r = -0.14$ and last name $r = -0.24$) of non-narcissist individuals.

Table 1. Showing the explicit and implicit scores of narcissists and non-narcissists

Scales		Means & SD		t-Test	Significance
NPI		18.76, 2.76			
		Narcissist	Non-Narcissist		
Explicit Self Esteem		20.48,5.73	17.28,3.10	2.48	Significant 0.01*
Implicit Self Esteem	First name	2.04,0.84	2.18,0.56	-0.7	Insignificant
	Last name	2.27,0.82	2.36,0.75	-0.4	Insignificant

*t(36)=2.48, p < 0.0

Discussion

The study investigates the explicit and implicit self-esteem of narcissists and non-narcissists, which was conducted on 90 male students.

After the analysis of data, it has been found that there is a convincing difference between the explicit self-esteem of narcissists and non-narcissists. Also, the first hypothesis has been proven to be true, i.e., In comparison to non-narcissists, narcissists have high explicit self-esteem. This type of self-esteem is defined as the awareness of one's own value, acceptance, and liking (e.g., Brown, 1993; Kernis, 2003; Rosenberg, 1965). It's probable that the understanding, which is based on analytical evaluations of self-absorbed feedback, is mostly to blame. Individuals having high explicit self-esteem are regarded to have positive views about themselves and are weak and unprotected to threats because of the worries and self-distrust linked with poor implicit self-esteem. This behavior of external magnanimity masking inner self-hatred is consistent with conventional narcissistic beliefs (Kernberg, 1970; Kohut, 1971; Morf and Rhodewalt, 2001; Raskin et al., 1991; Wink and Gough, 1990). High explicit self-esteem makes people more prone to narcissistic behaviors like self-enhancement and protective behaviour.

The second hypothesis has also been confirmed, i.e., there is no difference between narcissists and non-narcissists, in terms of implicit self-esteem. The findings show that the t value of implicit self-esteem is insignificant, in both first name (t(39)=0.7) as well as the last name (t(47)=0.4). Therefore, there is no difference between the implicit self-esteem of narcissists and non-narcissists scores. Self-evaluations that are not conscious, automated, or overlearned are thought to make up implicit self-esteem (Hetts and Pelham, 1999; Greenwald and Banaji, 1995). During a threatening interview, in 1999 Spalding and Hardin discovered that, Contrary to explicit self-esteem, uncontrollable behavior was predicted by implicit self-esteem (e.g., nonverbal anxiety). One of the most essential purposes of implicit self-esteem can be to buffer people against experiences that harm their sense of self (Dijksterhuis, 2004; Jones et al., 2002; Greenwald and Farnham, 2000; Spalding and Hardin, 1999; Shimizu and Pelham, 2004).

As can be seen, no correlation has been found between the implicit and explicit self-esteem of narcissists. Similarly, no correlation has been found between the implicit and explicit self-esteem of non-narcissist individuals. Therefore, the third and the fourth hypothesis, have not been confirmed as there is no correlation found in both cases, as per the obtained results.

Implications of the study

The study aims at understanding the implicit and explicit self-esteem of narcissists & non-narcissist. As the results conclude, narcissists have high explicit self-esteem, which explains the reason behind their overt grandiosity and extremely high positive self-regard. But at the same time, it can be understood that such people get easily offended and are fragile to challenges and threats because of their covered insecurities unknown to them. Considering the above findings, there is a need to provide introspective psychotherapy to people with narcissistic personalities so that they confront their weaknesses and insecurities leading to sound emotional and mental health.

References

Bosson, J. K., Brown, R. P., Zeigler-Hill, V., and Swann, W. B. (2003). Self-Enhancement Tendencies Among People With High Explicit Self-Esteem: The Moderating Role of Implicit Self-Esteem. Self and Identity, 2(3), 169–187. https://doi.org/10.1080/15298860309029

Gierczak, A. (2021). Relationships Between Explicit and Implicit Self-Esteem and the Need For Social Approval and Narcissism in Persons Holding Managerial Positions. Humanities and Social Sciences Quarterly. https://doi.org/10.7862/rz.2021.hss.27

Kernis, M. H. (2003). TARGET ARTICLE: Toward a Conceptualization of Optimal Self-Esteem. Psychological Inquiry, 14(1), 1–26. https://doi.org/10.1207/s15327965pli1401_01

Krizan, Z., and Herlache, A. D. (2017). The Narcissism Spectrum Model: A Synthetic View of Narcissistic Personality. Personality and Social Psychology Review, 22(1), 3–31. https://doi.org/10.1177/1088868316685018

Morf, C. C., and Rhodewalt, F. (2001). Unraveling the Paradoxes of Narcissism: A Dynamic Self-Regulatory Processing Model. Psychological Inquiry, 12(4), 177–196. https://doi.org/10.1207/s15327965pli1204_1

Nuttin, J. M. (1985). Narcissism beyond Gestalt and awareness: The name letter effect. European Journal of Social Psychology, 15(3), 353–361. https://doi.org/10.1002/ejsp.2420150309

Oura, S., Matsuo, K., Inagaki, T., Shima, Y., and Fukui, Y. (2017). Effect of explicit/implicit internal working models of attachment on frequency of experience in interpersonal life events: Using ST-IAT. The Proceedings of the Annual Convention of the Japanese Psychological Association, 81(0), 2C–004. https://doi.org/10.4992/pacjpa.81.0_2c-004

Raskin, R., Novacek, J., and Hogan, R. (1991). Narcissism, Self-Esteem, and Defensive Self-Enhancement. Journal of Personality, 59(1), 19–38. https://doi.org/10.1111/j.1467-6494.1991.tb00766.x

Rosenthal, S. A., and Hooley, J. M. (2010). Narcissism assessment in social–personality research: Does the association between narcissism and psychological health result from a confound with self-esteem? Journal of Research in Personality, 44(4), 453–465. https://doi.org/10.1016/j.jrp.2010.05.008

Stauner, N., Wilt, J., Grubbs, J., Pargament, K., and Exline, J. (2016). Neuroticism and Stressful Life Events Predict Religious and Spiritual Struggles. Personality and Individual Differences, 101, 517. https://doi.org/10.1016/j.paid.2016.05.306